ADVANCES IN X-RAY ANALYSIS

Volume 17

A Continuation Order Plan is available for this series. A continuation order will bring delivery of each new volume immediately upon publication. Volumes are billed only upon actual shipment. For further information please contact the publisher.

ADVANCES IN X-RAY ANALYSIS

Volume 17

Edited by
C. L. Grant
University of New Hampshire
Durham, New Hampshire

and

Charles S. Barrett, John B. Newkirk, and Clayton O. Ruud
Denver Research Institute
The University of Denver
Denver, Colorado

Sponsored by
University of Denver
Denver Research Institute
Metallurgy and Materials Science Division

SPRINGER SCIENCE+BUSINESS MEDIA, LLC

Proceedings of the Twenty-Second Annual Conference on Applications of
X-Ray Analysis held in Denver, August 22-24, 1973

Library of Congress Catalog Card Number 58-35928

ISBN 978-1-4613-9977-3   ISBN 978-1-4613-9975-9 (eBook)
DOI 10.1007/978-1-4613-9975-9

Copyright © 1974 Springer Science+Business Media New York
Originally published by Plenum Press, New York in 1974
Softcover reprint of the hardcover 1st edition 1974

PREFACE

The successful application of x-ray diffraction techniques and x-ray spectrometry depends in large measure on the availability of dependable standards and reference data. The preparation of such standards in the fields of metallurgy, geology, life sciences, and other disciplines is both costly and time consuming. As a result, the necessary standards for effective utilization of existing instrumentation are often not available. One of the purposes of the invited papers in this 22nd Annual Denver X-Ray Conference was to review the status of programs to prepare such standards and reference data. Simultaneously, it seemed appropriate to examine the role of sampling both in terms of standards and samples to be analyzed.

The first section of the invited papers focuses on the standards and reference data problems. In addition, many of the contributed papers offer information on this theme. The second topic in the invited papers considers the problem of sampling. If we recognize that analyses are conducted on samples which vary in size from several grams to a few micrograms or less, the magnitude of the random and systematic error components of sampling on the quality of results should be obvious. Many of the contributed papers in such fields as air pollution and similar disciplines speak clearly to the difficulty of obtaining "representative" samples.

The papers contained in this volume and the many lively discussions such as the panel discussion at the close of the first session of papers should stimulate further attention to this vital topic.

C. L. Grant

FOREWORD

The title STANDARDS AND SAMPLING best describes the subject chosen for emphasis at this 22nd Denver X-Ray Conference. The Denver committee was fortunate in having the cooperation of Professor C. L. Grant, University of New Hampshire, who served as invited co-chairman of the conference. Through his knowledge of the field Dr. Grant assembled a panel of six outstanding authorities in the field, each of whom gave a keynote lecture on a selected aspect of this broad subject. The invited keynote lecturers were:

1. H. H. Yolken, National Bureau of Standards, Wash., D.C.
   PROBLEMS IN THE STANDARDIZATION OF X-RAY MEASUREMENTS.

2. A. J. Gillieson, Department of Energy, Mines and Resources, Ottawa, Ontario, Canada.
   PRODUCTION OF STANDARD REFERENCE MATERIALS, THE ASSOCIATED PROBLEMS, AND THE PHILOSOPHY OF CERTIFICATION.

3. H. F. McMurdie, National Bureau of Standards, Wash., D.C.
   STANDARD DATA FOR IDENTIFICATION OF PHASES BY X-RAY DIFFRACTION.

4. R. Jenkins, Phillips Electronic Instruments, Mt. Vernon, N.Y.
   PROVISION, SUITABILITY, AND STABILITY OF STANDARDS FOR QUANTITATIVE X-RAY POWDER DIFFRACTOMETRY.

5. C. L. Grant, University of New Hampshire, Durham, N.H.
   INFLUENCE OF SAMPLING ON THE QUALITY OF ANALYSIS WITH EMPHASIS ON POWDERS.

6. R. K. Skogerboe, Colorado State University, Fort Collins, Colorado.
   VARIATION OF STANDARDS FOR SAMPLING REQUIREMENTS FOR COMPLEMENTARY ANALYSIS METHODS.

Written versions of these papers appear in the first part of this volume.

The remainder of the three-day conference was given to technical sessions made up of contributed papers. We owe our thanks to the following people who, with the keynote speakers, ably served as chairmen of the sessions:

Victor Buhrke, The Buhrke Company, Menlo Park, California.
D. M. Smith, Chairman, Department of Chemistry, University of Denver, Denver, Colorado.
B. L. Henke, Department of Physics, University of Hawaii, Honolulu, Hawaii.
H. F. McMurdie, National Bureau of Standards, Wash., D.C.

A special meeting on Aerosol Standard Samples was called and chaired by Dr. John Cooper of the Battelle Pacific Northwest Laboratories, Richland, Washington, and John Rhodes of Columbia Scientific Industries, Austin, Texas. This meeting presented results of an inter-laboratory cooperative program, in which analytical results obtained by different groups on identical specimens were compared and established a second program. The Denver X-Ray Conference is glad to provide a forum where results of this sort of program can be discussed and recorded. We hope to hear from this group as their results are achieved.

The preconference correspondence and the management of the myriad paper details of registration, etc., were handled by Mrs. Eithne Ruud, to whom the committee owes its sincere thanks. At the technical sessions themselves the projection of slides, manipulation of microphones and other important on-site details were performed by Mr. Tom McClelland, Mrs.Harovel Wheat, and Mr. Larry Apple. To all these supporting people we owe our gratitude for their cooperation in making possible a smoothly operating conference.

On request of the Editor of X-Ray Spectrometry, a brief history and description of the Denver X-Ray Conferences was prepared by the undersigned for that journal. It seems appropriate that we reprint that article here for the purpose of placing it with the series of Advances. We are grateful to the Editor and Publishers of X-Ray Spectrometry for permission to reproduce the article.

The Denver X-Ray Conferences

Every August for the past 21 years a large group of non-medical but x-ray oriented scientists and engineers have gathered in Denver, Colorado to discuss their common technical interests. This annual conference has come to be widely known as the Denver Conference on Applications of X-Ray Analysis.

The first organized meeting was arranged in 1952 by Professor James Blackledge and Mr. Merlyn Salmon, then of the Denver Research Institute, and Mr. Stephen Knight of Technical Equipment Corporation. Dr. I. Fankuchen of Brooklyn Polytechnic Institute and well known to the world of X-ray crystallography, was the featured lecturer. At this first meeting there were three local speakers and a total of 78 people in attendance. That year and for the next two years the X-ray conferences were held on the University of Denver campus with Fankuchen and about six speakers holding forth. Attendance dwindled from the original 78 to 58 in 1953. In 1954 the meeting was moved to a downtown Denver hotel. Yet only nine papers were offered and 50 people attended. Undaunted by these figures

Foreword

and convinced that X-ray analysis had a growing importance in national technology, Blackledge and his co-workers mounted a concerted effort to put the meeting on a national scale. Their efforts showed positive results the next year (1955) when 28 papers were heard by 78 people. Thereafter the value of the Denver X-ray conference was given increased national recognition and it grew steadily to the present three-day meeting with an annual attendance of 250 to 300 practising non-medical X-ray specialists. Featured speakers and attendees come from all parts of the globe. Also, hard cover volumes of the conference proceedings, entitled <u>Advances in X-Ray Analysis</u>, are to be found in technical libraries throughout the world. Following the policies established by the original founders, several people have carried on the annual organization of the X-ray conference. These include Dr. W. M. Mueller, Mr. Gavin Mallett, Mrs. Esther Marie Capps, and the late Mrs. Marie Fay, all of the University of Denver Research Institute.

The Denver conference has traditionally been self-supporting, though it has been officially sponsored and guaranteed by the University of Denver Research Institute. Efforts to keep down the cost to participants sometimes ran the conference into debt, resulting in the need for the University to bail it out. In 1959 financial aid from the Office of Naval Research was obtained as travel assistance for a few invited international speakers. In recent years, with the support of X-ray related industries who recognized this as a good opportunity to make effective contact with potential customers, the financial operation of the conference has been on a sound footing.

The technical policy by which each Denver X-ray conference is structured is set by the local committee, now consisting of Prof. C. S. Barrett, Prof. J. B. Newkirk and Dr. C. O. Ruud, all of whom are employed by the University of Denver. As the conference title implies, any high quality paper whose subject falls within the broad field of Applications of X-Ray Analysis is appropriate for presentation. Papers involving other radiation, such as electron diffraction and gamma rays, are also within the scope of the Denver conference. To bring focus to the meetings, about a year in advance of the meeting date, the local committee identifies a technical subject which appears to be ready for organized emphasis at the conference. Through direct contacts with persons who are active in that subject the committee selects an acknowledged leader in the subject and invites him to serve with them as Invited Co-chairman of the Denver Conference. Together, he and the local committee assemble a panel of known leaders in the selected field. The panelists then are invited to present keynote papers at the opening session of the conference. The subjects they cover are carefully chosen and interrelated so as to present as complete and balanced a picture as possible of the subject picked for emphasis.

Early in the calendar year, wide publicity is given to the next August meeting, including the subject to be emphasized, the names of the invited speakers and the titles of their talks. At this same time a general call for contributed papers is issued. The effect of this procedure is to stimulate the contribution of papers which deal with the amphasized subject and to promote increased attendance of technical people who share the common specialized interest. Consequently, the proceedings volumes stand as state-of-the-art summaries of certain techniques, applications, etc., involving X-rays or related radiation. Since 1966 the emphasized subjects and the respective invited co-chairmen have been:

Year	Emphasis	Invited Co-chairman
1966	X-ray Diffraction Topography	None
1967	Chemical Analysis by X-ray Fluorescence	H.G. Pfeiffer
1968	X-ray Diffraction - General	C.S. Barrett
1969	Low Energy X-ray Phenomena	B.L. Henke
1970	No subject emphasis	None
1971	X-ray Detection Methods	K.F.J. Heinrich
1972	X-rays in Environmental and Biomedical Applications	L.S. Birks
1973	Standards and Sampling Methods	C. L. Grant

Contributed papers which do not fall within the announced subject for emphasis are welcomed. In this way the Denver conference also serves as a forum at which new subjects may be discussed and where young or relatively unknown contributors can be heard. The only criteria for participation are an active interest in the subject matter and, if work is to be presented, that it be technically sound, original work of significance, and within the broad subject scope of the conference.

All papers which are orally presented at the Denver conference are also submitted in manuscript. Soon after the conference is held in August, the local committee and the invited general co-chairman read the manuscripts. Most of them are ultimately accepted for inclusion in Advances in X-Ray Analysis, and are coordinated for publication. By Spring of the following year the volume is available for distribution, free to the conference attendees and at moderate cost to others.

The detailed arrangements for the three-day meeting are made entirely by the local committee. Appropriate accommodations are made in a Denver hotel for the technical sessions and for pre-session meetings where speakers, chairmen and projectionists can coordinate a smooth presentation. Booths are made available near the meeting room where the latest commercial X-ray equipment is exhibited throughout the three-day conference period. At the close of the first day's technical sessions a social mixer is held which is traditionally sponsored by the commercial exhibitors. Other services to conferees include job and consulting referrals and a table where sample books, recently issued by such organizations as ASTM and U.S. Bureau of Standards, can be displayed.

In its nearly quarter century existence, the annual Denver conference on Applications of X-ray Analysis has apparently filled the need which was seen by its founders. In the foreseeable future, its continuing function will be to bridge the gap between pure X-ray science and the applied technologies which have their bases in X-ray and related phenomena.

CONTENTS

Preface v

Standard Reference Materials and Meaningful X-Ray Measurements,
    H. Thomas Yolken 1

Standard Reference Materials - Their Production and Use,
    A. H. Gillieson 16

Standard Data for the Identification of Phases by X-Ray
  Diffraction,
    Howard F. McMurdie 20

Provision, Suitability and Stability of Standards for
  Quantitative Powder Diffractometry,
    Ron Jenkins 32

Influence of Sampling on the Quality of Analyses with Emphasis
  on Powders,
    C. L. Grant and P. A. Pelton 44

Variation of Standards and Sampling Requirements for
  Complementary Analysis Methods,
    R. K. Skogerboe 68

Quantitative Analysis of Clay Minerals in Drilling Mud Solids,
    D. G. Feuerbacher and R. R. Clark, ... 75

Factors Limiting the Use of Standard Minerals in the X-Ray
  Diffraction Analysis of Clays
    Lowell A. Douglas 88

Role of Diffractometer Geometry in the Standardization of
  Polycrystalline Data,
    William Parrish 97

A New X-Ray Diffraction Method for Quantitative Multicomponent
  Analysis,
    Frank H. Chung 106

The Al Fe Be$_4$ Intermetallic Phase in Beryllium,
   I. Brower, E. C. Roberts and T. J. Bosworth 116

High Speed Retained Austenite Analysis with an Energy
Dispersive X-Ray Diffraction Technique
   A. P. Voskamp 124

Quantitative X-Ray Diffraction Phase Analysis of the Oxidation
of Steel by a Direct Comparison Method
   R. R. Biederman, R. F. Bourgault and R. W. Smith 139

Low Energy X-Ray and Electron Absorption within Solids
(100-1500 eV Region),
   Burton L. Henke and Eric S. Ebisu 150

X-Ray Fluorescence Analysis of Portland Cement Through the Use
of Experimentally Determined Correction Factors,
   C. H. Anderson, J. E. Mander and J. W. Leitner 214

Specimen Standards for X-Ray Spectrometric Analysis of
Atmospheric Aerosols,
   Bruce E. Artz and Henry Chessin 225

A Versatile X-Ray Fluorescence Method for the Analysis of
Sulfur in Geologic Materials,
   H. N. Elsheimer and B. P. Fabbi 236

Sampling and Standards in a Recycled World,
   H. E. Marr III, W. J. Campbell and D. L. Neylan 247

X-Ray Cross-Sections in Design and Analysis of Non-Dispersive
Systems,
   Benton C. Clark 258

Use of Multiple Standards for Absorption Correction and Quantitation with Frieda,
   P. S. Ong, E. L. Cheng and G. Sroka 269

Resin-Loaded Papers - A Versatile Medium for Sampling and
Standardization,
   Stephen L. Law and William J. Campbell 279

Chelating Ion Exchange Resins and X-Ray Fluorescence,
   Donald E. Leyden 293

Can Regression Equations be Optimized by Finagling X-Ray
Intensities,
   M. Fatemi and L. S. Birks 302

X-Ray Fluorescence Analysis of High-Temperature Super-alloys -
Calibration and Standards,
   K. F. J. Heinrich and S. D. Rasberry 309

Contents

Quantification of Sub-microgram Elemental Concentrations Using Micro-dot Samples,
James M. Mathiesen 318

A Rapid Direct X-Ray Fluorescence Method for Simultaneously Determining Brass Composition and Plating Weight for Brass-Plated-Steel Tire Cord Wires,
James Gianelos 325

Quantitative Nondispersive X-Ray Fluorescence Analysis of Highly Radioactive Samples for Uranium and Plutonium Concentration,
W. L. Pickles and J. L. Cate, Jr. 337

The Effects of Self-Irradiation on the Lattice of $^{238}(80\%)PuO_2$ III,
R. B. Roof, Jr. 348

The Effects of X-Ray Optics on Residual Stress Measurements in Steel,
Chester F. Jatczak and Harald H. Boehm 354

X-Ray Diffraction Residual Stress Analysis Using High Precision Centroid Shift Measurement Techniques - Application to Uranium - 0.75 Weight Percent Titanium Alloy,
W. E. Baucum and A. M. Ammons 371

An X-Ray Amorphous Scattering Investigation of the Corrosion of a Pottassium Silicate Glass $K_2O-3SiO_2$,
R. W. Gould and M. S. Hill 384

A Review of X-Ray Diffraction Methods for Diffusion Studies,
J. A. Carpenter, Jr. and D. R. Tenney 395

Pole Figure Random Intensity Calculation Using Powder Integrated Ratios,
Carlos Sergio Viana and Gustau Ferran 416

X-Ray Emission from Laser-Produced Plasmas,
C. M. Dozier, P. G. Burkhalter, B. M. Klein, D. J. Nagel, and R. R. Whitlock 423

Calculation and Measurement of Integral Reflection Coefficient Versus Wavelength of "Real" Crystals on an Absolute Basis,
D. B. Brown, M. Fatemi, and L. S. Birks 436

X-Ray Production Cross Sections for Ti, Co, Ge, Rb and Sn by 16-44 MeV Oxygen Ion Bombardment,
R. P. Chaturvedi, J. L. Duggan, T. J. Gray, C. C. Sachtleben, and J. Lin 445

Some Biomedical Applications of Charged-Particle-Induced X-Ray Fluorescence Analysis,
J. L. Campbell, A. W. Herman, L. A. McNelles, B. H. Orr, and R. A. Willoughby 457

Contents

Qualitative Analysis of the Kossel Back Reflection Pattern from Selected Semiconductors,
    Robert L. Fitzpatrick 467

An Experimental Evaluation of the Atomic Number Effect,
    L. Parobek and J. D. Brown 479

X-Ray Emission from Thin Film Materials,
    P. A. Stine, S. J. Hruska and G. L. Liedl 487

Auger Electron Emission Micrography and Microanalysis of Solid Surfaces,
    K. Hayakawa, H. Okano, S. Kawase, and S. Yamamoto 498

A Combined Photoelectron/X-Ray Fluorescence Spectrometer,
    H. K. Herglotz and D. R. Lynch 509

A Spherically Bent Crystal X-Ray Spectrometer with Variable Curvature,
    Donald L. Parker 521

Measurement of the X-Ray Sensitivity of Silicon Diodes in the Energy Region 1.8 to 5.0 KeV,
    Jacque J. Hohlfelder 531

Development of the High Performance "Solfa" On-Line Analyser to Measure Total Sulphur in Petroleum Distillates and Residual Fuels Using Non-Dispersive X-Ray Fluorescence,
    C. F. Gamage and W. H. Topham 542

Automatic Data Acquisition and Reduction for Elemental Analysis of Aerosol Samples,
    J. F. Harrison and R. A. Eldred 560

A Secondary-Source, Energy-Dispersive X-Ray Spectrometer and its Application to Quantitative Analytical Chemistry,
    R. P. Larsen and J. O. Karttunen 571

Author Index 585

Subject Index 587

STANDARD REFERENCE MATERIALS AND MEANINGFUL X-RAY MEASUREMENTS

H. Thomas Yolken
Institute for Materials Research
National Bureau of Standards
Washington, D.C. 20234

ABSTRACT

A review of the procedures and efforts at the National Bureau of Standards (NBS) to provide for meaningful measurements through the use of Standard Reference Materials (SRM's) is presented.

The examples of NBS standardization efforts for x-ray analysis range from basic metrology to applied environmental measurements. These examples include a determination of x-ray wavelength by a method which in part utilizes simultaneous x-ray and optical interferometry measurements of the atomic planes of near perfect silicon. In addition, Standard Reference Materials (SRM's) are being developed and applied to trace element analysis using x-ray fluorescence techniques. These efforts include development of SRM's for trace element analysis of air particulates. In another area, work is proceeding on the development of a silicon powder Standard Reference Material intended for x-ray diffractometer calibration. An effort to develop a suitable x-ray diffraction technique to determine the amount of quartz in mine dust is also underway. NBS efforts to provide SRM's for the calibration of electron microprobes and the validating of correction factor calculations are also described.

INTRODUCTION

For more than 65 years, the National Bureau of Standards (NBS) has been providing Standard Reference Materials (SRM's) to the Nation's scientific and technological laboratories to help ensure meaningful measurements. This paper is an attempt to describe the NBS view of meaningful measurements and the role Standard Reference Materials play in their realization, as well as to describe the current NBS efforts to develop and produce SRM's for the x-ray field.

ACCURACY AND PRECISION

The goal in any series of measurements is to produce meaningful data; this requires that the data be both accurate and precise.

Although most readers are aware of the difference between accuracy and precision, for sake of clarity, the following analogy is presented (figure 1), which illustrates the difference between these terms.

Three imaginary marksmen fire rifles at targets. In the top target, the marksman is inaccurate and imprecise. In the middle target, the marksman is quite precise, but

ACCURACY AND PRECISION

An inaccurate and imprecise marksman 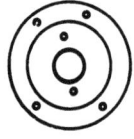 No statement of potential accuracy possible

An inaccurate but precise marksman  Potentially accurate Find source of systematic error

An accurate and precise marksman 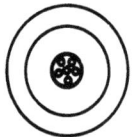 Accuracy cannot be attained until precision is first achieved

Bull's-eye corresponds to target value or 'true' value

Figure 1.

inaccurate. In the bottom target the marksman is both accurate and precise. The analogy in all cases is that the bull's eye corresponds to the target value, of course, or "true" value.

An example of the difference between accuracy and precision that is much closer to the x-ray field is the determination of the wavelength of x-rays which is important to many scientific and technological endeavors.

Early work by Siegbahn to measure the wavelength of x-rays relied on knowing the spacing of the (200) planes of rock salt, and then using this value to calculate wavelength from x-ray diffraction data. Siegbahn's knowledge of the "true" value of the spacing of the (200) planes of rock salt was obtained from its density, and atomic weight data together with Avogadro's number.

However, Siegbahn could measure wavelengths in terms of this spacing with much more accuracy than the spacing itself was known because of the uncertainties in the density, atomic weights and Avogadro's number. In other words, he could make relative wavelength measurements to six significant figures whereas the "true" spacing in absolute units was known only to four. In parallel with the case of the marksmen presented earlier, his shots were very close together but were a "barn door" away from the bull's eye. This difference of two orders of magnitude between "precision and accuracy" led to the decision to define arbitrarily the spacing of the (200) planes of rock salt in terms of X units. The reader is probably aware of the many difficulties that arose throughout the years in the x-ray field as a result of this difference between accuracy and precision.

It was later found that x-rays could be diffracted by ruled gratings thus providing a means of making wavelength measurements independent of crystal structure. The accuracy of this method, which was refined through the years, was about 20 to 30 parts in 10^6 in terms of primary optical wavelength.

An NBS program that has been underway for the last several years is that of Richard Deslattes and his coworkers [1,2]. Their initial efforts were to determine x-ray wavelength to 1 part in 10^6 in terms of primary optical wavelength. Their experiment was a two-step process. First, the simultaneous counting of x-ray and optical fringes from a silicon crystal and interferometer to determine crystal spacings and, second, the Bragg reflection from the crystal to determine x-ray wavelength.

The main parts of the apparatus used in the experiments were a single crystal of silicon-machined to form an x-ray and optical interferometer, and a uniquely machined ultra-small movement device. The x-ray interferometer was composed of the rows of atoms in the crystal and the optical interferometer is made of the faces of the crystal.

In the actual experiment, x-rays and laser light are used simultaneously to record superimposed interferometric tracings, as one part of the silicon crystal is slowly moved relative to the other part by the ultra-small movement device. Figure 2 shows an actual super-imposed interferometric tracing of the optical and x-ray fringes with approximately 1500 repeat distances for the (220) silicon plane for each passage of one-half optical wavelength for the helium-neon laser. From this experiment the lattice parameter of silicon can be related to the primary length standard, the wavelength of light. Following this by a Bragg reflection experiment on the same silicon crystal yields x-ray wavelength to one part per million accuracy. From this illustration, one could say that the marksman is getting much closer to the bull's eye, or stated another way, the systematic errors have been greatly reduced.

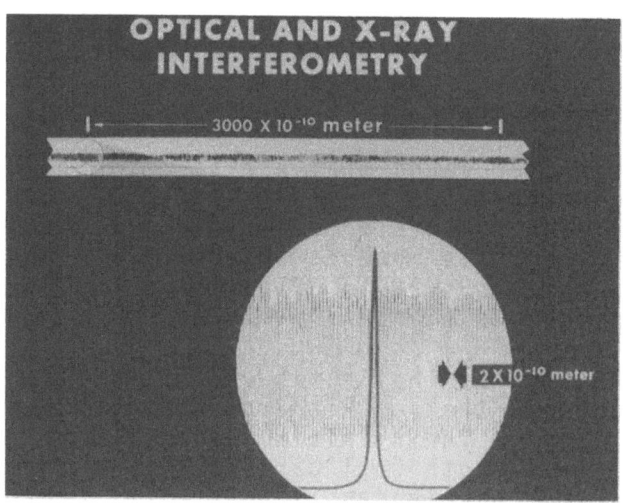

Figure 2.

Part of Deslattes' current work is concerned with directly determining Avagadro's constant to one part in a million accuracy. Avagadro's constant is currently known indirectly to only about 5 parts in 10^6 and directly to 20 parts in 10^6. To improve this situation, the spacing of the silicon atoms in a "perfect crystal" to 0.2 parts in 10^6 is being determined at NBS using x-ray and optical interferometry. These data, coupled with new, highly accurate determinations of both the density of the silicon crystal and the atomic weight of the silicon in the crystal now being determined at NBS should directly yield Avagadro's constant to one part in 10^6 accuracy.

DEFINITION AND DESCRIPTION OF NBS-SRM's

NBS-SRM's are well-characterized materials, with composition or physical properties certified by NBS, produced in quantity to: 1) Permit calibration "onsite by the user;" 2) Form the basis for equitable exchange of goods; 3) Tie measurements to compatible National and International Systems; 4) Permit evaluation of measurement techniques.

One of the most important facts built into most NBS-SRM's is known accuracy. The certification of an NBS-SRM is generally established by one or more of the following: (in order of preference)

1 - Use a reference method of analysis run independently by two or more analysts (at least one of whom is at NBS);
2 - Use two or more reliable independent methods (at NBS);
3 - Use an interlaboratory comparison system and allow several independent methods (NBS plus cooperating laboratories).

In the first case, a reference method (according to the NBS definition) is a method of known or proven accuracy, and its use assures the accuracy of the determination if personal bias is eliminated. This is why the measurements must be made, if possible, by two or more analysts.

The second way to certify an NBS-SRM, used when a reference method is not available, consists of using two or more independent methods. Each of these independent methods must have estimated systematic biases that are small relative to the accuracy goal set for certification.

The first two routes to certification are subject to critical scientific review through rigorous statistical analysis of the data obtained. This, however, is not the case for the third route, (interlaboratory or round-robin), which is subject to a somewhat less rigorous statistical analysis of the data. The valid proof of this route to certification is that, in a system under good quality control, it works.

Each of these three routes will now be discussed by presenting examples of the use of each to certify NBS-SRM's.

The first case is illustrated in Table 1 by the data for lead determined in a series of glass SRM's containing trace elements at levels from 0.02 ppm to 500 ppm. The lead content was determined by isotope dilution mass spectrometry, a recognized method of known and proven accuracy. The similar standard deviations obtained by the two analysts working 2000 miles apart is an important sign that the system is under control. The certified value of 426 ± 1 ppm for lead is believed to be a conservative figure.

Table I.

LEAD IN 500 PPM GLASS

Rod No.	Analyst 1	Analyst 2
2	426.5	---
13	426.2	---
18	425.6	425.9
48	426.1	426.0
56	426.9	425.0
66	426.0	425.4
78	426.2	425.6
106	425.7	---
Average	426.15	425.58
σ	± 0.41	± 0.40
95% L.E.	± 0.98	± 1.11

Table II.

CERTIFICATION BY TWO OR MORE INDEPENDENT METHODS

Cadmium in Bovine Liver

Sample	Concentration (μg/g)		
	ID-MS	Atomic Absorpt	Polarography
1	0.32 0.29	0.29	--
2	0.26 0.27	0.24 0.26	0.26
3	0.27 0.27	0.26 0.27	0.16*
4	0.28	0.24 0.27	0.28
5	--	0.30	0.28
6	0.26	0.26	--
$\bar{X}$	0.28	0.27	0.25
2σ	0.04	0.04	0.11

Range (all results) 0.24 – 0.32

*Outlier

Recommended Value 0.27 ± 0.04

The data in Table II illustrate the use of the second route used by NBS to certify SRM's; in this example three independent methods of analysis are used to determine cadmium in our bovine liver SRM. It should be noted that isotopic dilution mass spectrometry is not at present a reference method for the determination of cadmium.

The third route to NBS certification, which utilizes interlaboratory comparisons or round-robins, is used for various reasons and to a limited extent. In Table III the data are shown for a round-robin involving 6 different, highly competent laboratories in the determination of the carbon content of a steel SRM. Note that although 4 different methods on 6 different sample weights were used, the results obtained were highly satisfactory. Note also, however, that a previously issued NBS-SRM similar to the SRM under study was required for internal quality control in each of the labs. As discussed earlier, the round-robin approach should be used only in fields of science and technology that have obtained good quality control.

Table III.

CERTIFICATION BY CONSENSUS

SRM 337 – Basic Open-Hearth Steel

Carbon

ANALYST	METHOD/VARIATION (Note: Combustion step common to all)	% CARBON
1	Gravimetric – 1 g sample	1.08
2	Gravimetric – 3 g sample	1.06
2	Volumetric – 1 g sample	1.06
3	Gravimetric – Factor weight (2.73 g) sample	1.06
3	Gasometric – 1 g sample	1.07
4	Gravimetric – Half-factor weight (1.36 g) sample	1.06
5	Gravimetric – Half-factor weight (1.36 g) sample	1.08
5	Thermal Conductivity – 0.7 g sample	1.08
6	Gravimetric – 0.7 g sample	1.07

Mean 1.07

4 different methods
6 different sample weights

ALL LABS USED SRM 16d AS CONTROL

The dangers in using the round-robin approach are either that a field is not ready for it, or that the proper constraints may not be applied. The data in Table IV illustrate the danger in using the round-robin approach for a field that is not yet under quality control. Note the wide range of values reported by different labs for various elements in this orchard leaves SRM (this data was not used for certification). The center of the bull's eye is very elusive in this case.

Table IV.

CERTIFICATION BY CONSENSUS

A DANGER

Orchard Leaves

Element Determined	No. of Labs	Mean $\bar{X}$	Range	$S/\bar{X}$ %
P	10	0.20%	0.14 – 0.24	25
Al	10	222 μg/g	99 – 401	36
Fe	11	239 μg/g	151 – 367	28
Mo	5	5.4 μg/g	2.3 – 10.5	59

DEVELOPMENT OF SRM's FOR X-RAY INSTRUMENTATION

The remainder of the paper is devoted to describing some examples of ongoing efforts at NBS to develop, produce and certify new SRM's for calibrating various types of x-ray instrumentation.

A joint effort by the Surface Microanalysis Section (headed by Kurt Heinrich) in the Analytical Chemistry Division and the Glass Section (headed by Wolfgang Haller) in the Inorganic Materials Division is to develop small fiber and particle standards for calibration of electron microprobes and scanning electron microscopes. These proposed SRM's are to serve a different function than the microprobe SRM's that are currently available from NBS. Whereas, the existing SRM's are intended primarily as a check of or an aid to refining correction factor theory, the proposed small fiber SRM's are intended mainly for instrument calibration. This is necessary because correction factor theory is presently not applicable to particles smaller than the volume of specimen normally excited by the electron beam.

The preliminary work, just completed, utilized glass fibers of controlled size and composition. The chemical composition of the fibers can be produced to span a wide variety of compositions. Several elements can be present simultaneously with the concentration for any element of interest as high as 20 percent.

The fibers are produced by allowing molten glass to flow out of the bottom of a platinum crucible on to a revolving drum. The diameter of the glass fibers can be controlled by varying either or both the viscosity of the liquid glass, which is a function of the temperature or the speed of the drum. Typical fibers are from 5 to 30 micrometers in diameter.

The glass fibers are quite homogeneous both along their length and across their diameter. In addition, the composition of the fibers as determined by the microprobe is in agreement with conventional macroscopic analysis of bulk glass specimens made from the same melt of glass. This is an expected result as volatility is generally not a problem at the low melting points of the glasses used.

Future efforts in this area will most likely be aimed at producing glass fiber SRM's with compositions similar to various particulate air pollutants such as asbestos fibers.

Another example is the efforts of Stanley Rasberry and Kurt Heinrich to make existing NBS-SRM's more useful in x-ray fluorescence analysis, thereby, reducing the number of different SRM's needed for calibration of multi-element systems.

Their approach was to improve the methodology for the calibration of x-ray fluorescence spectrometers to account for interelement effects. The effects of secondary fluorescence and of absorption were considered separately in deriving their new empirical calibration equation. Figure 3 is a graphical summary of the interelement effects, in which relative intensity is plotted versus mass fraction concentration. Curve A is the expected result with no secondary fluorescence or adsorption present. The hyperbolic Curve B, for preferential absorption by the matrix is the same as previously derived empirical corrections. Curve C is the hyperbolic curve for preferential absorption by the analyte, while Curve D is their newly derived empirical expression for secondary fluorescence correction. The shape of this curve (D) is caused at low concentrations by an over abundance of exciter and at high concentrations by a lack of exciter.

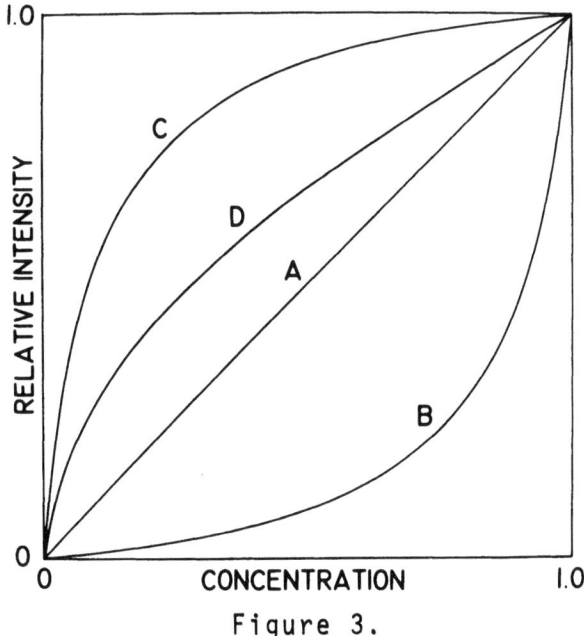

Figure 3.

This effort to better define these corrections reduces the number of SRM's of different concentrations needed for reliable calibration.

Another effort just beginning at NBS by John Taylor and coworkers is the development of an actual urban air particulate SRM, certified for chemical composition. X-ray fluorescence at this time is a very likely candidate method for the rapid, low cost field analysis of urban dust samples. The proposed air particulate SRM's, are useful in calibrating many types of methods of analysis. However, they should be particularly useful in the direct calibration of the proposed rapid field use of x-ray fluorescence instrumentation. In addition, they will be useful as an aid in evaluating other standardization methods and correction factors proposed for particulate analysis by x-ray fluorescence. An example of the latter use is shown in the data in Table V reported by Giauque and coworkers at the Lawrence Berkeley Laboratories (3). They utilized the NBS orchard leaves SRM to evaluate their calibration method and absorption corrections for x-ray fluorescence. The agreement appears to be excellent.

Table V.

ANALYSIS of NBS SRM 1571 ORCHARD LEAVES

	X-Ray fluorescence ppm	NBS ppm
Cr	2.5 ± 1.6	2.3
Mn	88.6 ± 2.2	91 ± 4
Fe	276 ± 8	300 ± 20
Ni	1.3 ± 0.4	1.3 ± 0.2
Cu	12.6 ± 0.6	12 ± 1
Zn	23.7 ± 0.8	25 ± 3
As	10.6 ± 0.8	14 ± 2
Br	9.3 ± 0.6	10
Rb	11.0 ± 0.8	12 ± 1
Sr	36.6 ± 1.2	37
Pb	45.4 ± 2.0	45 ± 3

Returning to the NBS effort, a contract has just been awarded by NBS to collect about 25 kilograms of urban dust in each of two cities, St. Louis and Los Angeles. The collection of the sample in St. Louis will take about 8 months; the collection apparatus (a large bag house that will collect particles smaller than about 10 μm) will then be moved to Los Angeles and the sample taken there over the next eight-month period.

The sample from each city will be blended separately and the homogeneity will then be checked using x-ray fluorescence analysis. Lead, cadmium, mercury, zinc, cobalt, nickel and beryllium are some of the metals most likely to be certified initially, utilizing several different measurement methods.

The next example of current NBS efforts is the work of Swanson, Hubbard and coworkers to produce a silicon SRM for x-ray diffraction. The proposed SRM is intended to be used both as an internal standard and as a check of goniometer accuracy. About 20 kilograms of high grade silicon powder (200 mesh) were blended and the homogeneity in terms of lattice parameters was determined for the whole lot of material by high precision x-ray diffraction work. The actual lattice spacings were next determined relative to a reference tungsten powder. The lattice constants of the tungsten were determined several years ago by Swanson and coworkers at NBS using a high-accuracy back-reflection camera with rigidly controlled atmosphere and temperature. The lattice constants for the silicon SRM will most likely be certified to about 30 ppm.

As an independent check of the method, a comparison was made between the lattice spacing of a silicon crystal as determined by x-ray interferometry measurements, and powder ground from the silicon crystal as determined by x-ray diffractometer measurements. Tungsten powder was used as an internal standard for the diffractometer measurements. Agreement was well within the proposed accuracy limits of certification.

A final example is work by Perloff and coworkers to assess and improve the quantitative analysis of alpha-quartz dust by x-ray diffraction. Alpha-quartz is a dangerous air pollutant in mines and other industrial environments. If this and other efforts supported by National Institutes for Occupational Safety and Health (NIOSH) show that x-ray diffraction is a viable approach to the rapid determination of alpha-quartz, then an SRM would be the next logical step.

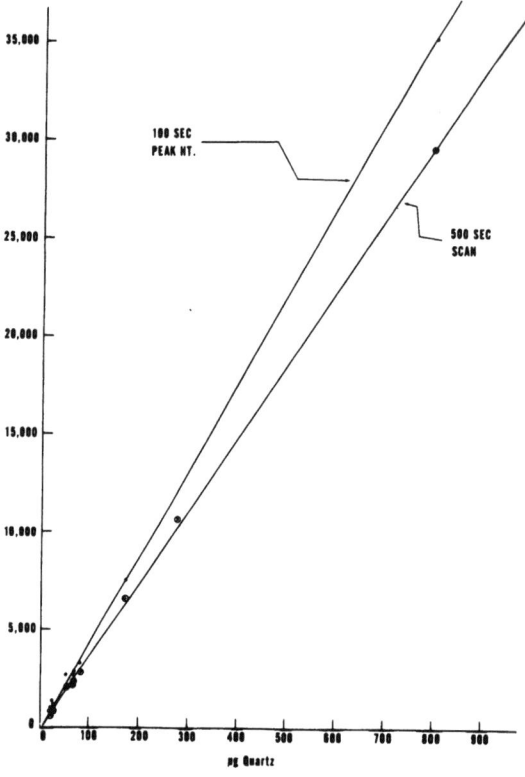

Figure 4.

Figure 4 shows the x-ray diffractometer responses of pure alpha-quartz versus the amount of an air-deposited sample on a silver-membrane filter. The lower limit of detection by x-ray diffractometer methods appears to be about 20 micrograms for pure alpha-quartz particles 10 micrometers and smaller; while the industrial safety laws call for detection of quartz in actual dust samples down to at least 100 micrograms of quartz. Proposed future NBS efforts will be aimed at investigating actual dust samples containing quartz to determine quantitatively the lower limit of detection when interferences are present, followed possibly by investigating candidate SRM materials.

The vignettes presented are examples of SRM-related x-ray standardization efforts currently underway at NBS. In closing, a brief glance into the crystal ball shows what may be in store for the future. I see that increased efforts in fluorescence and x-ray diffraction will come about as these techniques mature in the environmental field, and perhaps later in the clinical health field as they have done in the heavy industrial areas.

We can expect to see a continued and perhaps expanded effort at NBS to make a few SRM's go a long way. This might be done by both improvement in the calculation of various correction factors and in methods.

The last look into the crystal ball shows a group of SRM's under development for surface analysis techniques. These are not directly in the classical x-ray field, but in a somewhat related area -- photo-electron spectroscopy (ESCA) and Auger-electron spectroscopy. These SRM's will help answer such questions as: How much, and what valence of the analyte is present on the surface. This area will become increasingly important to a broad set of technologies, including: miniaturized integrated electronic circuitry, heterogeneous catalysis, corrosion prevention, and adhesion.

REFERENCES

1. Richard D. Deslattes, "X-Ray/Optical Interferometry and X-Ray Fundamental Constants Measurements," Proceedings of the International Conference on Precision Measurement and Fundamental Constants, National Bureau of Standards, August 1970, NBS Spec. Publ. 343, U. S. Government Printing Office, Washington, D. C. 20234, 1971.

2. J. H. Sanders and A. H. Wapstra, "Atomic Masses and Fundamental Constants," Proceedings of the Fourth International Conference on Atomic Masses and Fundamental Constants held at Teddington, England, September 1971, Plenum Press, New York 1972.

3. Robert D. Giauque, Fred S. Goulding, Joseph M. Jaklevic, and Richard H. Pehl, "Trace Element Determination with Semiconductor Detector X-Ray Spectrometers," Anal. Chem. $\underline{45}$, 1, 671-681 (April 1973).

NOTE: Information on NBS-Standard Reference Materials may be obtained by writing to: The Office of Standard Reference Materials, Room B311, Chem. Bldg., National Bureau of Standards, Gaithersburg, Maryland, 20234.

STANDARD REFERENCE MATERIALS - THEIR PRODUCTION AND USE

A. H. Gillieson
Department of Energy, Mines and Resources
Mines Branch, Mineral Sciences Division
Ottawa, Ontario, Canada

ABSTRACT

   The purpose and the value of standard reference materials (S.R.M.) are defined, and their increasing world importance is emphasized. The desirable features of an S.R.M. are outlined and the procedures necessary to meet, as far as possible, the necessary requirements in S.R.M. production are discussed. The method and organization of certification for homogeneity and composition are detailed, as well as the statistical procedures to evaluate the certification analysis. The present status and future plans for the production of S.R.M.'s by the Canadian Standard Reference Materials Project, Mines, Branch, Department of Energy, Mines and Resources, are outlined.

INTRODUCTION

   Standard reference materials are prepared primarily to assist the analyst in checking the correctness of his methods and results; they are compositional standards, rather than, say, dimensional standards such as are used for length or time. Unfortunately, they are misused to a certain extent inasmuch as they are sometimes used in place of secondary (laboratory) standards. For reasons of economy, standard reference materials (S.R.M.) should be used to establish the concentration of certain elements in secondary standards. The secondary standards are then used in the laboratory as required, whereas the S.R.M. are held in reserve to be sampled and compared to new secondary standards as stocks of old ones become depleted.

The actual cost of producing an S.R.M. is several times the amount charged the purchaser, yet its price must lie within the purchasing power of the small analytical laboratory. The Mines Branch is responsible for the production of both mineralogical and metallurgical S.R.M. In the United States, in general, mineralogical standards are prepared by the U.S. Geological Survey and metallurgical standards by the National Bureau of Standards. The United Kingdom has produced a variety of S.R.M.; of late, East Germany, Tanzania, and South Africa have come into the picture, supplying standards to satisfy domestic and export requirements.

An S.R.M. must be similar in composition to the material being processed but not necessarily identical; in addition, sufficient funds must be available to underwrite its production. We, at the Mines Branch, are very fortunate that several divisions have conventional size reduction, blending, and sampling equipment and supporting analytical laboratories.

We began our project with two basic raw materials, the first being a typical Sudbury ore characterized by its copper, nickel and iron content. We still have a supply of this standard (original weight about 500 lb.) and we have extended its analysis.

The second material prepared and assayed for reference was a syenite rock from the Bancroft area. Although initially it seemed destined to appeal more to geologists than to chemists, it has been sought after by mineral analysts as well, judging from the orders received.

The resulting supply of S.R.M. from this rock was quickly depleted and, within six years of the first sampling, another rock lot was taken. Mines Branch mineralogists were quick to point out that the replacement rock was a most inferior mineralogical specimen indeed. We replied by saying that it made little difference to us because our primary interest was in its chemical analysis hence its suitability as an S.R.M. The standards are to be used to establish the accuracy of analytical methods, that is, how we are to measure their worth, primarily, and their effectiveness in mineral and metallurgical chemical analysis.

CERTIFICATION OF S.R.M.

When a material judged to be suitable for an S.R.M. is first obtained, one most significant virtue, from the standpoint of chemical analysis, is its trace element content. So you must prepare the material in a manner so as to add an absolute minimum of these same trace elements to the sample.

The research scientists in the Mineral Processing Division came up with the suggestion that we grind the material autogenously; i.e. that the ore be used as the grinding medium. In most instances, fist-sized chunks of ore are loaded into the mill and, as they collide with and abrade one another, are ground. The amount of contamination is thus kept to a minimum.

Size of sample is our next consideration, i.e., the particle size to which the bulk sample is ground. In general, we agree that the ground pulp should pass through a 200-mesh Tyler screen; small enough for good homogeneity, minimum electrostatic agglomeration, and easy dissolution of sample.

Homogeneity is all-important in the preparation, handling, analysis and ultimate use of S.R.M. If you start off by considering the homogeneity of a sample with respect to trace elements, you had best forget about it entirely. The sample material, especially if it is a rock, usually contains a wide variety of minerals in several states of distribution. So we run a check on the distribution of one or more of the major constituents of the sample material. The method of analysis is usually non-destructive such as x-ray fluorescence which permits accurate determination of major constituents in a large number of samples. In some instances such as in the preparation of S.R.M. to be used in the calibration of an optical emission spectrometer, we have gone so far as to separately analyze each standard to ensure accuracy. Such an individualized procedure is the exception to the system usually employed.

To avoid segregation, the S.R.M. must be sampled for analysis as soon as possible after preparation for use; this is an essential point.

In order to establish analytical values for the elements of interest, we invited certain laboratories to collaborate with us, somewhat to our mutual advantage. But such an undertaking is becoming increasingly more difficult. The number of laboratories in the world that are interested, capable, and have the time to spare is limited. In addition, the number of S.R.M. in preparation in the world is increasing; therefore we are beginning to reach a state of diminishing returns, as far as collaboration on analysis is concerned.

Apparently the National Bureau of Standards has not entirely forsaken the round-robin approach to certification of S.R.M. but, in some instances, it will accept as reliable assays those in good agreement derived from two entirely different analytical procedures performed in their own laboratories. The most prominent features of this method are that it is self-checking and very precise.

AVAILABLE S.R.M.

I should like to briefly outline what the Mines Branch has to offer by way of standard reference materials, how and where they may be obtained.

Standard reference materials were first produced in Canada in 1958 through the initiative of Dr. Jan Hurwitz -- at that time working in the Mines Branch. As the result of a cooperative undertaking between manufacturers, the Mines Branch, and several universities, three phosphor-bronze disc S.R.M. and, later, four commercial-purity copper rod S.R.M. were certified, and these constitute the Non-Ferrous Group of S.R.M.

The non-metallic S.R.M., the second group, were prepared through the cooperation of the Geological Survey of Canada, the Mines Branch, the mining industry, the universities, and consultant analysts. Initially a Sudbury sulphide ore and a Bancroft syenite rock were certified, and these have been followed by three platinum-group ores, a molybdenum ore, a zinc-tin-copper-lead ore, and a copper-molybdenum ore.

The Geological Survey of Canada has produced three ultramafic-rock S.R.M. which are representative of ore-bearing rock, although only one contains sufficient elements of economic importance to be classified as an ore.

In preparation are three ferrous-alloy "setting-up" S.R.M., a zinc-lead ore, a gold ore, an iron ore, and a copper-nickel-iron sulphide ore.

Plans have been made for the production of four Canadian soil S.R.M., two tungsten ores, a cerium-iron alloy, a blast-furnace iron, and a blast-furnace slag.

CONCLUSION

Our plan, at the Mines Branch, is to endeavour to produce S.R.M. that will be of use to the Canadian mining industry. Some of those that I have mentioned were suggested by ourselves, but most of the standards in use in Canada were produced at the request of certain segments of industry; e.g., the platinum group and all our metallic standards. Our intention is to have standard reference samples representing all typical Canadian ores.

As these standards become available, we receive requests from literally every country around the world, primarily because there are relatively few S.R.M. available and because relatively few countries are producing them.

STANDARD DATA FOR THE IDENTIFICATION OF PHASES BY X-RAY DIFFRACTION

Howard F. McMurdie

Associateship of the Joint Committee on Powder
Diffraction Standards, National Bureau of Standards
Washington, D.C. 20234

ABSTRACT

The identification of crystalline phases by x-ray diffraction, either by powder or single crystal techniques requires a dependable body of reference data. It is not only necessary to have data on each phase which are accurate and complete, it also is desirable to have data on as wide a range of compounds as possible, and to have the data organized in such a manner as to be readily usable. The outstanding compilations which approach these goals are the Powder Diffraction File and Crystal Data.

The Powder Diffraction File, published by the Joint Committee on Powder Diffraction Standards has data covering about 22,500 phases, both organic and inorganic. These data are of various degrees of accuracy as is indicated by symbols. The File is continuously being improved by the addition of evaluated data from the general literature and by data produced by supporting projects, the principal one being the Joint Committee Associateship at the National Bureau of Standards.

To be noted in the File with a star, and to be truly considered standard data a powder pattern must be complete in the sense of including all reflections above the minimum "d" spacing covered, both weak lines and those with large "d" spacings. Since the best test of a pattern is its own internal consistency, the reflections must all have hkl's assigned and must show a good agreement between the spacings observed and those calculated from a refined cell, and they must be consistent with the known space group. This agreement can be best obtained by the use of an internal standard and a computer program. The intensities should be measured by a method which minimizes the effect of crystal orientation.

The PDF is provided with search procedure manuals arranged on a scheme of the strongest lines to help in locating data matching that from an unknown. A computer program for rapid searching is available. A recent development is the inclusion of a "reference intensity" to aid in estimating the quantitative analysis of mixtures.

Crystal Data is a compilation now in the third edition made at the National Bureau of Standards and published by the Joint Committee on Powder Diffraction Standards. It contains data on the unit cell parameters of over 24,000 phases. These data are arranged by crystal system and axial ratios to simplify identification of phases from unit cell data obtained from single crystal cameras.

Both of these large compilations are also important reference sources for crystallographic information giving structural information and literature references.

INTRODUCTION

The x-ray diffraction method of chemical analysis does not have the glamor of more modern methods, but it is used increasingly and has several strong advantages over other techniques. It is quick, uses a small sample, is nondestructive, and most important gives results in terms of crystalline phases rather than elements. It depends, however, on having available a dependable source of standard reference data, for comparison with data from unknowns. Only laboratories with very limited applications can depend on their own reference standards. The need is for a very large file of accurate reference data, and a quick simple way to relate data from unknowns to the standards. Two compilations approach these desirable goals. The Powder Data File (PDF), for use with powder diffraction, contains data on over 22,500 phases, both inorganic and organic. It is coordinated to several searching schemes and has annual additions and improvements. Crystal Data (CD) now available in a third edition, has unit cell parameters, some physical properties and bibliographic information on about 24,000 phases, organic and inorganic, and has these arranged (in such a way as) to facilitate identification of phases from single crystal data. Both of these compilations are published by the Joint Committee on Powder Diffraction Standards (JCPDS). More details on these compilations will be given below.

JOINT COMMITTEE ON POWDER DIFFRACTION STANDARDS (JCPDS) (1)

The JCPDS, under various names, has been in existence for about 23 years, created originally as a joint function of the American Society for Testing and Materials (ASTM) and the American Crystallographic Association (ACA). It is now an independent incorporated non-profit group, still retaining relationships with the ASTM and ACS but with additional sponsoring organizations including the American Ceramic Society, the Mineralogical Society of America and other foreign and domestic groups involved in solid state research. It has about 40 members including representatives of its sponsoring groups and other scientists both in the USA and abroad interested in crystallographic problems, and it maintains a full time managerial and secretarial staff at its permanent headquarters in Swarthmore, Pa. It also employs a small full time research group at the National Bureau of Standards (NBS), and several part time editors. The JCPDS collect new powder data from the general literature which are then evaluated by its editorial staff. New data are also produced by grants in aid to crystallographic workers in various parts of the world and by the associateship working at NBS with NBS cooperation. Each year data on about 2000 phases are added to the file, these being partly data on phases not previously covered and partly improved patterns replacing earlier, less reliable data. Each year new index books covering the full operational file are issued in which data can be located using the strongest intensities, or alphabetical listings.

The data in the PDF are classified as to quality in the judgment of the editors. Data of a high degree of reliability from well characterized samples are marked by a star. Other data where the cell parameters are given and which have accurate hkl indexing are indicated by an i (for indexed). Many patterns in the file have intensities measured by a qualitative method, are unindexed or are from samples whose purity is open to question. These patterns have no quality indication, but in many cases are the best available published powder data for that "phase". Certain patterns are marked with a 0. These are of questionable validity and usefulness, because of lack of information, possibility of being multiphase or because of lack of completeness. It is, of course, the aim and desire of the JCPDS to replace poor data as other information becomes available and in fact the percentage of starred patterns has greatly increased over the years.

CRITERIA FOR ADEQUATE POWDER REFERENCE STANDARDS

For a pattern to be fully adequate as a reference standard to be used for phase identification it should have the following qualities:

1. It should be made on material known to be of the stated composition and to be a single phase. This should be ascertained by chemical analyses, known preparation ingredients or optical examination.

2. An important aspect of a set of powder diffraction data are their own internal consistency. The "d" spacing data should give reasonable refined unit cell parameters and the observed and calculated diffraction angles for the various hkl's should agree to with 0.1° (2θ). Lines of strong or medium intensity should be resolved if greater than 0.2°(2θ) apart.

3. The pattern should be complete as to lines of low 2θ angle and to include lines as weak as one on a relative intensity scale of 100. It is not important for identification to have a pattern which includes the back reflection region.

4. The intensities should be measured by a quantitative method and the closest attention should be paid to the elimination of errors resulting from particle size and preferred orientation.

It should be self evident that patterns should meet these specifications. However most of those available at the present time, and many new published patterns do not meet these standards.

It is desirable to have reference data of the highest possible quality even if poor data is obtained from an unknown because of sample characteristics or instrumentation. Poor data from an unknown can often be successfully compared with excellent reference data, but problems will occur if either poor or excellent data from an unknown are compared with poor reference patterns.

Let us examine some types of unsatisfactory data. Table I shows how the available data on $SrCl_2 \cdot 2H_2O$ was improved by recent work, particularly in completeness. The old data was obtained in 1942 with Cr radiation on a Debye-Scherrer camera and was the only data published on the phase until very recently. The new information was the result of diffractometer work using Cu radiation at NBS. A worker using the old data as a reference standard and obtaining a good or excellent pattern from his sample, would conclude by comparison that his sample contained a second phase. He would conclude this by seeing the two strong low angle reflection lines, and also by finding two lines near 3.20A instead of one. On the other hand a worker obtaining a poor pattern from his sample would have no difficulty arriving at a satisfactory identification using the new data. A second example of an incomplete pattern is given in Table II for $MgCl_2 \cdot 6H_2O$, which shows a number of important missed lines. In this case the earlier work was done with Mo radiation before 1938.

TABLE I

Comparison of old and new data for $SrCl_2 \cdot 2H_2O$

Monoclinic
Low angle portion only

New dÅ	I	hkl	Old dÅ	I
5.63	60	$\bar{2}00$		
4.52	75	$\bar{1}11$		
3.964	35	111	3.96	70
3.238	35	310	3.24	60
3.229	30	002		
3.206	100	$\bar{3}11$-020	3.20	100
2.966	4	$\bar{1}12$		
2.871	20	021	2.87	55
2.816	2	400		
2.784	45	220	2.78	70

TABLE II

Comparison of old and new data for $MgCl_2 \cdot 6H_2O$

Monoclinic
Low angle portion only

New dÅ	I	hkl	Old dÅ	I
5.77	17	110	5.8	15
4.263	20	$\bar{1}11$		
4.101	100	111	4.10	100
3.955	30	$\bar{2}01$		
3.708	10	201		
3.556	30	020	3.57	15
3.068	2	021		
3.032	5	002		
2.981	35	310	2.98	20
2.883	65	220	2.88	50
2.740	35	$\bar{3}11$		
2.728	35	$\bar{1}12$	2.72	44
2.661	8	$\bar{2}02$		
2.643	90	112	2.65	75

The availability in recent years of programs for computer refinement of cell parameters from powder data and the ability to calculate the "d's" of the lines allowed by the cell parameters and the space group have made it possible to evaluate powder data more critically. The test of the accuracy of measurement of a pattern is a comparison of calculated and observed "d's" to ascertain whether the observed data give a satisfactory refinement for the unit cell, commensurate with the space group. However, much poor data is still being published. Table III present recent published data for Ni_2FPO_4 done with Fe radiation. The original article reports "d" spacings and gives unit cell parameters. The table indicates the poor relationship between the observed and calculated values expressed in degrees $(2\theta)(Fe)$. This poor fit casts doubt on the choice of cell, purity of phase or accuracy of measurement.

A particularly interesting example of recent data of questionable validity is shown in Table IV for γNa_2SO_4 (at 250°C). Differences of up to 1.9° (2θ) lead to a complete lack of faith in the cell, phase purity and ability to measure patterns.

Another type of pattern which produces poor data for identification is shown in Table V. This is for a cubic compound, $Li_4B_7O_{12}Br$. Here the data seems to have been used primarily to obtain an accurate cell size by extrapolation to 180° (2θ). This data, only partially given here, may or may not have given a good cell, but the errors of up to a third of a degree (2θ) in the front part of the pattern, the important part for identification, make it poor for an identification reference. These data were obtained by diffractometer and one assumes the alignment of the equipment or sample, was poor.

The accuracy of "d" spacing and the inclusion of extra lines can be checked by computer, if a space group and cell information are known, but the evaluation of intensities is a more difficult problem. There is no simple test, but one knows that peak heights will cover a wide range of relative intensities and very strong lines will not occur in the back reflection region because of the geometric factors involved. When such strong lines are indicated in a pattern it is usually a result of high absorption in the front part of a Debye-Scherrer pattern. Of course the intensities can be checked by calculation in cases where accurate structural information is available. Experimentally the greatest sources of error in powder intensity measurement are large crystallite size and lack of random crystal orientation, the latter effect being particularly pronounced when using a diffractometer because of the packing usually involved.

TABLE III

Comparison of Observed and Calculated Data for Ni_2FPO_4

Monoclinic
Low angle portion only

hkl	dÅ		Δ2θ° (Fe)	
	Cal	Obs.		
002	4.605	4.526	+	.43
121	4.457	4.345	+	.65
201	4.299	4.251	+	.30
220	4.242	4.200	+	.27
211	4.070	3.867	+	1.47
1̄31	3.783	3.786	−	.05
310	3.652	3.583	+	.60

TABLE IV

Comparison of Observed and Calculated Data for γNA_2SO_4 (at 250°)

Hexagonal
Low angle portion only

hkl	dÅ		Δ° (2θ) (Fe)	
	Cal	Obs.		
100	4.67	4.33	−	1.89
101	3.95	3.95		0.00
002	3.658	3.607	−	.45
102	2.880	2.867	−	.18
110	2.705	2.705		0.00
111	2.533	2.596	+	1.15
200	2.334	2.348	+	.30

TABLE V

Comparison of Observed and Calculated Data for $Li_4B_7O_{12}Br$

(cubic, $a° = 12.20$)

Partial pattern only

hkl	dÅ		$\Delta 2\theta$ (Cu)
	Cal	Obs.	
200	6.02	6.10	.19
220	4.26	4.31	.23
222	3.49	3.52	.23
400	3.02	3.05	.29
420	2.70	2.73	.37
422	2.47	2.49	.30
12 0 0	1.017	1.017	---

NBS FELLOWSHIP METHODS

Data of excellent quality in regard to "d" spacings and intensities can be produced by various methods. Certainly Guinier cameras and diffractometers are both capable of fine results. At NBS we routinely use a diffractometer operated at a scanning rate of 8 min/degree (2θ) and with an internal standard usually W or Ag for "d" spacings. The use of an internal standard gives corrections which can be applied throughout the pattern to compensate for various instrumental and mounting errors and for mismeasurement due to paper or film shrinkage. Our use of Ag or W is not fully satisfactory because the first line of Ag is at 38° (2θCu) and that of W at 40°. NBS in cooperation with the JCPDS is now completing work on measuring the pattern of a sample of Si which will be distributed by NBS as a Standard Reference Material for use as an internal standard in powder diffraction studies. NBS uses a crystal monochromator in the diffracted beam and thus eliminates β radiation. We are also able to use Cu radiation on all compositions.

Our 2θ data and known or estimated unit cell parameters are the input for a least square refinement (2). The final refinement must give a difference in 2θ values of not greater than 0.04° over the range reported or the data are not released. A poorer fit is considered to raise questions as to phase purity, or choice of cell parameters.

The optimum mounting in the diffractometer for the measuremen of "d" spacings is one which is tightly packed so that diffraction is from a thin dense layer. However, this is the poorest possible condition for intensity measurements since it promotes preferred orientation. Therefore separate patterns are produced for intensity determinations, the sample being drifted into a cell employing no packing or smoothing. It is also important that the sample be of very fine particle size, and where grinding results in broadened lines, a rotating holder is useful. The problem of overcoming orientation is a complex one, each material requiring special techniques. Methods of detecting orientation problems and of eliminating them are discussed in an article by B. Post (3). At NBS we make at least three intensity patterns of each material, which must show reasonable agreement. In a number of cases we have compared our experimental results with those calculated from published structure studies and found good agreement.

We report peak height relative intensities. We realize that integrated intensities are more scientifically defensible and are not affected by degree of crystalline perfection, but for identification the peak heights are more practical. NBS has published a chart which experimentally compares the two kinds of intensity measurements for materials of average crystal perfection (4).

The patterns prepared by the NBS Associateship are, of course, published in the Powder Diffraction File, but they are also published in an NBS Monograph (5) series issued annually. Here we include other relevant information on structure type, optical data, polymorphism, and available pertinent references.

REFERENCE INTENSITY

The NBS Associateship, in cooperation with Dr. J. Visser of the Netherlands, have developed a Reference Intensity Value (I/I cor), to relate the intensities of the strongest peaks of different materials. This is useful as an aid in estimating the relative quantities of materials in mixtures. The value relates the intensity of the strongest peak of a material to the strongest peak of αAl_2O_3 (corundum) in a 50-50 mixture by wt. Al_2O_3 which was chosen because of its chemical stability, and its ready availability. The I/I cor value is reported on almost all new compounds studied at NBS and by an increasing number of other workers.

CALCULATED POWDER PATTERNS

In some cases accurate cell parameters and atomic positions

are known from single crystal studies but a usable powder pattern is not available. In such cases, and where the material is unstable or is difficult to obtain or handle, a pattern is calculated by NBS or others and the data added to the PDF. These patterns are designed by a "c" for calculated. This, of course, is done only where the atomic positions are fixed, or the single crystal structure has given a low R factor. Both the integrated and peak intensities are reported for these patterns. The latter requiring a calculation which produces data simulating the resolution of a good diffraction pattern (6).

INDEXING SYSTEMS

No body of data is useful without practical indexes and there is considerable thought and work on this problem by a special subcommittee on the JCPDS. At present there are three principle printed indexes issued annually covering the full inorganic operational file. The best known, called the "Hanawalt" is based on three permuted entries for each phase using the three strongest lines of the front reflection portion of the pattern. The "d" spacings are in groups with the position within the group ordered on the second permuted line. There is an overlap in the groups equal to approximately twice the expected experimental error.

The second index, known as the "Fink" was devised to place less importance on accurate intensities. Here the eight strongest lines are used with six entries for each phase, and placed in order of decreasing "d's" rather than on intensity. This index is most useful when intensities are distorted or with electron diffraction data.

A third index, called the "Davey-KWIC" is an alphabetical chemical word listing with parts of the chemical name permuted (i.e. Potassium chloride and Chloride, potassium). An alphabetical mineral list is also provided.

The organic data is searched similarly with a Hanawalt type and alphabetic indexes. There is a continuing demand for more specialized indexes and for the sale of cards for special groups of compounds such as poisons or paint pigments. A numerical index of minerals is being prepared as a trial to test the practicality of such special listings.

Besides the book indexes, the numerical and chemical data of the file are on magnetic tapes for searching by computer. Tapes can be leased, or searching can be done utilizing a program set up by the JCPDS using a central tape and ones own computer interface. Programs for searching are constantly being improved.

A few years ago an earlier version of a search program was discussed at the Denver Conference (7).

The complete file is also available on microfiche for use with a viewer. This is particularly useful where storage space is at a premium.

CRYSTAL DATA

Crystal Data is a compilation based on unit cell parameters of over 24,000 crystal phases, both organic and inorganic. This compilation is now available in the third edition, published and distributed by the JCPDS. These data are arranged by crystal system and then by cell size for the cubic phases, and by axial ratios for the other systems. If one has an unknown crystal of suitable size, it is simple to obtain cell parameters by a single crystal diffraction camera or the axial ratios from optical goniometry, and to identify it by the use of this compilation. The publication gives, in most cases, other pertinent information such as space group, density, crystal habit, cleavage, twinning forms, optical data, known structural relationships and relevant literature references. This compilation, newly published in two volumes, (organic and inorganic) is complete with name and formula indexes, and has extremely good coverage of the crystallographic literature up to 1967. The data were assembled and evaluated by members of the staff of NBS and Cambridge University with help and advice from a number of special experts. Dr. J.D.H. Donnay, who was the editor of the first two editions was also the general editor of this edition with the cooperation of Dr. H. Ondik of NBS. The material is on computer tapes and the data have been checked by computer for internal consistency. The tapes will be available for distribution.

Work is continuing on the compilation and updating supplements will be issued at suitable intervals in the future.

Both of these compilations, the Powder Diffraction File and Crystal Data, are of considerable value beyond their use as reference data for identification. The thousands of references, their wide coverage, and the additional information on crystallographic properties and preparation make them major sources of information on the solid state.

CONCLUSION

We do not know what improvements and changes will be made in the techniques of phase identification by x-ray diffraction in future years. But we may expect increased resolution, improved monochromatization and extended ability to detect weak lines. It may be that the development and improvement of the energy dispersive method, with its attendant ability to obtain data quickly will extend automation of analysis. We, however, can be sure that we will need better reference data. Much of our present data is not up to present standards for this purpose, and even our best may seem very inadequate in a generation.

REFERENCES

1. Joint Committee on Powder Diffraction Standards, a non profit Pennsylvania Corporation with offices at 1601 Park Lane, Swarthmore, Pennsylvania 19081.

2. H.T. Evans, Jr., D.E. Appleman and D.S. Handwerker, "The least-squares refinement of crystal unit cells with powder diffraction by an automatic computer indexing method," (abs), Am. Cryst. Assoc. Annual Meeting, Cambridge, Mass. Program 42-43 (1963).

3. B. Post, "Laboratory Hints for Crystallographers," Norelco Reporter 20, 8 (1973).

4. National Bureau of Standards Monograph 25, Section 1, "Standard X-Ray Diffraction Powder Patterns," U.S. Government Printing Office, Washington, D.C. 20402 (1962).

5. National Bureau of Standards Monograph 25, Sections 1-11, "Standard X-Ray Diffraction Powder Patterns," Issued Annually, U.S. Government Printing Office, Washington, D.C. 20402 (1962-1973).

6. D.K. Smith, Norelco Reporter 15, 57 (1968). See also Lawrence Radiation Laboratory, Livermore, California, UCRL-7196, UCRL-50262, UCRL-70078 and UCRL-70674.

7. G.S. Johnson and V. Vand, "Computerized Multiphase X-Ray Powder Diffraction Identification System," in Advances in X-Ray Analysis, Vol. 11, p. 376, Plenum Press, New York, N.Y. (1968).

PROVISION, SUITABILITY AND STABILITY OF STANDARDS FOR QUANTITATIVE POWDER DIFFRACTOMETRY

Ron Jenkins

Philips Electronic Instruments

Mount Vernon, New York 10550

INTRODUCTION

Standards required for quantitative powder diffractometry are essentially of three types as shown in Table 1 - external instrument standards for the initial and long term check of instrument alignment (i.e., an "instrument" standard), external analytical standards, which are similar in composition to the specimens to be analysed, for the preparation of calibration curves for "in-type" analysis; and internal analytical standards for the minimization of problems due to variations in specimen absorption.

The ideal equipment alignment standard should be free from preferred orientation, stable and available in large quantities, since it should serve the dual role of providing a continuing indication of the performance of a given diffractometer, as well as allowing absolute intensity comparisons to be made between different diffractometers.

The selection of an internal or external analytical standard will of course be determined by the sample matrix range to be covered, but in all cases some pre-treatment is required of both the samples to be analysed and the analytical standards. It is useful to differentiate between the sample in its raw (as-received) state and that treated portion of the sample (or standard) that is eventually placed in the diffractometer. In this paper the former will be referred to as the "sample" and the latter as the "specimen". Two assumptions are frequently made in quantitative powder diffractometry, namely that the specimen has essentially the same phase composition as the sample, and that this phase composition remains unchanged during the calibration and analysis procedure.

Whether or not these assumptions are valid will depend very much upon the chemistry of the phases making up the sample. Under certain circumstances, the assumptions are certainly not true. In this paper mention will be made of polymorphic changes before irradiation due to grinding with the added introduction of strain and amorphism. There will also be some discussion of changes during irradiation due to the photoelectric effect of the X-rays and/or the influence of the atmospheric conditions.

TABLE 1

STANDARDS UTILIZED IN X-RAY POWDER DIFFRACTOMETRY

Type of Standard	Composition	Main Purpose
External instrument standard	Almost any material that can be reproducibly formed into a stable disc	Initial and long term check of instrument alignment and performance
External analytical standard	As close as possible to that of the samples to be analyzed	Preparation of calibration curves for "In-Type" analysis
Internal analytical standard	Of similar absorption and particle size to that of the sample range to be analyzed	Minimization of absorption problems

EXTERNAL INSTRUMENT STANDARDS

The essential function of an instrument standard is to provide a stable reference sample that can be measured to judge how well a diffractometer has been aligned, to recheck the performance of the diffractometer at any required point in time and to act as a means of comparing the absolute performance of different instruments. Most equipment manufacturers have their own ideas on what material should be utilized to fulfill the above requirement and will generally supply the requisite material in a suitable form. Standards for diffractometer alignment have traditionally included silicon, gold and various silicates.

A few years ago, the Philips X-Ray Powder Diffractometer User Group in Britain, investigated various alternative materials which might be used as instrument standards, and among those found to be most generally suitable were hot pressed alumina and Arkansas stone. Since the latter material is cheaper and more readily available, this was chosen as the basic reference standard for all

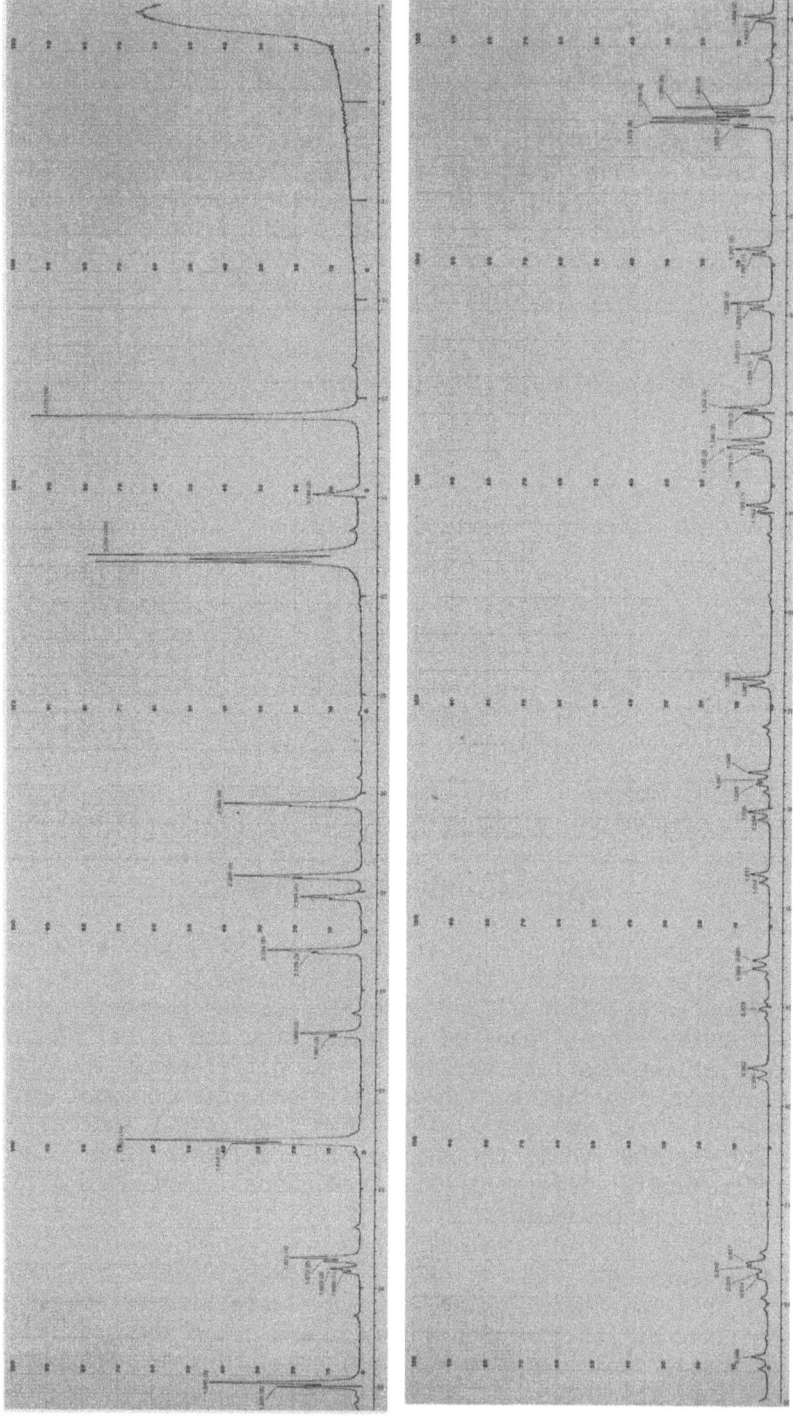

Figure 1. The diffraction of Arkansas stone.

members of the user group. The same material has also been adopted for all new powder diffractometers supplied by Philips in the U.S.A. Arkansas stone is essentially a very fine grained alpha quartz and its complete diffractogram is shown in Fig. 1.

This particular diagram was obtained using a computer controlled diffractometer and the digital to analogue converter has an overflow feature built in. This scales down by a factor of ten any signal in excess of full scale on the recorder, eg, the 3.36Å reflection in the diffractogram. Current practice in our own laboratory is to measure the absolute peak intensity of the 3.36Å reflection as an intensity check, and to use the separation of the lines of the quartz quintuplet at 67-69°2θ as a means of checking resolution.

The Arkansas stones are cut into sections 2 x 1 x 0.2 cms and cemented into standard aluminum holders. Each standard is cross-referenced against a master standard and only those within ± 10% of the original master standard are employed. A recent batch of 100 of these standards were cut and mounted in the manner described and all but two were found to lie within the required ±10% limits. We have found these standards to be free from orientation problems, very stable - they can be washed with soap and water if necessary, and particularly durable.

The need for a good universal instrument reference can be judged from results of several round robin tests with these standards among users where intensity differences of several hundred percent have been reported under nominally identical equipment conditions.

ANALYTICAL STANDARDS

Internal or external analytical standards are generally chosen to reflect the concentration and/or absorption range of samples to be analysed. In the cases of both analytical standards and samples to be analysed, some form of sample preparation is required to produce a specimen ready for analysis. This preparation may include such steps as grinding, sieving, drying and compacting and during the course of these treatments, the sample may undergo certain changes.

Table 2 lists some of the possible changes which may occur during the preparation of the specimen or indeed during the actual analysis. By far the greatest potential source of problems is in grinding the sample where the introduction of strain and amorphism is common and where under certain circumstances, polymorphic and compositional changes may also occur. Needless to say, incomplete sieving of a ground heterogeneous powder can lead to further complications. Decomposition and polymerisation may occur during the

TABLE 2
POSSIBLE CAUSES FOR COMPOSITIONAL VARIATIONS BETWEEN

THE AS-RECEIVED SAMPLE AND THE PREPARED SPECIMEN

1. <u>Induced by grinding</u>

   a) Amorphism
   b) Strain
   c) Decomposition (e.g., loss of CO_2)
   d) Polymorphic change
   e) Solid state reaction

2. <u>Induced by irradiation</u>

   a) Polymerization
   b) Decomposition

3. <u>Special problems</u>

   a) Hydration, carbonation, etc., due to atmospheric conditions
   b) Loss of water in vacuum
   c) Decomposition at high temperature

---

irradiation of the specimen particularly where weak ligands are present in the compounds making up the material under investigation. Finally, the atmosphere immediately surrounding the specimen during analysis may create problems of loss or gain of water of hydration, carbon dioxide, etc. Of these three sources of difficulty, the analyst really has control only of the third, where the use of controlled atmospheres can be employed.

POLYMORPHIC CHANGES INDUCED BY GRINDING

The fact that polymorphic changes may be induced by grinding has long been recognized, but the problem may be more common than is generally recognised by many practicing diffractionists. Table 3 gives a short list of some of the cases reported in the literature and includes examples of polymorphic changes, decomposition and formation of solid solutions. Several of the indicated polymorphic changes are reversable, for example the transformations between calcite and aragonite, and litharge and massicot.

Mills used for reducing the particle size of samples for powder diffractometry are generally of the ball mill (eg, the Wig-L-Bug) or disc mill (eg, Shatterbox) types. The stress system generated in these mills is a combination of hydrostatic and shear stresses [7] of which the shear stress is generally the more important. Further, although local heating may certainly accelerate any solid state interaction, the major factor in any phase change is purely a mechanical one [16,17]. It has been estimated [8] that the pressure component in a Wig-L-Bug is between 10-20 Kbars

Table 3 - Reported Polymorphic Changes and Solid State Reactions Resulting from Grinding*

Initial Phase	Final Phase	Reference
Calcite ($CaCO_3$ rhombohedral)	Aragonite ($CaCO_3$ orthorhombic)	Burns & Bredig[1], Schrader and Hoffman[2], Dandurand[3]
Vaterite ($CaCO_3$ hexagonal)	Calcite ($CaCO_3$ rhombohedral)	Gregg[4], Northwood and Lewis[5]
Litharge (PbO, tetragonal)	Massicot (PbO orthorhombic)	Lewis et al[6], Senna & Kuno[7]
Lead dioxide (PbO_2, tetragonal)	Lead dioxide (PbO_2, orthorhombic)	Dachille & Roy[8]
Kaolinite ($Al_2(OH)_4 Si_2O_5$, triclinic)	Mullite ($3Al_2O_3 \cdot 2SiO_2$, orthorhombic)	Takahashi[9]
Boehmite (γ-AlO(OH), orthorhombic)	γ-Alumina (γ-Al_2O_3, hexagonal) plus Corundum (α-Al_2O_3 rhombohedral)	Panis[10]
Wurtzite (ZnS, hexagonal)	Sphalerite (ZnS, cubic)	Gregg[4]
KCl plus KBr	solid solution	Vegard & Hauge[11]
Antimony plus Bismuth	solid solution	Dandurand[12]
Massicot (PbO plus Sulphur	PbS plus SO_2	Lin[13]
$ZnCO_3$ $CdCO_3$	Loss of CO_2 to form basic carbonates	Burton[14]
Montmorillonite ($Al_2Si_4O_{10}(OH)_2$)	Al_2O_3, MgO, etc.	Bloch[15]

*Note: it should be pointed out that some of the indicated phase changes are reversable.

(one bar = 10^6 dynes/cm² ≃ 1 atmosphere) and one would intuitively assume that that generated in a disc mill is considerably in excess of this value. This should be considered in the light of the fact that many of the listed phase transformations occur at much less than 10 Kbars, for example the calcite-aragonite transformation will commence at 3 Kbars and the lead dioxide transformation at 9 Kbars.

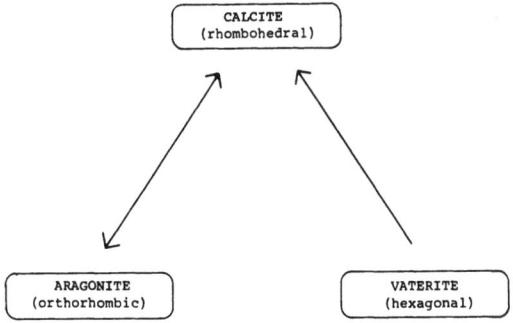

Figure 2. Phase transformation in calcium carbonate.

Fig. 2 indicates the possible phase transformations found in calcium carbonate, in which three polymorphic forms are involved, namely, calcite, aragonite and vaterite. The arrows in the figure indicate the possible phase changes which are induced by grinding.

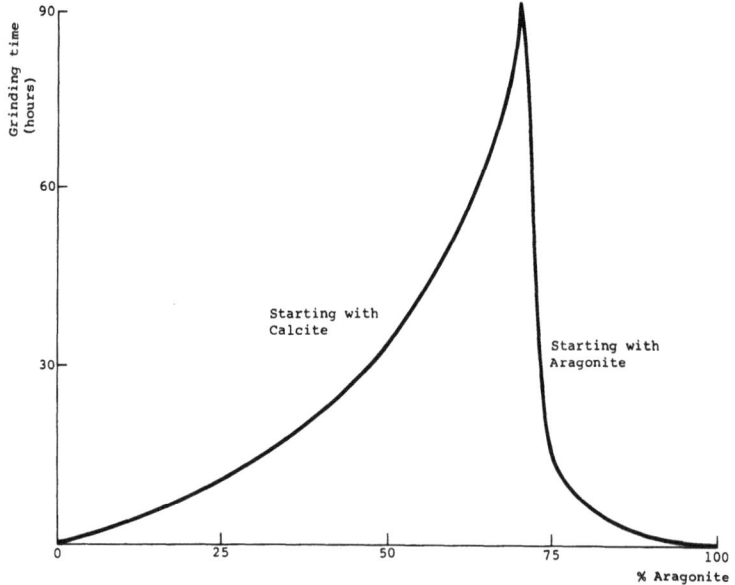

Figure 3. Effect of grinding on calcite and aragonite.

Fig. 3 indicates the degree of conversion of calcite and aragonite as a function of grinding time. It will be seen that whichever of the pure phases is used as a starting material a roughly 70/30 equillibrium mixture is eventually obtained. Of particular importance to the diffractionist is the degree of transformation which might occur within the normal period of time used to prepare the sample. The grinding time might be typically 2-20 minutes, depending upon the initial and required fineness of the material to be analysed. In the case of calcite/aragonite, the maximum transformation would probably not exceed a few percent in a normal grinding time, but in the case of a less stable system such as massicot/litharge, Fig. 4, the possible transformation could be several tens of percent. It is also important to note that the effect of the phase changes on the diffractogram can be amplified where the various phases may have different packing efficiencies.[8]

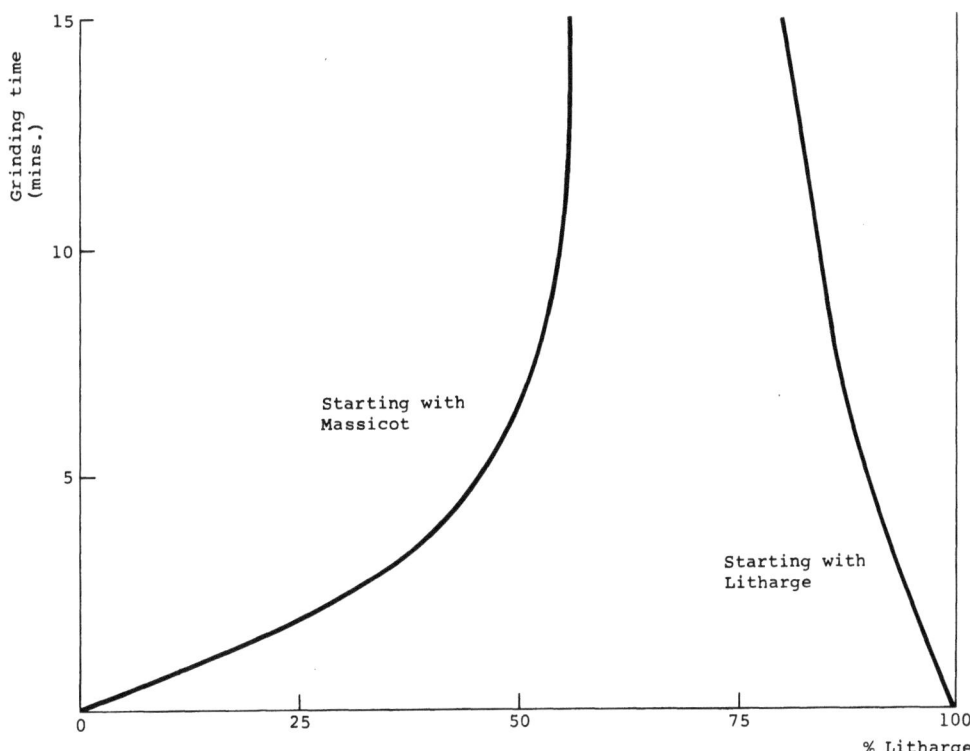

Figure 4. Effect of grinding on litharge and massicot.

CHANGES IN THE SPECIMEN DURING ANALYSIS

Specimen changes during the actual collection of count data may occur as a result of the effect of atmosphere surrounding the specimen, or the photoelectric effect of the X-radiation. Loss or gain of water or carbon dioxide from a specimen is not uncommon although this can be minimized where necessary, by covering the specimen with a plastic film during irradiation. In cases of, for example, extreme hygroscopy, the inconvenience caused by amorphous scatter from the film is more than counterbalanced by specimen stability. Where the specimen is for any reason damp, the photo-electric effect of the X-rays may cause the production of ozone, leading in turn to oxidation.

Significant decomposition of a specimen due to bond rupture is relatively uncommon and there is, to my knowledge, no equivalent case to that of the complete decomposition of tetra-ethyl lead, found in X-ray spectrometry. Here, it will be remembered that when working in dilute solutions of tetra-ethyl lead in gasoline, $Pb-C_2H_5$ bonds are completely ruptured causing the deposition of a lead complex on the cell window. Fig. 5 illustrates the effect of this on X-ray intensity and it will be seen that the X-ray intensity (in an inverted optics spectrometer) increases initially by a rate of about 1% per minute. Part of this effect is due to the lowering of specimen density by heating as will be seen from the dotted line in Fig. 5.

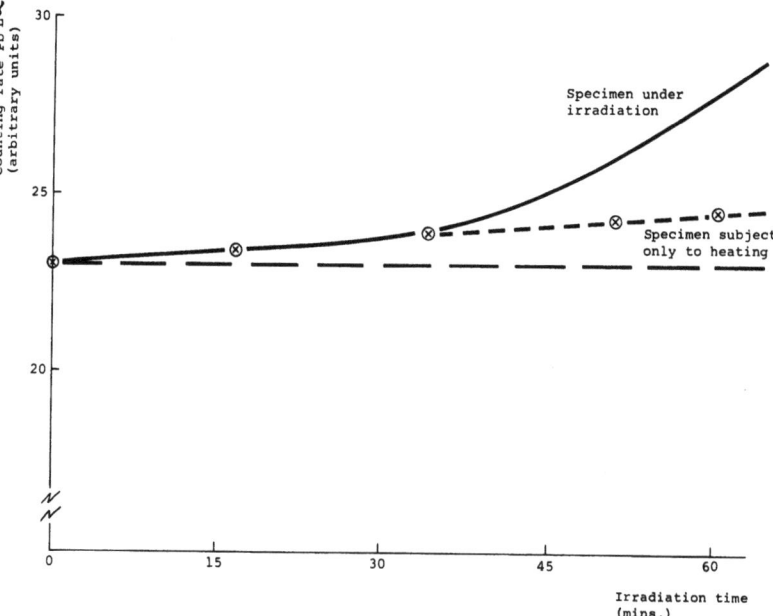

Figure 5. Decomposition of tetra-ethyl lead due to X-radiation.

One example of specimen changes due to bond rupture that is common to both X-ray diffraction and X-ray spectrometry is to be found in the analysis of polymeric films. As an example, Fig. 6 shows a region of the infra-red spectrum of an unstabilised polyvinyl chloride film, before and after irradiation with X-rays[19].

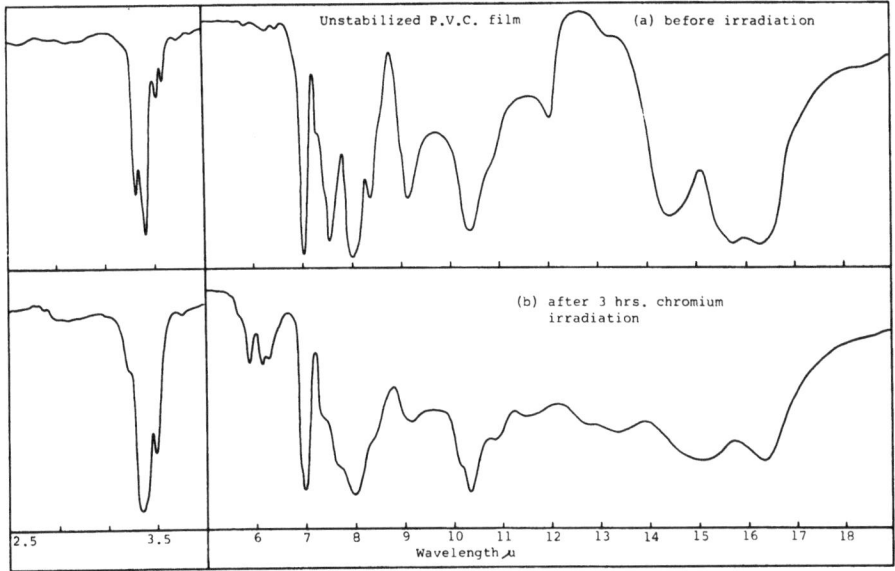

Figure 6. Effect of X-radiation on polyvinyl chloride.

A marked change in physical properties is observed following irradiation, and the irradiated sample is by comparison, brittle and shows a decrease in solubility and flow characteristics. A gradual fall-off in chlorine content is also observed during irradiation. Comparison of the IR spectra indicate significant changes at 1450 cm^{-1}, 970 cm^{-1} and 3030 cm^{-1}. The first of these is attributed to loss of Cl from a - CHCl- group to give a -CH$_2$- group, and the latter two to the formation of a trans -CH=CH- group. Both of these can be taken as evidence of additional cross linking, yielding a product which is significantly lower in chlorine content than the original.

CONCLUSIONS

From what has been described, it will be apparent that the diffractionist must be wary in the interpretation of data from prepared specimens, particularly when dealing with unfamiliar materials, or in cases where significant sample preparation is required. Where some doubt exists as to whether specimen preparation has brought about changes to the sample, an alternative preparation procedure should be sought and its results compared with the original method. In all cases, reliable equipment standards should be regularly employed to confirm equipment performance, since the first step in explaining an unexpected result is to be able to repeat the measurement and confirm that it was not an instrumental artifact.

REFERENCES

1. Burns, J.H., and Bredig, M.A., J. Chem. Phys., 25 (1956) 1281

2. Schrader, R., and Hoffman, Br., Z. Chem., 6 (1966) 388

3. Dandurand, J-L., C.R. Acad. Sci. Paris, Ser. D (1970) 881

4. Gregg, S.J., Chem. Ind., 11 (1968) 611

5. Northwood, D.O., and Lewis, D., Amer. Miner., 53 (1968) 2089

6. Lewis, D., Northwood, D.O., and Reeve, R.C., J. Appl. Cryst., 2 (1969) 156

7. Senna, M. and Kuno, H., J. Am. Ceram. Soc., 54 (1971) 259

8. Dachille, F. and Roy, R., Nature, 186 (1960) 34

9. Takahashi, H., 6th Nat. Conf. Clays and Clay Minerals, Pergamon: New York Vol. 2 (1959) 279

10. Panis, A., C.R. Acad. Sci. Paris, Ser. D (1970) 1057

11. Vegard, L. and Hauge, Th., Z. Physik, 42 (1927) 1

12. Dandurand, J-L., C.R. Acad. Sci. Paris, Ser. D (1970) 808

13. Lin, I.J., Israel J. Earth Sci., 20 (1971) 41

14. Burton, T.G., Trans. Inst. Chem. Engrs., 44 (1966) 37

15. Bloch, J.M., Bull. Soc. Chim. Fr., (1950) 774

16. Tomashevskii, E.E., Soviet Physics - Solid State, $\underline{12}$ (1971) 2588

17. Snow, R.H., Powder Tech., $\underline{5}$ (1971/2) 351

18. Silk, S.T. and Lewin, S.Z., Advances in X-Ray Analysis, $\underline{14}$ (1970) 29

19. Squirrell, D.C.M., private communication.

INFLUENCE OF SAMPLING ON THE QUALITY OF ANALYSES WITH EMPHASIS ON POWDERS

C. L. Grant and P. A. Pelton

Center for Industrial and Institutional Development

University of New Hampshire, Durham, NH 03824

INTRODUCTION

Sampling is a necessary part of the chemical analysis of particulate matter where the objective is to characterize bulk properties since it is usually undesirable or impossible to test an entire lot. The sample must be a miniature replica of the bulk material at least in respect to those features being tested. In other words, the sample must be representative to permit extrapolation from the sample to the bulk.

There is error associated with sampling particulate material, but this sampling error is only one component of the total error associated with an analytical result. Sample preparation and determination of the property being tested are two other major sources of error. These three errors combine as the squares of their standard deviations, i.e., their variances, to produce the total analytical error. Very often, sampling error is the largest of the three and, therefore, contributes a disproportionately large share to the total error.

The preparative and determinative errors for most analytical procedures are readily estimated, and most investigators report the size of these errors. Clearly, a total analytical program could be planned more intelligently if the sampling error could also be predicted. Some work has been done toward this end, but there is a need for considerably more investigation if analysts are to reliably predict sampling error in advance of analysis. Similar considerations apply for standards especially for synthetic powder standards.

SAMPLING THEORY

If the particles of material are identical with respect to all chemical and physical properties, then sampling is very simple. Any sample taken will be representative. However, for most particulate systems, the particles do differ both chemically and physically; and we must sample accordingly. The degree of chemical nonuniformity can range from only slight differences between particles such as a "pure" material with a near ubiquitously distributed contaminant to particles that are distinctly different such as gold particles in mineral sand. The most common physical variables are particle size and shape.

In the analysis of large lots of particulate material, two sampling operations are generally required. First, a sample of the bulk material is obtained and submitted for laboratory study. Since this sample is generally too large for direct analysis, it must be further subdivided by an acceptable sampling scheme in the laboratory. The error associated with sampling either bulk or laboratory size samples is due to particle segregation and random particle distribution.

Numerous articles have been written concerning the problems of sampling bulk quantities (10 Kgm and more). Bicking (1) presents a comprehensive review of the problems of sampling particulate material in railroad cars, fertilizer in bags, and ore deposits in situ. He also describes some of the tools and procedures used to extract samples from bulk such as the "thief" sampler, the sample riffler, and others. Taschler (2) discusses the use and description of sampling equipment in more detail.

For laboratory sampling, many of the same tools can be used. However, Allen and Khan (3), in a comparison of several devices, found that the spinning riffler is most effective in producing representative samples for small lots of material that can be completely processed. In fact, Charlier and Goossens (4) show that, using a spinning riffler, unmixed materials are sampled just as effectively as are completely mixed materials. A spinning riffler is a series of sampling cups arranged on the perimeter of a disc and located under a delivery shute. A schematic is shown in Figure 1. A new type of suspension sampler for separating solid particles dispersed in a slurry is described by Burt, et al. (5).

Segregation occurs according to particle size, shape, density, and mixture flow characteristics (6-9). Under the influence of gravitational forces, the smallest and/or most free flowing and/or most dense particles will settle downward. (Spherical particles are more free flowing than are angular particles.) Segregation is enhanced by movement of the total particle system. Such motion may be of small magnitude such as that due to natural vibration.

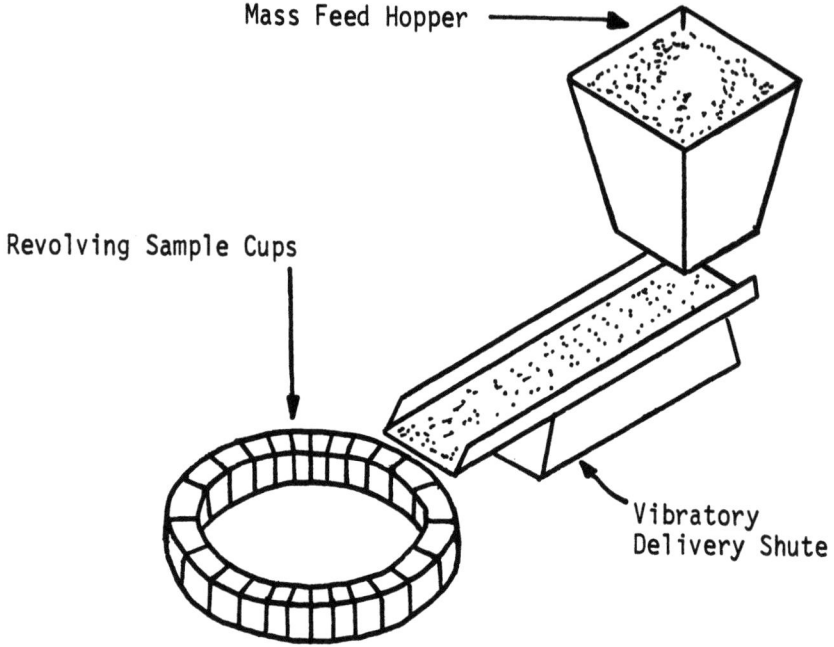

Figure 1. Schematic of Spinning Riffler

The effect of random particle distribution is more subtle. To understand this effect, apply binary statistics to an ideal mixture of 5,000 white beads and 1,000 red beads (10). From this mixture, samples of 24 beads are extracted. Despite the fact that the mixture may have been homogenized as completely as possible, there will not always be four red beads in each sample. Because of the random manner in which the particles are extracted, one could expect between zero and eight red beads per sample if we only consider 95% of the samples. This effect is based on the number of particles extracted per sample, i.e., the greater the number, the smaller the effect. Therefore, it can be generally stated that random particle distribution is of greatest influence when extracting small laboratory size samples having few particles. On the other hand, particle segregation exerts its greatest influence for bulk samples which cannot be conveniently mixed or processed through a device that eliminates particle segregation.

Visman (11) applies the above reasoning in developing his general sampling equation as given below--

$$S^2 = A/W + B/N \qquad (1)$$

where S^2 = the total variance for the system
   A = sampling constant (random variance)
   B = sampling constant (segregation variance)
   W = size of the gross sample
   N = number of samples collected

The random variance constant A is estimated by taking a series of small samples where the sampling variance is primarily due to random effects, and it is assumed that the segregation variance is negligibly small. Similarly, the segregation variance constant B is estimated by taking a series of large samples where the sampling variance is due primarily to segregation effects; and it is assumed that random effects are negligibly small.

Further, W and N in Equation 1 are in the nature of operating variables that can be manipulated within certain limits by the sampler. For instance after A and B have been determined and for a given S^2 and N, the sample weight, w, can be determined by substituting the equality Nw = W into Equation 1. This substitution gives--

$$w = A/(NS^2 - B) \qquad (2)$$

Conversely, by specifying w, N can be determined. Therefore, the number and size of samples can be calculated for a specified sampling variance.

Visman's method is most applicable for the continuous analysis of similar lots of bulk material. However, the method is not as helpful in estimating the expected sampling error for laboratory size samples.

Since segregation of particles in laboratory sampling can be largely eliminated, we need to emphasize errors resulting from random particle distribution. Benedetti-Pichler (10) considers this situation; and for a binary population, he applies the Bernoulli Theorem which yields the equation--

$$\sigma_p = \frac{d_A d_B}{d^2} (P_A - P_B) \sqrt{\frac{p(1-p)}{n}} \qquad (3)$$

where σ_p = the absolute standard deviation of the percent of component x in a mixture of particles A and B
   d_A = the density of the A particles
   d_B = the density of the B particles
   d = the weighted average density assuming all particles have the same volume

P_A = the percent of the component x in the A particles
P_B = the percent of the component x in the B particles
p = the fractional number of the total particles which are type A
1-p = the fractional number of type B particles
n = the total number of particles

This equation is idealized by assuming only two species of particles. However, this restriction can be overcome by considering that the components are either rich in x or poor in x. Equation 3 is also idealized by assuming that all particles are the same size.

Equation 3 is inconvenient because the relative numbers of particles for the different species are seldom known. However, Wilson (12) modified the equation to employ terms of particle size, sample weight, concentration, and density. Thus, for the simple case of two distinct particulate species of the same particle size, we have--

$$S_E = (t_1 - t_2)\sqrt{W_1 W_2 \left(\frac{d_1 d_2}{d}\right)\left(\frac{V}{w}\right)} \qquad (4)$$

where S_E = the absolute standard deviation of the concentration of element E
t_1 = the weight fraction concentration of element E in species 1
t_2 = the weight fraction concentration of element E in species 2
W_1 = the weight proportion of species 1
W_2 = the weight proportion of species 2
d_1 = the density of species 1
d_2 = the density of species 2
d = the weighted average density
V = the volume of the individual particles
w = the weight of sample taken

Consider the determination of trace elements in powder samples. If the element under consideration is a minor ingredient of one or more of the major constituents, its relative sampling error will be comparatively small because (t_1-t_2) will be very small. If on the other hand the trace ingredient is present as a major constituent of the species of minor abundance, its relative sampling error will be comparatively large. The latter situation would pertain to elements in certain mineral mixtures such as zirconium in beach sands where the zirconium is almost exclusively associated with individual zircon particles, gold in mineral sands, and metals in synthetic powder mixtures used to calibrate analytical systems for trace analysis.

Further, if we assume none of the minor ingredient is present in species 2, then t_2 equals zero. Also, if we express the sampling error, C_E, as the percent relative standard deviation, then Equation 4 becomes--

$$C_E = \frac{100 S_E}{t_1 W_1} = 100 \sqrt{\frac{W_2}{W_1}\left(\frac{d_1 d_2}{d}\right)\left(\frac{V}{w}\right)} \quad (5)$$

We now see that the relative sampling error varies directly with the square root of W_2, d_1, d_2, V, and inversely with the square root of W_1, d, w. However, d is approximately equal to d_2 because most of the sample is composed of species 2 particles, so they cancel each other; and changes in W_2 can be ignored because the term, W_2/W_1, is largely controlled by changes in W_1 for low concentration analytes. Therefore, the significant aspects of Equation 5 are that C_E varies directly with the square root of the density of the minor species, d_1, and the size of the particles, V, and inversely with the square root of the concentration of the minor species, W_1, and the weight of the sample taken, w. Thus, for a given particle system with a given analyte concentration, we must either decrease the particle volume (by grinding) or increase the sample weight in order to decrease the sampling error.

For the case of more than two mineral species and with the simplifying assumption that all species have the same density, Wilson derived the following equation--

$$S_E = \sqrt{(t_1-t_2)^2 \frac{W_1 W_2}{n} + (t_2-t_3)^2 \frac{W_2 W_3}{n} + (t_3-t_1)^2 \frac{W_3 W_1}{n}} \quad (6)$$

For this study, our interest is primarily to determine trace elements which are major components of species in minor abundance. Therefore, we can assume a binary system in which everything other than the species of interest is the second component. This assumption does not introduce serious error and allows us to use Equation 5.

Two assumptions made in deriving Equation 5 are that both analyte and matrix particles are spherical and of uniform size. Neither of these ideals is met in real samples; and, therefore, we must know more about the effect of violation of these assumptions. Knowledge of the particle size distribution is required in order to estimate the average particle size. In general, this distribution varies with the type of material and the method of grinding and sieving. Herdan (13) states that the particle size distribution for many ground minerals is log-normal, suggesting that a volume corresponding to the geometric average could be used. Benedetti-Pichler states that the particle volume can be estimated from the mesh size of the sieve that just passes all the sample. However, this can cause the sampling error to be seriously overestimated in some cases. For samples of wide particle size distribution, Wilson (12) recommends that the average volume, V, be replaced by a weighted average volume, $\overline{V}$, calculated from--

$$\overline{V} = \sum_{h=1}^{k} g_h v_h \qquad (7)$$

where k = the number of groups of different particle size
v_h = the average volume of the individual particles of each group
g_h = the fraction by weight of the h group

One major drawback to the use of any estimate of particle volume is that, after grinding, different species will not likely show the same particle size distribution due to differences in hardness and brittleness. For rock samples, minerals such as zircon, chromite, and magnetite which are harder than the bulk materials may be concentrated in the larger fractions. Since these coarse particles have the greatest influence on the standard deviation of sampling, such segregation could cause a serious discrepancy between expected and observed sampling errors.

The assumption of spherical particles can also lead to problems. Some species such as micas tend to fragment as flakes or platelets in which case the relationship between volume and particle diameter established for spheres no longer holds.

Lastly, we must recognize the tacit assumption that the bulk material is homogeneous with respect to the sampling unit. If this is not the case, the observed sampling error will be larger than predicted because a systematic error due to particle segregation will be superimposed on the random error, unless a sample splitting device such as the spinning riffler is fully capable of eliminating segregation effects.

A sampling equation intended for the same purpose as Equation 5 is given in Kolthoff and Elving (14)--

$$s = \sqrt{bp} \qquad (8)$$

where s = relative sampling standard deviation
$b = \dfrac{\text{weight of largest individual particle (assumed cubic)} \times 100}{\text{weight of sample}}$
p = weight percent of the analyte

This equation, in contrast to Equation 5, predicts sampling errors which are proportional to the analyte concentration, i.e., as the analyte concentration increases, the sampling error increases. This, of course, is illogical; and, since Equation 8 was taken from Davies (15) which is now out of print and was unavailable, we must conclude that it was copied incorrectly. For low analyte concentration, Equation 8 written as $s = \sqrt{b/p}$ is nearly identical to Equation 5 since $b = (100W_2d_1V)/w$ and $p = 100W_1$. This further confirms the suspicion that Equation 8 was copied incorrectly.

Ottley (16) presents a sampling equation developed by Pierre M. Gy that is similar to Equation 5--

$$s = \sqrt{\frac{Cd^3}{M}} \qquad (9)$$

where s = relative sampling standard deviation
      d = nominal sieve opening in centimeters that retains 5-10% of particles, i.e., approximately the diameter of the largest particles assumed to be cubic
      M = weight of sample in grams
      C = a sampling constant

The sampling constant, C, is equal to--

$$f \times g \times 'l' \times m \qquad (10)$$

where f = a shape factor that ranges from 0.2 for gold ore particles to 1.0 for spherical particles
      g = a particle size distribution factor with values from 0.25 for most natural particle systems to 1.0 for systems where the particles are all of the same size
    'l' = a liberation factor with values from 0 to 1.0. For an analyte distributed uniformly throughout the matrix, 'l' = 0; and, for an analyte contained only in the minor species particles, 'l' = 1.0
      m = a mineralogical composition factor that is calculated from the equation--

$$m = \frac{1-a}{a}[(1-a)r + at] \qquad (11)$$

where r and t are the average densities of the analyte and matrix respectively, and a is the average analyte content expressed as a fraction.

If we apply the same assumptions to Equation 9 that were used in the development of Equation 5, i.e., all particles are spherical and the same size and all analyte particles are contained in the minor species particles, then Equation 9 reduces to the following--

$$s = \sqrt{\frac{md^3}{M}} \qquad (12)$$

For analyte concentrations less than 1%, Equation 11 can be approximated by $m = r/a$ and Equation 12 becomes--

$$s = \sqrt{\frac{rd^3}{aM}} \qquad (13)$$

In comparing Equations 5 and 13, it can be seen that W_2/W_1 is essentially equal to $1/a$; d_1d_2/d (for low analyte concentration) is equal to r; d^3 equals V because cube-shaped particles are assumed; and M equals w. Therefore, Equation 13 multiplied by 100 (to give percent relative sampling standard deviation) is nearly identical to Equation 5 for trace element analysis except for the difference in assumed particle shape.

EXPERIMENTAL

Several assumptions were made in deriving the Wilson Equation in its most useful form (Equation 5). Briefly, they are (1) all particles are spheres; (2) all analyte and matrix particles are the same size, designated V; for wide particle size distributions, a weighted average particle volume, $\overline{V}$, can be determined and used as a legitimate estimate of V; (3) the analyte is contained only in the minor species particles; and (4) the bulk being sampled is homogeneous.

Purpose of Investigation

The object of this study is to experimentally test Equation 5 by (1) studying a nearly ideal particulate system with these assumptions optimized, and (2) studying more typical particulate systems whose properties do not meet some of the idealized assumptions.

Assumptions 1 and 2 are of principal interest since most particulate samples whether natural or synthetic contain nonspherical particles of a wide size distribution. This is particularly true for comminuted material. McCabe and Smith [17] state--

> The product (of comminution) always consists of a mixture of particles ranging in size from a definite maximum to a submicroscopic minimum—unless they are

smoothed by abrasion after crushing, comminuted particles resemble polyhedrons with nearly plain faces and sharp edges and corners. The number of major faces may vary but is usually between four and seven. Particles may be compact, with length, breadth, and thickness nearly equal, or they may be platelike or needlelike. A compact grain with several nearly equal faces can be considered to be spherical—.

The particulate systems studied, therefore, will be primarily designed to demonstrate the effect of changes in particle geometry and particle size distribution. The experimental sampling error associated with each particulate system will be compared with the theoretical values predicted by Equation 5.

DESIGN OF EXPERIMENT

The particulate systems used throughout this work are synthetic, i.e., not natural systems and were specially prepared to study the validity of the assumptions made in the derivation of Equation 5. First, three systems that each contained spherical analyte and matrix material were studied. Two of these had the same narrow particle size distribution; but the range of particle diameter varied between the systems, i.e., one contained particles that were -100, +140 mesh and the second contained particles that were -200, +230 mesh. A third system contained particles that were -325 mesh, representing a relatively wide size distribution. Clearly, Assumptions 1 and 2 are approached for these systems; and, since Assumptions 3 and 4 can be closely approximated, Equation 5 was tested under quite ideal conditions. Also, by using three different particle sizes, this effect can be studied. Three similar systems were prepared from angular analyte and angular matrix material. These systems were used to show the effect of nonspherical particle geometry. A system of spherical analyte and matrix material with very wide particle size distribution was also studied to simulate what might be encountered in a typical comminuted material. Finally, a system was prepared that contained platelet analyte and spherical matrix material of narrow particle size distribution. This system will show the effect of a particle geometry that is significantly less spherical than are the angular particles previously considered.

Figure 2 is a flow diagram that shows the various steps in the preparation and analysis of each of these eight particulate systems. First, the analyte and matrix materials were screened using U. S. Standard sieves to give the desired particle size distributions. Next, the densities of the various fractions were determined experimentally by displacement of methyl ethyl ketone in LeChatelier

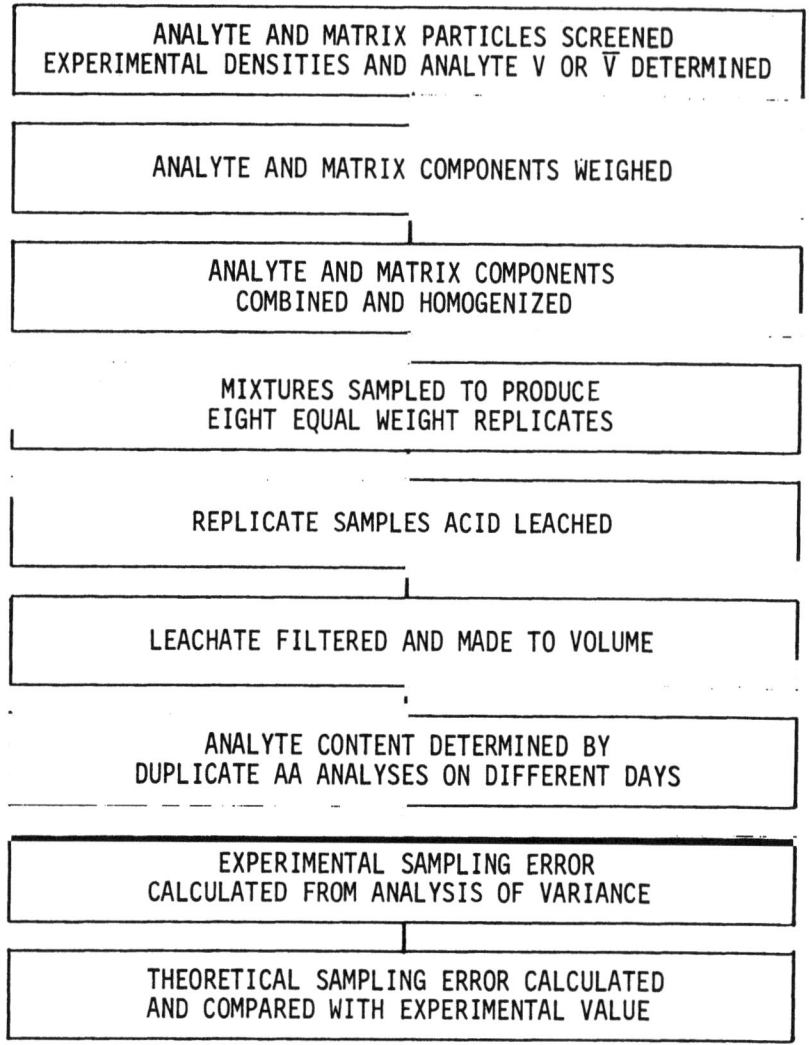

Figure 2. Experimental Design Flow Sheet

flasks. This was necessary because some voids were present in spherical particles. The appropriate estimate of particle volume, V or $\bar{V}$, is then determined for each particulate system by methods that will be described in the section on results and discussion. Portions of analyte and matrix components were carefully weighed using a semimicro balance, combined, and homogenized using an off-axis blender. After blending, each mixture was sampled using a Jones-type riffler. Eight approximately equal weight samples were produced for each mixture sampled. For the first six sets of spherical and angular particles, "grab samples" were also taken by simply removing samples from each mixture with a laboratory spatula. This was done to study the ability of this frequently used method to generate reproducible samples.

After sampling, each replicate was acid leached to dissolve the analyte. The inert matrix was separated from the acid solutions by filtration, water washed, made to volume, and analyzed for the analyte content by expanded scale atomic absorption (AA) techniques. Two separate AA determinations were made for each solution on separate days using independent calibration curves. The experimental percent relative standard deviation, i.e., the sampling error, was determined for each particulate system by conducting an analysis of variance on each set of data. Finally, theoretical sampling errors were calculated using Equation 5 and compared with the experimental values.

Plasma heated zircon was used as the spherical matrix material, and natural crystalline zircon was used as the angular matrix material. These forms of zircon were employed because they are inert to the analyte acid extraction after purification by acid leaching and, therefore, would not interfere with the analyte analysis. Also, these materials were readily available. Spherical and angular iron particles were used as the analytes. These materials were selected because they were also readily available but, more importantly, because iron is easily acid extracted from an inert zircon matrix and is easily quantitatively analyzed by AA techniques. Copper was used as the platelet analyte material because it, too, could be acid extracted and quantitatively analyzed by AA. Also, the copper platelets could be prepared from copper spheres that were available.

Density was also a factor in the selection of iron and copper as the analyte materials. For sampling errors of the desired size, i.e., significantly greater than the analytical error, for the particle sizes, analyte concentration, and sample weight to be studied, Equation 5 predicts the need for an analyte with a density of 7 g/cm^3 or greater. Iron with a density of 7.86 g/cm^3 and copper with a density of 8.92 g/cm^3 both met this requirement.

Equipment and Materials

Angular iron and zircon was the feed material used to manufacture spherical particles. The iron was typical iron grit, and the zircon was a typical Australian product. The spherical iron and zircon was manufactured at Humphreys Corporation, Bow, New Hampshire by passing the angular materials through a large direct current plasma system to form spherical molten droplets. Particles in the range of 200 down to 10 or less micrometers generally show excellent sphericity (18).

Copper platelets with an average diameter of 185 μm and an average thickness of 25.3 μm were used for this study. These platelets were made by pressing copper spheres between two-inch diameter 404 stainless steel plates under a 10,000 pound load.

A schematic of the off-axis particle blender used to homogenize the sample mixtures is shown in Figure 3. It is simply a laboratory stirring motor equipped with a set of reducing gears, a large pulley attached to dampen effects of voltage changes, and a Variac for speed control. The sample vial is placed in the jaws of an extension clamp which is attached perpendicular to the drive shaft of the stirrer. The position of the vial and its speed of rotation are adjusted so that the sample tumbles slowly from one end of the sample vial to the other.

A Jones-type sample splitter was employed to subdivide samples. Reagent quality acid with acceptably low iron concentration was used for the acid extraction. All glassware was acid washed to insure low blanks. Atomic absorption was used to measure iron concentration in the extracted solution.

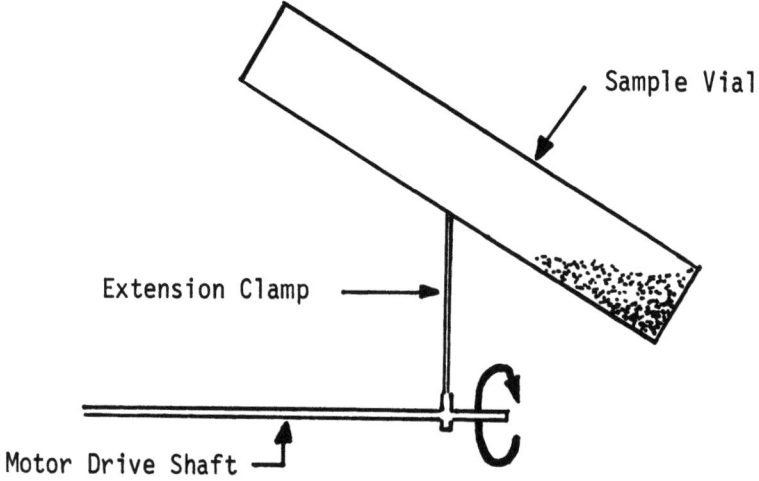

Figure 3. Schematic of Particle Blender

RESULTS AND DISCUSSION

Both iron and zircon spherical and angular particles were screened using U. S. Standard sieves of 100, 140, 200, 230, and 325 mesh. Of the resulting fractions, the -100, +140 mesh, -200, +230 mesh, and -325 mesh fractions for each particle geometry and each material were selected for experimentation. Results of these studies have been reported (19).

The average volumes, V, and the corresponding average diameters of the spherical and angular iron particles for each size fraction were estimated. For the -100, +140 mesh spherical particles, two weighed samples were counted microscopically and gave average counts of 92 and 96 particles/mg. The average particle volume, V, and corresponding average diameter were then calculated using the experimentally determined density (7.54 g/cm^3). For the -100, +140 mesh angular particles, average counts of 136 and 142 particles/mg. were found. The average particle volume, V, and average diameter were calculated using the experimentally determined density (7.28 g/cm^3). As expected, the average volumes for the angular particles were smaller than for the spherical particles of the same screen size.

The -200, +230 mesh particles for each geometry were too small to count microscopically. Therefore, the same size distribution as the -100, +140 mesh particles was assumed. The average volume, V, and the corresponding average diameter of the particles were then calculated based on this assumption, using linear interpolation. Since the size range is very narrow in this case, the error caused by this process should be small.

For the -325 mesh spherical particles, the weighted average diameter and weighted average volume, $\overline{V}$, were calculated from a particle size distribution obtained with a Coulter Counter. The distribution was integrated using 10 μm intervals according to Equation 7. A similar Coulter Counter distribution for the -325 mesh angular particles could not be obtained so it was assumed to be the same as for the spherical particles. This assumption should be reasonably valid since the angular material was used to manufacture the spherical material except that the angular particles probably have a slightly smaller average volume than the spherical particles of the same mesh size. In general, the average diameter is close to the upper boundary of the range of diameters for each size fraction since the large particles contribute proportionately greater sample weight than the small particles. It was suggested by Benedetti-Pichler (10) that the upper boundary particle volume could be used without significant error in making theoretical calculation. While this may be true for narrow size distributions, the data reported in (19) showed that a significant error could result if this practice were followed.

Average volumes and diameters were not determined for the zircon fractions since that information is not needed to make the theoretical calculations. However, they should be close to those found for the iron.

After screening, the spherical iron fractions were assayed and found to be nearly 96% iron. There was also 3% carbon; and, by optical emission spectrographic techniques, small amounts of manganese, nickel, and silicon were found. The zircon fractions were acid leached to remove any iron to insure that the analyte was contained only in the minor species particles. The adequacy of this process was verified experimentally.

Mixtures of iron and zircon of the same size distribution were prepared with an iron concentration of approximately 3,000 ppm by careful weighing of the appropriate amounts of each. The total amount of iron needed for each sample was small. However, special precautions were not taken to achieve maximum weighing accuracy since the parameter of interest was variation from sample to sample within a mixture rather than minor differences in the absolute iron content of the various mixtures. The experimentally determined average iron concentration for each mixture was used for the theoretical calculations.

The mixtures were homogenized in an off-axis blender. There is no guarantee that homogeneity was achieved, but it should have been approached. Each mixture was split repeatedly in a small Jones-type sample riffler. This operation was continued until eight approximately 100 mg. samples were produced. Each of these fractions was carefully weighed to the nearest 0.0001 grams using a semimicro balance and placed in separate beakers for iron extraction. To deduce the benefit of the Jones-type riffler, each of the mixtures was also "grab sampled," i.e., eight approximately 100 mg. samples were removed from the vials using a laboratory spatula and treated like the riffled samples.

In this manner, twelve sets of eight approximately 100 mg. samples, each of accurately known weight, were produced for evaluation by extraction and analysis. Two separate iron determinations were made for each solution on separate days using independent calibration curves. These duplicate measurements, after removing the systematic effects due to days, provide an estimate of the random error associated with reading the absorbance units from the AA and from constructing and reading the calibration curves. Unfortunately, there is no convenient way to determine the sample preparation error associated with the weighing, leaching, and filtering operations required by the analytical methodology. However, previous work indicates that this error is only about ±0.2% relative.

An analysis of variance was performed on each set of data. A typical analysis for the -200, +230 mesh riffled spherical particles is shown in Table 1.

TABLE 1

Analysis of Variance for -200, +230 Mesh Riffle Sampled Spherical Particles

Source of Variation	Sum of Squares	Degrees of Freedom	Mean Square	Quantity Estimated by Mean Square
Between Samples	555,286	7	79,326	$\sigma_a^2 + n\sigma_s^2$
Within Samples	895	7	128	σ_a^2
Block	121	1		
Total	556,302	15		

σ_a^2 = true analytical mean square

σ_s^2 = true sampling mean square

n = number of replicate measurements on each sample

Starting at the bottom of the table, the total variation (the total corrected sum of squares) is the summation of the block, within samples and between samples effects. The block effect relates to any systematic difference between the two separate sets of analyses that exist because the data were not completely randomized. The within samples mean square represents the random variability in the duplicate analytical measurements for each sample of a set and provides an estimate of the true analytical variance, σ_a^2. It must be understood that this estimate does not include the sample preparation variability because duplicate analyses were made on aliquots of the same solution rather than on two separately prepared solutions of the same sample.

The between samples mean square, M_s, combines the analysis mean square, σ_a^2, and the sampling mean square, σ_s^2, as shown in the table, where n is the number of replicate analytical measurements on each sample. The quantities, σ_a^2 and σ_s^2, cannot be determined directly because they are parameters of a total population; but they can be estimated by S_a^2 and S_s^2. Therefore, by substitution,

$$M_s = S_a^2 + nS_s^2 \qquad (14)$$

and

the experimental sampling error, S_s, can then be calculated from the equation--

$$S_s = \sqrt{\frac{M_s - S_a^2}{n}} \quad (15).$$

As reported elsewhere (19), comparison of the experimental and theoretical sampling errors for the spherical riffled samples showed generally good agreement although the experimental value for the -325 mesh material was twice as high as predicted (4.5% versus 2.4%). The agreement between the experimental and theoretical sampling errors for the spherical grab samples was not as good, possibly due to the ease of segregation of spherical particles leading to large experimental sampling errors. For angular riffled particles, the experimental values were consistently larger than predicted by theory. The somewhat improved agreement for angular grab samples is explained on the basis of increased shear forces between angular particles which tend to prevent segregation (20).

Spherical Particles of a Wide Particle Size Distribution

In practical situations, samples of a wide particle size distribution are more commonly encountered than are samples of narrow particle size distribution. Therefore, a synthetic mixture of wide particle size was prepared and studied. The mixture consisted of two grams each of five zircon particle size distributions (-100, +140 mesh; -140, +200 mesh; -200, +230 mesh; -230, +325 mesh; and -325 mesh) added to a mixture of approximately 30 mg. of iron spheres containing equal portions of each of the five size ranges specified for the zircon. The total mixture was blended in the off-axis mixer, and eight samples of about 0.15 grams were obtained using the Jones-type riffler. The samples were treated in the same manner as for the previous work. The sample weights and iron concentrations are presented in Table 2.

From an analysis of variance, an experimental percent relative sampling standard deviation of 7.6% was obtained. All terms required to calculate the theoretical percent relative sampling standard deviation were known except a weighted average particle volume, $\bar{V}$. Three different methods of estimating $\bar{V}$ were used to determine their effect on the relative sampling error predicted by Equation 5. First, as suggested by Benedetti-Pichler (10), the volume of the largest particles of the distribution was taken as $\bar{V}$. The theoretical percent relative sampling standard deviation was 17.2. Clearly, the agreement between this predicted value and the experimental value of 7.6% is not very good. Next, $\bar{V}$ was estimated using Equation 7 where the volume of the largest particles

TABLE 2

Sample Weights and Iron Analyses of Riffle Sampled
Spherical Particles of a Wide Particle Size Distribution

	Iron Content, ppm	
Sample Weight, g	1st Analysis	2nd Analysis
0.16200	2809	2747
0.14800	2838	2796
0.17770	2870	2800
0.16155	2909	2863
0.15281	3198	3166
0.13300	2613	2584
0.15910	3072	3017
0.14715	2548	2506
Ave. 0.15516	2857	2810

for each of the five particle size distributions was used as the individual average particle volume for that fraction, i.e., the volume of a 149 μm diameter particle was used as the average particle volume for the -100, +140 mesh fraction. For $\overline{V}$ generated in this manner, the predicted percent relative sampling standard deviation was 9.6 which is in much better agreement with the experimental percent relative sampling standard deviation. Lastly, the experimentally determined average particle volumes or weighted average particle volumes were used in Equation 7 to calculate $\overline{V}$. For the -100, +140 mesh size range, an experimental value was obtained. For $\overline{V}$ calculated in this manner, Equation 5 gives a theoretical percent relative sampling standard deviation of 8.7. This final approach to the calculation of $\overline{V}$ provides the best agreement with the experimental percent relative sampling standard deviation and is the recommended procedure when the necessary data are available.

Although the lowest predicted value of the sampling error is slightly greater than the experimental value, the difference is too small to conclude that Equation 5 overestimates the sampling error for a wide particle size distribution. In fact, considering the limited amount of data, the agreement between the theoretical and the experimental estimate is quite good using the recommended procedure for calculating $\overline{V}$.

Platelet Particles of Narrow Size Distribution

Platelet particles are frequently encountered in natural systems; however, the chief interest here is to measure the effect of a particle geometry that differs from a sphere more than the angular particles previously studied. Copper platelets of -100, +120 mesh were used for this study. A representative sample of these platelets was examined under a Leitz microscope, and an average diameter of 185 μm with a range of 140-214 μm was observed. The average thickness was 25.3 μm with a range of 15-42 μm. The range of diameters (140-214 μm) is greater than expected for material that is passed by a 100 mesh sieve (149 μm sieve opening) and is retained on a 120 mesh sieve (125 μm sieve opening). This discrepancy occurs because the sieve openings are square, and platelets go through these openings on the diagonal as shown in Figure 4.

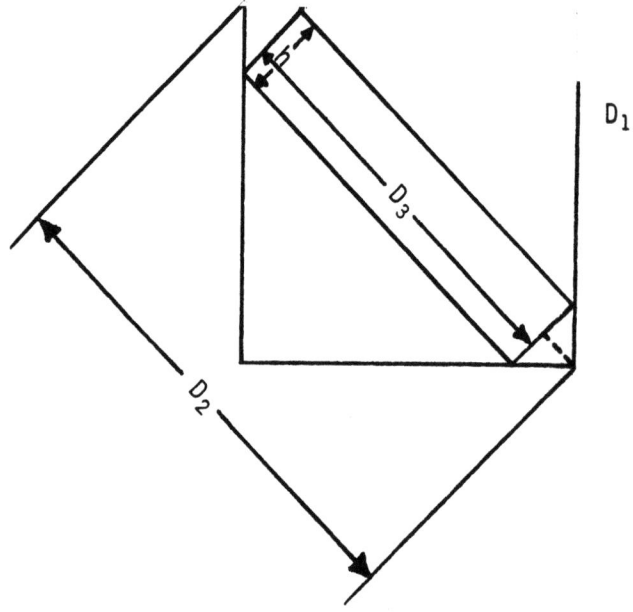

h = platelet height or thickness
D_1 = sieve opening
D_2 = sieve opening diagonal
D_3 = platelet diameter

Figure 4. Diagram of a Platelet Particle Passing through a Sieve Opening

For this paper, a platelet is defined as a particle whose thickness, h, is no greater than 0.414 times its diameter, D_3. This is a reasonable definition since platelets with a larger thickness to diameter ratio approach the shape and volume of a sphere. From this definition, the diameter of a platelet that will just pass a sieve opening, D_1, can be calculated by

$$D_3 = (1.414\ D_1) - h \tag{16}$$

Applying Equation 16 to platelet particles of 15-42 μm thickness, the corresponding range of diameters for these particles that could be passed by a 100 mesh sieve and retained on a 120 mesh sieve is 135-196 μm. This range of diameters still is slightly less than the range estimated microscopically. Probably the major cause for this minor discrepancy is that Equation 16 assumes right angles at the platelet edges; whereas, these copper platelets have slightly rounded edges. Also, the sieve openings may be larger than the standard value given for each mesh size due to wear. Both of these effects would cause the range of platelet diameters to be shifted to larger values than predicted by Equation 16.

A mixture of copper platelets and spherical dissociated zircon was prepared, homogenized in an off-axis blender, and split with the Jones-type riffler repeatedly to produce eight approximately 0.14 gram samples. After weighing, each sample was extracted with concentrated nitric acid and analyzed by AA in a manner similar to that used for the iron-zircon mixtures. The sample weights and copper concentrations are presented in Table 3.

TABLE 3

Sample Weights and Copper Analyses for a Riffle Sampled Copper Platelet-Spherical Zircon Mixture of a Narrow Particle Size Distribution

Sample Weight, g	Percent Copper Content	
	1st Analysis	2nd Analysis
0.13092	1.54	1.61
0.13916	1.54	1.58
0.13525	1.51	1.59
0.13745	1.56	1.61
0.14618	1.49	1.56
0.15475	1.63	1.69
0.13880	1.35	1.39
0.14482	1.68	1.73
Ave. 0.14092	1.54	1.60

An analysis of variance was performed on these data, and an experimental percent relative sampling standard deviation of 6.3 was obtained. As in the previous work, all terms required to calculate the theoretical percent relative sampling standard deviation were known except the average particle volume, V; and, as previously, three different estimates of V were used.

The first estimate (1.73×10^{-6} cm^3) was obtained by assuming the mixture contained sieved spherical particles with a diameter equal to the largest sieve opening as suggested by Benedetti-Pichler (10). The corresponding calculated theoretical percent relative sampling standard deviation was 8.3. The agreement between this value and the experimental value was much better than expected. Estimating the volume of a platelet by assuming it to be a sphere ought to lead to a large overestimation of the average particle volume and, therefore, a large overestimation of the sampling error. However, for a given square sieve opening and for both platelet and spherical particles that just pass through this opening, the maximum platelet diameter (assuming thin platelets) will be much greater than the maximum sphere diameter. Therefore, the expected platelet volume overestimation caused by assuming spherical particles is partially compensated.

The second estimate of V was also based on the assumption of spherical particles but with an average diameter corresponding to the same particle size distribution as the -100, +140 mesh iron spheres. The theoretical percent relative sampling standard deviation using this value was 7.8. While this value is closer to the experimental value, the improvement over the value obtained for the first estimate of V is small, verifying that Benedetti-Pichler's recommendation works satisfactorily for very narrow size distributions of platelet particles.

The final estimate of V (6.78×10^{-7} cm^3) was made using the microscopically measured average platelet thickness (25.3 µm) and diameter (185 µm). The corresponding calculated theoretical percent relative sampling standard deviation was 5.2. Although this estimate is smaller than the experimentally determined sampling error of 6.3%, it should be the best estimate of the true sampling error. Recall that sampling error is dependent on the number of particles being sampled. For a given sample weight, analyte concentration, and density, an estimate of the analyte particle volume that is greater than the actual volume will result in an underestimation of the number of analyte particles and, therefore, an overestimation of the sampling error. The converse is, of course, also true. Therefore, the best estimate of the true sampling error is obtained when the best estimate of the true average analyte particle volume is used.

The difference between a predicted sampling error of 5.2% and one of 7.8% (obtained by the second estimate of V) or even one of 8.3% (obtained by first estimate of V) is not that great when attempting to estimate an experimental sampling error. For many cases, either the first or second method of estimating V would be adequate. However, for very small sample weight, low analyte concentration, large analyte density, and, most importantly, very thin platelets, the difference between the predicted sampling error for the first two methods of estimating V, which are based on screen sizes, and the method employing the true average volume could be quite large. For example, a platelet of 10 μm thickness and 200 μm diameter will just pass a 100 mesh sieve and has a volume of 3.14×10^{-7} cm^3. The volume of a sphere that just passes a 100 mesh size is 17.3×10^{-7} cm^3. The ratio of the sampling errors predicted using these two volumes is $\sqrt{17.3/3.14} = 2.3$. In other words, if the predicted sampling error were 5.0% using $V = 3.14 \times 10^{-7}$ cm^3, then the predicted sampling error would be 11.5% using $V = 17.3 \times 10^{-7}$ cm^3. Therefore, for situations where many repeat analyses are to be made, it may be advantageous to microscopically analyze the particulate system to obtain the best possible estimate of the average analyte particle volume.

ACKNOWLEDGMENT

Thanks are due for permission to reproduce sections of the material on Sampling Theory and Table 1 which also appear in an article in ASTM Special Technical Publication 540.

REFERENCES

1. C. A. Bicking, "The Sampling of Bulk Materials," Materials Research and Standards, MTRSA 7, 95-116 (1967).

2. A. F. Taschler, "Sampling of Solids," in Technical Report 3939, Quality Assurance Directorate, Picatinny Arsenal, Dover, N. J. (1969).

3. T. Allen and A. A. Khan, "Critical Evaluation of Powder Sampling Procedures," Chem. Eng. (London) 238, 108-112 (1970).

4. R. Charlier and W. Goosens, "Sampling a Heterogeneous Powder Using a Spinning Riffler," Powder Technology 4, 351-359 (1970/1971).

5. M. W. G. Burt, C. A. Fewtrell, and R. A. Wharton, "A Suspension Sampler for Particle Size Analysis Work," Powder Technology 7, 327-330 (1973).

6. J. J. Fischer, "Solid-Solid Blending," Chemical Engineering 67, 107-128 (1960).

7. J. F. Van Denburg and W. C. Bauer, "Segregation of Particles in the Storage of Materials," Chemical Engineering 71, 135-142 (1964).

8. H. Campbell and W. C. Bauer, "Cause and Cure of Demixing in Solid-Solid Mixers," Chemical Engineering 73, 179-185 (1966).

9. J. C. Williams, "The Mixing of Dry Powders," Powder Technology 2, 13-20 (1968/1969).

10. A. A. Benedetti-Pichler, "Theory and Principles of Sampling for Chemical Analysis," in W. M. Berl, Editor, Physical Methods in Chemical Analysis, Vol. 3, p. 183-217, Academic Press (1956).

11. J. Visman, "A General Sampling Theory," Materials Research and Standards, MTRSA 9, 9-13, 51-66 (1969).

12. A. D. Wilson, "The Sampling of Silicate Rock Powders for Chemical Analysis," Analyst 89, 18-30 (1964).

13. G. Herdan, "Attainment of a Specified Fineness," in Small Particle Statistics, p. 229-246, Butterworths (1960).

14. E. B. Sandell and P. J. Elving, "Principles and Methods of Sampling," in I. M. Kolthoff, P. J. Elving, and E. B. Sandell, Editors, Treatise in Analytical Chemistry, Part 1, Vol. 1, p. 67-97, Wiley (1959).

15. O. L. Davies, Statistical Methods in Research and Production, Oliver and Boyd (1947).

16. D. J. Ottley, "Gy's Sampling Slide Rule," World Mining, 20-24 (August 1966).

17. W. L. McCabe and J. C. Smith, "Size Reduction," in Unit Operations of Chemical Engineering, p. 201-241, McGraw-Hill (1956).

18. P. H. Wilks, P. Ravinder, C. L. Grant, P. A. Pelton, R. J. Downer, and M. L. Talbot, "The Commercial Production of

Submicron ZrO_2 Via Plasma," in Symposium on the Application of Electric Discharge Chemistry, San Francisco, American Institute of Chemical Engineers (1971).

19. C. L. Grant and P. A. Pelton, "The Role of Homogeneity in Powder Sampling," in ASTM STP 540, Sampling, Standards, and Homogeneity, in press.

20. K. Ridgway and R. Rupp, "The Mixing of Powder Layers on a Chute: The Effect of Particle Size and Shape," Powder Technology 4, 195-202 (1970/1971).

VARIATION OF STANDARDS AND SAMPLING REQUIREMENTS FOR COMPLEMENTARY ANALYSIS METHODS

R.K. Skogerboe

Department of Chemistry, Colorado State University

Fort Collins, Colorado 80521

ABSTRACT

The requirements imposed on the standards and samples used for any analytical technique are determined by the sample utilization characteristics of each technique as well as the purpose of the analysis. A practical comparison of said sample utilization characteristics for several analytical techniques will be examined to define limitations imposed on the standards/sampling aspects of analysis.

INTRODUCTION

The requirements imposed on the samples and standards used for analysis are particularly stringent at trace concentration levels. Trace elements or trace species are almost invariably distributed in a material in a non-uniform manner to some degree. Witness to this fact may be based on the continued development and use of the ion, electron, and laser microprobe instrumentation which has occurred over the past decade. If sample heterogeneity did not exist, the use of such instruments for the analysis of discrete microsamples would be the only justification for their existence.

Three primary considerations influence the requirements that must be imposed on the samples and standards used. 1) The physical form of the material to be analyzed clearly affects the requisites involved. While there are some prominent exceptions, solution samples are normally homogeneous and present the least problems. The case for powder samples has been discussed by Dr. Grant. Problems associated with continuous solids (e.g., metals

and alloys) will be considered below. 2) Certainly the sample/
standards requirements are strongly influenced by the type of
information required in trace characterization, e.g., bulk,
surface, or local concentrations of trace constituents. There is
considerable interest in all of these types of analyses; they
will be discussed below. 3) A limiting influence is often imposed
on the sampling/standards criteria by the analytical technique(s)
selected for the particular analysis of interest. The impositions
involved may be broadly defined as originating from sample utilization characteristics of the method(s). In order to demonstrate
this, consider the more common analytical problem, i.e., bulk
analysis.

SAMPLING/STANDARDS REQUIREMENTS FOR BULK ANALYSIS

If the determination of a bulk or average composition is of
interest, the accuracy of the analysis will depend on the repre-
sentational integrity of the samples and standards. In this
context, integrity refers to the accuracy with which the sample
represents the population (or universe) from which it is drawn
and the accuracy with which the relevant concentrations in the
standards are known. That the general compositional identity of
the standards should approximate or simulate that of the samples
is strongly implied if interferences are to be avoided. In such
systems, accurate representation is best obtained if the materials
under investigation are homogeneous. On a practical basis, a
homogeneous material is defined as one in which the unit which
determines heterogeneity (or segregation) is considerably smaller
than the size of the analysis sample, i.e., the aliquot used for
the actual measurement. Grains or particles may be the units defining segregation of the trace element from the matrix in solids.
Ions, molecules, or complexes may be the defining units in
solutions. A homogeneous material is thus one for which the
probability of including the same number of trace constituent
units in each analysis aliquot approaches unity. Obviously,
as the size of the unit determining segregation gets larger, the
sample size required to maintain this unit probability also gets
larger. In other words, for a particular material, the degree of
heterogeneity observed by the analytical technique gets greater
as the sample size decreases.

A general indication of this effect can be obtained by examin-
ing data representative of different analytical techniques. Table
I presents a comparison with appropriate comments. Techniques
which require that the samples be presented to the measurement
system as a solution typically use large samples; the samples are
homogenized during dissolution; and precise bulk analysis are
generally obtained. Techniques which are capable of analyzing
a sample directly without prior sample treatments such as dissolution or fusion are often at a disadvantage because sample

homogenization is not obtained. In general, these techniques also utilize less sample in the measurement step and the results reflect heterogeneities in the sample to a larger extent than those which use larger samples. The extreme case is that in which microprobe techniques are used for bulk analyses. Surely a major portion of the variation observed is due to sample inhomogeneity. While one may argue that probe techniques were not designed for such analyses, an examination of the literature readily indicates that they are being used for that purpose, e.g., the ion microprobe. If such usage is to continue, the availability of standards which are homogeneous on a microscale is essential.

Recognizing that there is a sampling problem and a shortage of homogeneous standards for bulk analyses, numerous approaches can be used to alleviate the problem. Prior homogenization of the samples or standards by chemical or physical means can be carried out. While dissolution, fusion, separation, and preconcentration methods may be used for this purpose and are essential for some types of analyses; such methods add to the cost of the analysis and serve as additional potential sources of error. In effect, direct analyses which do not require prior sample treatment are to be preferred on at least a practical basis. For many direct analysis techniques, "artificial" homogenization approaches have been used widely. For example, many spectrometric techniques have long relied on spinning samples within the measurement section of the instrument as a means of homogenization. Rotation of a sample simply allows sampling a greater total area and tends to average out some of the random variations due to heterogeneity. While such methods have some limitations, they can be readily used to reduce the problem and increase the accuracy of bulk analysis. A related technique is based on the use of longer measurement (instrument) time constants to integrate out random signal variations. While such techniques have found extended use, every experienced analyst can cite examples for which they have failed.

It should be noted that the sample utilization characteristics of any technique are determined, (not only by the amount of sample used for a measurement), by the physical nature of the sample and by the process involved in producing the analytical signal used for measurement. These factors are implied by the comments presented in Table I.

It has been pointed out that methods relying on the use of a sample as a solution suffer from the disadvantage that contamination or loss of analytical species may occur during the dissolution process. Similar effects may also be associated with other sample preparation methods such as fusion and ashing. At the same time, such preparative steps may be beneficial in that samples are frequently homogenized during the process. A third factor of importance must be considered. Analytical errors often originate

TABLE I. Relation Between Bulk Analysis Precision for Analytical Techniques and the Sample Sizes Used

Technique	Typical Sample Size, mg	Typical Measurement Precision Range, %	Comments
Colorimetry	100 - 1000	1 - 5%	Homogenization by dissolution.
Flame spectrometry methods	100 - 1000	1 - 5%	Homogenization by dissolution.
Neutron Activation	10 - 1000	1 - 10%	Separations to increase specificity add to error.
Emission Spectrography	1 - 100	5 - 30%	Specific excitation medium used affects precision of results.
X-ray Fluorescence	10 - 100	5 - 30%	Homogenization by pre-treatment often used to improve precision.
Solids Mass Spectrometry	0.1 - 10	5 - 30%	Sample homogeneity very important.
Microprobe techniques	0.01 - 1	5 ≈ 50%	Sample homogeneity critical for precise bulk analyses.

from matrix effects, i.e., the presence of one or more constituents in the samples causes a change in the magnitude of the analytical signal measured for another constituent. Such matrix effects frequently depend on the concentration ratio of analyte to interferent. The magnitude of the effects may also depend on the compositional identity (compound form) of the analyte and/or the interferent. Consequently, the analyte/interferent concentration ratios and/or the compound forms of the species may have to be closely matched for both samples and standards. Certainly, this is most often impossible. Sample preparation methods involving dissolution, fusion, ashing, separation, or some other form of chemical or physical treatment consequently may offer one or more advantages over the direct analysis methods. The preparative method used may serve to destroy the physical and chemical dissimilarities between samples and standards thereby eliminating matrix effects. Such preparative methods may also be used as means for separating the analyte and the interferent species. Finally, the use of some physical and chemical methods of preparation may make it easy to prepare synthetic standards of high integrity. In many instances, these advantageous possibilities far outweigh the time and effort required and the other disadvantages associated with the destructive preparation techniques. Numerous examples to support these contentions can be found in the literature.

Emission spectroscopy, spark source spectrometry, and x-ray fluorescence may be considered to be closely related and competitive as well as complementary techniques. Generally, these techniques are used for multielement determinations on each sample and the majority of these analyses are made directly on solid sample materials. In many respects, the sample and standards requirements are accordingly quite similar. One primary difference exists. Emission and mass spectrometry both must rely on vaporization of the sample during the measurement step. One need not be widely versed in thermodynamics to realize that the vaporization process will almost universally be highly dependent on the physical and chemical characteristics of the samples. Small changes in these characteristics can clearly affect both the rate and the extent of vaporization thereby affecting the analytical signals measured. X-ray fluorescence should not be generally subject to these problems and the attendant errors that may be observed if the samples and standards differ appreciably. Alternatively, instances occur for which selective vaporization has been used to separate analytical species from matrix interferents. In such instances the x-ray method is at a disadvantage. As a result, it is difficult to assign a generally distinct advantage to any one of these competing techniques insofar as sampling and standards requirements are concerned. Such assignments must be based on each specific analytical problem.

MICROPROBE TECHNIQUES

The sampling/standards problems for microprobe techniques are strikingly unique in comparison to those associated with bulk analyses. In a sense, the probe techniques are most appropriate for the determination of heterogeneity (segregation). Their design features are oriented toward analyses on a highly localized scale in all three spatial dimensions. The electron and ion microprobe systems are capable of providing spatially resolved analyses in the low to fractional micron range; a spatial resolution of about 10 microns is about the best that can be achieved with the laser probe system. If one considers the use of such systems for the quantitative determination of minor or trace constituents, it is intuitively obvious that their greatest potential lies in locating and defining heterogeneity and/or in analyzing microsamples. The preparation of standards on a microscale or bulk standards with a certified level of heterogeneity is clearly a momentous task. While the author has considered for several years possible means by which such standards can be prepared, no obvious solutions to the problem are apparent. For this reason, it is fortunate that the majority of microanalysis problems which can be handled by the probe techniques actually require only relative concentration profiles. If it were not for this, the practitioners of microprobe analyses would be in serious trouble.

Microprobe analyses are almost exclusively direct analyses. For this reason, the requirement that standards and samples be closely matched is an implicit if not explicit requirement. Exactly how close the match must be will depend on the particular sample type and on the type of information to be provided by the analysis.

CONCLUSIONS

In choosing an analytical technique for any determination, two general considerations must be evaluated and brought together. The first deals with the nature of the problem. The question to be answered must be exactly determined and a research plan developed. This should define the sampling strategy, the number of samples to be collected, and their variety. The plan must designate the species to be determined, the type of information required, and the levels of accuracy and precision needed. With this information, the attributes of the analytical techniques available can be considered. The technique selection may be based on considerations involving sensitivity, precision, accuracy, type of information provided, sampling/standards requirements, and economy. The appropriate matching of these two factors is the essential role of the professional analytical chemist. While many erroneous results have been blamed on the inavailability of adequate standards, it is apparent that this blame has often been falsely placed. Rather, the analyst has failed to recognize the limitations and/or problems common to certain techniques. In effect, he failed to adequately evaluate the considerations mentioned above. It is true that a greater variety of standard reference materials is needed. The realist recognizes that, in spite of the efforts of the National Bureau of Standards, fulfilling all requests for standards is a task that will never be completed. A practical compromise must be to seek a limited number of standard reference materials generally representative of the most important sample matrices. Beyond this, laboratories must take the initiative in establishing the integrity of any analytical methods. This may be best accomplished by participation in critical intermethods and interlaboratory comparisions. By selecting materials for which good standards do not exist, the intercomparisons can be used to establish these materials as standards. Why not make this a goal of the next intercomparison that you participate in?

QUANTITATIVE ANALYSIS OF CLAY MINERALS

IN DRILLING MUD SOLIDS

D.G. Feuerbacher and R.R. Clark

Dresser Industries

Oilfield Products Division

Houston, Texas

ABSTRACT

Quantitative analysis of six clay minerals commonly found in drilling mud solids is studied by x-ray diffraction using the method of internal standards. Standard clay samples are used with three internal standards to derive calibration curves from which four synthetic mud solids are analyzed. If careful sample preparation and handling are employed, reasonably accurate and reproducible results are obtained.

INTRODUCTION

Clay minerals constitute a vitally important ingredient of oil well drilling fluids, and their detection and analysis, both qualitative and quantitative, is integral to a successful drilling operation. Clays are not only added to the drilling fluid to build viscosity, gel strength, and fluid loss control, but are also added continuously to the mud from bit cuttings. (See Table I.)

TABLE I

USE AND OCCURRENCE OF CLAYS IN DRILLING FLUIDS

Clay Mineral	Occurrence, Use
Sodium Montmorillonite (bentonite)	Viscosity improver, gelling and suspending agent, fluid loss agent
Calcium Montmorillonite	Dilutant in bentonite clays; poorer viscosity improvement
Attapulgite	Viscosity builder in salt water drilling fluids
Muscovite	Native clay – component of slates
Illite	Native clay – major component of shales
Kaolinite	Native clay

X-ray diffraction is a commonly used technique for qualitative analysis of clay minerals, and has also been successfully used for quantitative measurements[1,2]. Other methods which have been tried, with varying degrees of success, for quantitative analysis of clays are electron microscopy[3], X-ray fluorescence[4,6], atomic absorption[7], UV-visible spectroscopy[8,9], thermal analysis[10], and X-ray macroprobe analysis[11]. Any technique used for analysis of clays must, however, overcome the limitations of rather poor and variable crystallinity, tendency to orient, varying particle size distributions, difficulty of obtaining pure standards, low sensitivity to detection, different degrees of hydration, etc., all of which clays impose on the analyst.

One of the most difficult problems in clay analysis by X-ray diffraction is sample preparation. Some researchers feel that a completely random-oriented sample is required for successful analysis[12], while others claim that this is impossible to achieve with clays, so that highly oriented or partially oriented samples must be used[13].

Once the method of sampling has been decided, there are basically four variations in the method of quantitative X-ray diffraction analysis:

1. Direct measurement of peak height or area for a particular reflection and comparison with a pre-run calibration curve[14,15].

2. The method of known additions, in which sequential additions of the mineral being analyzed are made, and the results extrapolated to zero addition[15].

3. The method of external standards, in which a standard mineral is compared with the unknown mixture by measuring both intensity of reflection and mass absorption coefficients[15,16].

4. The method of internal standards, in which the mineral being determined is compared on a weight ratio and intensity ratio basis with a pure, crystalline substance which was added to the sample in a known concentration[15].

The last method has been used successfully for the determination of calcite[17], quartz[18,19], hematite[20], feldspar[21], and clays[22,23]. Most workers use as internal standards substances which are in the cubic system and have rather low mass absorption coefficients, but one study[22] used platy internal standards ($ZnCl_2$ and pyrophyllite) which was felt would orient to the same degree as the clays.

Our study on clays in drilling mud solids sought to achieve a method of quantitative analysis which was selective, rapid, and reasonably accurate and sensitive. We felt that attempts to achieve completely non-oriented clay samples have been unsuccessful and have also resulted in a loss of sensitivity; hence we chose a rather highly oriented sample, either a pressed pellet or a thin film on a glass slide. The method of internal standards was used because of the relative freedom from matrix effects.

A total of six clay minerals, those most commonly found in drilling fluid solids, were chosen for analysis. Structure and formula data on these clays is given in Table II.

EXPERIMENTAL

Sample Preparation and Standards

Sources of the six clay standards used are given in Table III. With the exception of the kaolinite, and muscovite, all standard clays had one or more impurities. These impurities, along with their concentrations (found by the method of known additions), are given in Table IV.

Two standards were used for sodium montmorillonite, a commercial Wyoming bentonite, and a synthetic montmorillonite. The synthetic clay was prepared by Mr. Stan Alford of the Oilfield Products Division, Dresser Industries, Inc. by a process of dispersion of an impure sodium montmorillonite, settling, dialysis of the dispersed layer, and ion exchange with a sodium-saturated cation exchange resin.

Internal standards employed were sodium fluoride, sodium chloride, and nickel(ous) oxide. These were chosen on the basis of their cubic (non-orienting) structure and low X-ray mass absorption coefficients.

All samples and standards were dried at $105°C$ for at least 6 hours, and then ground to <325 mesh in a Pitchford Selective Particle Size Grinder[24]. All clay standards prepared in this way were then analyzed by Mr. Morris Cordova of the Oilfield Products Division, Dresser Industries, Inc. on a Coulter Model T Particle Counter. All samples showed a rather symmetrical particle size distribution with peaks ranging between 2 microns and 25 microns.

Mixtures of clays, samples, and internal standards were weighed on an analytical balance and then mixed on the Pitchford Grinder (utilizing mixing action only) for at least one minute. The samples were treated in two ways:

1. Approximately 2 grams of the powder mixture was pressed to 24,000 psi in a 1¼" X-ray die to form a thin pellet.

TABLE II

STRUCTURE AND FORMULA DATA ON CLAYS

CLAY	GENERAL FORMULA	SYSTEM	LATTICE PARAMETERS		
			A	B	C
Sodium Mont.	$(Al_{3.34}Mg_{0.66})Si_8O_{20}(OH)_4Na_{0.66}$	Monoclinic	5.17	8.95	12.5 [1]
Calcium Mont.	$(Al_{3.34}Mg_{0.66})Si_8O_{20}(OH)_4Ca_{0.33}$	Monoclinic	5.18	9.0	15.0 [2]
Attapulgite	$Mg_5Si_8O_{20}(OH)_2 \cdot 8\ H_2O$	Monoclinic	5.22	18.06	12.75
Muscovite	$KAl_2(Si_3Al)O_{10}(OH)_2$	Monoclinic	5-20	9.00	20.03
Illite	$K_{1-1.5}Al_4Si_{7-6.5}Al_{1-1.5}O_{20}(OH)_4$	Monoclinic	5.2	9.0	20.0
Kaolinite	$Al_2Si_2O_5(OH)_4$	Triclinic	5.15	8.95	7.39

NOTES
1. Lattice parameter for one interlamellar water layer
2. Lattice parameter for two interlamellar water layers

TABLE III

CLAY STANDARDS SOURCES

CLAY	SOURCE
Kaolinite	Hydrite Flat D, Lot 1071 Georgia Kaolin Company
Attapulgite	Production Salt Gel Floridin Company
Muscovite	Stoneham, Maine; Ward's Natural Science Est. Inc.
Illite	API Clay Mineral Standard Project # 49; Illite # 36 Morris, Illinois
Calcium Montmorillonite	Clark Clay, Gonzales, Texas Southern Clay Products, Inc.
Sodium Montmorillonite	a. Purified (see test) b. Bear Creek bentonite Bear Creek, Wyoming

TABLE IV

IMPURITIES IN CLAY STANDARDS

CLAY	IMPURITY	CONCENTRATION	CLAY PURITY
Kaolinite	none	-	100%
Attapulgite	Dolomite	5%	88%
	Quartz	6%	
	Calcite	1%	
Illite	Quartz	14%	84%
	Kaolinite	2%	
	Feldspar	trace	
Muscovite	none	-	100%
Clark Clay	Cristobalite	1%	95%
	Oligoclase	2%	
	Gypsum	1%	
	Quartz	1%	
Bear Creek Clay	Quartz	1%	68%
	Oligoclase	8%	
	Illite	20%	
	Cristobalite	trace	
	Gypsum	1%	
	Calcite	2%	
Pure Na Mont.	Cristobalite	5%	95%

2. To analyze for montmorillonites, part of the sample (2 grams) was placed in a controlled humidity (40% relative humidity) chamber for at least 8 hours. This has the effect of hydrating the relatively faster hydrating calcium montmorillonite to give a basal 001 spacing of 15 Å, while leaving the slower hydrating sodium montmorillonite at 10 - 12 Å. In this way the calcium montmorillonite can be analyzed separately.

Another part of the sample, 200 mg., was treated with 10-12 drops of ethylene glycol to form a slurry. This slurry was then spread on a cavity 2.0 x 4.5 cm formed by masking tape on a standard microscope slide. The slide was then placed in a 105°C oven and dried until the surface of the sample was just slightly moist. The slide was then analyzed by X-ray diffraction. This treatment gave total montmorillonites in the sample; sodium (and mixed-layer) montmorillonites were calculated by subtracting calcium from total montmorillonites.

X-RAY MEASUREMENTS

All diffraction measurements were made on a General Electric Model XRD-5 Diffractometer modified with Ortec solid-state electronics. A Cu-target tube was used for all scans.

Integration of all pertinent lines was done with an Infotronics CRS-208 Automatic Digital Integrator (if there were no interfering peaks and a reasonably flat baseline) or a Hruden planimeter if there was interference or a changing baseline.

RESULTS

The method of internal standards relies on the mass absorption coefficients cancelling out of the intensity ratio equations, so that

$$I_p/I_s = K\, Wt_p/Wt_s$$

where I_p, Wt_p = integrated intensity, weight of component P in the sample
I_s, Wt_s = integrated intensity, weight of internal standards
K = constant which depends on p and s, but is independent of the matrix

K can be found by calculating the slope of the graph of intensity ratios vs weight ratios. In our calculations a modified least squares linear fit is assumed, with the intercept forced to be the origin.

Table V gives the results of these calculations; percent deviation is the average deviation of data points from the linear fit. Table VI lists the slope values for kaolinite in several matrices; the closeness of these slopes verifies the validity of the internal standards approach.

A total of four synthetic drilling mud solids was then mixed and analyzed by the method of internal standards. The results are given in Table VII.

Since most of the clay standards used were not pure (see Table IV), the value of the clay concentration found should be multiplied by the concentration of that clay in the standard. These values are given in Table IV on the right side of the page.

In order to determine relative sensitivity of our sampling and analytical technique to the six various clays, small concentrations of each clay were analyzed in two different matrices: Kaolinite and commercial amorphous lignosulfonate. The result of this study is given in Table VIII.

Precision (reproducibility) of results is not included in tabular form, but was on the order of $\pm 5\%$ for kaolinite and muscovite, and $\pm 10\%$ for the other clay minerals studied.

DISCUSSION

It is immediately obvious that there are several problems associated with the quantitative analysis of clays by X-ray diffraction. In Table V, for example, some of the percent deviation values are quite high, and in Table VII the percent errors for several clays, particularly in Sample 4, are extremely high. (This sample and consequent results were included to show the results of poor sample mixing; even though this sample was mixed on a Spex Mixer-Mill for 30 minutes, the reagent barium sulfate still tended to "ball up" and caused inhomogeneous powder distribution). It also can be seen that the method is much more highly sensitive to some clays, e.g. kaolinite and muscovite. This may have to do with their degree of crystalinity as well as particle size distribution, degree of preferred orientation, etc. This sensitivity is reflected in the size of the slopes in Table V, with illit showing smaller slope values than kaolinite or muscovite.

In actual practice the analysis would proceed as follows: first the dried, ground sample would be scanned to identify the clay minerals present. Then the suitable clay standards and internal standards would be chosen and the method of internal standards used as above. Since there are so many varieties of

TABLE V

RESULTS OF INTERNAL STANDARD CALIBRATIONS

CLAY	INTERNAL STANDARDS					
	NaCl		NiO		LiF	
	$K^{(1)}$	% Dev.$^{(2)}$	K	% Dev.	K	% Dev.
Kaolinite	1.06	6.0	2.03	1.9	1.98	4.2
Muscovite	1.04	4.0	4.03	20.4	–	–
Illite	0.155	8.9	0.224	6.6	0.190	22.4
Attapulgite	0.323	25.4	0.83	6.0	–	–
Ca Mont.	–	–	3.05	7.8	2.70	9.5
Total Mont.	–	–	3.65	8.5	3.22	11.2

NOTES
(1) K = Slope of I clay/I Int. Std. vs Wt. Clay/Wt. Int. Std.
(2) % Deviation = Average deviation of data points from least square linear fit.

TABLE VI

EFFECT OF MATRIX ON SLOPE VALUES
CLAY: KAOLINITE
INTERNAL STANDARD: NiO

Matrix	Slope (K)
Calcium Montmorillonite	1.77
Attapulgite	2.03
Illite	1.81
Muscovite	1.76

TABLE VII
ANALYSIS OF SYNTHETIC MUD SOLIDS

SAMPLE	COMPOSITION	FOUND	% ERROR
1	40% Attapulgite	39.6	1.0
	10% Kaolinite	9.3	7.0
	30% $BaSO_4$	-	-
	10% NaCl	-	-
	10% LiF	-	-
2	50% Bear Creek Mont.	47.9	4.2
	30% Illite	28.7	4.3
	10% Kaolinite	8.6	14.0
	10% NiO	-	-
3	40% Bear Creek Mont	41.3	3.2
	30% Clark Mont.	25.3	15.7
	10% Muscovite	7.8	22.0
	10% Kaolinite	7.5	25.0
4	5% Bear Creek Mont	6.3	26.0
	10% Illite	16.8	68.0
	5% Kaolinite	10.3	106.0
	70% $BaSO_4$	-	-
	10% LiF	-	-

TABLE VIII
MINIMUM DETECTABLE CLAY CONCENTRATIONS
IN KAOLINITE AND LIGNOSULFONATE MATRICES

CLAY	SENSITIVITY (Wt%) in Matrix[1]	
	Kaolinite	Lignosulfonate[2]
Kaolinite	0.3	0.3
Muscovite	0.8	0.5
Attapulgite	3.0	3.0
Illite	3.0	3.0
Sodium Montmorillonite	1.0	3.0
Sodium Chloride [4]	<0.1	<0.1

NOTES
(1) Sensitivity is the lowest concentration of clay required to give a major (usually 001) peak height of at least 2 x residual noise level.
(2) Lignosulfonate is Swedish Spersene, a product of Dresser Industries, Inc., Oilfield Products Division.
(3) Matrix used was attapulgite.
(4) Used as a comparison only

even one type of clay (e.g., kaolinite), however, much care must be taken to thoroughly identify each clay present and to use that clay as the standard (or calibration) in quantitative methods.

There is also the problem in the separation of peaks in order to calculate areas. In addition to this, problems in particle size distributions, sample mixing and preparation, instrument stability and reproducibility, non-reproducible degrees of particle orientation, interaction of clays with internal standards, etc. all contribute to the errors involved in this technique.

This laboratory is in the process of further refining our capabilities for clay mineral analyses, and we feel that the following precautions and methods will improve the accuracy and reproducibility of further quantitative studies:

1. Examination of several lines for each clay instead of just the strongest line; perhaps computer programming might be of help here.

2. Use of a spinning sample holder to minimize particle distribution effects.

3. Careful accumulation and characterization of clay standards.

4. Extra care in preparation and handling of samples, particularly with regard to grinding, mixing, and controlled hydration or complexation.

5. Use of other analytical techniques (e.g. DTA, electron microscopy, etc.) to supplement x-ray data.

ACKNOWLEDGEMENTS

The authors wish to thank Dresser Industries and Mr. Keith Wagner for allowing us the equipment and time to finish this study. We would also like to thank our wives, Loma and Kay, for their help and patience.

REFERENCES

1. Johns, W.D., Grim, A.E., and Bradley, W. Jr.
   Journal of Sedimentary Petrology 24, 242 (1954)
2. Norrish, K., and Taylor, R.M., Clay Minerals Bull. 5, 98 (1962)
3. Babitsyn, B.K., and Ushatinskii, I.N. Chemical Abstracts, 78 168 (1973)

4. Nicolas, J., Dorrillet, B., and Quintin, M., Chemical Abstracts 69, 766 (1968)
5. Bentelspacher, Hans, and Rietz, Egbert, Chemical Abstracts 71, (1969)
6. Jabikh, A.A., Chemical Abstracts 69, (1968)
7. Davey, J., Chemical Abstracts 76, 596 (1972)
8. Bailey, G.W. and Karickhoff, S.W., Anal. Lett. 6, 43 (1973)
9. Karickhoff, Samuel W., and Bailey, George W., Clays and Clay Minerals 21, 59 (1973)
10. Davis, C.E., and Holdridge, D.A., Clay Minerals 8, 193 (1969)
11. Hermes, O.D. and Ragland, P.C., Amer. Mineral, 52, 493 (1967)
12. Thomson, A.P., Duthie, D.M.L., and Wilson, M.J., Clay Minerals 9, 345 (1972)
13. Shaw, H. Jr., Clay Minerals 9, 349 (1972)
14. Davis, C.E., J. Sci. Res. Counc. Jam. I, 98 (1970)
15. Brown, G., "The X-Ray Identification and Crystal Structures of Clay Minerals," Jarrold and Sons Ltd., Norwich (1961)
16. Plesch, R., Chemical Abstracts 78, 618 (1973)
17. Jacob, C., Chemical Abstracts 78, 600 (1973)
18. Novosel-Radovis, Vera, Chemical Abstracts 77, 258 (1972)
19. Millet, J. and Hommey, R., Chemical Abstracts 78, (1973)
20. Novosel-Radovic, Vera, Chemical Abstracts 77, 669 (1972)
21. Weber, Jr. Larque, P., and Feurer, R., Chemical Abstracts 78, (1973)
22. Mossman, M.H., Freas, D.H., and Bailey, S.W., Clays Clay Minerals 15, 441 (1966)
23. Takat, Tibor, and Peter, Mrs. Tibor, Chemical Abstracts 76, 247 (1972)
24. Pitchford, A.H., Norelco Reporter 7 (1960)

FACTORS LIMITING THE USE OF STANDARD MINERALS IN THE X-RAY DIFFRACTION ANALYSIS OF CLAYS[*]

Lowell A. Douglas

Dept. of Soils and Crops, Rutgers University

New Brunswick, New Jersey 08903

ABSTRACT

In the quantitative x-ray diffraction analysis of clays, "standard" clays are compared with "unknowns" assuming both standards and unknowns have similar diffraction characteristics. This assumption is valid in certain cases involving kaolinites, chlorites, vermiculites, and where standards have been obtained from the unknown clays. In other cases; especially those involving smectites, illites, and some chlorites and vermiculites; lack of homogeneity within mineral species and inability to measure this inhomogeneity in mixed mineral suites limits the accuracy obtainable by diffraction methods.

INTRODUCTION

The intensity of an x-ray diffraction pattern is proportional to the material producing it, when appropriate considerations are made for absorption effects (1). This fact has enabled mineralogists to develop clay mineral quantitative analysis schemes (2,3,4,5, etc.) Hofmann (3) and Moore (5) showed that with modern x-ray equipment high reproducibility was obtained when the unknowns were mixtures of standard minerals. Hofmann (3) reported poor reproducibility when naturally occuring mixtures were used in an interlaboratory comparison. This paper discusses some properties of clay minerals that affect their use as standards.

[*] Journal series paper of the New Jersey Agricultural Experimental Station.

PARTICLE SIZE

Some clay minerals occur over a very wide range of particle sizes. Kaolinite is often found in < 0.2 um fractions, but is sometimes found as sand size material (6). Vermiculite and chlorite are found over a wider range of particle sizes. Smectites are always limited to <0.2 um fractions, and sometimes limited to <0.08 um fractions (4). Illite's particle size distribution is not as distinct because of nomenclature problems. Illite is a discredited mineral name (7), the preferred name being that of the specific mica (s) involved. Where the specific mica is not known the material should be called mica. It is relatively easy to assign a specific mica name to particles 10 um in diameter. The problem is much more involved when studying smaller size particles. I will use the term illite to refer to any 10 A., clay-size mica, to emphasize the fact that the true nature of clay-size mica is often more conjecture than fact.

Several investigators (2,8,9) have claimed that oriented clay samples can not be used in quantitative x-ray diffraction analysis. Oriented samples are usually prepared by pouring a clay - water suspension on a glass slide and allowing the water to evaporate. In this process large particles settle faster than small particles producing a sample in which the smaller particles are preferencially exposed to the x-ray beam. On the other hand, in the clay (< 2 um) size range, smaller particles are poorer diffracters of x-rays (Figure 1). Moore (5) showed that oriented samples are as useful as unoriented samples in quantitative studies, the important point being that the sample preparation methodology must result in a reproducible diagnostic sample. Thus, the investigator has a choice between preparing oriented samples sensitive to small amounts of fine clays or unoriented samples that will be more sensitive to larger size clay minerals. With either method the particle size distribution of the "standard" minerals should be similar to the particle size distribution of the unknowns.

CLAY CRYSTALLINITY

The ability of a material to diffract x-rays (structure factor) depends on (among other considerations) the atomic scattering of a single atom relative to that of an electron. The International Tables for x-ray Crystallography lists atomic scattering factors for copper radiation at low angles including: lithium 2.21, sodium 9.76, potassium 16.73, magnesium 10.5, iron 23.68, silicon 12.16, aluminum 11.23, and water 8.87. Substitutions common in clay minerals, as magnesium/iron, silicon/aluminum, magnesium/potassium, etc.,

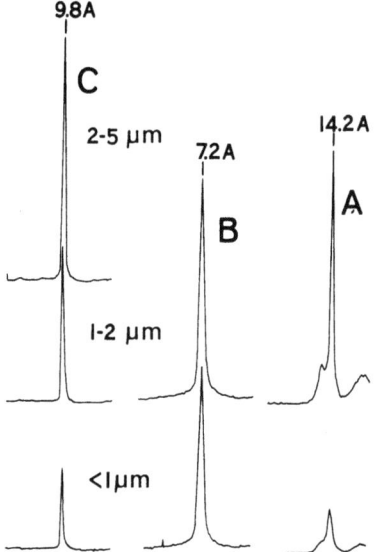

Figure 1. Diffractograms of: top 2 - 5 um, middle 1 - 2 um and bottom <1 um; A, vermiculite; B, kaolinite; and C, biotite.

may cause considerable change in the intensity of x-ray diffraction lines.

Illite

Grim, et al. (10) proposed the name illite for "the mica-type mineral occuring in argillaceous sediments". This (10) and subsequent (11, 12) discriptions of illite have led investigators to consider illite as a slightly altered muscovite.

The data of Gaudette, et al. (11) show considerable variation in the x-ray diffractograms of the samples used to characterize illite (Table 1). Warshaw (13) showed that the intensity of the 00 3 diffraction line of illite is very sensitive to the potassium content of the mineral.

It has long been recognized that it is virtually impossible to distinguish between clay - size illite and glauconite, in mixed suites, by X-ray diffraction; even with their contrasting chemical formulas, $(OH)_2 Al_2 Si_{4-x} Al_x O_{10} K_x$ and $(OH)_2 (Fe, Al, Mg)_2 Si_{4-x} Al_x O_{10} K_x$ with octahedral iron plus magnesium four to five times octahedral aluminum in glauconite. Two recent studies (14,15) have considerably altered the popular generalizations regarding illite. The reader must refer to Weavers (14) and Hower and Mowatt's (15) papers for details, but in essence they have shown that much illite is a diocta-

Table 1. Relative intensity of diffraction lines of standard illites. The most intense line among all samples = 100. (Data recalculated from Gaudette et al. (11).

Sample Source	Diffraction Lines			
	001	002	003	005
Beavers Bend	100	30	84	27
Marblehead	81	46	89	41
Rock Island	97	35	89	30
Fithian	81	19	43	11
Grundite	89	16	46	11

hedral mica more like glauconite than muscovite. Until more is learned about how to distinguish between different illite types in mixed mineral suites, the investigator will have a relatively large disadvantage in attempting to choose a similar "standard" illite for diffraction work.

Kaolinite

Kaolinite probably presents the smallest problem when one wants to select a "standard" kaolinite. Among the clay minerals, kaolinite shows a remarkable constancy of lattice parameters and chemical composition (16). Murray (17), and Murray and Lyons (18), pointed out that although most kaolinites are highly crystalline, some varieties have poor crystallinity and corresponding variations in their diffraction patterns. The criteria of Murray and Lyons (18) may be used to identify the crystallinity of kaolinites and to select an appropriate "standard" kaolinite.

Chlorite and Vermiculites

Chlorites and vermiculites present a dichotomy to the diffractionist. These minerals often occur both as micro and megascopic minerals in zones of mineralization or metamorphism (19). In these cases pure specimens of the chlorite on vermiculite under study can often be used as a standard for preparing analytical data. Bailey (20) and Brindley and Gillery (21) have presented methods for using d-spacings and 001 line intensities to indicate the chemical composition of chlorites, enabling one to compare unknown and standard chlorites. On the other hand in vermiculite the magnesium-iron ratio varies, with associated changes in structure factor.

Chlorite and vermiculite are usually thought to be trioctahedral. Chlorites (21) and vermiculites (22) in soils and sediments are often dioctahedral minerals. Trioctahedral chlorite includes a brucite layer and trioctahedral vermiculite includes a layer composed of magnesium octahedrally coordinated to water. In the dioctahedral analogues, ideally aluminum substitutes for the magnesium in interlayer position. Unfortunately, there is considerable evidence that mixed aluminum - magnesium - hydroxide - water interlayers, with associated change in structure factor, may be rather common (Table 2). Identification of these minerals involves the collapse of the 14A diffraction line to 10 to 11.5A when the sample is heated (22,23). However, there is a considerable temperature range over which these minerals collapse (24) probably indicating the variable nature of the interlayer. In clay-size material there is no explicit way to define the nature of these interlayers. The selection of standard dioctahedral chlorites or vermiculites presents the added problem that samples of desired mineral purity are very rare.

Smectites

Smectites include all those expanding clay minerals formerly included under the group, montmorillonite (7). Ross and Hendricks (25) listed the smectites along with the ions in octahedral position, as: montmorillonite, aluminum and magnesium; beidellite, aluminum; nontronite, iron; hectorite, magnesium and lithium; and sapronite, magnesium and aluminum. In spite of these compositional differences it is very difficult to distinguish these minerals by x-rays (26). Rodriguez Gallego et al., (27) found the diffractograms of several montmorillonites of variable chemical composition to be very different. This could not be due to chemical composition according to calculations of structure factors. Greene-Kelly (28) described a method that is diagnostic for montmorillonite as opposed to other minerals of the group.

Table 2. Interlayer variations in clay-size chlorites and vermiculite

Mineral	Chlorite		Vermiculite	
Octahedral layer	di	tri	di	tri
Interlayer	gibbsitic	brucite	$Al + H_2O + OH$ or $Al + Mg + H_2O + OH$	$Mg + H_2O$

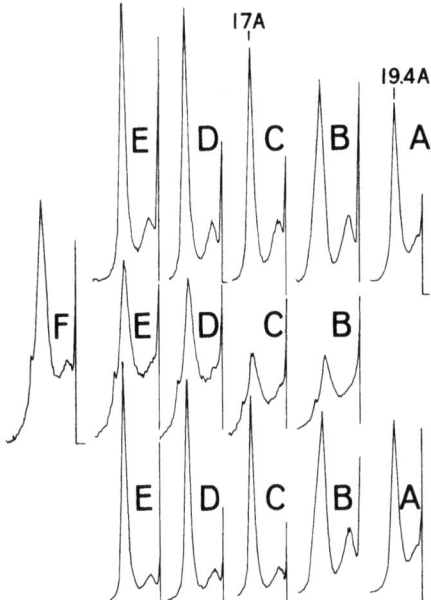

Figure 2. Diffractograms of ethylene glycol treated smectites. A, sample on Millipore filter and immediately x-rayed. B, on glass slide and air dry. C,D,E,F sample in vacuum desicator for 12, 36, and 58 hours and 2 weeks. Top, Umiat bentonite (29); middle, < 2 um saprolite; bottom, Wyoming bentonite. Top and middle, full scale = 200,000 cpm, bottom, full scale = 400,000 cpm.

The 001 line is usually used for quantitative identification of smectites. In mixtures including chlorite or vermiculite (also 14-15 A spacings) the smectite is usually expanded with either ethylene glycol or glycerol. The diffraction characteristics of expanded smectites depend on experimental methodology. Figure 2 shows diffractograms of three smectites that were treated with ethylene glycol. "A" were suspended in a ten percent ethylene glycol solution and sedimented (with vacuum) on Millipore filters, and immediately x-rayed. After sitting two hours the 19.4 A line shifted to 17 A. "B,C,D,E, and F" are of smectites that were also suspended in ten percent ethylene glycol and then poured on a glass slide. "B" diffractograms were made sixteen hours later, the slide was then placed in a vacuum desicator (over free ethylene glycol) and rescanned at the stated time intervals. The intensity of the 17 A line increased with increased time in the desicator.

CONCLUSIONS

Methods are available that yield high precision, but often questionable accuracy in quantitative x-ray diffraction analysis of clays. The major obstacle to high accuracy is the large amount of variation, both chemical and structural, within most clay mineral species. Higher accuracy can be obtained when standard clays are extracted from the samples under investigation (4) or when pure standard clays can be obtained nearby, as is often done in clay mining operations or in metamorphic environments. Often the investigator is interested in trends rather than the precise clay in a given sample. The high precision of present methods is adequate for these studies provided the clays under investigation have developed under comparable conditions and from similar parent minerals.

REFERENCES

1. H. P. Klug and L. E. Alexander, X-ray Diffraction Procedures. Wiley, New York (1954).

2. G. W. Brindley, "Quantitative Analysis of Clay Mixtures," in G. Brown, Ed., The X-ray Diffraction and Crystal Structures of Clay Minerals, p. 489-516. Mineralogical Society, London (1961).

3. U. Hofmann, "Summary of Clay-Mineral Studies in Germany, 1954 and 1955", in A. Swineford, Ed., Clay and Clay Minerals, Proceedings of the Fourth National Conference. p. 79-90. National Academy of Sciences - National Research Council, Washington, (1956).

4. R. J. Gibbs, Quantitative X-ray Diffraction Analysis Using Clay Mineral Standards Extracted from the Samples to be Analyzed, Clay Min. 7, 79-90 (1967).

5. C. A. Moore, Quantitative Analysis of Naturally Occuring Multicomponent Mineral Systems by X-ray Diffraction. Clays and Min. 16, 325-326 (1968).

6. W. H. Grant, Kaolinite Stability in the Central Piedmont of Georgia, Clays and Clay Min. 13, 131-140 (1965).

7. G. W. Brindley, "Discussions and Recommendations Concerning the Nomenclature of Clay Minerals and Related Phyllosilicates", In S. W. Bailey, Ed., Clays and Clay Minerals, Proceedings of the Fourteenth National Conference, p. 27-34, (1966).

8. G. W. Brindley and S. S. Kurtossy, Quantitative Determination of Kaolinite by X-ray Diffraction, Amer. Min. 46, 1205 - 1215, (1961).

9. R. J. Gibbs, Error Due to Segregation in Quantitative Clay Mineral X-ray Diffraction Mounting Techniques, Amer. Min., 50, 741-751, (1965).

10. R. E. Grim, R. H. Bray and W. F. Bradley, The Mica in Argillaceous Sediments. Amer. Min. 22, 813 - 829 (1937).

11. H. E. Gaudette, J. L. Eades and R. E. Grim, "The Nature of Illite", in W. F. Bradley and S. W. Bailey, Eds., Clay and Clay Minerals. Proceedings of the Thirteenth National Conference. p. 33-48, Pergamon Press, Oxford, (1965).

12. D. E. Lapham and M. J. Jaron, Rapid, Quantitative Illite Determination in Polycomponent Mixtures, Amer. Min. 49, 272-276, (1964).

13. C. M. Warshaw, "Experimental Studies of Illite", in A. Swineford, Ed., Clay and Clay Minerals, Proceedings of the Seventh National Conference, p. 303-316, Pergamon Press, London, (1959).

14. C. E. Weaver, Relations of Composition to Structure of Dioctahedral 2:1 Clay Minerals, Clay and Clay Min., 16, 51-61, (1968).

15. J. Hower and T. C. Mowatt, The Mineralogy of Illites and Mixed - Layer Illite/Montmorillonites, Amer. Min., 51, 825-853, (1966).

16. G. W. Brindley, "Kaolin, Serpentine, and Kindred Minerals", in G. Brown, Ed., The X-ray Identification and Crystal Structures of Clay Minerals, p. 51-131, Mineralogical Society, London (1961).

17. H. H. Murray, Structural Variations of Some Kaolinites in Relation to Dehydrated Hulloysite, Amer. Min., 39, 97-108, (1954).

18. H. H. Murray and S. C. Lyons, "Correlation of Paper-Coating Quality with Degree of Crystal Perfection of Kaolinite", In. A. Swineford, Ed., Clay and Clay Minerals, Proceedings of the Fourth National Conference, p. 31-44, (1956).

19. G. Millot, Geology of Clays, Springer - Verlog, 425 p., New York, (1970).

20. S. W. Bailey, Determinations of Chlorite Compositions by X-ray Spacings and Intensities, Clay and Clay Min., 20, 381-388. (1972).

21. G. W. Brindley and F. H. Gillery, X-ray Identification of Chlorite Species, Amer. Min., 41, 169-186, (1956).

22. G. Brown, The Dioctahedral Analogue of Vermiculite, Clay Min. Bull., 2, 64-69, (1953).

23. C. M. Warshaw and R. Roy, Classification and a Scheme for the Identification of Layer Silicates, Bull. Geol. Soc. Amer., 72, 1455-1492, (1961).

24. J. C. Hathaway, "Studies of Some Vermiculite - Type Clay Minerals", In, W. O. Milligan, Ed., Clay and Clay Minerals, Proceedings of the Third National Conference, p. 74-86, National Academy of Sciences-National Research Council, Washington, (1955).

25. C. S. Ross and S. B. Hendricks, Minerals of the Montmorillonite Group, U. S. Geological Survey Proff. Paper 205-B, p. 23-79, (1945).

26. D. M. C. MacEwan, "Montmorillonite Minerals", in G. Brown, Ed., The X-ray Identification and Crystals of Clay Minerals, p. 143-207, Mineralogical Society, London (1961).

27. M. Rodriguez Gallego, J. M. Martin Vivaldi, and J. M. Martin Pozas, Analisis Cuantitativo de Filosilicatos de la Arcilla por Difraccion de Rayos - X III. Influencia de la Sustitucion insomorfica y de la Cristalinidad. Anal. Quimica (Espano) 65, 25-29, (1969).

28. R. Greene - Kelly, Identification of Montmorillonoids, J. Soil Sci., 4, 233-237, (1953).

29. D. M. Anderson and R. C. Reynolds, Umiat Bentonite, an Unusual Montmorillonite from Umiat, Alaska, Research Report 223, 11p., U.S. Army Cold Regions Research & Engineering Lab., Hanover (1967).

ROLE OF DIFFRACTOMETER GEOMETRY IN THE STANDARDIZATION OF

POLYCRYSTALLINE DATA

William Parrish

IBM Research Laboratory

San Jose, California 95193

ABSTRACT

This paper reviews the factors in diffractometer geometry most likely to cause large errors in intensity measurements due to imperfect specimen preparation, particularly preferred orientation. It is recommended the same specimen be transferred to two or more diffractometer geometries in which the specimen has different orientations to obtain sets of complementary data. Three geometries are illustrated: θ-2θ scanning with reflection and transmission specimens, and Seeman-Bohlin with reflection specimen.

INTRODUCTION

The quality of the specimen preparation is probably the most important single factor in determining the precision of X-ray powder diffractometer data. Diffraction angles are derived from intensity measurements which in turn are determined by many factors related to the type of specimen and its relation to the diffractometer geometry. It takes a great deal of time, patience and experience to prepare good specimens for X-ray analysis and "perfect" or "ideal" powder specimens are probably not often achieved.

There are many possible definitions of a "perfect" specimen. For example, comparing experimental and computed powder patterns would require the structure factors calculated from the observed integrated intensities to agree with those calculated from accurate single crystal data to some acceptable reliability factor. To achieve high precision the crystallites must be completely randomly oriented, small enough to avoid intensity errors arising from the statistics of the crystallite size distribution but not so small as to cause line broadening, and free of strain or other defects. It also assumes that the chemical composition and the atomic distribution are the same in the bulk powder sample as in the tiny single crystal used for the crystal structure analysis.

Powder diffractometry covers a wide spectrum of problems of varying difficulties ranging from absolute measurements with an accuracy of 1% (1) to simple phase determinations and the experimental and specimen preparation work necessary depends on the information desired. The perfect specimen definition and its realization would be too restrictive for most applications and in practice the acceptable quality need only be sufficient to meet the requirements of the particular analysis. The preparation of powder data standards obviously requires a considerably greater amount of characterization and care in preparation than specimens prepared for routine analyses. The purpose of this paper is to review those factors in the diffractometer geometries which are most likely to cause errors due to improper specimen preparation, particularly preferred orientation. It will be shown that it is often advisable to transfer the same specimen to two or more diffractometer geometries in which the specimen has different orientations to obtain sets of complementary data.

DIFFRACTOMETER GEOMETRIES

A number of diffractometer geometries are listed in Table I. The θ-2θ scanning diffractometer with incident divergent beam, reflection specimen and parallel slits for limiting the divergence normal to the focusing plane was first described in 1949 (2) and is by far the most widely used geometry for polycrystalline specimens. (This is often referred to as "Bragg-Brentano parafocusing" but neither used parallel slit collimators which are essential to obtain good line profiles and high intensities from a thin long line source.) A wide angular range ($\approx 0°$ to $165°2\theta$) is accessible and the intensity and resolution can be varied by selection of the divergence and receiving slit apertures. The specimen rotates at one-half the angular speed of the receiving slit so that the incident and diffracted beams make the same θ-angles to the specimen surface at all 2θ's. Hence the relative intensities can be measured directly without absorption corrections except for the special case of very low absorption (3). If peak intensities are measured corrections should be made for the Kα doublet separation.

In θ-2θ scanning the incident beam irradiates a constant specimen width in the axial direction (normal to the focusing plane) at all diffraction angles but the irradiated specimen length L in

Table I. Diffractometer Geometries

Scanning Type	Specimen Type	Monochromator Position	Designation
θ-2θ	Reflection	None	
θ-2θ	Reflection	Diffracted beam	S(R) M(R)
θ-2θ	Transmission	Diffracted beam	S(T) M(R)
Seeman-Bohlin	Reflection (stationary)	Incident beam	M(R) S(R)
θ-2θ	Reflection	Incident beam	M(R) S(R)
θ-2θ	Transmission	Incident beam	M(R) S(T)
Seeman-Bohlin	Transmission (stationary)	Incident beam	M(R) S(T)

the focusing plane varies with 2θ (4). For a reflection specimen L = αR/cos θ (α = angular aperture in radians, R = goniometer radius) and increases rapidly as the specimen reaches small values of θ where it is nearly parallel to the incident beam. α must be selected with respect to the smallest 2θ to be measured to prevent the incident beam from exceeding the specimen length and thereby causing extraneous scattering and incorrect relative intensities. For a transmission specimen sin θ replaces cos θ, L becomes the limiting factor at high 2θ's and larger apertures can be used at small 2θ's.

S(R) M(R). The addition of a focusing monochromator in the diffracted beam beyond the receiving slit makes it possible to eliminate specimen fluorescence. The use of a cylindrically bent highly oriented pyrolytic graphite monochromator (5) results in an increase of intensity by a factor of about two over the conventional θ-2θ diffractometer because of the elimination of the beta filter and the diffracted beam parallel slit collimator. However, the rocking angle is large (0.4° or more) and the resolution is not as good as that obtained with less mosaic monochromators; unfortunately the latter have much lower intensities. The geometrical factors relating to the specimen are the same as for the diffractometer without monochramator.

S(T) M(R). The θ-2θ geometry is also used with the specimen in transmission by rotating the specimen 90° from its reflection geometry position (6). The thin specimens required are no more difficult to prepare than the reflection specimens. The incident beam continues to diverge after diffraction from the specimen and a focusing reflection monochromator is required to refocus the beam on the detector window. To avoid long path lengths the crystal must be cut at an angle to the reflecting planes which is not possible with graphite, and quartz cut at about 3° to 10.1 is commonly used. The scanning range is limited to a maximum of approximately 85°(2θ). The geometry is similar to that of the Guinier camera and has two important advantages over the camera: specimen fluorescence is eliminated and the detector permits direct quantitative intensity measurements. The Kα doublet is focused so that the components overlap at around 30°2θ. Corrections are required for the doublet separation which increases at higher angles but at a different rate and not necessarily in $2K\alpha_1/K\alpha_2$ ratio that is found with the reflection specimen geometry.

Both geometries could be used with the monochromator before the specimen: M(R) S(R), M(R) S(T). The disadvantage of these arrangements is that fluorescence might be excited by the characteristic line and the angular aperture is fixed by the monochromator and cannot be varied.

M(R) S(R). In the Seeman-Bohlin geometry the specimen is in a fixed position, the incident beam makes a fixed angle γ with the specimen surface and the irradiated area remains constant (7). The detector is linked to face the specimen as it rotates around a fixed focusing circle. An incident beam focusing monochromator makes it possible to move the specimen closer to the source slit to reduce γ to smaller values than can be obtained using the focal line of the X-ray tube as the source. A small γ of around 5° is advantageous in studying thin films because the path length of the incident beam in the specimen is t/sin γ (t = thickness of film) thereby enhancing the intensities. For powder specimens somewhat larger values of γ are desirable to reduce the shadowing effect (see below). The lowest scanning angle is determined by γ and the highest by mechanical restrictions; typical values for γ =·10°, R = 175 mm are 40° to 200° (4θ). Reflections occur from planes with different inclinations β to the specimen surface so that θ = β + γ. The relative intensities must be corrected for the varying receiving slit aperture and other factors (7).

SPECIMEN ERRORS

Radial displacement of the specimen surface from the goniometer axis of rotation is the most common cause of errors in angle measurements and systematic errors in lattice parameter determination (8). The errors may arise from improperly prepared specimen holders or specimen surfaces, failure to shim slide smears or incorrectly machined or worn banking surface on the diffractometer. If the surface is displaced a distance S the entire observed profile is shifted from its correct value to larger or smaller angles depending on the direction of the displacement but the intensities and line profile shapes are not modified if the displacements are small.

In $\theta-2\theta$ geometry the shift $\Delta 2\theta$ radians is $\pm(2S/R)\cos\theta$ for a reflection specimen. For a transmission specimen $\cos\theta$ is replaced by $\sin\theta$ and it is possible to eliminate the error by repeating the measurement after rotating the specimen 180° around the goniometer axis (9). The value free of displacement error lies midway between the two measurements. For a reflection specimen in Seeman-Bohlin the shift $\Delta 4\theta$ radians is $\pm(S/R)\sin 2\theta/\sin\gamma \sin(2\theta - \gamma)$. The shift is greater than for $\theta-2\theta$ geometry and increases rapidly as θ approaches γ.

If the specimen mount is tilted and displaced the line profiles will be modified and the intensities and angles will be incorrect. In both $\theta-2\theta$ and Seeman-Bohlin incorrect 2:1 setting of as little as 1° may cause systematic errors in the relative peak intensities at small 2θ's; the error decreases with increasing 2θ.

The flat specimen aberration can be reduced by use of a curved specimen whose radius fits the focusing circle at the smallest 2θ to be measured, $R(FC) = R/2\sin\theta$ (10). If only a small amount of powder is available it may be glued to the front or back of a thin beryllium foil mounted on the curved portion of the specimen holder. Flexible specimens whose curvature varies continuously during the scan are more difficult to prepare.

Surface roughness or microabsorption may cause large relative intensity errors due to the shadowing of the incident beam. The maximum errors occur in $\theta-2\theta$ scanning at small 2θ's and in Seeman-Bohlin aligned for small γ's. Since there is no way to correct these errors the roughness must be avoided in the preparation of the specimen surface. Transmission specimens are less prone to error from this effect because the angle of incidence is larger and the thin specimen is less likely to be influenced by surface inequalities. A somewhat similar effect may arise by the inclusion of relatively large crystallites in a fine grained matrix. The large grains cause shadowing and if oriented properly excessively large reflection intensities.

Rapid rotation of the specimen in its own plane during scanning is essential to reduce the effect of crystallite size statistics on the relative intensities and to eliminate in-plane preferred orientation (8). The absolute intensities vary with the crystallite size and packing density of the specimen.

Specimens with very low absorption require corrections of the observed relative intensities because different proportions of the primary beam are transmitted through the specimen during the scan

(3). Thick transparent specimens give asymmetrically broadened line profiles shifted from their correct position.

PREFERRED ORIENTATION

When the crystallites in the specimen are not randomly oriented the observed relative intensities vary with the diffractometer geometry. In the θ-2θ geometry only those crystallites oriented nearly parallel to the surface of a reflection specimen and perpendicular to the surface of a transmission specimen will be in a position to reflect as shown in Fig. 1. The alignment necessary for a reflection to occur is dependent on the mosaic character of the crystallites, and the angular apertures of the incident beam in the focusing and the axial planes; in practice this must be within a few degrees. The specimen mount and the rotating specimen device should be designed to allow interchangeability of the same specimen preparation between diffractometers set up for reflection and transmission patterns. Comparison of these patterns with each other, and with a computed pattern, if available, reveals far more information than can be obtained with only one of the patterns.

If the powder has a pronounced morphological feature it may be very difficult to prepare a specimen without preferred orientation. This is shown in Fig. 2 for the clay mineral dickite which is platy and tends to orient with its basal planes parallel to the specimen surface. The upper chart obtained with the specimen in reflection shows enhanced 00ℓ intensities while the same specimen in transmission exhibits enhanced h00 and hk0 intensities. Comparison of the intensities from both diffractometer geometries is necessary to obtain a more complete set of data.

The top pattern in Fig. 3 was computed for a random bismuth powder. The lower two patterns were obtained from a 300Å thin film of bismuth evaporated on a glass substrate. The large degree of preferred orientation is shown by the presence only of strong 00ℓ reflections in the lower θ-2θ reflection pattern. The middle pattern was obtained with the same specimen in a Seeman-Bohlin diffractometer and shows many of the reflections missing from the θ-2θ pattern but their relative intensities are markedly different from those of the computed pattern. The high index planes 014, 105, 017 are much stronger and the low index planes 102, 110 and 022 much weaker. Such data are useful in studying the growth and physical properties of thin films.

The use of a pole figure device to obtain additional data on preferred orientation should also be tried when the method is applicable (11).

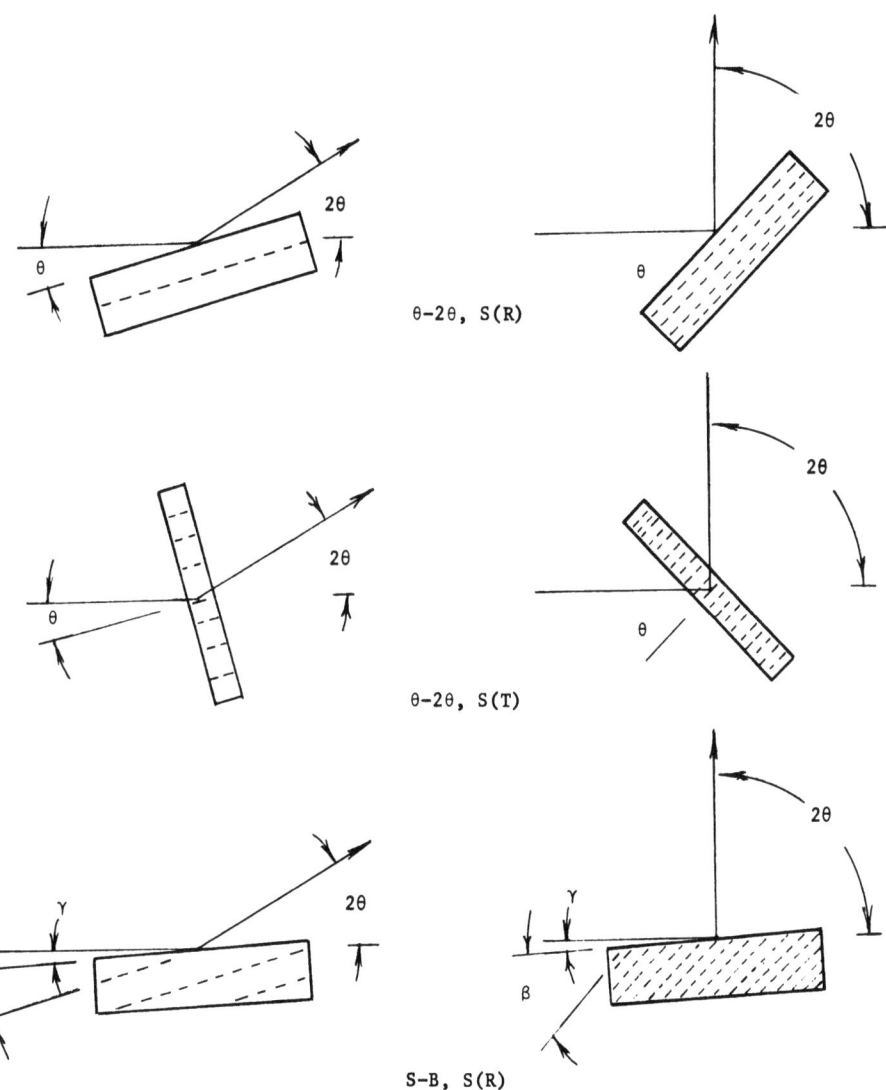

Fig. 1. Specimen orientation for three diffractometer geometries. Diffraction is possible only from planes nearly parallel to the reflection specimen surface in θ-2θ (top), from planes nearly normal to the transmission specimen surface in θ-2θ (middle), and from planes inclined different amounts to the reflection specimen surface in Seeman-Bohlin geometry (bottom). Left column for θ = 15°, right column for θ = 45°.

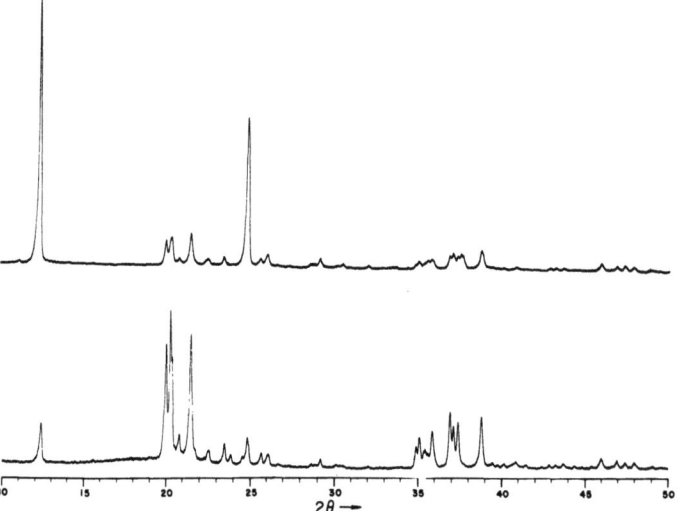

Fig. 2. θ-2θ diffractometer patterns of highly oriented dickite specimen in reflection (top) and transmission (bottom).

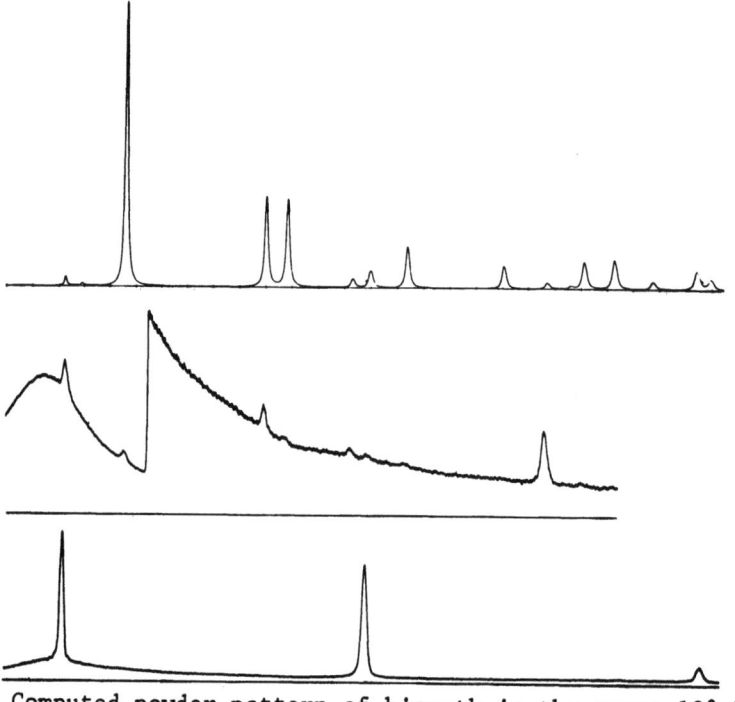

Fig. 3. Computed powder pattern of bismuth in the range 18°-73° (2θ), CuKα (top). Thin (300Å) film of bismuth on glass substrate with Seeman-Bohlin diffractometer (middle) and same film in reflection with θ-2θ diffractometer (bottom). The sharp drop of intensity in the middle pattern is due to a change in scaling factor.

REFERENCES

1. T. Paakkari, P. Suortti and O. Inkinen, "A Contribution to the Powder Intensity Project of the International Union of Crystallography," Ann. Acad. Sci. Fennicae A.VI.345, 3-29 (1970).

2. W. Parrish, "X-Ray Powder Diffraction Analysis: Film and Geiger Counter Techniques," Science 110, 368-371 (1949).

3. A. J. C. Wilson, "Mathematical Theory of X-Ray Powder Diffractometry," Philips Technical Library (1963).

4. W. Parrish, M. Mack and J. Taylor, "Determination of Apertures in the Focusing Plane of X-Ray Powder Diffractometers," Jour. Sci. Instr. 43, 623-628 (1966).

5. R. W. Gould, S. R. Bates and C. J. Sparks, "Application of the Graphite Monochromator to Light Element X-Ray Spectroscopy," Appl. Spectr. 22, 549-551 (1968).

6. P. M. deWolff, "Focussing Monochromators and Transmission Techniques," Norelco Reporter XV, 44-49 (1968).

7. W. Parrish and M. Mack, "Seeman-Bohlin X-Ray Diffractometry. I. Instrumentation. II. Comparison of Aberrations and Intensity With Conventional Diffractometer," Acta Cryst. 23, 687-700 (1967).

8. W. Parrish, "X-Ray Analysis Papers," Centrex Publ. Co., Eindhoven (1965).

9. P. M. deWolff, "Diffractometer Measurements of Low-Order Powder Reflections," Acta Cryst. 13, 835-837 (1960).

10. W. Parrish, "X-Ray Diffractometry Methods for Complex Powder Patterns," in H. van Olphen and W. Parrish, Editors, X-Ray and Electron Methods of Analysis, p. 1-35, Plenum Press (1968).

11. A. Segmüller and J. Angilello, "Automatic Pole Figure Evaluation," Jour. Appl. Cryst. 2, 76-80 (1969).

A NEW X-RAY DIFFRACTION METHOD FOR QUANTITATIVE MULTICOMPONENT ANALYSIS

Frank H. Chung

Sherwin-Williams Research Center

Chicago, Illinois 60628

ABSTRACT

A unified matrix-flushing theory and its practical applications for quantitative multicomponent analysis by X-ray diffraction are presented.

In this method, a fundamental "matrix-flushing" concept is introduced; the calibration curve procedure is shunted; the matrix (absorption) effect is totally eliminated; all components, crystalline or amorphous, can be determined.

INTRODUCTION

X-ray diffraction analysis is a standard technique used in industrial laboratories for quality control and routine analysis. It is often the only technique available for distinguishing polymorphic structures or for analyzing solid solutions (1). However, the X-ray diffraction technique is hampered by a matrix (absorption) effect which makes quantitative X-ray diffraction analysis rather tedious.

An X-ray diffraction method for quantitative multicomponent analysis has been developed in which the calibration curve procedure is shunted and a more fundamental "matrix-flushing" concept is introduced. This concept provides an exact relationship between intensity and concentration free from matrix effects. The unified matrix-flushing theory and its applications are presented below.

UNIFIED MATRIX-FLUSHING THEORY

The intensity (energy per second) of a reflection (hkℓ) in a powder diffraction pattern is derived from the well known X-ray diffraction theory and is described elsewhere (2, 3). The intensity equation is given below:

$$I(hk\ell) = \left(\frac{I_0 e^4 \lambda^3 d}{32 \pi m^2 c^4 r}\right) \left(N^2 p F^2\right) \left(\frac{1 + \cos^2 2\theta}{\sin^2 \theta \cos \theta}\right) TAV \qquad (1)$$

where: I = Intensity of X-rays diffracted by (hkℓ) plane.
I_0 = Intensity of primary X-rays.
e, m = Charge and mass of electron.
λ = X-ray wavelength.
d = Slit width of detector.
c = Velocity of light.
r = Specimen-to-detector distance.
N = Number of unit cells per unit volume.
p = Multiplicity.
F = Structure factor.
θ = Bragg angle.
T = Temperature factor.
A = Absorption factor.
V = Volume of powder in the beam.

This complete intensity equation can be conveniently separated into six factors: (1) a constant factor for a particular diffractometer, (2) a structure factor characteristic of the diffracting sample, (3) an angle factor known as the Lorentz-polarization factor, (4) a temperature factor to correct for thermal vibration of atoms, (5) an absorption factor which attenuates the diffracted X-rays, and (6) the volume of powder in the primary X-ray beam.

For a specific (hkℓ) reflection from a component i in an X-ray powder diffraction pattern of a mixture, the first four factors are constant, K_i. The absorption of X-rays, just like the absorption of visible light, follows the well-known exponential law. For infinite specimen thickness, we have the fifth factor:

$$A = \int_0^\infty e^{-\mu_t s} ds = \frac{1}{\mu_t} \qquad (2)$$

where μ_t and s are the linear absorption coefficient and thickness of the total sample. The last factor V, in case of a mixture, indicates the volume fraction V_i of component i.

Let X_i and ρ_i be the weight fraction and density of component i; the complicated intensity equation (1) can be reduced to:

$$I_i = \frac{K_i}{\rho_i} \cdot \frac{X_i}{\mu_t} = k_i X_i \tag{3}$$

where $\frac{K_i}{\rho_i}$ is a characteristic constant of component i, and k_i is a factor containing the linear absorption coefficient of the total sample. At this moment, k_i is constant for a very small variation in X_i. Later on, it will be shown that k_i in the ratio k_i/k_j is constant for any X_i; the matrix factor μ_t is cancelled.

For the quantitative X-ray diffraction analysis of a mixture of n components, we have n unknowns (X_i, i = 1 to n) which must satisfy the following (n + 1) nonhomogeneous equations:

$$\left. \begin{array}{l} I_i = k_i X_i, \quad i = 1 \text{ to } n \\ \\ \sum_{i=1}^{n} X_i = 1 \end{array} \right\} \tag{4}$$

This situation can be most conveniently treated in terms of matrix algebra as follows:

$$\begin{pmatrix} k_1 & 0 & 0 & \cdots & 0 \\ 0 & k_2 & 0 & \cdots & 0 \\ 0 & 0 & k_3 & \cdots & 0 \\ \cdots & \cdots & \cdots & \cdots & \cdots \\ 0 & 0 & 0 & \cdots & k_n \\ 1 & 1 & 1 & \cdots & 1 \end{pmatrix} \begin{pmatrix} X_1 \\ X_2 \\ X_3 \\ \vdots \\ X_n \end{pmatrix} = \begin{pmatrix} I_1 \\ I_2 \\ I_3 \\ \vdots \\ I_n \\ 1 \end{pmatrix} \tag{5}$$

If the above matrix equation, $KX = I$, has a solution, the solution is unique if and only if the rank of the K matrix is equal to the rank of the (K, I) matrix.

In order to satisfy this condition, the unique solution of equation (5) has the following simple and symmetrical form:

$$X_i = \left(\frac{k_i}{I_i} \sum_{i=1}^{n} \frac{I_i}{k_i} \right)^{-1} \tag{6}$$

Note the unusual property of this unique solution is that the weight fraction of any component in a multicomponent system is expressed in terms of ratios like I_i/I_j and k_i/k_j (i,j = 1, 2, ..., n). The use of an intensity ratio makes it immune to many sources of errors; the use of a k-ratio makes it free from matrix effect. But what is k_i? How can it be measured?

From equation (4) the following relationship between intensity and concentration holds true for any pair of components in a multicomponent system:

$$\frac{I_i}{I_j} = \frac{k_i}{k_j} \cdot \frac{X_i}{X_j} \tag{7}$$

hence
$$\frac{k_i}{k_j} = \frac{I_i}{I_j} \quad \text{at} \quad \frac{X_i}{X_j} = 1 \tag{8}$$

or
$$\frac{k_i}{k_j} = \left(\frac{I_i}{I_j}\right)_{50/50} = \text{slope} = \text{constant} \tag{9}$$

If a compound j is a universal reference material such as corundum (α-Al$_2$O$_3$), then $k_j = k_c = 1$, where the subscript c stands for corumdum, hence (4, 5):

$$k_i = \left(\frac{I_i}{I_c}\right)_{50/50} \tag{10}$$

This is the definition of "Reference Intensities" as given in the Powder Diffraction File (6) by the Joint Committee on Powder Diffraction Standards.

One interesting point is that the relationship between intensity and concentration of any two components as prescribed by equation (7) is independent of the existence of other components in the sample, which means equation (7) holds true whether the universal reference material corundum is indeed a component of the sample or not.

The foregoing deduction can be summarized in the following theorem: The plot of intensity ratio (I_i/I_j) to the weight ratio (X_i/X_j) of any two components is a straight line passing through the origin with a slope equal to the ratio of corresponding Reference Intensities (k_i/k_j). This intensity-concentration relationship between each and every pair of components in a multicomponent system is not perturbed by the presence or absence of other components.

The above derivation assumes that all the components in the mixture are crystalline and identified. However, in many situations some components in the mixture are amorphous and/or unidentified, quantitative data are often requested only for identified components or components of interest. In these cases, equation (6) cannot be applied, but equation (7) always holds true. In order to make use of this equation a flushing agent must be added into the sample to flush out the matrix effect. The flushing agent may be any pure compound not present in the sample.

Let the weight fraction of the flushing agent and the original sample be designated X_f and X_o respectively, that is:

$$X_f + X_0 = X_f + \sum_{i=1}^{n} X_i = 1 \tag{11}$$

$$\frac{I_i}{I_f} = \frac{k_i}{k_f} \cdot \frac{X_i}{X_f} \tag{12}$$

$$X_i = X_f \frac{k_f}{k_i} \cdot \frac{I_i}{I_f} \tag{13}$$

Equation (13) can be used for quantitative multicomponent analysis and in addition one more simplification can be realized. Since corundum (α-Al$_2$O$_3$) has been chosen for Reference Intensities by the JCPDS for its purity, stability, and availability, it is convenient to choose the same corundum as a flushing agent for the same good reasons. Consequently, $k_f = k_c = 1$, (k_c = relative intensities of corundum) from equation (13) we obtain:

$$X_i = \frac{X_c}{k_i} \cdot \frac{I_i}{I_c} \tag{14}$$

This is the working equation for quantitative multicomponent analysis. It represents a straight line passing through the origin with a slope equal to X_c/k_i. It is free from matrix factors. No previous information relating to the approximate concentration ranges of various components sought is required. Intensity ratios from a single scan are the only experimental data needed. Since intensity ratios from the same scan are used the errors due to instrumental drift and sample preparation are minimized. Note that the working equation (14) prescribes the slope of calibration curve for every component in the sample, thus it is not necessary to actually work out calibration curves.

Another interesting application of this matrix-flushing concept is that it can be used to detect and quantify the amorphous content in a mixture. When corundum is chosen as the flushing agent, substituting equation (14) into equation (11), we get:

$$\sum_{i=1}^{n} \frac{I_i}{k_i} = \frac{X_0}{X_c} \cdot I_c \tag{15}$$

where X_0/X_c is the weight ratio of the original sample to flushing agent corundum.

Equation (15) affords a means to experimentally check the correctness of this concept, to appraise the reliability of intensity data, and to predict and assay the presence of amorphous materials in a sample. That is to say the experimental data may fall into one of the following three cases:

$$\sum_{i=1}^{n} \frac{I_i}{k_i} \gtreqless \frac{X_0}{X_c} \cdot I_c \tag{16}$$

where: > indicates wrong data,
= indicates all components are crystalline,
< indicates the presence of amorphous material.

For a binary system, an "auto-flushing" phenomenon exists. No flushing agent is needed. One component automatically serves as a flushing agent for the other component and vice versa. Because:

$$\left. \begin{array}{c} X_1 + X_2 = 1 \\ \frac{I_1}{I_2} = \frac{k_1}{k_2} \cdot \frac{X_1}{X_2} \end{array} \right\} \tag{17}$$

hence

$$X_1 = \frac{1}{1 + k \cdot \frac{I_2}{I_1}} \tag{18}$$

where

$$k = \frac{k_1}{k_2} = \left(\frac{I_1}{I_2}\right)_{50/50} \tag{19}$$

The slope k is simply the corresponding intensity ratio of a 50/50 mixture of the same two components, hence the quantitative analysis of a binary system can be performed without using the Reference Intensities at all. Of course, equation (18) is the result of degeneration of equation (6) when n = 2.

The soundness and usefulness of this matrix-flushing method for quantitative multicomponent analysis are demonstrated in the experimental section.

EXPERIMENTAL

In order to illustrate the simplicity and usefulness of the matrix-flushing theory, eight synthetic samples were prepared with pure chemicals of analytical reagent grade or better. The analytical procedure and the instrumental conditions used are described elsewhere (4, 5).

For the sake of clarity, the experimental data and their interpretation are divided into three categories:

I. All Components Are Crystalline and Identified:

In this situation, the X-ray diffraction pattern of the original sample can be interpreted directly and quantitatively without using a flushing agent or an internal standard. Equation (6) is used to calculate the percentage composition from intensity data. The data of two examples No. 1 and No. 2 are given in Table 1. All intensity data refer to the strongest reflections of corresponding components.

Table 1. Intensity and Composition Data

Sample No.	Composition (grams)		Intensity I_i, c/s	Ref. Int. k_i	% Composition	
					Known	Found
1	ZnO	0.2236	610	4.35	9.87	9.2
	NiO	0.5454	1412	3.81	24.06	24.5
	CdO	0.6588	3303	7.62	29.07	28.7
	KCl	0.8386	2207	3.87	37.00	37.7
2	ZnO	1.8901	5968	4.35	41.49	41.3
	KCl	1.0128	2845	3.87	22.23	22.1
	LiF	0.8348	810	1.32	18.32	18.5
	Al_2O_3	0.8181	599	1.00	17.96	18.1

II. Some Components Are Amorphous And/Or Unidentified:

In this case, a known quantity of flushing agent (α-Al_2O_3) was added to the sample. The sample was then ground to a homogeneous fine powder before the X-ray diffraction experiment was run. The working equation (14) was applied to each component sought to calculate its weight percentage.

In many practical situations the composition of an unknown sample is only partially determined. Even when all the peaks in the X-ray diffraction pattern are accounted for, one is still not sure whether there is any amorphous material present. Therefore, it is safe to use the flushing agent for the following reasons: First, the true values of X_c and k_c of the flushing agent are absolutely known; Secondly, the weight fraction X_i is dependent upon I_i and k_i only, independent of any X_j, I_j and k_j where $j \neq i$; Thirdly, the weight fraction of a component sought is independent of the presence or absence of any other component, crystalline or amorphous. The data of two examples No. 3 and No. 4 are listed in Table 2. Note that Sample No. 3 is identical to Sample No. 2 except that the Al_2O_3 is an unknown component in Sample No. 2, but is a flushing agent in Sample No. 3.

Table 2. Intensity and Composition Data

Sample No.	Composition (grams)		Intensity I_i, c/s	Ref. Int. k_i	% Composition	
					Known	Found
3	ZnO	1.8901	5968	4.35	41.49	41.1
	KCl	1.0128	2845	3.87	22.23	22.0
	LiF	0.8348	810	1.32	18.32	18.4
	Al_2O_3 Flushing	0.8181	599	1.00	17.96	----

Table 2 (Cont.)

Sample No.	Composition (grams)		Intensity I_j, c/s	Ref. Int. k_j	% Composition	
					Known	Found
4	NiO	1.0743	4162	3.81	51.28	48.4
	CdO	0.1495	1160	7.62	7.14	6.7
	KCl	0.5410	2404	3.87	25.82	27.5
	Al_2O_3 Flushing	0.3304	356	1.00	15.77	----

It is found that even the intensity data read directly from the diffraction pattern on a strip-chart provide fairly good accuracy.

The authenticity of the matrix-flushing theory is further scrutinized by equation (15). Using the data of example No. 3, we have:

$$\Sigma \frac{I_j}{k_j} = \frac{5968}{4.35} + \frac{2845}{3.87} + \frac{810}{1.32} = 2721 \text{ (Experiment)}$$

$$\frac{X_0}{X_c} \cdot I_c = \frac{82.04}{17.96} \times 599 = 2736 \text{ (Theory)}.$$

Another interesting feature of this matrix-flushing theory is that it can be applied to detect and determine the total amorphous material in a sample.

Sample No. 5 contains an amorphous silica (silica gel) and Sample No. 6 contains an amorphous resin (Goodyear VPE). Their intensity data are presented in Table 3.

Table 3. Intensity and Composition Data

Sample No.	Composition (grams)		Intensity I_j, c/s	Ref.Int. k_j	% Composition		$\Sigma \frac{I_j}{k_j}$	$\frac{X_0}{X_c} \cdot I_c$
					Known	Found		
5	ZnO	0.9037	4661	4.35	34.43	36.4		
	$CaCO_3$	0.7351	2298	2.98	28.00	26.2		
	SiO_2 Gel	0.4234	0	----	16.13	15.9		
	Al_2O_3 Flushing	0.5629	631	1.00	21.44	----	1842	2312
6	ZnO	0.8090	4948	4.35	40.81	38.7		
	CdO	0.2825	3337	7.62	14.25	14.9		
	Resin (VPE)	0.4057	0	----	20.46	21.9		
	Al_2O_3 Flushing	0.4854	719	1.00	24.48	----	1575	2218

Normal scans of these two samples do not indicate the presence of amorphous material. However, the intensity data from the matrix-flushing method show a large intensity imbalance which indicates the presence of amorphous materials. By the use of material balance, the total amorphous content found in each case is in good agreement with the respective amount of silica gel and VPE resin actually put into the sample.

III. Binary Systems:

For a binary system no flushing agent is needed. The data of two examples, No. 7 and No. 8, are shown in Table 4. Equation (18) is used to obtain the % composition.

Table 4. Intensity and Composition Data

Sample No.	Composition (grams)		Intensity I_i, c/s	Ref. Int. k_i	% Composition	
					Known	Found
7	KCl	2.4530	5371	3.87	74.90	74.6
	LiF	0.8219	604	1.32	25.10	25.4
8	ZnO	1.4253	6259	4.35	71.22	72.1
	TiO$_2$ (R)	0.5759	1461	2.62	28.78	27.9

There are two different ways to obtain the slope k: $k = k_1/k_2$ or $k = \left(\frac{I_1}{I_2}\right)_{50/50}$, the latter requires less work, gives better accuracy, and needs no Reference Intensities. For Sample No. 7:

$$k = \frac{k_{KCl}}{k_{LiF}} = \frac{3.87}{1.32} = 2.93 \quad \text{or} \quad k = \left(\frac{I_{KCl}}{I_{LiF}}\right)_{50/50} = \frac{5583}{1846} = 3.02$$

An unusual example of auto-flushing is the determination of crystallinity of polymers (7) where one component is amorphous while the other component is a poor crystalline material. The slope of the straight line in this reference verifies equation (19) remarkably well.

CONCLUSION

Three working equations, (6), (14) and (18), for quantitative multicomponent analysis are derived from the well established complete intensity equation (1). No assumption or approximation is made in the derivation. Since no calibration curves are required for this method, the work involved for quantitative X-ray diffraction analysis is greatly simplified.

ACKNOWLEDGMENT

The author wishes to thank Mr. Richard W. Scott for his stimulating discussion and warm encouragement.

REFERENCES

1. F. H. Chung and R. W. Scott, "Vacuum Sublimation and Crystallography of Quinacridones". J. Appl. Cryst. 4, 506-511 (1971)

2. C. W. Bunn, "Chemical Crystallography", p. 223, Oxford, London (1961)

3. L. V. Azaroff, "Elements of X-Ray Crystallography", p. 202, McGraw-Hill, New York (1968)

4. F. H. Chung, "Quantitative Interpretation of X-Ray Diffraction Patterns of Mixtures. I. Matrix-Flushing Method for Quantitative Multicomponent Analysis". J. Appl. Cryst. in press (1973)

5. F. H. Chung, "Quantitative Interpretation of X-Ray Diffraction Patterns of Mixtures. II. Adiabatic Principle of X-Ray Diffraction Analysis of Mixtures". J. Appl. Cryst. in press (1973)

6. L. G. Berry, Ed., "Inorganic Index to the Powder Diffraction File", p. 1421, JCPDS (1972)

7. F. H. Chung and R. W. Scott, "A New Approach to the Determination of Crystallinity of Polymers by X-Ray Diffraction", J. Appl. Cryst. 6, 225-230 (1973)

THE AlFeBe$_4$ INTERMETALLIC PHASE IN BERYLLIUM

I. Brower, E. C. Roberts and T. J. Bosworth

The Boeing Company

Seattle, Washington   98124

ABSTRACT

X-ray diffraction analysis has been used to determine the nature and relative concentration of the phases present in diffusion welded joints of fairly high purity beryllium. The analyses were done in conjunction with the major objective of the work which was to correlate the types and amounts of phases in the weld joint with the weld processing variables.

The beryllium used contained less than one percent by weight of alloying elements. Those elements present in major amounts were aluminum, less than 0.10 weight percent (less than 0.033 atom percent) and iron, less than 0.15 weight percent (less than 0.023 atom percent).

A phase concentration technique was used to obtain the quantities of phases required for positive identification. X-ray powder patterns obtained from the residues provided detailed diffraction data of which those for the phase AlFeBe$_4$ are of particular interest. These data are compared with the findings of other research investigations.

Experimental procedures are discussed in detail and a summary of the test results in included.

INTRODUCTION

Commercially pure beryllium has been found to be an alloy and a mixture of a number of elements and chemical compounds. Among the former, elemental aluminum and silicon have been detected and, among the latter, beryllium oxide and beryllium carbide have been reported, (1). These substances have been detected by x-ray diffraction studies which most commonly have been made on solid specimens.

In addition to these, however, there also are found two beryllium intermetallic phases, e.g., $AlFeBe_4$ (2), (3), and $FeBe_{12}$ (3)*. In the first of these, four aluminum atoms apparently replace four of the beryllium atoms in the twenty-four-atom cell of the phase $FeBe_5$ (5). Both this latter and the ternary phase are isomorphous and not greatly different in density and lattice parameter. Such ternary complexes can also form in which the iron is replaced by chromium, manganese, or nickel (2), but these latter three are not dominant impurities in beryllium.

The occurrence of one or the other of these phases has been found to be dependent upon thermal exposure with the $FeBe_{12}$ being preferentially formed at temperatures above 850°C (1560°F) while the $AlFeBe_4$ is formed at lower temperatures in such processes as solution treatment and aging (1). There is also a dependency on the amount and distribution of aluminum. As long as aluminum is present in the matrix, $AlFeBe_4$ will form in preference to $FeBe_{12}$. However, aluminum tends to segregate to the grain boundaries of beryllium and thus high solution temperatures with their greater diffusion rates will tend to deplete the matrix of this element. For example, long-time solution annealing commercial purity beryllium above 1970°F will tend to segregate aluminum to the point that $FeBe_{12}$ can form in quenched specimens. Under other conditions of alloys, where aluminum is somewhat randomly distributed, $FeBe_{12}$ will not form until sufficient $AlFeBe_4$ has formed and depleted the matrix of aluminum (6).

Electron microscopic studies have also shown that the $AlFeBe_4$ shows a definite preference for interface development particularly at grain boundaries (3). See Figure 1.

BACKGROUND

The present studies are an offshoot of work involving the development of macroscopic interfaces via the process of diffusion welding of beryllium. The major effort was a study to develop welding parameters as a function of beryllium oxide content and specimen thickness (7). The procedures used

---

*Controversy exists regarding the composition and crystal structure of this phase. The stoichiometry used here is from the most recent binary constitution diagram (4).

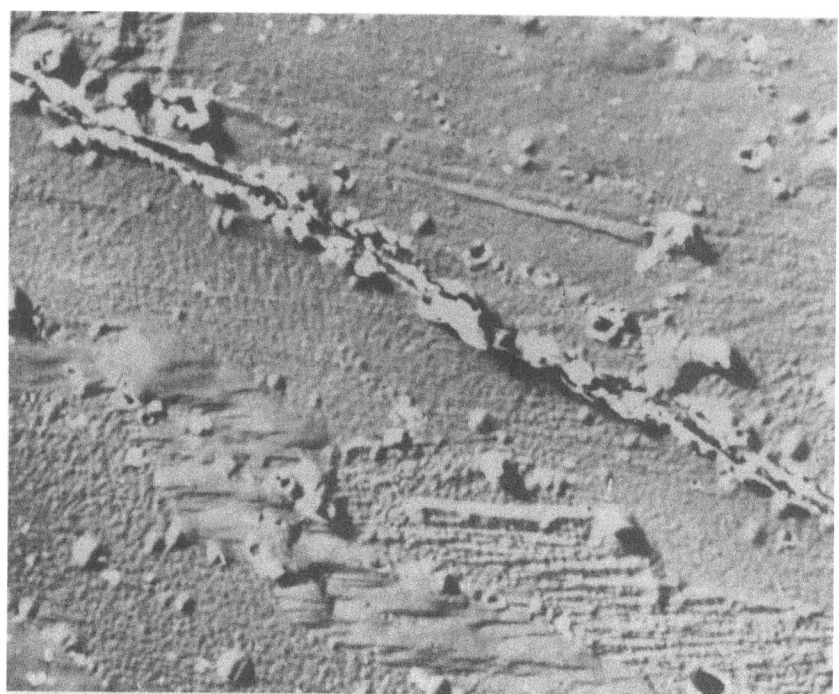

Figure 1. Electron Micrograph of Cross-Section of Diffusion Weld Made at 1450°F. (Mag. 6000X)

can briefly be described as follows:

A pair of beryllium sheet specimens approximately 0.5 inch x 1.0 inch by 0.060 inch thick and having the composition shown in Table 1 were chemically cleaned, then loaded to a specific pressure in a vacuum chamber (10^{-4} to 10^{-5} torr). The temperature of the specimens was then raised to 1400°, 1450° or 1500°F and held for differing periods of time from thirty minutes to three hours.

TABLE I. Chemical Composition (Weight Percent) of Commercially Pure Beryllium Sheet

Be	Beo	Fe	Si	Al	Mg	C
98.7	0.97	0.14	0.04	0.07	0.03	0.11

EXPERIMENTAL PROCEDURE

After completion of the diffusion welding exposure the welded specimens were broken apart by cleaving along the joint interface. Each half specimen was then placed in a diffractometer and a run made using copper $K\alpha$ radiation.

These exposures soon showed that one or more phases were present in the fractured interface in even greater quantities than the beryllium oxide. Because these initial samples had very irregular fracture surfaces, it was difficult to draw any conclusions on the nature of the phases on the basis of the diffractometer runs. On the other hand it was important that a more specific identification be made because of the possible contribution of the presence of a phase or phases to the strength of the interface bond. On the basis of the temperature the phase $AlFeBe_4$ was expected but the effect of the bonding pressure on the phase development was unknown.

To improve the phase identification a first attempt was made at scraping the fractured surface with a tungsten carbide tool and retaining all of the material so removed. This material was then prepared and exposed in a 114.6mm Debye-Scherrer camera using copper $K\alpha$ radiation. Some improvement was noted, e.g., the elemental aluminum diffraction lines were clearly defined but pinpointing the specific substances responsible for the remaining diffraction lines was not possible.

The next step was to utilize a phase extraction technique thus allowing a concentration of the particles of interest and a removal of the beryllium from the sample. The small chips and scraping provided by the procedure described above were again accumulated from individual specimens fractured along the joint interface. The material removed was limited to that within 0.003 inch of the diffusion interface. This mechanically removed material was chemically reacted with an anhydrous methanol-iodine (20%) solution by heating in a petri dish to about 70°C. At the proper temperature a vigorous reaction occurs. When the reaction stops, the mixture is allowed to cool. The exhausted solution is then removed with an eye dropper taking care to retain the solid material in the dish. New solution is then added and the process repeated until the residue left in the petri dish no longer reacts with fresh iodide solution. This procedure removes elemental beryllium down to trace amounts and completely eliminates the aluminum.

The residual particles obtained by this procedure were of two types: (1) fine, light-colored particles that floated to the surface of the extraction solution and (2) coarser particles ranging in color from white to light gray to dark gray.

Following careful washing in absolute methanol, particles of each type were accumulated on a vaseline-coated glass fiber and exposed to copper Kα radiation in a 114.6mm Debye-Scherrer camera.

RESULTS AND DISCUSSION

Diffraction analysis shows the floating residue to be made of extremely fine beryllium oxide, beryllium, and silicon with possibly some silicon dioxide and beryllium carbide. Many very weak lines from unidentifiable substances were present in some samples in addition to the lines related to the above materials.

The darker residue was found to clearly contain the following substances in decreasing order of concentration: $AlFeBe_4$, beryllium oxide, silicon, silicon dioxide and beryllium. The approximate ratio of concentration is: 50:35:5:3:1. Analyses of specimens diffusion welded at 1400° and 1500°F for three hours showed these substances in the same relative concentrations with remarkable consistency. Debye-Scherrer patterns are shown in Figure 2. Additionally, it was surprising to find how closely the interplanar spacings from these samples matched.

The phase concentration technique provided a considerably greater range of diffraction lines for the compound $AlFeBe_4$ than heretofore has been reported. Table II shows the interplanar spacing information obtained by very careful measurement of the diffraction lines followed by film shrinkage correction. The d-values obtained are shown in comparison with those obtained by Carrabine and Paine (8), and Rooksby (1). Also shown are those calculated on the basis of an a_o value of 6.054 Å. This was determined by plotting the lattice parameter values determined from the corrected film measurements against the function ($cos^2 \theta/\sin\theta + cos^2\theta/\theta$) and extrapolation to a zero value of this function (9). (See Figure 3). Considerable random scatter is encountered even though only the higher angle lines are used. This is attributable to the large amount of superposition due to the presence in the extracted residue of the aforementioned elements and compounds. For this reason it is difficult to give proper relative intensity values.

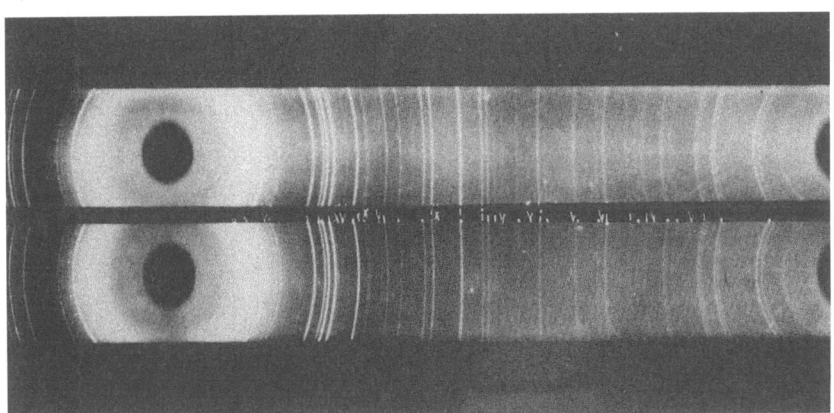

Figure 2. Debye-Scherrer Patterns of Extracted Residues from Samples Diffusion Welded at 1400°F for Three Hours (Top) and 1500°F for Three Hours

TABLE II. Interplanar Spacing Data For the Phase Al Fe Be$_4$

hkl	This Work d meas, Å	This Work d calc, Å	Carrabine & Paine (8) d, Å	Rooksby (1) d, Å
111	3.489	3.495	3.49	3.50
200	3.015	3.027	3.01	
220	2.135	2.140	2.14	2.142
311	1.823	1.824	1.83	1.826
222	1.739	1.748	1.74	
400	1.509	1.513	1.51	1.514
331	1.387	1.389	1.39	
420	1.348	1.354	1.35	
422	1.234	1.236	1.24	1.237
511,333	1.163	1.165		1.166
440	1.069	1.070		1.071
531	1.022	1.022		
600,442	0.990	0.982		
620	0.957	0.957		
533	0.922	0.923		
622	0.912	0.913		
444	0.873	0.874		
711,551	0.848	0.848		
640	0.840	0.840		
642	0.809	0.809		
731,553	0.788	0.788		

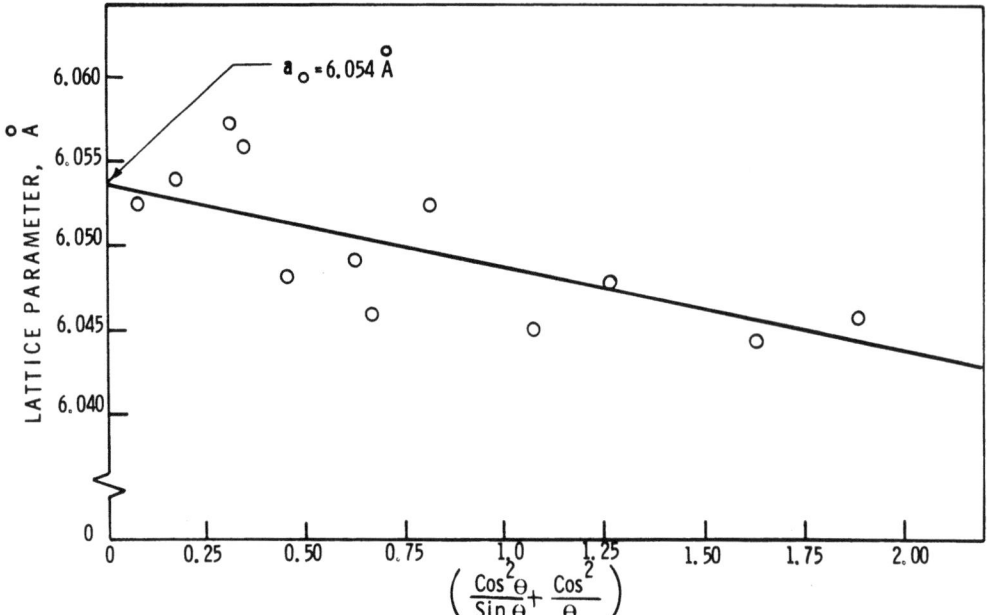

Figure 3. Extrapolation of Lattice Parameters Calculated from Measured d-Values Against $(\cos^2 \theta/\sin \theta + \cos^2 \theta/\theta)$

This a_o value can be compared with those obtained by Carrabine, 6.06 Å, and Rooksby, 6.057 Å. These are all remarkably close. Considerable confidence in the fact that this is indeed Al Fe Be$_4$ also is taken from the chemical analysis of the beryllium shown in Table I and from the a_o values of the isomorphous phases prepared and investigated by Carrabine and Paine (2):

Phase	a_o, Å
Al Ni Be$_4$	6.01
Al Mn Be$_4$	6.11
Al Cr Be$_4$ (cubic)	6.12

A calculation of the theoretical density of the Al Fe Be$_4$ phase made on the basis of a 24 atom unit cell (similar to Fe Be$_5$ with which it is isostructural) shows it to be 3.56 g/cc.

SUMMARY

A phase extraction technique has been used to remove the bulk of the matrix material and thus provide a concentrated residue from which the interplanar spacings of the intermetallic phase Al Fe Be$_4$ have been determined. Utilizing this data a new value for the lattice parameter and a theoretical density value have been obtained.

ACKNOWLEDGMENT

The authors express their thanks for the electron microscopy work to Mr. C. D. Smith of The Boeing Aerospace Company Metallurgical Laboratory.

REFERENCES

1. H. P. Rooksby, "Intermetallic Phases in Commercial Beryllium", Jl. of Nuc. Mat. 7:205, 1962.

2. J. A. Carrabine, "Ternary Al Mn Be_4 Phases in Commercially Pure Beryllium", Jl. of Nuc. Mat. 8:278, 1963.

3. F. J. Fraikor and V. K. Grotzky, "Grain Boundary Precipitation in Sheet Rolled from Beryllium Ingots", Trans AIME 239:2008, 1957.

4. F. A. Shunk, "Beryllium-Iron" Constitution of Binary Alloys, Second Supplement, McGraw-Hill Book Company, New York, 1969, p. 160.

5. L. Misch, "Strukturen intermetallischer Verbindungen des Berylliums mit Kupfer, Nickel und Eisen", Zeitschrift für Physikalische Chemie, B29:42, 1935.

6. F. J. Fraikor and A. W. Brewer, "Precipitation in Quenched and Aged Beryllium Ingot Sheet", Trans. ASM 61:784, 1968.

7. T. J. Bosworth, "Diffusion Welding of Beryllium: Part 1 - Basic Studies", The Welding Journal, Welding Research Supplement 57:579s, 1972.

8. J. A. Carrabine and R. M. Paine "Aluminum Beryllium Iron", ASTM Data Card 18-6, 1965.

9. B. D. Cullity, Elements of X-ray Diffraction, Addison-Wesley Publishing, Reading, Mass. (1956) p. 330.

HIGH-SPEED RETAINED AUSTENITE ANALYSIS WITH AN ENERGY

DISPERSIVE X-RAY DIFFRACTION TECHNIQUE.

A. P. Voskamp

SKF European Research Centre B.V.

Jutphaas, Netherlands

SUMMARY

A time saving method has been applied for the determination of retained austenite.

The method involved is based on the approach of energy dispersive X-ray diffraction analysis. With this approach, polychromatic radiation from the X-ray tube is used and diffraction maxima will occur at a fixed angle 2θ in as many wavelengths or energies as "d" values are present.

Giessen and Gordon published the first application of this method to powder diffraction analysis in 1968 for the identification of crystal structures. As the determination of retained austenite is a quantitative type of analysis, based upon identification of the crystal structure, the new approach should also be applicable in principle.

With almost 100 samples of unknown austenite content, experiments have been carried out both with the conventional and the energy dispersive X-ray diffraction technique. The results obtained are closely comparable and the retained austenite values together with the errors are shown.

For these measurements, experiments have been carried out with the energy dispersive technique to determine the relation between the known concentration of retained austenite in a number of standards and the intensity correction factors (R). The results obtained from these experiments have shown good reproducibility of the intensity correction factors.

(b) The effect of additional line broadening due to internal stresses upon the FWHM of an energy peak in an FAS spectra is small.

(c) The background in the FAS spectra due to the Compton scattering is acceptably low.

In figure 1., using the FAS technique, an intensity against energy plot is shown for a hardened 52100 steel containing 18.3% retained austenite.

The total analysis time was ten minutes and the diffraction lines from (110) α Fe up to the (321) α Fe are shown.

In figure 2., using the conventional technique, an intensity versus 2θ plot is shown with the same number of diffraction lines. The total time to perform this second analysis was 90 minutes.

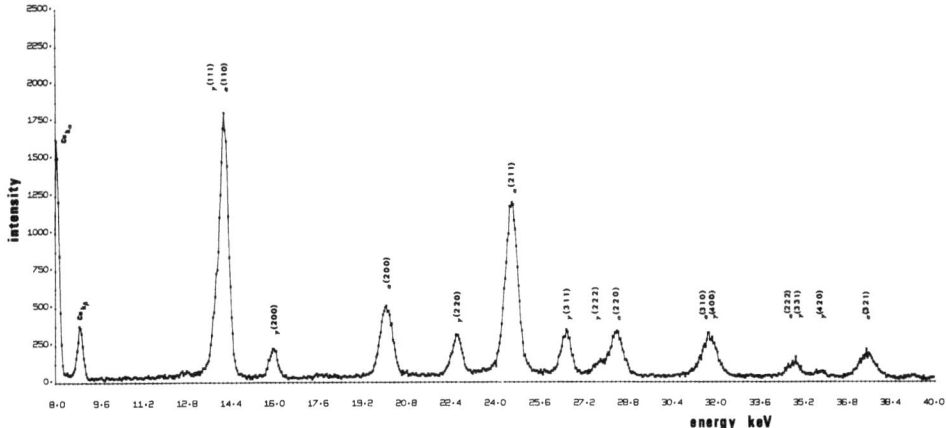

Figure 1. FAS diagram 52100 steel 18.3% retained austenite. Cu tube 50KV 24 mA 20 eV/ch 2θ = 25.00° 10 min. ¼°/0.2 mm.

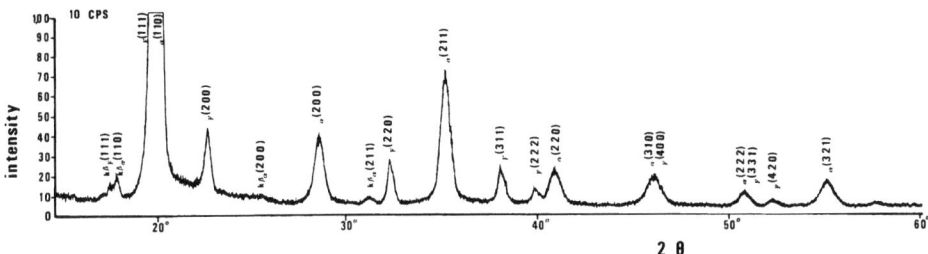

Figure 2. Diffractogram 52100 steel 18.3% retained austenite. MoKα/Zr filter ½°/min. 50 KV 24 mA ¼°/0.2 mm.

It is evident that, in addition to a nine-fold reduction in analysing time, the peak to background ratio for the FAS technique is increased with no Kβ lines in the background. Comparison of the two diagrams shows that the resolution obtainable is almost equal.

In order to carry out quantitative analysis for the determination of retained austenite with the FAS technique, the angle 2θ at which the diffraction spectra must be collected has to be initially decided upon because many parameters can contribute to the optimum value. The following are important:

a. Resolution obtainable.

b. Intensity distribution in the incident beam.

c. Interference with fluorescence lines.

d. X-ray penetration in the sample.

a. Resolution obtainable.

Besides the influence from the sample itself on the FWHM of a diffraction line, the peak to background ratio and the FWHM are strongly dependent upon the choice of slits. This is shown in the plot presented in figure 3, where the FWHM and the peak to background ratios are given as a function of the width of the divergence and receiving slits. The minimum FWHM obtainable, shown in figure 3, is approximately 380 eV and is almost the same value as the FWHM of the detector at 22.8 KeV. This relatively high value is inherent in this type of detector. In the range of 10 to 30 KeV there is an almost linear increase of the FWHM with an increase in photon energy. This effect is shown in figure 4. Presently available detectors with 140 eV resolution at 5.9 KeV show values somewhat below the plotted HM line but still around the 300 eV level for 22.8 KeV.

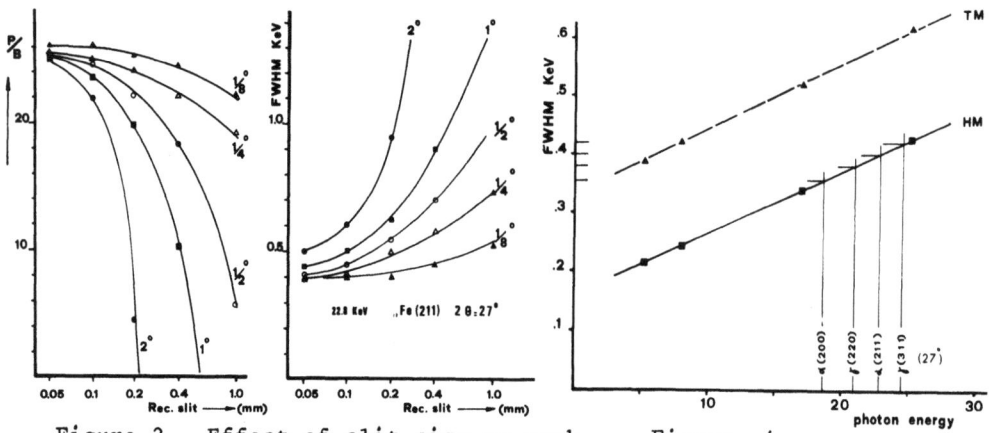

Figure 3. Effect of slit size on peak to background ratio P/B and FWHM.

Figure 4. Detector resolution

b. Intensity distribution in the incident beam.

The intensity distribution in the polychromatic incident beam is a photon energy dependent factor. If we consider only a few lines in the energy spectra with eV values very close to each other, this factor will have an almost negligible influence on the intensities obtainable. However, if we are interested in more diffraction lines in the d range of 0.6 to 2.0 Å and in a range of 2θ from 20 to 30 degrees, the polychromatic radiation band from 14 to 38 KeV is involved, which cannot be considered as constant. Therefore, the intensity distribution must be known for that part of the X-ray spectra.

With a conventional spectrometer or with an energy dispersive detector placed in the primary beam, it is possible to obtain the spectral distribution of an X-ray tube. Both methods were applied to the determination of the spectra from a copper tube at several take-off angles. The results obtained for an angle of 3 degrees is in close agreement with the spectra published by Gilfrich and Birk (13). Up to now spectral intensity measurements have been carried out for a copper tube and the spectra obtained, shown in figure 5, was used for the calculations of the intensity correction factors. However, the intensity obtainable is less than with a tungsten tube. Experiments have shown a two-fold increase in the net diffracted intensities in the FAS spectra on steel samples using a tungsten tube. Therefore, a tungsten tube is to be preferred due to the increased count-rate obtained.

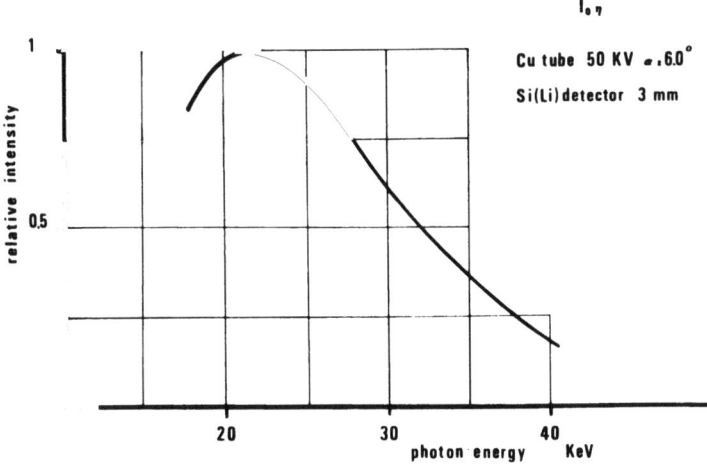

Figure 5. Relative intensity of the X-ray spectra of a copper tube using a Si(Li) 3 mm. detector.

c. Interference with Fluorescence Lines

Since relatively high X-ray energies are used it is obvious that the elements present in the sample will fluoresce. Some of the materials under investigation were alloyed with nickel, chromium and molybdenum. In order to avoid interference of the diffraction lines by the fluorescence lines in the FAS spectra, it was necessary to select a particular 2θ value.

In figure 6, the fluorescence energies are plotted against the atomic number. From this it is shown that the energy range for FAS between 18 - 25 KeV, can be disturbed by the K fluorescence energies of elements with an atomic number between 40 and 50. Since the fluorescence lines have a fixed energy in the spectra at any chosen 2θ value, and since the diffraction lines are variable with 2θ, it is possible to arrange the fluorescence and diffraction spectra such that no overlapping occurs.

In figure 7, the photon energies for the α and γ iron diffraction lines in FAS are plotted as a function of 2θ, with the K energies for molybdenum also being indicated.

It is common in austenite analysis by X-ray diffraction to determine the intensities of at least two α- and two γ-diffraction lines. These lines are mostly the α(200), γ(220), α(211) and γ(311). In order to avoid overlapping it is necessary to choose a 2θ angle below 25° or such that the molybdenum Kα and Kβ lines are approximately equal in distance on the energy scale from the (200) line. This condition is satisfied at 2θ = 27°.

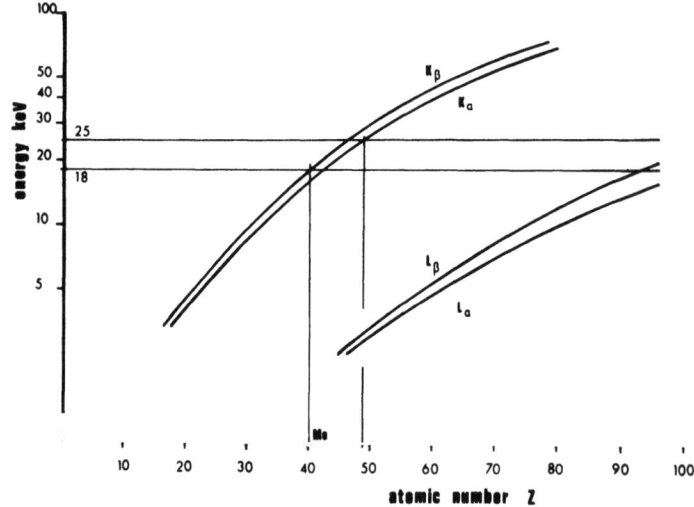

Figure 6. Moseley-law diagram for selected X-ray spectral lines.

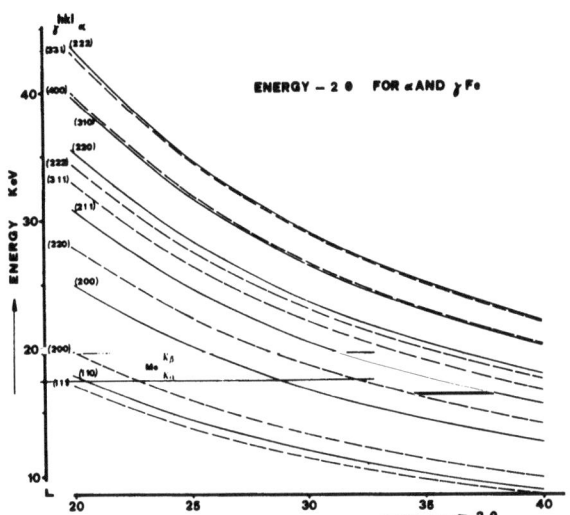

Figure 7. Bragg's Law diagram for the FAS method.

d. X-ray Penetration in the Sample

The main reason for the occurrence of large differences in analysing volume fraction for various diffraction planes is the variation in penetration of the photon energy. This is especially pronounced at high energy levels; the effect is shown in figure 8.

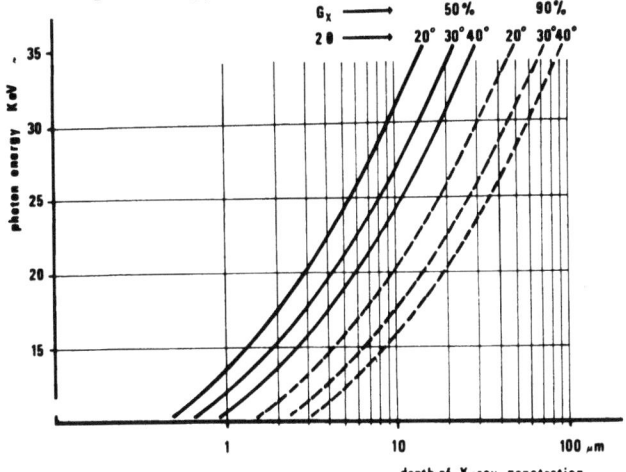

Figure 8. X-ray penetration in low alloyed steels for several fixed 2θ angles and two fractions G_x of the total diffracted intensity.

After consideration of those parameters previously mentioned, the 2θ angle of 27 degrees is favoured for retained austenite analysis. All the following experiments for the determination of retained austenite were carried out at this angle.

Compton Scattering

Another aspect of the FAS technique which has an influence on the absolute intensities and also on the peak to background ratio is the amount of Compton scattering. This gives rise to an increased background count in the energy spectra from samples with low atomic numbers. This effect is shown in figure 9, where the intensities against energy are given for six different diffraction spectra between gold and carbon. Although from gold to iron the effect is almost negligible, below iron the effect is increased until with carbon the intensity distribution is mainly due to Compton scattering.

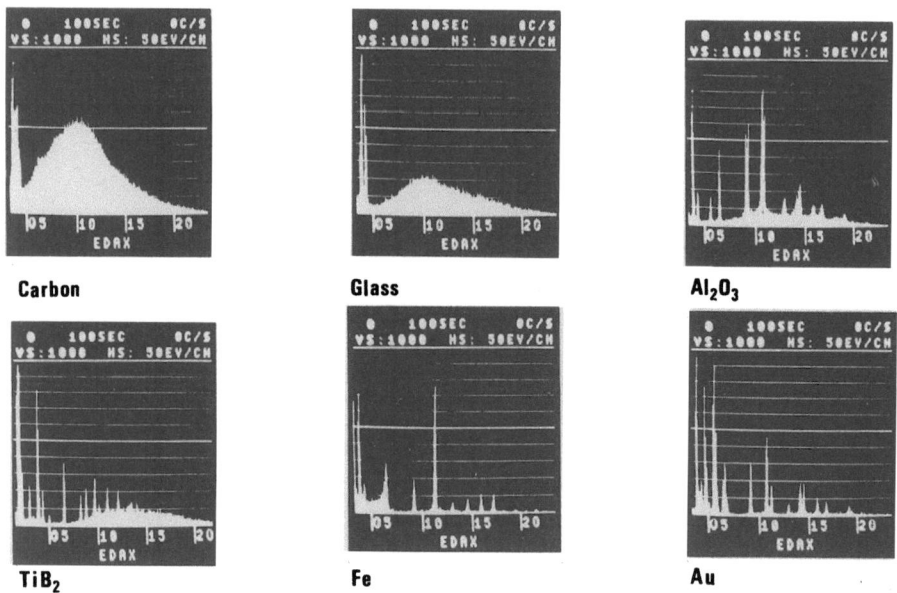

Figure 9. Increased background counts due to Compton Scattering.

As far as the scattering efficiency for different photon energies is concerned, experiments have been carried out to determine the influence on the intensity distribution by several scatterers. This work is not yet completed but it has been shown experimentally that for an iron scatterer, this correction is almost constant for the photon energy range which is involved.

Count-rate

Of the many factors which are needed to determine the optimal conditions for FAS analysis, one of the most important is the obtainable count-rate. Since the count-rate for a diffraction line in the FAS spectra depends on many parameters, extra attention has to be paid to the reliability of the results obtained for a particular analysing condition.

In order to measure the performance of the FAS method, experiments were carried out both with the FAS method and the CSM under the same geometrical conditions. Net intensities were measured for the $\alpha(211)$ and $\gamma(311)$ reflections on a sample containing 18.3% austenite.

For the CSM, the scan rate was 0.5 degrees per minute and the background intensities were measured on both sides of each reflection for a time of 100 sec. The total time to achieve a net intensity was respectively 600 and 400 sec. for the $\alpha(211)$ and $\gamma(311)$ reflections. The integration time for the FAS method was 400 sec.

In order to compare the results of both methods, the net intensities obtained were calculated for 100 sec. The results are presented in Table 1.

Method 50KV 24mA	div. slit	rec. slit	$I_{net}\ \alpha(211)$ 100 sec	$I_{net}\ \gamma(311)$ 100 sec
MoK$\alpha_{1,2}$	$\frac{1}{4}°$ x 12mm	0.2 x 12mm	6,700	1,500
Cu tube $\alpha=6.0°$	$\frac{1}{4}°$ x 12mm	0.2 x 12mm	2,500	500
W tube $\alpha=6.0°$	$\frac{1}{4}°$ x 12mm	0.2 x 12mm	6,000	1,200
Mo K$\alpha_{1,2}$	$1°$ x 12mm	0.4 x 12mm	51,000	11,000
W tube $\alpha=6.0°$	$1°$ x 12mm	0.4 x 12mm	30,000	6,000

Table 1. Obtainable count-rates for retained austenite measurements on a sample with 18.3% austenite.

From Table 1 one can find that for small irradiated sample areas the count-rate for the FAS method is comparable with the CSM if a tungsten tube is used.

Although the CSM yields a higher intensity for larger irradiated areas the FAS method under these conditions will also yield a reliable count in almost the same count time.

Intensity Calculations

The intensity calculation for diffraction lines of polycrystalline material is also based upon the Debije-Scherrer intensity formula for the FAS technique.

In general, working with a diffractometer, for a single-phase powder specimen with random crystal orientation the well-known exact expression for the diffraction intensity is given by Cullity (14).

$$I_{\lambda_1} = I_{o\lambda_1} \frac{e^4}{m^2 c^4} \frac{\lambda_1^3 A}{32\pi R} \frac{1}{v^2} \frac{1}{2\mu} F H \frac{1 + \cos^2 2\theta}{\sin^2\theta \cos\theta} e^{-2M} g\eta \quad \{1\}$$

For the retained austenite analysis the net diffracted intensities of a number of lines which are involved may be taken as relative to each other. Therefore, and because of the fixed angle 2θ under which the intensities are measured, the formula must be modified by the following factors:

a. For intensity calculations and intensity ratio determinations by a fixed angle θ the Lorentz and Polarisation factor will not be required.

b. The contribution of λ for different d values is not constant, therefore the intensity must be integrated over all values $\Delta\lambda$ of λ that can contribute to the peak. From the Bragg equation one finds $\Delta\lambda = \lambda \cot\theta \Delta\theta$. $\Delta\theta$ is constant for a particular experiment and $\cot\theta$ is constant. Therefore $\Delta\lambda$ depends directly on λ, but according to the Bragg equation λ is a linear function of d for a fixed angle θ; therefore it follows that $\Delta\lambda$ is proportional to the d value.

c. As long as there are no absorption edges near the wavelengths used, λ exp.3 is proportional to μ. Therefore they cancel each other out in the formula.

d. $\frac{e^2}{m^2c^4}$ A 32 mR can be considered as constant for a particular diffractometer.

e. Air absorption is negligible for the relatively high photon energies.

The intensity formula will be reduced by this number of factors and will show up as:

$$I_{\lambda_1} = I_{o\lambda_1} \eta \frac{F^2}{v^2} H e^{-2M} d \quad \{2\}$$

where

I_{λ_1} = integrated intensity per unit length of diffraction line

$I_{o\lambda_1}$ = intensity of incident beam

v = volume of unit cell

F = structure factor

H = multiplicity factor

e^{-2M} = temperature factor

η = counting efficiency of the detector

d = displacing.

The correction factor R is the ratio of the calculated intensities for the diffraction lines involved.

$$R = \frac{I\ \alpha(hkl)\ calc.}{I\ \gamma(hkl)\ calc.} \qquad \{3\}$$

The retained austenite can be found with the equation given by Lucas and Nützel (15)

$$X_\gamma = \frac{100}{\frac{I\alpha(hkl)obs.}{I\gamma(hkl)obs.} \cdot R + 1} \qquad \{4\}$$

Intensity Correction Factors (R) and Results

To determine the intensity correction factors (R), retained austenite standards with known concentrations were used. An advantage of this procedure is the fact that the correction can be found without knowing the intensity distribution of the polychromatic incident beam, and even without knowing the interaction of the incident beam with the sample. Experiments have been carried out with ten samples each having a known austenite content in the range of 5 to 20%. In total each sample was measured ten times by the FAS technique and ten times with the CSM. The austenite content determined by the CSM was calculated according to {4} using the correction factors (R) given by Faninger and Hartmann (16). From the results obtained it was possible to find experimentally the correction factors for the FAS method.

In order to check these experimentally obtained values, the corrections were determined from first principles. With the intensity formula {2}, it was possible to determine the intensity ratios for any pair of diffraction lines. With the exception of the relative intensity distribution in the polychromatic energy range, the corrections required are to be found in the International Tables for X-ray Crystallography (17).

Table 2 presents the intensity corrections experimentally obtained and following the theoretical approach. The results obtained from the theoretical approach and the experimentally determined ratios can be seen to be closely comparable.

(R)	$\dfrac{\alpha(200)}{\gamma(220)}$	$\dfrac{\alpha(200)}{\gamma(311)}$	$\dfrac{\alpha(211)}{\gamma(220)}$	$\dfrac{\alpha(211)}{\gamma(311)}$
EXP.	2.03	2.48	0.74	0.90
THEOR.	1.95	2.43	0.71	0.83

Table 2.   Intensity Correction Factors

To prove the statistical reliability of the FAS method for the quantitative intensity measurements, two samples with differing amounts of austenite were examined by both techniques for a total of 40 times each. The average austenite content and the 1σ values were respectively: $18.3 \pm 0.4\%$ and $5.2 \pm 0.6\%$ for the CSM and for the FAS method, $18.1 \pm 0.5\%$ and $5.3 \pm 0.6\%$.

With the results obtained concerning the intensity correction factors and the statistical reliability, retained austenite analyses were carried out on a number of hardened steel samples. In Table 3 the results are presented for samples which were examined by both FAS and CSM. The values presented are the average from four measurements. It is evident that the austenite content within the limits of the associated errors for the two methods is closely comparable.

	MO K Radiation Time: 2400 sec.		Polychromatic Radiation Time: 400 sec.	
Sample	R.A.	Abs. Error	R.A.	Abs. Error
A1	21.3%	0.9%	22.4%	0.8%
A2	19.3	0.7	20.4	0.9
A3	18.4	0.8	17.8	0.7
A4	17.7	0.3	17.5	0.6
A5	14.3	0.5	14.3	0.6
A6	9.4	0.4	9.0	0.4
A7	8.7	0.5	8.1	0.3
A8	8.2	0.2	8.0	0.5
A9	7.3	0.2	8.1	0.4
A10	5.9	0.4	5.9	0.3

Table 3: Results of retained austenite measurement.

CONCLUSIONS

X-ray diffraction under fixed angle scattering conditions using the Si(Li) detector for spectra analysis is applicable to retained austenite analysis. The results obtained with the FAS technique have shown good agreement with measurements carried out with the conventional scanning method using molybdenum Kα radiation.

It is possible within approximately six minutes to carry out a quantitative analysis.

For the determination of retained austenite by FAS for a range of low alloyed carbon steels, there is an optimum angle $2\theta = 27°$.

There are no extra manpower requirements or additional controlling units required for diffractometer settings on the new method.

The whole operation for a retained austenite determination is greatly simplified and is therefore highly suitable for quality control measurements. However, even in research investigations, this method can save considerable time.

ACKNOWLEDGEMENTS

The author wishes to express appreciation to Mr. A. Moggré of the EDAX International Inc. for the continued support of this work in providing the detector and periphery electronics. Special thanks are extended to Mr. R. J. Smalley and Dr. H. G. Nützel for critical comments on the manuscript.

REFERENCES

1. R. M. Brugger, R.B. Bennion, T.G. Worlton and E.R. Peterson "Neutron Diffraction Studies of Samples at High Pressure" Proceedings of a panel on research applications of Repetitively Pulsed Reactors and Boosters, Dubna USSR 1966. International Atomic Energy Agency, Vienna 1967 P 35.

2. R.L. Heath,"The Application of High Resolution Solid State Detectors to X-ray Spectrometry - A Review", Advances in X-ray Analysis, Vol.15, 1971.

3. B. Buras, J. Chwaszczewska, S. Szarras and Z. Szmid, "Fixed Angle Scattering Method for X-ray Crystal Structure Analysis", Institute of Nuclear Research, Report 894/11/PS Warsaw, 1968.

4. B.C. Giessen and C.E. Gordon, "New High-Speed Technique Based on X-ray Spectrography", Science, Vol 159, March 1968.

5. C.J. Sparks Jr. and D.A. Gedcke, "Rapid Recording of Powder Diffraction Patterns with Si(Li) X-ray Energy Analysis System: W and Cu Targets and Error Analysis", Advances in X-ray Analysis, Vol. 15, 1971.

6. G.W. Martin and A.S. Klein, "A Complete Instrumental System for Energy Dispersive Diffractometry and Fluorescence Analysis", Advances in X-ray Analysis, Vol 15 1971.

7. W. Lin, "A Rapid Fluorescence and Energy Powder Pattern Analysis System", Advances in X-ray Analysis, Vol. 16, 1972.

8. J.C. Nutter, "A Non-dispersive On-stream X-ray Diffractometer for the Cement Industry", Cement Technology, March/April, 1972.

9. R.E. Ferrell Jr., "Applicability of Energy-dispersive X-ray Powder Diffractometry to Determinative Mineralogy", The American Mineralogist, Vol. 56, 1971.

10. H. Cole, "Bragg's Law and Energy Sensitive Detectors", Journal for Applied Crystallography 3, 405, 1970.

11. P. Banerjee and P. Charbit, "Rapid X-ray Diffraction Investigations by Means of a Si(Li) Semiconductor Detector", Siemens Review, October 1971.

12. E. Laine, I. Lähteenmäki and M. Kantola, "Adaptation of Solid State Detector in X-ray Powder Diffractometry", X-ray Spectrometry 1, 1972.

13. J.V. Gilfrich and L.S. Birks, "Spectral Distribution of X-ray Tubes for Quantitative X-ray Fluorescence Analysis", Analytical Chemistry, Vol 40, June 1968.

14. Cullity, "Elements of X-ray Diffraction", Addison-Wesley Published Company Inc., 1956.

15. G. Lucas and H. Nützel, "Einverbessertes Verfahren zur röntgenographischen Bestimming des Restaustenits in gehärteten-Stählen", Siemens-Z 35, (1961), S445.

16. G. Faninger und U. Hartmann, "Physikalische Grundlagen der Quantitativen röntgenographischen Phasenanalyse", Harterei Technische Mitteilungen, 27 (1972).

17. International Tables for X-ray Crystallography. Kynoch Press, Birmingham.

QUANTITATIVE X-RAY DIFFRACTION PHASE ANALYSIS OF THE OXIDATION OF STEEL BY A DIRECT COMPARISON METHOD

R. R. Biederman, R. F. Bourgault and R. W. Smith

Worcester Polytechnic Institute

Worcester, Massachusetts 01609

ABSTRACT

A direct comparison x-ray method using crystal structure information has been developed for rapid quantitative analysis of the oxide phases present in oxidized steels. This procedure is similar to that used for the analysis of retained austenite in hardened steels and requires no external or internal standards for accurate oxide determinations. Successful application of this technique has been achieved for a variety of steel oxide mixtures and excellent correlation with microstructural analysis results has been found. Accurate determinations have been made of the volume fractions of Wustite (FeO), Magnetite (Fe_3O_4) and Hematite (Fe_2O_3) for a wide range in steel carbon content and transformation treatments. Many oxide mixtures that differ over a wide range in the amounts of each of the three oxides phases have been measured successfully. The procedure is simple, the method direct and the analysis accurate.

INTRODUCTION

A typical oxide scale in the condition that it formed on the steel is shown in Figure 1. This scale contains three crystallographically distinct phases in different regions of the scale:

Hematite (Fe_2O_3) - outer light layer with "icicle" character inward growth

Magnetite (Fe_3O_4) - dispersed cuboid shaped phase

Wustite (FeO) - columnar grained darker matrix phase

Figure 1. Typical Oxide Scale 300X
60.7v/oWustite, 29.4v/oMagnetite, 9.9v/oHematite

Numerous cracks and voids are present in the scale also. While quantitative metallographic methods (1) can be used to determine the amounts of each phase present, it is often a time consuming and arduous task of questionable accuracy. X-ray diffraction techniques offer potential for greater accuracy and rapid analysis especially if careful weighing and standards can be avoided.

An x-ray diffraction method for the quantitative analysis of multiple phase mixtures by direct comparison (2,3) has been used successfully for several years. This method is based on the fact that the integrated intensity of a diffracting condition is proportional to the volume fraction of the phase present. In a randomly oriented sampling, only one reflection from each phase has to be considered for complete analysis. However, if texture is present in the sampling, more reflections from every phase must be analyzed for an accurate analysis. This complicates the analysis but analytical methods are available (4) to correct for texturing effects and analysis by direct comparison is still feasible. If at all possible, it is best to minimize preferred orientation effects. This was achieved in this study of steel oxide scales by pulverizing the scale and randomly dispersing the oxide in powder form.

THEORY

The integrated intensity for each Bragg diffraction condition from a phase in a polycrystalline specimen is:

$$I_{(hkl)} = \left(\frac{I_o e^4}{m^2 c^4}\right) \left(\frac{\lambda^3 A}{32 \pi r}\right) \left(\frac{1}{V^2}\right) (F^2)(L-P)(p) \left(\frac{\exp^{-2M}}{2\mu}\right) \tag{1}$$

$$I_{(hkl)} = (\quad K \quad) \left(\frac{R}{2\mu}\right)$$

The first two parentheses reflect parameters that depend only on the incident x-ray conditions and the geometry of the x-ray experiment. These parameters are constant for a specific geometry and power condition. The remaining terms in the integrated intensity expression are conveniently lumped into a single factor, R, which relates to the crystallography of the phase under analysis. The various parameters included in this R factor are: V - the volume of the unit cell, F - the structure factor, L-P - the combined Lorentz factor and the polarization factor, μ - average linear absorption coefficient, $\exp^{-2M}$ - temperature factor and p - multiplicity factor. Calculation of R depends only on the Bragg angle, theta, and crystallographic information for each phase for the particular (hkl) planes of interest.

For a mixture of the three oxide phases, Wustite (W), Magnetite (M) and Hematite (H) the integrated intensity relationship for constant experimental conditions is:

$$I_W : I_M : I_H = R_W C_W : R_M C_M : R_H C_H \tag{2}$$

and the total volume fraction of oxide phases is:

$$C_W + C_M + C_H = 1 \tag{3}$$

The volume fraction of each phase, C_W, C_M and C_H is readily calculated by simultaneous solution of equations (2) and (3) provided all R factors are known and the integrated intensity can be experimentally measured. The solution expressions are presented below in a convenient form for direct analysis.

$$C_W = \frac{1}{1 + \dfrac{R_W I_M}{R_M I_W} + \dfrac{R_W I_H}{R_H I_W}}$$

$$C_M = \frac{1}{1 + \dfrac{R_M I_H}{R_H I_M} + \dfrac{R_M I_W}{R_W I_M}} \tag{4}$$

$$C_H = \frac{1}{1 + \dfrac{R_H I_W}{R_W I_H} + \dfrac{R_H I_M}{R_M I_H}}$$

These expressions apply for oxide mixtures that exhibit no preferred orientation. Random orientation is most easily achieved by dispersing a mixture of fine oxide powder.

R FACTOR ANALYSIS

The determination of R factors for each diffracting condition requires an understanding of the crystallography of each phase. Table 1 summarizes the positions and atom types within the unit cell for Wustite (FeO), Magnetite (Fe_3O_4) and Hematite (Fe_2O_3). Wustite forms a cubic unit cell structure analogous to that of NaCl. Magnetite forms a cubic cell structure similar to NaCl also, but with 56 atoms in the unit cell in comparison to 8 for FeO or NaCl. Hematite forms a trigonal or rhombohedral cell structure which is isomorphous to that of hexagonal Al_2O_3. The positions listed in

Table 1 for Hematite are referred to the standard hexagonal axes.

TABLE 1
ATOM POSITIONS IN OXIDE UNIT CELLS

WUSTITE STRUCTURE (FeO), SPACE GROUP Fm3m - Fe located at center (m3m)   a = 4.307A   ASTM CARD 6-0615

Positions of: Fe
0   0   0 ,   0 1/2 1/2,   1/2 0 1/2,   1/2 1/2 0

Positions of: O
1/2 1/2 1/2,   1/2   0   0 ,   0 1/2   0 ,   0   0 1/2

MAGNETITE STRUCTURE   (Fe_3O_4), SPACE GROUP Fd3m - O located at center ($\bar{3}$m), a = 8.396A   ASTM CARD 11-614

Positions of: O
0   0   0 ,   0 1/2 1/2,   1/2 0   1/2,   1/2 1/2 0
0   1/4 1/4,   0   3/4 3/4,   1/2 1/4 3/4,   1/2 3/4 1/4
1/4 0   1/4,   1/4 1/2 3/4,   3/4 0   3/4,   3/4 1/2 1/4
1/4 1/4 0 ,   1/4 3/4 1/2,   3/4 1/4 1/2,   3/4 3/4 0
1/2 1/2 1/2,   1/2   0   0 ,   0   1/2   0 ,   0   0   1/2
1/2 1/4 1/4,   1/2 3/4 3/4,   0   1/4 3/4,   0   3/4 1/4
1/4 1/2 1/4,   1/4 0   3/4,   3/4 1/2 3/4,   3/4 0   1/4
1/4 1/4 1/2,   1/4 3/4 0 ,   3/4 1/4 0 ,   3/4 3/4 1/2

Positions of: Fe   ($2/3 Fe^{3+}$ - - $1/3 Fe^{2+}$)
1/2   0   1/4,   1/2 1/2 3/4,   0   0   3/4,   0   1/2 1/4
0   1/4 0 ,   0   3/4 1/2,   1/2 1/4 1/2,   1/2 3/4 0
3/4 0   0 ,   3/4 1/2 1/2,   1/4 0   1/2,   1/4 1/2 0
1/4 1/4 1/4,   1/4 3/4 3/4,   3/4 1/4 3/4,   3/4 3/4 1/4
1/8 7/8 1/8,   1/8 3/8 5/8,   5/8 7/8 5/8,   5/8 3/8 1/2
3/8 5/8 3/8,   3/8 1/8 7/8,   7/8 5/8 7/8,   7/8 1/8 3/8

HEMATITE STRUCTURE   (Fe_2O_3), SPACE GROUP R$\bar{3}$c Fe located at center ($\bar{3}$)   a = 5.032A;   c = 13.74A   ASTM CARD 13-534

Positions of: Fe
0   0   0 ,   1/3 2/3 2/3 ,   2/3 1/3 1/3
0   0   1/2,   1/3 2/3 1/6 ,   2/3 1/3 5/6
0   0   1/6,   1/3 2/3 5/6 ,   2/3 1/3 1/2
0   0   2/3,   1/2 2/3 1/3 ,   2/3 1/3 0

Positions of: O
2/3 0   1/12,   0   2/3 3/4 ,   1/3 1/3 5/12
0   2/3 1/12,   1/3 1/3 3/4 ,   2/3 0   5/12
1/3 0   1/4 ,   2/3 2/3 11/12,   0   1/3 7/12
0   1/3 1/4 ,   1/3 0   11/12,   2/3 2/3 7/12
1/3 1/3 1/12,   2/3 0   3/4 ,   0   2/3 5/12
2/3 2/3 1/4 ,   0   1/3 11/12,   1/3 0   7/12

Specific x-ray diffraction pattern information is available from the inorganic x-ray diffraction card file (5, 6, 7) for all three oxides.

Six independent diffracting conditions, two from each oxide phase, were used in all oxide determinations. Since only three reflections, one from each phase, are required for any analysis, the eight combinations possible with six independent reflections allowed an evaluation of preferred orientation effects in each specimen. Table 2 presents the structure factor analysis for the six diffracting conditions considered. The atomic scattering

TABLE 2
STRUCTURE FACTOR ANALYSIS

104)H; (113)H; (111)W; (200)W; (220)M; (400)M

$F_{(104)H} = f_{Fe^{3+}} (3 - 5.196i)$; $F^2_{(104)H} = 35.95 f^2_{Fe^{3+}}$

$F_{(113)H} = f_{O^{2-}} (15.588)$; $F^2_{(113)H} = 247 f^2_{O^{2-}}$

$F_{(111)W} = 4 (f_{Fe^{2+}} - f_{O^{2-}})$; $F^2_{(111)W} = 16(f_{Fe^{2+}} - f_{O^{2-}})^2$

$F_{(200)W} = 4 (f_{Fe^{2+}} + f_{O^{2-}})$; $F^2_{(200)W} = 16(f_{Fe^{2+}} + f_{O^{2-}})^2$

$F_{(220)M} = 8 (f_{Fe^{3+}})$; $F^2_{(220)M} = 64 f^2_{Fe^{3+}}$

$F_{(400)M} = 8 (f_{Fe^{2+}}) + 32 (f_{O^{2-}})$; $F^2_{(400)M} = 64 (f^2_{Fe^{2+}}) + 1024 (f^2_{O^{2-}}) + 512 (f_{Fe^{2+}})(f_{O^{2-}})$

factors reflect the charge on the respective ions in each structure. A numerical summary of all the parameters considered in the R factor expression and the calculated R factors are presented in Table 3 for the six reflections used in this analysis. Temperature factors have been omitted for this factor is essentially constant for the reflections considered. Numerical values of all other parameters were obtained from the International Tables for X-ray Crystallography (8). The ratio of R factors for any pair of diffracting conditions from the same phase is proportional to the corresponding integrated intensity ratio for that phase. If the integrated intensity ratio is different from the calculated R factor ratio, preferred orientation is present in the sample and the analysis must be corrected (4) for this effect.

TABLE 3

R FACTOR INTENSITY ANALYSIS FOR IRON OXIDES

FOR CHROMIUM RADIATION

HKL	2θ	$\sin\theta$	$\dfrac{\sin\theta}{2.291}$	$f_{Fe^{2+}}$	$f_{Fe^{3+}}$	$f_{O^{2-}}$	F^2	P	L-P	v^2	R
(111)W	54.87	0.462	0.202	18.48	-	5.87	2544	8	7.018	6383	22.37
(200)W	64.28	0.533	0.2325	17.48	-	5.35	8339	6	4.936	6383	38.69
(220)M	45.43	0.386	0.1685	-	18.95	-	22983	12	10.860	350287	8.55
(400)M	66.25	0.547	0.2385	17.49	-	5.25	93754	6	4.636	350287	7.44
(104)H	50.40	0.425	0.1855	-	18.50	-	12304	12	8.612	90546	14.04
(113)H	62.72	0.520	0.227	-	-	5.31	6967	12	5.243	90546	4.84

Combining the information calculated in Table 3 and equations (4) result in expressions for the volume fraction of each oxide phase that depend only on the experimentally determined integrated intensity for the respective diffraction conditions. For example, considering the (200)W, (220)M, and (104)H reflections, the expressions for oxide analysis are presented in Table 4. Accurate

TABLE 4

VOLUME FRACTION PHASE ANALYSIS USING (200)W; (220)M; AND (104)H INTENSITIES

$$C_W = \dfrac{1}{1 + 4.525\left(\dfrac{I_{(220)M}}{I_{(200)W}}\right) + 2.756\left(\dfrac{I_{(104)H}}{I_{(200)W}}\right)}$$

$$C_M = \dfrac{1}{1 + 0.609\left(\dfrac{I_{(104)H}}{I_{(220)M}}\right) + 0.221\left(\dfrac{I_{(200)W}}{I_{(220)M}}\right)}$$

$$C_H = \dfrac{1}{1 + 0.363\left(\dfrac{I_{(200)W}}{I_{(104)H}}\right) + 1.642\left(\dfrac{I_{(220)M}}{I_{(104)H}}\right)}$$

oxide analysis is dependent only on accurate integrated intensity determination.

EXPERIMENTATION AND RESULTS

The theory presented was tested on over 200 oxide scales from various combinations of steels, oxide treatments and oxidizing atmospheres. Each oxide scale was stripped from the steel by uniaxial tensile loading and the stripped scale was pulverized to a powder of approximately 325 - 400 mesh. The powder was suspended in a 2% solution of parlodion in amyl acetate and dispersed onto a glass slide. Enough oxide powder was used to assure that the irradiated specimen area was always greater than the projected x-ray beam size.

Diffraction patterns were obtained by standard diffractometer techniques using vanadium filtered chromium Kα radiation. No attempt was made to monochromatize the x-ray beam. The diffractometer scanned the specimen at a rate of 2°/minute. All other experimental x-ray diffractometer parameters were adjusted to maximize the peak to background ratio for optimum chart readability.

Examples of typical diffractometer traces for the range containing the six reflections used in this analysis are shown in Figures 2, 3, 4.

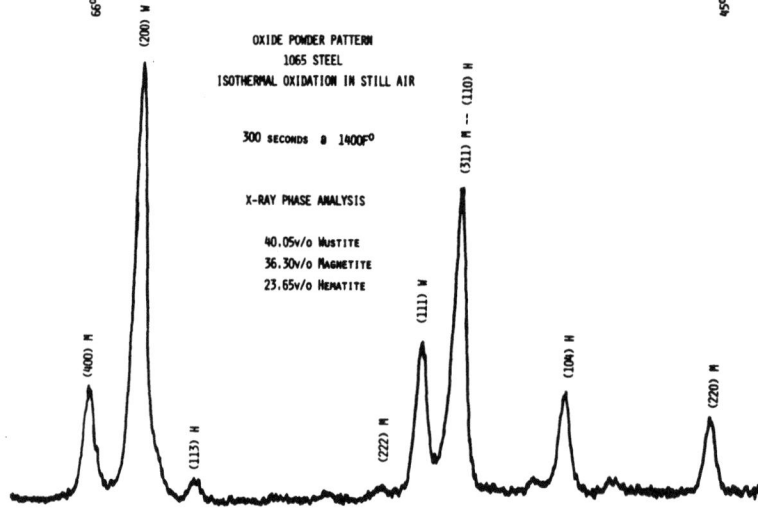

Figure 2. Oxide Powder Pattern 1065 steel Isothermal oxidation in still air 300 seconds at 1400F

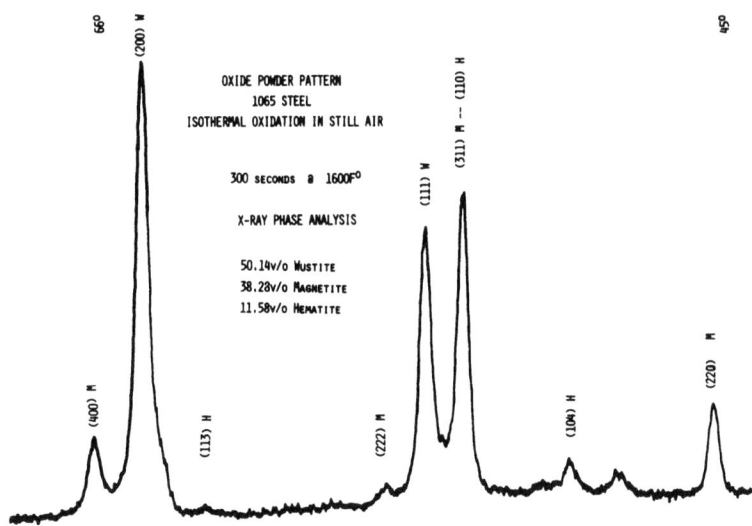

Figure 3. Oxide Powder Pattern 1065 steel Isothermal oxidation in still air 300 seconds at 1600F

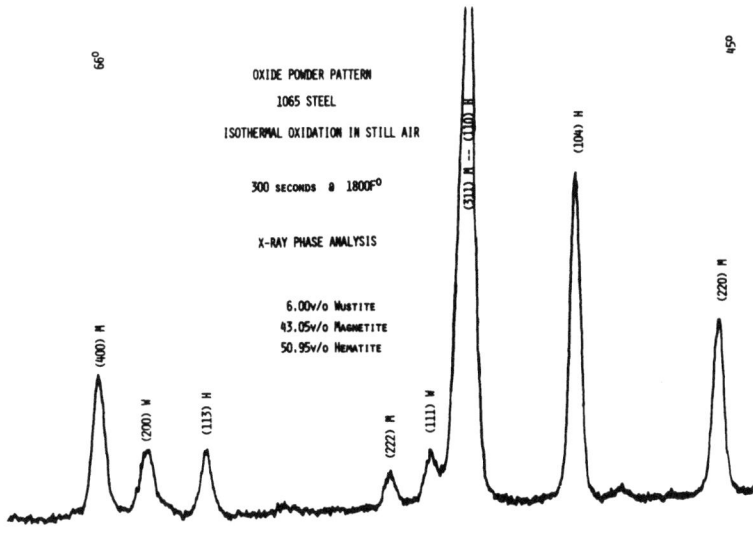

Figure 4. Oxide Powder Pattern 1065 steel Isothermal oxidation in still air 300 seconds at 1800F

These patterns were obtained from 1065 steel scales formed by isothermal oxidation in still air at 1400F, 1600F and 1800F respectively. The integrated intensity or area under each diffraction peak was evaluated by planimeter. The average phase analysis of the oxide scale is shown in each figure for the (200)W, (400)M, (104)H and (200)W, (220)M, (104)H combinations. A complete analysis of all eight combinations of reflection for the 1065 steel oxidized isothermally for 300 seconds at 1600F is presented in Table 5.

TABLE 5

EXAMPLE SCALE ANALYSIS - ISOTHERMAL OXIDATION IN AIR

1065 STEEL HELD AT $1600°F$ FOR 300 SECONDS

INTEGRATED INTENSITY (IN^2)	INTENSITY COMBINATION	ANALYSIS v/oW	v/oM	v/oH
$I_{(220)M}$ = 0.23	(400)M; (200)W; (104)H	49.22	39.41	11.37
$I_{(400)M}$ = 0.22	(220)M; (200)W; (104)H	51.05	37.16	11.79
$I_{(104)H}$ = 0.12	(400)M; (111)W; (104)H	49.63	39.07	11.30
$I_{(113)H}$ = 0.03	(220)M; (111)W; (104)H	51.45	36.85	11.70
$I_{(111)W}$ = 0.84	(220)M; (200)W; (113)H	52.76	38.39	8.85
$I_{(200)W}$ = 1.43	(400)M; (200)W; (113)H	50.81	40.68	8.51
	(220)M; (111)W; (113)H	53.15	38.08	8.77
	(400)M; (111)W; (113)H	51.22	40.31	8.47
	AVERAGE OF ALL EIGHT	51.16±2	38.74±1.9	10.10± 1.7

DISCUSSION

The experiments performed indicate that the volume fraction of Wustite, Magnetite and Hematite can be determined by the presented x-ray method to a reasonable degree of accuracy. Higher accuracy is attainable by more accurate measurement of integrated intensity. The procedure is simple, the method direct and the analysis accurate provided a random oxide orientation can be obtained.

REFERENCES

1. R. T. Dehoff and F. N. Rhines, "Quantitative Microscopy", McGraw-Hill (1968).

2. B. L. Averbach and M. Cohen, "X-ray Determination of Retained Austenite by Integrated Intensities", Trans. AIME 176, p. 401 - 415 (1948).

3. B. D. Cullity, "Elements of X-ray Diffraction", p. 388 - 400, Addison-Wesley Publishing Co. (1956).

4. R. Gullberg and R. Lagneborg, "X-ray Determination of the Volume Fraction of Phases in Textured Materials", Trans. AIME 236, p. 1482 - 1485 (1966).

5. ASTM CARD 6-0615 - Wustite - Inorganic X-ray Powder Data File, American Society for Testing Materials.

6. ASTM CARD 11-614 - Magnetite - Inorganic X-ray Powder Data File, American Society for Testing Materials.

7. ASTM CARD 13-534 - Hematite - Inorganic X-ray Powder Data File, American Society for Testing Materials.

8. "International Tables for X-ray Crystallography", Vol. II & III, The Kynoch Press (1962).

LOW ENERGY X-RAY AND ELECTRON ABSORPTION WITHIN SOLIDS

(100-1500 eV Region)

Burton L. Henke and Eric S. Ebisu

University of Hawaii

Honolulu, Hawaii 96822

ABSTRACT

Quantitative analysis by x-ray fluorescence and photoelectron and Auger electron analysis can be effectively extended through a precise knowledge of the total and subshell photoionization cross sections. Light element and intermediate element analysis, as based upon K and L series fluorescence respectively, involve x-ray interactions in the low energy region. Optimized analysis for essentially all the elements by x-ray induced photoelectron and Auger electron spectroscopy involves both x-ray and electron interactions in the low energy region. Unfortunately, theory and measurement for interaction cross sections in this 100-1500 eV region are difficult, particularly for the heavier elements. Nevertheless, recent advances in experimental and computerized-theoretical techniques for the determination of low energy interaction coefficients do permit establishing appreciably more complete tabulations of cross sections than are currently available in this energy region.

In this paper, the types of interaction cross section data that are needed for quantitative x-ray and electron analysis are defined. Such data that are available from experiment and from theory are reviewed and compared. Some newer techniques for the measurement of cross sections are discussed. And finally, new "state of the art" tables are presented for the mass absorption coefficients of all of the elements and of some special laboratory materials. These are tabulated specifically for twenty-six of the most commonly applied characteristic wavelengths in the 8-110 Å region and are based upon the best currently available theoretical and experimental data.

COMPARISON — THEORY AND EXPERIMENT

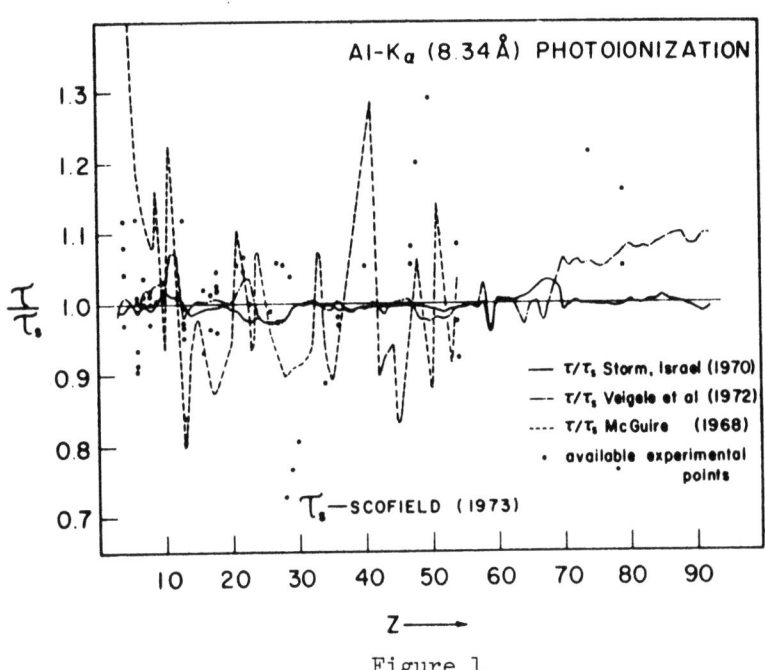

Figure 1

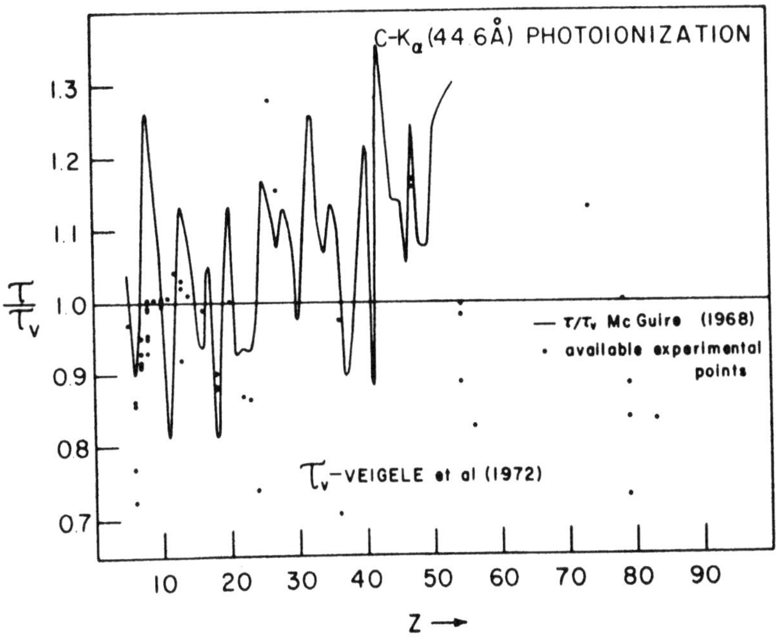

Figure 2

I. LOW ENERGY INTERACTION COEFFICIENTS - MEASUREMENT AND THEORY

A. Total Photoionization Cross Sections

Among the most efficient methods for quantitative structure and chemical analysis that have been developed to this time are those based upon x-ray fluorescence analysis, electron microprobe analysis, scanning electron microscopy, and most recently, low energy electron spectroscopy. It is generally recognized that the most severe limitation to the application of these methods to quantitative analysis is the inadequacy of the currently available photoionization and electron ionization cross section data.

In order to begin to meet this need the author and his co-workers have combined the available theory and experimental data to develop working, "state of the art" tables for the total photoionization cross sections in the ultrasoft x-ray region. These works appeared in 1957, 1967, 1969, (1,2,3) and another is published here.

Fortunately, there has been some excellent theoretical work accomplished in recent years (4,5,6,7,8) as effectively assisted by the availability of the new, large computers. Nevertheless, the theoretical developments are severely handicapped by the lack of good experimental data with which to test the choice of models and in order to be able to extend the present theory to the calculation of the true partial photoionization cross sections which accurately distinguish the multiple excitation and ionization events. To illustrate this need for more experimental data, Figs. 1 and 2 present most of the available experimental data points on total photoionization cross sections for the two most widely and precisely measured low energy x-ray wavelengths, viz. Al-Kα (8.34 A) and C-Kα (44.7 A) respectively, and relative to the recent theoretically calculated cross section values. To give a more complete view of the situation, the available experimental data for the elements of atomic number up through Z=54 (Xe), and for the ultrasoft x-ray region are presented in Fig. 3 with the very recent and complete theoretical curves of Veigele et al (1972) (9) for the elements above Z=18 (Ar), and with the semi-empirical curves of Henke, et al (1969) (3) for the elements Z=18 (Ar) and below. In Fig. 3 the atomic cross sections are presented dimensionlessly as relative to the square of the wavelength in order to illustrate the universal quality of the shape and the progression of the curves versus atomic number which do not overlap except for the high-Z, long wavelength regions. (This universality has been useful in our combining theoretical and experimental data.) It must be pointed out that about one-third of the reported experimental data had to be rejected in the presentations of Fig. 3 as "way off", improbable points. Up-to-

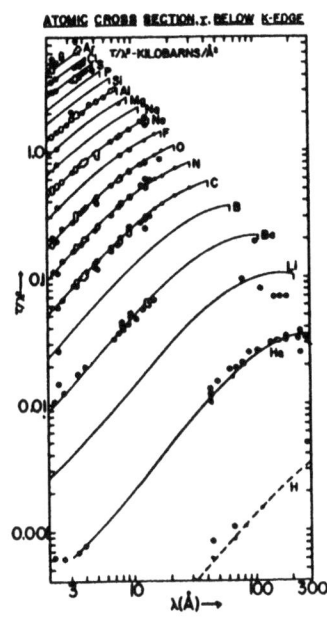
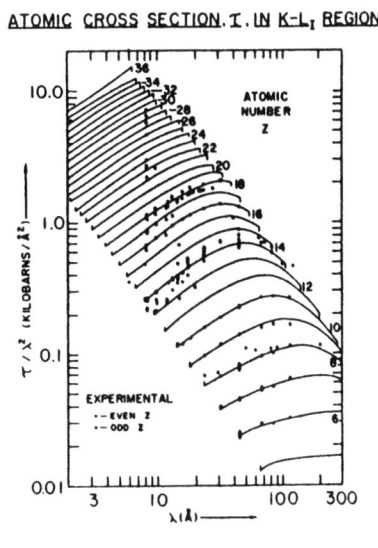
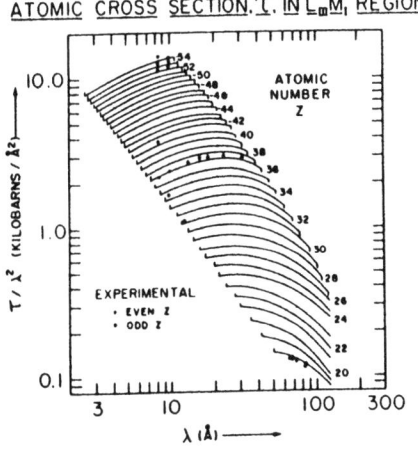
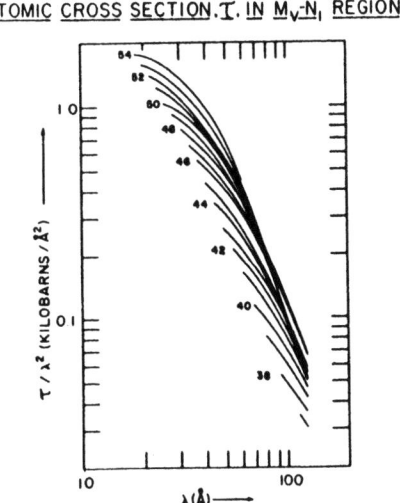

Figure 3

date bibliographies of sources of measured, ultrasoft x-ray absorption coefficients have recently been reported by Hubbell (10) (National Bureau of Standards) and by this author (11).

The most reliable and precise photoionization cross sections in the ultrasoft region are those for systems in the gas state for which the mass thickness of the absorption path can be determined more accurately than that for thin solid film systems. In 1967 the author and his co-workers (2) were able to add to the available data more than one hundred new coefficients as measured on reagent grade gases for characteristic source wavelengths in the 8-110 A region. These measurements on the monatomic rare gases, presented in Fig. 4, demonstrate very good agreement with the calculated curves for the total photoionization cross sections by Veigele, et al (1972) (9). Also the agreement with present theory for the measured molecular gas systems was well within experimental error. Kennedy and Manson (1972) (7) have also calculated the photoionization cross sections for the rare gases and in the low energy x-ray region. Their theoretical curves also fit the experimental data of Fig. 4 very well (11). This remarkable ability for the present theory to fit the experimental data on the rare gases except for the high Z and longest wavelengths, has been established also by the measurements of Wuilleumier (12) as based upon continuous x-ray source measurements.

It is interesting to note that an independent electron model, using a self-consistent-field theory, was used to calculate the total photoionization cross sections which yield good agreement with the experimental values for the rare gases. Apparently the subshell photoionization cross sections as calculated by the "independent electron" model include to some extent the multiple excitation and ionization effects (see below) which have been measured by Carlson, Krause and Wuilleumier (13,14,15) to be of the order of 20% for the rare gases. Fadley (16) suggested that the present theory (using the "sudden approximation" model) is yielding true total photoelectric cross sections, which include multiple ionization and excitation effects. Therefore these <u>should</u> fit the experimental attenuation measurement in the ultrasoft x-ray region for which elastic and inelastic photon scattering losses are insignificant. An important associated question that needs to be investigated is whether or not the solid state absorption can accurately be described by free atom absorption theory as applied to the ultrasoft x-ray region. If solid state effects are present, these must be within the present experimental errors.

The present limitation on the precision of measurement of the total photoionization cross sections for atoms in the solid state is the inability to construct very thin, uniform samples of known mass thickness. This problem is particularly difficult in the

ULTRASOFT X-RAY PHOTOIONIZATION
THEORY AND EXPERIMENT

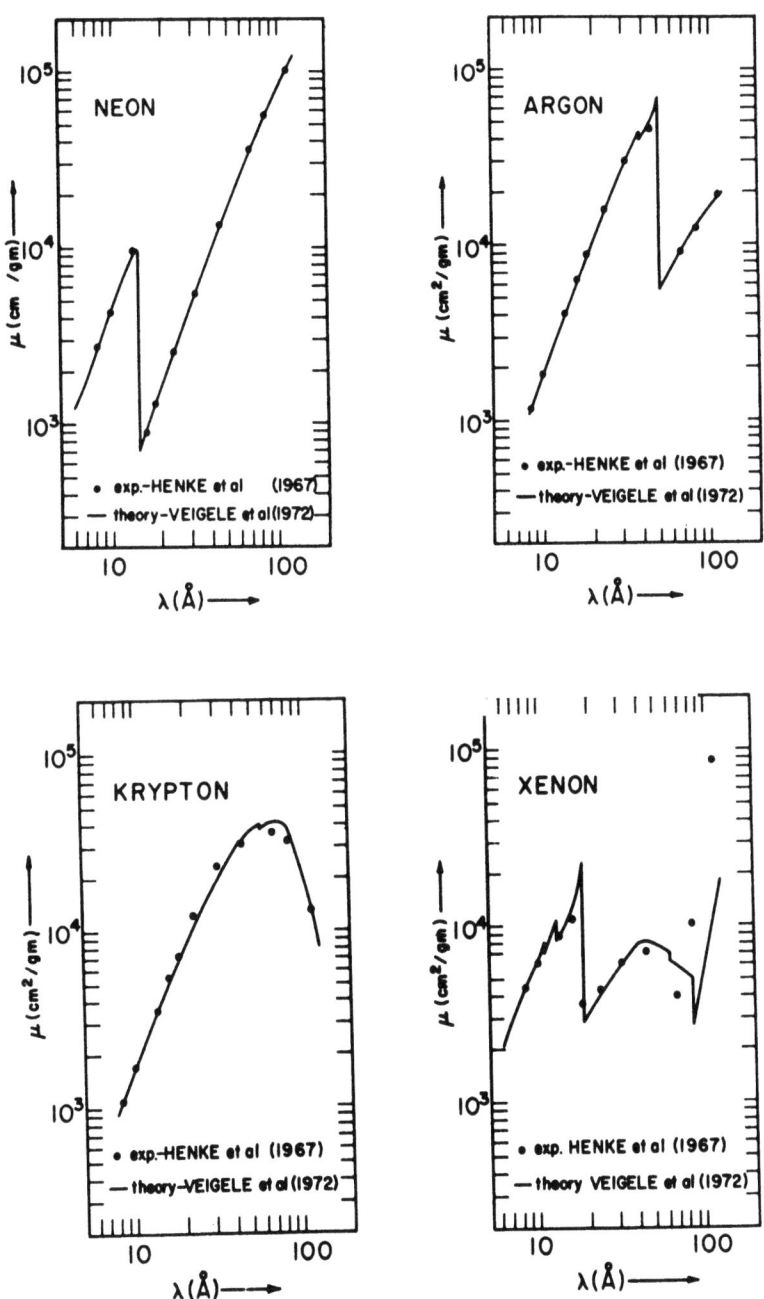

Figure 4

heavy element region for which the need for precise absorption measurements is very great as has been noted above. Thin film deposition techniques can yield satisfactory absorption samples in the submicron thickness range only for a very few selected systems.

Several methods are proposed here for heavy element cross section measurement. The author (17) has recently reviewed ultrasoft x-ray reflection theory as applied to the interpretation of the characteristic reflected intensity, R, versus grazing incidence angle, ϕ, in the region of the total reflection cut-off as a function of the optical constants of the material. These constants, in turn, have then been used to yield the photoionization cross section values for the surface material. Examples of theoretically fitted experimental data are shown in Fig. 5 for a thick evaporated gold surface. Such solid state measurements upon any smooth surface are characteristic of effective sample depths of the order of one hundred angstroms (17).

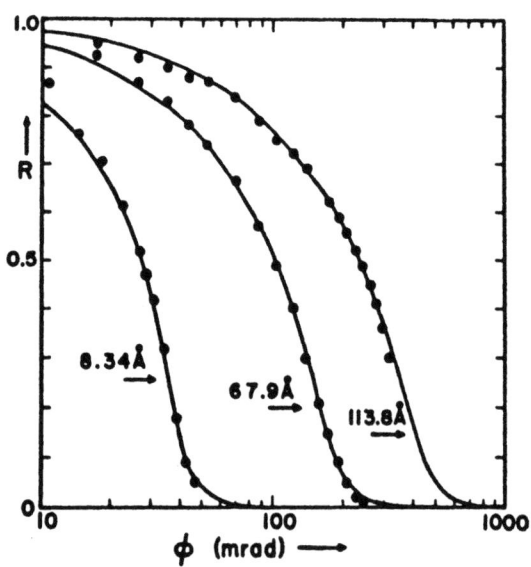

Figure 5

Another technique that has been developed in this laboratory particularly for heavy metal atoms is illustrated in Fig. 6. Here the Langmuir-Blodgett method for constructing uniform, multimolecular layer systems is applied to "lock" into a known number of stearate molecular layers a known area density of cations of the type under measurement. As an example of the method, layers of lead stearate were built upon a thin, calibrated substrate of polypropylene. Because the mass thickness is well defined, and the cross sections in the ultrasoft region for H, C and O are well known, the absorption cross sections for the lead atoms can be determined from the composite film absorption. These measurements are shown here with the theoretical values of Veigele, et al (1972) (9).

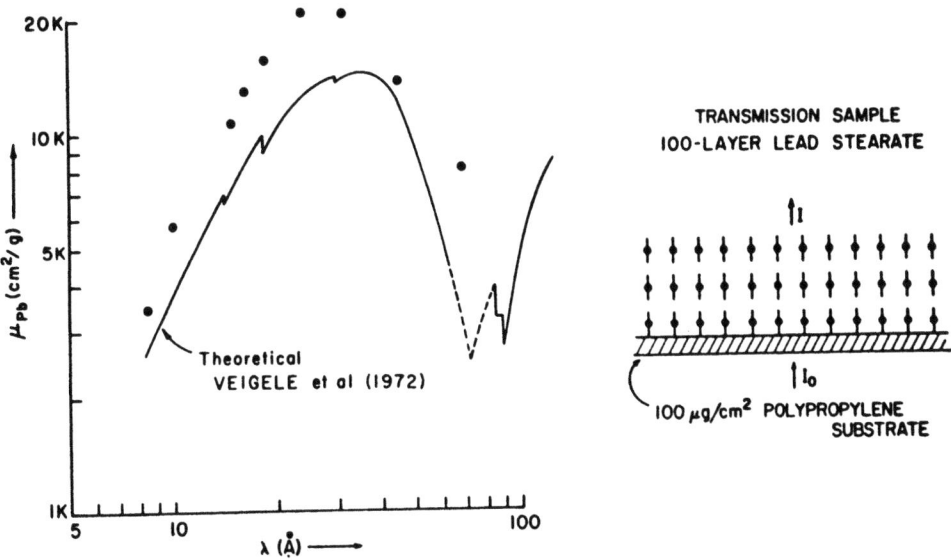

Figure 6

Some systems can be deposited as uniform films of the required thickness in the submicron region on top of solid substrates. Such samples can be deposited upon one half the area of an oversize analyzing crystal surface (2" x 3") so that a Bragg reflection intensity can be measured off the uncovered crystal surface, and then off the covered portion as shifted into the beam. These readings yield the required absorption information as measured for a path length of known Bragg-angle geometry within the deposited films.

The ultrasoft x-ray spectrographic system that has been developed for such measurements as described above is shown in Fig. 7. A clean, high-power source of ultrasoft, characteristic x-ray lines has been specially designed for this work and is closely coupled to a fluorescent secondary radiator. The fluorescent sources are mounted in an eight position holder and yield about twenty standard monochromatic line radiations (K, L, or M) distributed throughout the 5-200 A region. In this fluorescent mode, continuum background is eliminated. The interchangeable x-ray anode is chosen to provide intense line radiation just on the high energy side of the desired fluorescent lines in order to optimize excitation efficiency. Specially constructed, synthetic organic crystals of 2d-spacings up to 160 A are used in Bragg reflection to isolate the characteristic wavelength being measured from adjacent series lines. Finally, residual background, including electronic, is reduced to a minimum by pulse height discrimination with thin-window, subatmospheric pressure flow proportional counters.

B. Partial Photoionization and Electron Ionization Cross Sections

In the study of the low energy x-ray interactions within matter, in addition to the measurement of the transmitted x-ray intensity, as discussed above, one may also measure the associated excited emissions, namely the fluorescent photons, the Auger electrons and the photoelectrons. These signals are directly proportional to the partial photoionization cross sections of the atomic subshells involved in the initial ionization. It would be from such measurements that information might be derived for the subshell cross sections. The Auger electron and photoelectron signals are inversely proportional to the total electron ionization cross sections for the material. The electron spectroscopic measurement of the emitted photoelectron, "no loss" lines can be used as a means to investigate both the partial photoionization cross sections and the electron ionization cross sections.

A typical low energy electron spectrum is shown in Fig. 8 which was measured from an evaporated BaF_2 sample as excited by Al-Kα (1487 eV) photons. The relatively sharp lines are charac-

PHOTOIONIZATION CROSS SECTION MEASUREMENT
(100-1000 eV Region)

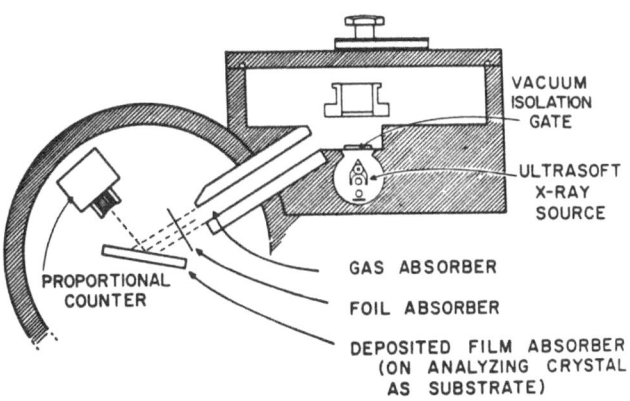

Figure 7

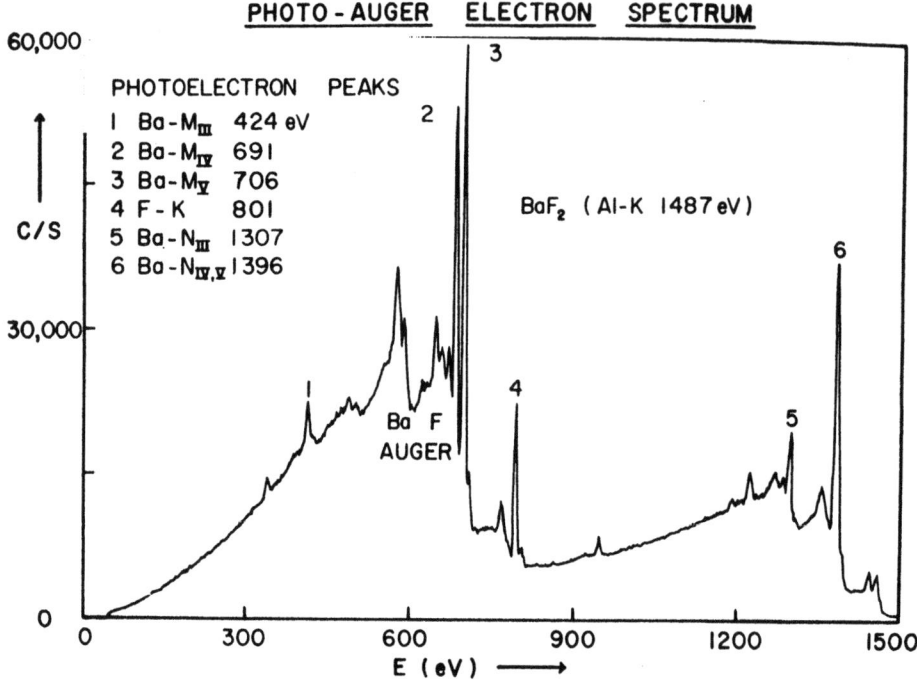

Figure 8

teristic of those electrons which leave the sample surface with no energy loss and are the most amenable to precise measurement and interpretation.

In order to suggest that the intensities of these no-loss lines do "measure" the ratio of the partial photoionization cross section to the total electron ionization cross section, a model for the generation of the photoelectron line intensity is presented below. This is included in some detail because there is relatively little presented in the literature to date on the development of quantitative photoelectron intensity analysis.

Shown in Fig. 9 is the essential geometry of the model adopted here for the generation of an unscattered, no-loss photoelectron signal (the spectrographic photoelectron intensity "peak"). Monoenergetic, parallel x-radiation of wavelength λ is incident upon a plane surface at an angle ϕ, refracting into a homogeneous, isotropic sample with angle ϕ'. The <u>linear</u> x-ray absorption coefficient for this radiation within the sample is equal to μ. In a sample layer at depth z, and thickness dz, photoelectrons are generated within an effective sample area which is the projection of the limiting slit area, A_0. The effective atomic cross section

MODEL FOR
NO-LOSS PHOTOELECTRON INTENSITY, N

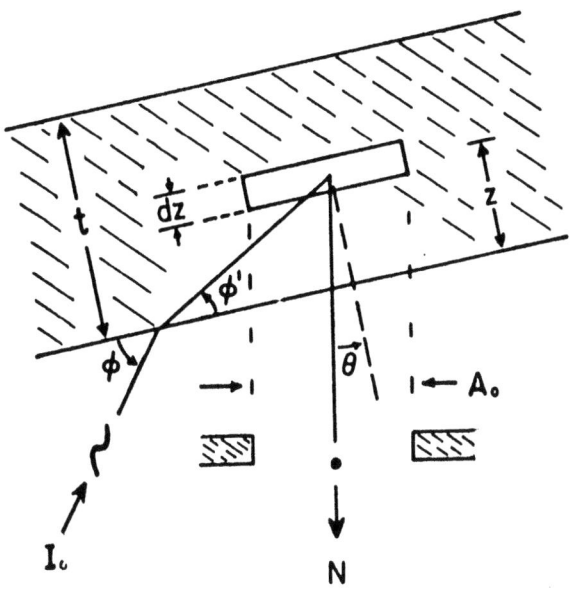

Figure 9

for the creation of q-type photoelectrons is τ_q. The number per unit volume of atoms within the sample which can emit q-type photoelectrons is ρ_q. [Note: If the angular distribution of the particular q-type photoelectron being measured is essentially isotropic, τ_q is equal to the partial photoionization cross section for either a single or multiple ionization process, $\tau_{n,\ell}$ (e.g., τ_{1s}, τ_{2s}, $\tau_{2p1/2}$, etc.)]. For a nonisotropic distribution, τ_q is equal to the differential cross section multiplied by 4π, i.e., $4\pi(d\tau_{n,\ell}/d\omega)\psi$, where $d\omega$ is the differential solid angle and ψ is the angle between the incident, unpolarized x-rays and emitted, no-loss photoelectrons being measured.

For the photoelectron energies of interest here, viz., in the 100-1000 eV region, elastic scattering and back reflection or refraction of the photoelectrons are considered to be negligible. (Such effects are considered here to be of second-order and to be included later when more precise experimental data are available.) The fraction of the photoelectrons that originate at depth z and pass through the surface can be expressed as $\exp(-\varepsilon z/\cos\theta)$. ε is a linear electron attenuation coefficient, characteristic of the electron energy and of the sample and is related to the mean free path and to the electron ionization cross section. Finally, the small, effective solid angle around the emission angle θ, subtended at the sample by the effective entrance slit annulus of the spectrograph, is defined here as ω.

The contribution, dN, to the no-loss photoelectron signal may now be written as

$$dN = S(1-R)\rho_q \tau_q \frac{\sin\phi}{\sin\phi' \cos\theta} \exp(-hz) dz$$

where $\quad S = I_0 A_0 \left(\frac{\omega}{4\pi}\right)$ (an instrumental constant)

and $\quad h = \left(\frac{\mu}{\sin\phi'} + \frac{\varepsilon}{\cos\theta}\right)$

R is the Fresnel coefficient for x-ray reflection, i.e., the fraction of the I_0 unpolarized intensity that does not enter the sample due to reflection. R differs from zero value only in the small angle region.

Integrating for a thick sample yields

$$N = S(1-R) \left(\frac{\rho_q \tau_q}{\varepsilon}\right) \frac{\sin\phi}{\sin\phi' + \frac{\mu \cos\theta}{\varepsilon}}$$

For the large angle ϕ region, and for $\theta = 0$ (the usual geometry in photoelectron spectroscopy), this becomes

$$N = S \left(\frac{\rho_q \tau_q}{\varepsilon}\right)$$

because, for large ϕ, $\sin\phi \simeq \sin\phi' \gg \mu/\varepsilon$. In this case, the x-rays penetrate much farther than the effective escape depths of the photoelectrons; consequently, x-ray absorption effects become negligible.

This prediction that the photoelectron signal, N, is independent of sample angle ϕ to the x-ray beam (except for the grazing incidence region) has been experimentally verified (17). And because of this independence of the photoelectron signal with sample surface "tilt", N is also relatively insensitive to surface structure.

The linear electron attenuation coefficient, ε, may be defined by

$$\varepsilon = \Sigma \rho_i \sigma_i = (1/\Lambda)$$

where ρ_i is the number of atoms of type i per unit volume, σ_i is their electron ionization cross section, and Λ is the mean free path for the q-type photoelectrons being measured.

We may write a relation for the photoelectron "no-loss" signal, N_m, in terms of a total electron ionization cross section, σ, characterizing a molecular unit of the sample

$$N_m = S z_q \frac{\tau_q}{\sigma},$$

in which z_q is the number of atoms within the molecular unit which can emit the q-type photoelectrons being measured. σ is the total electron ionization cross section.

As mentioned earlier, the photoionization cross section, τ_q, as used in these expressions is defined in terms of the differential cross section by

$$\tau_q = 4\pi \left(\frac{d\tau_{n,\ell}}{d\omega}\right)_\psi.$$

It has been suggested (7) that for the low energy photoelectron emission, describable by electric dipole processes, the energy dependence of the photoelectron angular distribution may be included very simply through a parameter, β, by the relation (for unpolarized incident x-rays),

$$\left(\frac{d\tau_{n,\ell}}{d\omega}\right)_\psi = \left(\frac{\tau_{n,\ell}}{4\pi}\right)[(1-\beta/2) + \frac{3}{4}\sin^2\psi].$$

For s-type photoelectrons, $\beta=2$, and the angular distribution is independent of electron energy and is given by

$$\left(\frac{d\tau_{n,s}}{d\omega}\right) = \left(\frac{3}{8\pi}\right)\tau_{n,s} \sin^2\psi.$$

And according to recent theoretical calculations (18), for the p-type electrons and for the 100-1000 eV region, a typical value of β lies between 1.0 and 1.5.

In Fig. 10 the low energy photoelectron angular distributions are plotted for the s-type and typical p-type photoelectrons. As has been noted by Samson (19) and indicated here, if the photoelectrons are measured at 54.7° (or 125.3°) the differential cross section that would result is simply proportional to the total cross section and thus requires no energy correction or knowledge of the value for β.

In the model analysis for the photoelectron intensity, N, as measured in the "no-loss" peak, the product of the effective sample area, A_0, and the solid angle fraction, $\omega/4\pi$, is determined by the

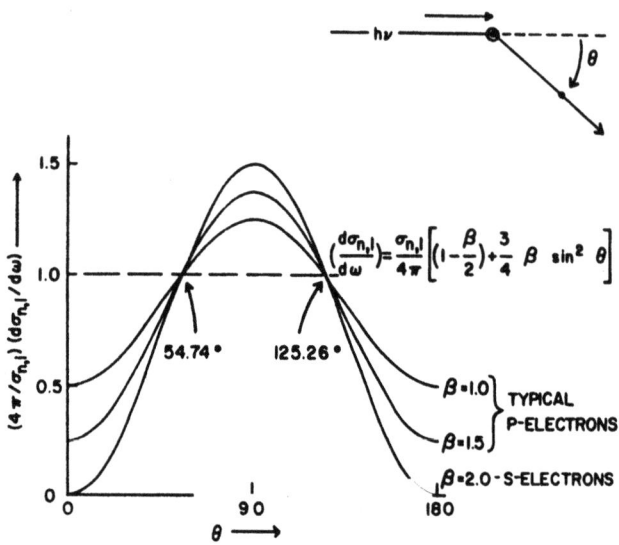

Figure 10

entrance slit geometry of the spectrograph. For the double-focussing, hemispherical analyzer that is used in this work, it is easy to show (20) that this product is equal to $KR^2(\Delta E/E)^2$, where K is an instrumental constant, the size parameter, R, is chosen here to be the mean radius of the hemispheres, and $\Delta E/E$ is the relative energy resolution and is also fixed by the slit geometry.

Often, rather than varying the hemisphere potential difference, the acceptance energy E of the spectrograph is fixed (thus fixing the absolute energy resolution, ΔE) and the energy spectrum is scanned by imposing a retarding or accelerating field between the sample and the entrance slit. In the presence of this field, the product of the effective sample area, A_0, and solid angle fraction, $\omega/4\pi$, must change by the factor E/E_0, where E is the electron spectrograph acceptance energy (electron energy at the entrance slit) and E_0 is the electron energy at the sample. This correction may be established by an application of the Lagrange-Helmholtz brightness law (21).

Therefore, we may finally estimate the measured photoelectron intensity, N, in the no-loss peak as follows:

$$N = \{KI_0 R^2 (\Delta E)^2/E\} \frac{z_q \tau_q}{E_0 \sigma(E_0)} \text{ (for large } \phi\text{)}$$

where τ_q is equal to 4π times the differential photoionization cross section that is appropriate for the angular region in which the photoelectrons are being measured (as noted above) and the factor in brackets becomes an instrumental constant for the sample-potential-scan mode of energy analysis.

It is important to emphasize here that the partial photoionization cross section, τ_q, is that for the single electron process for which the ejected photoelectron leaves with all of the available energy-viz.-that of the photon minus the binding energy of the q-type electron. There is a significant probability, however, that other processes may be involved which may share some of this available energy and produce satellite structure on the low energy side of the normal photoelectron line. These multiple excitation and ionization processes, generally defined as shake-up and shake-off processes, are depicted schematically in Fig. 11 for a photoelectron spectrum as, for example, from a rare gas sample. The relative probabilities for a particular shake-up event can be measured from the area underneath the satellite line as compared to that under the main peak and have been shown to be of the order of twenty percent (13, 14, 15).

The shake-up and shake-off processes, as an intrinsic possible part of the photoionization event, are excited by the sudden change in the central potential associated with the ejection of the q-electron and the subsequent loss of its shielding. In the "sudden approximation" calculation for the photoionization cross sections this simultaneity is assumed in that the velocity of the ejected electron is considered to be large as compared to orbital electron velocities. It has been experimentally determined by photoelectron spectroscopy (see, for example, 13, 14, 15) that the ratio of the multiple excitation cross sections to that for the single electron photoionization cross section is a constant, independent of the photon energy, when the "sudden approximation" assumption obtains and that the relative multiple excitation cross sections approach zero as the photon energy approaches threshold energy for the q-excitation ionization.

The extensive calculations of photoionization cross sections which are currently available (for which the sudden approximation model is generally used) contain implicitly, it is believed here (16), multiple excitation ionization effects--except for the region in which this model is not applicable, viz. near the absorption edges. As discussed in Sect. II below, we would therefore expect that present theory should yield total photoionization cross sections which describe those determined by photon absorption measurements. It follows, however, that the present theoretically calculated tables for the partial photoionization cross sections need to be modified by subtracting out the contributions due to the multiple excitation effects in order to yield the τ_q values as needed to quantitatively predict the normal photoelectron line intensities as discussed above.

There is a considerable need at this time for more detailed theory which does permit the explicit evaluation of the multiple electron contributions to the photoionization cross sections and thus yield the precise determination and tabulation of the probability for the single photoelectron ejection process, τ_q.

To date, there are very few direct measurements of the partial photoionization cross sections, and those for the total electron ionization within solids are nearly non-existent for the low energy region. Experiment and theory are much more difficult for the description of the electron ionization (hence for low energy electron transport) than for the photoionization cross sections.

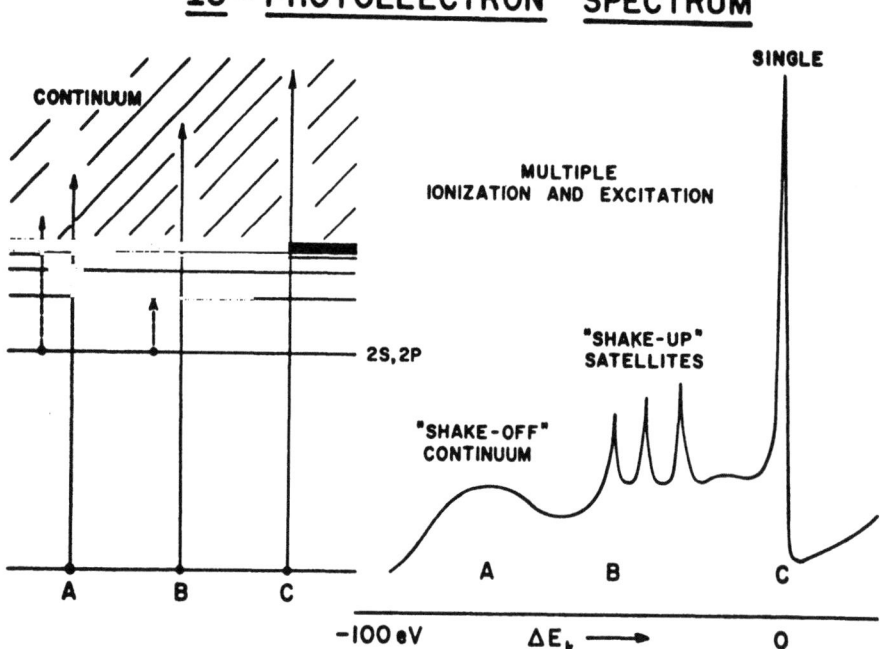

Figure 11

Available experimental cross sections are very imprecise because of the extreme difficulties involved in defining the required sample systems. Fortunately the product $E\sigma(E)$ is a slowly varying function with energy for $E>100$ eV as is shown in the logarithmic plot of Fig. 12 for the available data for gold (17, 22).

Unlike the case for photoionization cross sections, the free-atom theory (as the Bohr-Bethe theory) for electron ionization does not apply for solids in the low energy region, (for example at one kilovolt). Much closer are the predictions of the plasmon theory for energy loss as developed by Pines and Bohm (23), by Ferrell (24) and by other (25, 26). A typical result of this theory (that of Ferrell) may be written as

$$E\sigma(E) = \frac{E_p}{2a_0 N} \ln\left[\frac{4E}{E_p}\right]$$

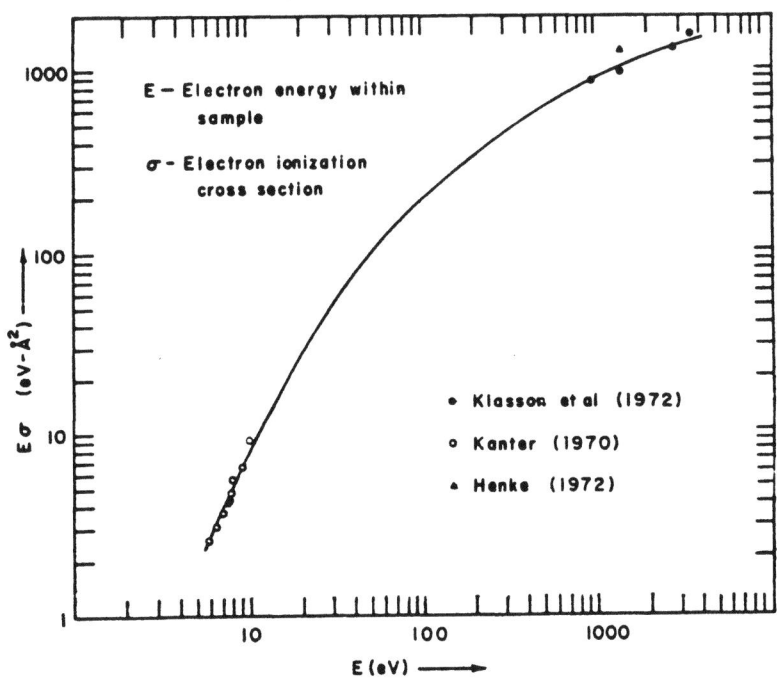

Figure 12

where E_p is the characteristic plasmon energy loss value, a_0 is the Bohr radius and N is the number of atoms per unit volume of sample. Very similar expressions have been proposed for nonmetals, which include single electron as well as collective electron losses (27).

As a sample test of the application of the photoelectron intensity model described above, we have compared the experimental intensity ratios for the 1-s electrons from carbon to that of oxygen in Mylar and of that for carbon to that of fluorine in Teflon to those predicted by the model equation. Here we have assumed that the $E\sigma(E)$ values are constant for the given sample because the energy differences are not large. And we have used subshell, 1-s photoionization cross sections as determined by the author (11) by graphical subtraction of experimental total photoionization cross section data. (These cross sections for C, O and F 1-s levels were later calculated by Scofield (28) to be within one percent of those experimental values). As listed in Fig. 13, the agreement between

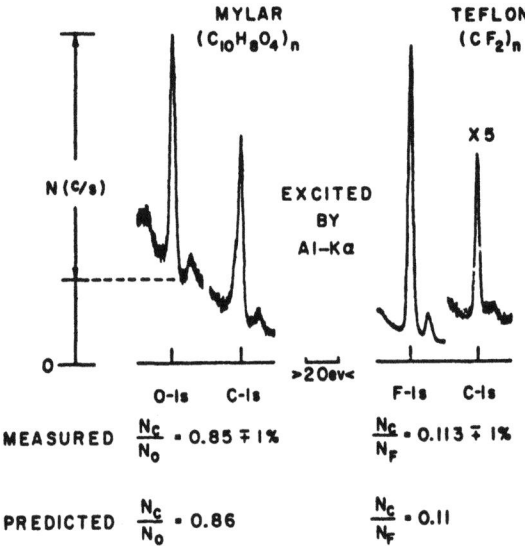

Figure 13

the predicted and experimental ratios is very good. In this preliminary work, we have not been able to obtain such good agreement with the p, d, and f state electrons and heavier element samples. It is planned that when the partial photoionization cross sections are known with sufficient accuracy, such measurements may then be made over large energy ranges in order to investigate the energy dependence of the $E\sigma(E)$ function experimentally.

Another method for measuring the total electron ionization cross section has been recently described by this author (17). It is based upon the measurement of the sharp enhancement of the photoelectron signal from a given atomic level when the exciting photons impinge upon the sample at a small grazing angle, near that for the critical angle for total reflection of the x-ray beam. In this critical angle region, the x-ray beam is refracted along the sample surface and is thus concentrated within the effective escape depth for the ejected photoelectrons.

The electron spectrograph that has been developed for this research program is shown in Fig. 14. It is a second generation instrument based upon one built by the author in 1960 (29, 30), utilizing the double focussing property of hemispherical analyzer plates at the 180° position. In order to achieve relatively high photoelectron signal levels (which scale as the square of the instrument size as indicated earlier) large (20" mean diameter) hemispheres were chosen for the second instrument. These were constructed at the Los Alamos Scientific Laboratories in 1963. A high power, ultrasoft x-ray source is closely coupled to a sample chamber illuminating one of eight sample on a rotating holder at a grazing angle around 20°. The electrons under measurement leave the sample normally. An ion bombardment cleaning station and sample heating is provided in the sample chamber. The slits are usually set for about 0.3% relative energy resolution, yielding 0.3 eV resolution in the sample-potential-scan mode with the analyzer set to accept 100 eV electrons. The system is evacuated by synthetic zeolite traps and titanium sublimation and sputter-ion pumps to about 10^{-7} torr. In order to indicate spectrographic resolution, speed and freedom from carbon contamination, a recording of the Au-$4f_{5/2}$ - $4f_{7/2}$ doublet is shown in Fig. 14 as excited by aluminum foil filtered Mg-Kα (1254 eV) photons. The scan time was 2.5 minutes for an x-ray source setting of 200 ma and 6 kv.

II. ESTABLISHING A "STATE OF THE ART" TABLE OF ULTRASOFT X-RAY ABSORPTION COEFFICIENTS

It has recently become possible to extend this laboratory's ultrasoft x-ray absorption coefficient tables of the light elements to the medium and heavy element region. This is because of the

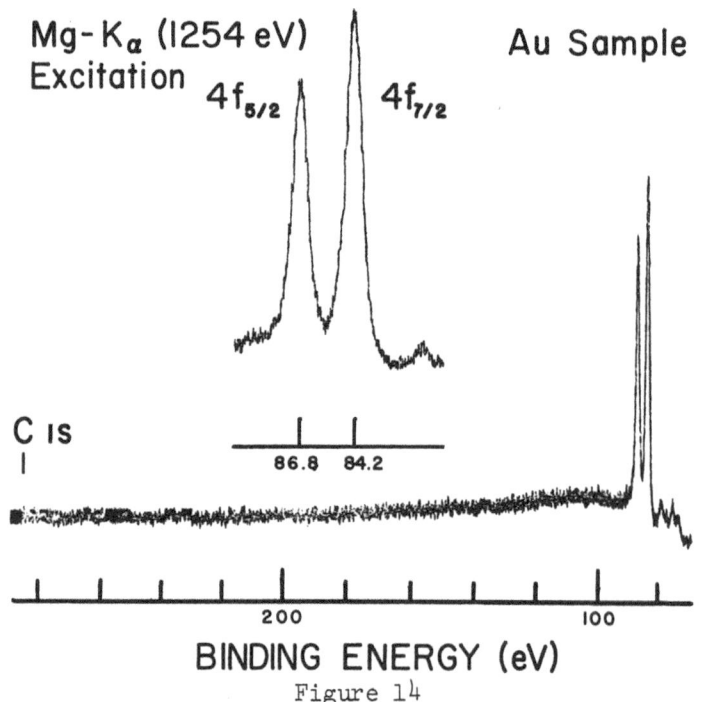

Figure 14

availability of new experimental data, and more important, of the complete theoretical calculations for all elements and for the low energy region as presented by Henry, et al (8) and Veigele, et al (9). By combining the information present in the available data and theory, a tentative, or "state of the art" set of tables have been developed and are presented here.

Because most applications of such absorption tables in ultra-soft x-ray spectroscopy are for certain characteristic wavelengths, and because the experimental measurements are usually most precise for line radiation courses, these tables are presented for twenty six of the most commonly applied characteristic wavelengths in the 8-110 A region. The particular atomic sources of these characteristic wavelengths and the particular wavelength values assigned to each for this work are presented in Table 1.

TABLE I

Line Source	Wavelength (Angstroms)	Line Source	Wavelength (Angstroms)
Al-$K\alpha_{1,2}$	8.34	Ti-$L\alpha_{1,2}$	27.4
Mg-$K\alpha_{1,2}$	9.89	Ti-$L\ell$	31.4
Na-$K\alpha_{1,2}$	11.9	N -$K\alpha$	31.6
Zn-$L\alpha_{1,2}$	12.3	C -$K\alpha$	44.7
Cu-$L\alpha_{1,2}$	13.3	W -$N_{V,VII}$	58.4
Ni-$L\alpha_{1,2}$	14.6	Mo-$M\zeta$	64.4
Co-$L\alpha_{1,2}$	16.0	B -$K\alpha$	67.6
Fe-$L\alpha_{1,2}$	17.6	Nb-$M\zeta$	72.2
F -$K\alpha$	18.3	Zr-$M\zeta$	82.1
Mn-$L\alpha_{1,2}$	19.4	S -$L\ell$	83.4
Cr-$L\alpha_{1,2}$	21.6	Y -$M\zeta$	93.4
Mn-$L\ell$	22.3	Sr-$M\zeta_1$	109.
O -$K\alpha$	23.6	Be-$K\alpha$	114.

In Appendix I is presented a nearly complete set of experimental absorption coefficients as measured with characteristic line source radiations. The corresponding literature references are cited along with others for data that has been reported only as graphs or as measured with continuous radiation sources (such as synchrotron radiation sources). This bibliography has been based upon one developed by Hubbell in 1971 (10) and upon an extensive literature search conducted at this laboratory. The author would like to apologize for any possible omissions in this listing and would appreciate being informed of such reported ultrasoft x-ray absorption coefficients that have not been included here.

Approximately 250 of the nearly 400 measurements listed are for the light elements through Ar-18. These light element data yield a net deviation, $\bar{d}$, of only +0.7% from the values as interpolated from the semi-empirical tables of Henke, et al (1970) (3). Therefore it was considered unnecessary to attempt to improve the previous fits at this time, and the light element tables presented here in Appendix II are based upon logarithmic interpolations from the author's previous tables (3) for the light elements He-2 through Ar-18. (For the Z-ranges 2-10 and 11-18, the net deviations from Veigele (9) are -1% and 6% respectively. The corresponding deviations from Henke (3) are -0.5% and 2%.)

For the medium and heavy elements it was decided to place the greatest weight upon interpolations from the calculated results of Veigele, et al (8.9) because the spread in the present experimental data for the medium and heavy elements is considerable, as has been indicated in Fig. 2. Nevertheless, it was considered significant to raise all of the theoretical calculated values of Veigele, et al, by 2% throughout the medium and heavy element region in order to minimize the net deviation, $\bar{d}$, of the experimental values from the averaging table values. It is important to note here that if the present single electron theoretical models that have been used to calculate low energy cross sections do not include to some extent the multiple excitation and ionization effects, the calculated cross section values would need to be adjusted upward by appreciably more than 2% in order to match total photoionization cross section measurements. This increase of the calculated values of 2% resulted in net deviation value of -0.4% for the medium and heavy elements. (In calculating this net deviation value, experimental data points were rejected which were more than 100% off the averaging curves.) The net deviation of the experimental data from the table values presented here for the entire range of atomic numbers is +0.31%. The net deviations, $\bar{d}$, along with the associated standard deviations, σ, for different ranges in atomic number are summarized in Table II.

Also included in the mass absorption coefficient tables of

TABLE II

ANALYSIS OF EXPERIMENTAL-MINUS TABLE VALUES (d)
OF APPENDIX I

Range in Z	No. of Exp. Values, n, (for d<100%)	$\bar{d} = \dfrac{\Sigma d}{n}$	$\sigma = \sqrt{\dfrac{\Sigma d^2}{n}}$
2-18	253	0.66	± 7.7
19-36	41	-5.9	±18.
37-54	52	0.75	±28.
55-94	32	4.8	±36.
19-94	125	-0.41	±28.
2-94	378	0.31	±17.

Appendix II are those calculated for certain compound materials which have been found to be particularly useful in this laboratory as thin film substrates or windows and as proportional counter gases for ultrasoft x-ray analysis. These are the light element materials Formvar, Parlodion, Polypropylene, Mylar, Kimfoil, Aluminum Oxide, Quartz, P-10, Methane and Propane.

Finally, presented in Appendix II, are the self-absorption coefficients for the $K_{\alpha 1,2}$ radiations of Be-4 through Si-14 and for the $L_{\alpha 12}$ radiations of Ti-22 through Ge-32. Also listed with these values are the corresponding mass absorption coefficients extrapolated to the high and low side of the K and the L_I absorption edges and their ratios, r_{KL} and r_{LM}. (In the μ_{LI}, r_{LM} values, the extrapolation of the absorption coefficient curve for which only the M-shells and higher orbitals are excited is taken to the L_I edge.)

III. THE STATE OF THE ART

In the development of absorption coefficient tables as based upon principally experimental data (as for the light element tables presented here) it is generally assumed that chemical and solid state effects can be neglected. And similarly this is assumed in the application of such tables to the prediction of compound absorption coefficients as simple sums of atomic cross sections.

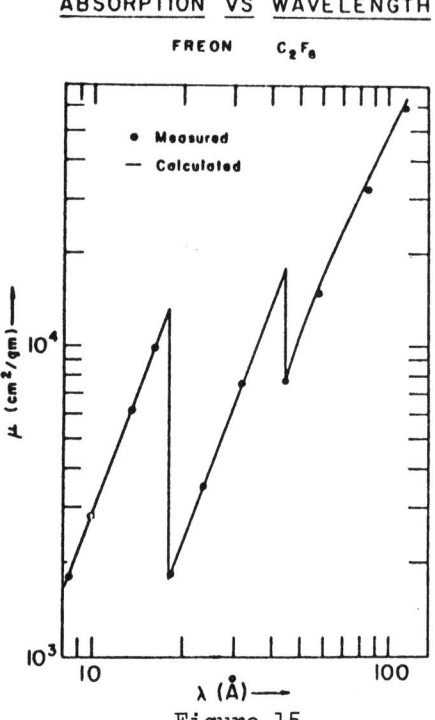

Figure 15

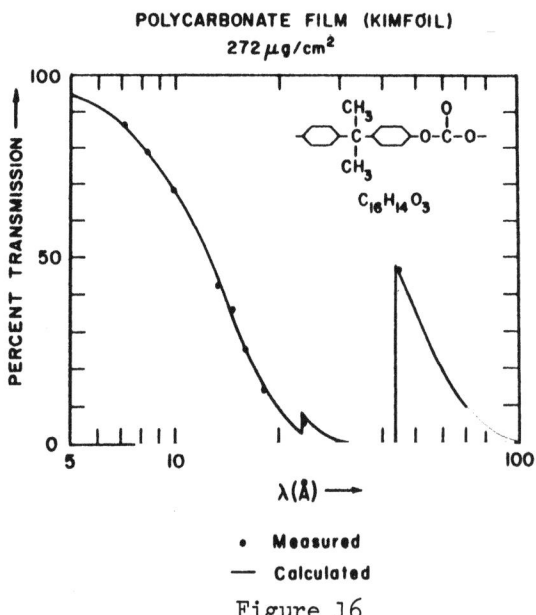

Figure 16

This assumption seems reasonably good for the light element region as is indicated in Fig. 15 in which the experimentally measured mass absorption coefficients for the molecular system Freon (C_2F_6) are compared to those calculated from the tables presented in Appendix II. The relatively good agreement with the calculated absorption for the solid film system, Kimfoil ($C_{16}H_{14}O_3$) is indicated in Fig. 16 in which the experimental and calculated values for the film transmission vs. wavelength are compared.

As discussed above, the medium and heavy element experimental data are not sufficient at this time to establish cross section tables without a strong reliance upon available theory. This situation is indicated in the listing of experimental data of Appendix I, in the standard deviation values presented in Table II, and in Figs. 17 and 18 in which the experimental absorption coefficients by two independent methods are compared to the theoretical values from Veigele, et al (9) for gold-79 and for lead-82. For gold, the transmission measurements of Lukirskii (31) and of Jaegle (32), and the total reflection measurements of Ershov (33) are presented. For lead, the transmission measurements of Lukirskii (34) and those presented earlier in this paper as deduced from lead stearate multilayer absorption are presented. Such disagreement can be attributed to experimental difficulties, to chemical and crystalline effects in condensed materials, and to the inadequacy of present theory. There is a considerable need expressed here for more and improved experimental measurements in the medium and heavy element region. And there is a considerable need for more detailed theoretical calculations which clearly identify and predict quantitatively the role of multiple excitation and ionization effects and, ultimately, the role of chemical and solid state effects upon photoionization in the low energy region.

Finally, the present "state of the art" seems to be approaching a point at which precise and detailed descriptions of the <u>partial</u> photoionization cross sections will be obtained both experimentally and theoretically and thus provide an important new dimension for the extension of quantitative x-ray and electron analysis. As described above, photoelectron spectroscopy may well be an effective method for the determination of multiple and single electron photo excitation and ionization cross sections, of electron ionization cross sections, and of photoelectron angular distributions--all of which may be of critical value in the development of more precise theoretical models and for the prediction of solid state effects in the low energy x-ray and electron interactions.

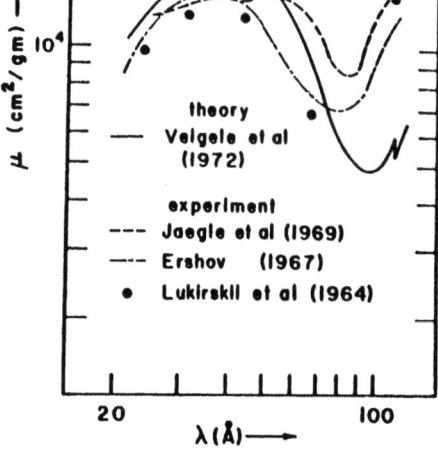

Figure 17

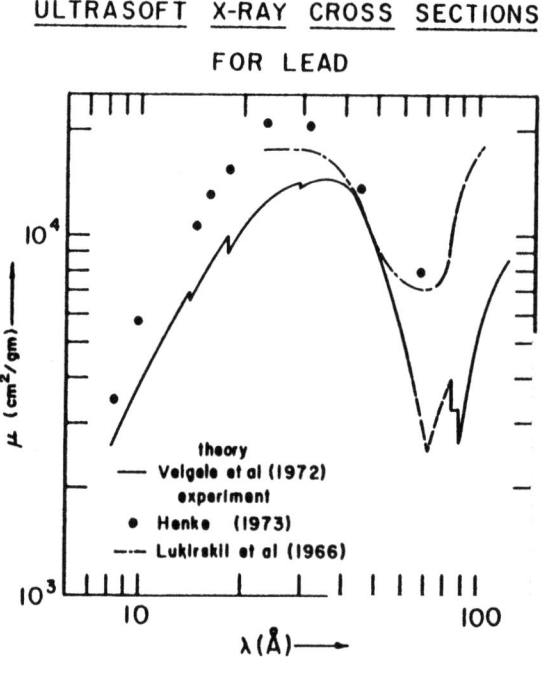

Figure 18

ACKNOWLEDGEMENTS

We gratefully acknowledge the important assistance in this work of Susan D. Holmes and Ronald H. Ono. This research program is supported under Grant No. AFOSR 72-1274 by the Air Force Office of Scientific Research.

REFERENCES

1. B. L. Henke, R. White, and B. Lundberg, J. Appl. Phys. 28, 98 (1957)
2. B. L. Henke, Norelco Reporter, Vol. XIV, No. 3-4 (1967)
3. B. L. Henke and R. L. Elgin, in Henke, Newkirk and Mallett, Editors, Advances in X-Ray Analysis, Vol. 13, p. 639-665 Plenum Press (1970)
4. U. Fano and J. W. Cooper, Rev. Mod. Phys. 40, 441 (1968)
5. E. Storm and H. I. Israel, Nucl. Data Tables 1, 565 (1970)
6. E. J. McGuire, Phys. Rev. 175, 20 (1968)
7. D. J. Kennedy and S. T. Manson, Phys. Rev. A5 227 (1972)
8. E. M. Henry, C. L. Bates, and W. J. Veigele, Phys. Rev. A6, 2131 (1972)
9. W. J. Veigele, Atomic Data Tables 5, 51 (1973)
10. J. H. Hubbell, Atomic Data 3, 241 (1971), also see Natl. Bur. Std. Report No. NSRDS-NBS 29 (1969)
11. B. L. Henke, in R. W. Fink, et al, Editors, Inner Shell Ionization Phenomena and Future Applications, (Technical Information Division of the U. S. Atomic Energy Commission, Oak Ridge, Tennessee, 1973)
12. F. Wuilleumier, Phys. Rev. A6, 2067 (1972)
13. M. O. Krause, Phys. Rev. 177, 151 (1969)
14. T. A. Carlson, W. E. Moddeman, and M. O. Krause, Phys. Rev. A1, 1406 (1970)
15. F. Wuilleumier and M. O. Krause, in D. A. Shirley, Editor, Electron Spectroscopy, p. 259, American Elsevier Publishing Co., Inc. (1972)
16. Charles Fadley - Private Communication (submitted to Chemical Physics Letters)
17. B. L. Henke, Phys. Rev. A6, 94 (1972)
18. H. Brysk and C. D. Zerby, Phys. Rev. 171, 292 (1968)
19. J. A. R. Samson, Phil. Trans. Roy. Soc. Lond. A268, 141 (1970)
20. B. L. Henke, in Barrett, Newkirk and Mallett, Editors, Advances in X-Ray Analysis, Vol. 12, p. 480, Plenum Press (1969)
21. J. C. Helmer, Am. J. Phys. 34, 222 (1966)
22. M. Klasson, et al, Physica Scripta 5, 93 (1972)
23. D. Pines, Rev. Mod. Phys. 28, 184 (1956)
24. R. A. Ferrell, Phys. Rev. 101, 554 (1956)
25. J. J. Quinn, Phys. Rev. 126, 1453 (1962)

26. A. W. Blackstock, R. H. Ritchie, and R. D. Birkhoff, Phys. Rev. 100, 1078 (1955)
27. A. van der Ziel, Phys. Rev. 92, 35 (1953)
28. J. H. Scofield, Technical Report No. 51326 (1973), University of California Radiation Laboratory
29. B. L. Henke, in W. M. Mueller, Editor, Advances in X-Ray Analysis, Vol. 5, p. 285, Plenum Press (1962)
30. B. L. Henke, X-Ray Optics and X-Ray Microanalysis, p. 157 Academic Press, Inc. (1963)
31. A. P. Lukirskii, et al, Opt. and Spectrosc. 16, 168 (1964)
32. P. Jaegle, et al, Phys. Rev. 188, 30 (1969)
33. O. A. Ershov, Opt. and Spectrosc. 22, 252 (1967)
34. A. P. Lukirskii, T. M. Zimkina, and S. A. Gribovskii, Sov. Phys. Solid State 8, 1525 (1966)

APPENDIX I

Line (Å)	Exp.		Table	% Dev.	Line (Å)	Exp.		Table	% Dev.
Be (4)					C (6)	(continued)			
8.34	0.199	(CW)	0.179	11.4	67.60	5.570	(De)	6.456	-13.7
	0.206	(O)		15.3		6.550	(He)		1.5
	0.192	(O**)		7.4		6.800	(M**)		5.3
9.89	0.327	(CW)	0.301	8.8		6.592	(We)		2.1
	0.340	(O)		13.1	72.20	6.560	(De)	7.494	-12.5
	0.318	(O**)		5.8	82.10	8.440	(De)	10.050	-16.0
13.30	0.770	(O)	0.745	3.4	83.40	10.350	(He)	10.350	0.0
	0.722	(O**)		-3.1	114.00	21.200	(He)	20.810	1.9
B (5)					N (7)				
44.70	33.000	(KS)	37.020	-10.9	8.34	1.150	(B)	1.107	3.9
						1.121	(He)		1.3
C (6)						1.109	(W)		0.2
8.34	0.656	(A)	0.718	-8.7	9.89	1.825	(B)	1.789	2.0
	0.670	(B)		-6.7		1.800	(He)		0.6
	0.720	(He)		0.2		1.796	(W)		0.4
	0.725	(O)		0.9	12.30	3.430	(B)	3.205	7.0
	0.650	(Se)		-9.5	13.30	3.836	(A)	4.022	-4.6
	0.711	(W**)		-1.0		4.530	(B)		12.6
	0.806	(We)		12.2		4.040	(He)		0.4
9.89	1.063	(A)	1.174	-9.5	14.60	5.300	(B)	5.069	4.6
	1.090	(B)		-7.2	16.00	6.550	(He)	6.496	0.8
	1.167	(He)		-0.6	18.30	9.100	(He)	9.161	-0.7
	1.235	(O)		5.2	23.60	17.160	(De)	17.310	-0.9
	1.085	(Se)		-7.6		17.200	(He)		-0.6
	1.156	(W)		-1.5	31.60	1.730	(He)	1.637	5.7
11.90	2.082	(We)	1.984	4.9	44.70	3.850	(D)	3.903	-1.4
12.30	2.030	(B)	2.147	-5.4		3.830	(De)		-1.9
	1.975	(Se)		-8.0		3.940	(He)		0.9
13.30	2.170	(A)	2.714	-20.0		3.800	(K)		-2.6
	2.550	(B)		-6.0		3.790	(M)		-2.9
	2.740	(He)		1.0	64.40	9.160	(De)	9.334	-1.9
	2.580	(O)		-4.9	67.60	10.060	(De)	10.570	-4.8
	2.510	(Se)		-7.5		10.270	(He)		-2.8
	2.756	(We)		1.5		10.900	(M)		3.1
14.60	3.200	(B)	3.457	-7.4	72.20	12.130	(De)	12.230	-0.8
16.00	4.470	(He)	4.451	0.4	82.10	15.950	(De)	16.550	-3.6
17.60	5.591	(We)	5.786	-3.4	83.40	17.000	(He)	17.070	-0.4
18.30	6.400	(He)	6.366	0.5	114.00	36.500	(He)	34.810	4.9
23.60	12.000	(De)	12.380	±3.1					
	12.200	(He)		-1.5	O (8)				
	12.993	(We)		5.0	8.34	1.615	(B)	1.597	1.1
31.60	25.400	(He)	25.490	-0.4		1.604	(He)		0.4
	23.803	(We)		-6.6		1.540	(O)		-3.6
44.70	1.930	(De)	2.373	-18.7		1.585	(W)		-0.8
	2.280	(He)		-3.9	9.89	2.520	(B)	2.533	-0.5
	2.280	(O)		-3.9		2.540	(He)		0.3
	2.300	(M**)		-3.1		2.480	(O)		-2.1
	2.535	(We)		6.8		2.540	(W)		0.3
64.40	5.140	(De)	5.755	-10.7	(continued)				

APPENDIX I

Line (Å)	Exp.		Table	% Dev.	Line (Å)	Exp.		Table	% Dev.
O (8)	(continued)				Ne (10)	(continued)			
12.30	4.340	(B)	4.476	-3.0	13.30	9.770	(He)	9.190	6.3
13.30	5.456	(A)	5.601	-2.6		9.800	(B)		6.6
	5.500	(B)		-1.8		8.500	(A)		-7.5
	5.560	(He)		-0.7	14.60	0.750	(B)	0.715	4.8
	5.340	(O)		-4.7	16.00	0.897	(He)	0.918	-2.3
16.00	8.850	(He)	8.877	-0.3	18.30	1.310	(He)	1.301	0.7
17.60	10.000	(A)	11.300	-11.5	23.60	2.600	(He)	2.582	0.7
18.30	12.620	(He)	12.390	1.9	31.60	5.540	(He)	5.617	-1.4
23.60	1.440	(He)	1.200	20.0	44.70	13.630	(He)	13.570	0.4
	1.320	(De)		10.0		13.100	(D)		-3.5
31.60	2.550	(He)	2.527	0.9		13.700	(De)		1.0
44.70	6.250	(He)	6.044	3.4	64.40	34.610	(De)	31.800	8.8
	6.150	(O)		1.8	67.60	35.900	(He)	35.760	0.4
	5.650	(M)		-6.5		39.090	(De)		9.3
	5.765	(D)		-4.6	83.40	56.800	(He)	54.770	3.7
	6.000	(K)		-0.7	114.00	102.000	(He)	97.520	4.6
	5.800	(A)		-4.0					
	6.020	(De)		-0.4	Na (11)				
64.40	14.830	(De)	14.550	1.9	44.70	17.300	(KS)	18.460	-6.3
67.60	16.500	(He)	16.530	-0.2					
	16.250	(M)		-1.7	Mg (12)				
	17.010	(De)		2.9	8.34	4.260	(BU)	4.284	-0.6
72.20	19.570	(De)	19.160	2.1		4.120	(BU)		-3.8
82.10	26.310	(De)	25.980	1.3	9.89	0.510	(BU)	0.489	4.3
83.40	26.500	(He)	26.800	-1.1	11.90	0.900	(BU)	0.815	10.5
114.00	56.000	(He)	53.950	3.8	13.30	1.340	(BU)	1.110	20.7
					14.60	1.800	(BU)	1.411	27.6
F (9)					16.00	2.180	(BU)	1.818	19.9
8.34	2.030	(He)	2.037	-0.3	17.60	3.000	(BU)	2.364	26.9
	2.035	(O)		-0.1	23.60	6.900	(BU)	5.174	33.4
9.89	3.140	(He)	3.198	-1.8	44.70	31.400	(KS)	24.900	26.1
	3.140	(O)		-1.8					
13.30	6.850	(He)	6.941	-1.3	Al (13)				
16.00	11.000	(He)	10.910	0.8	8.34	0.399	(LC)	0.403	-1.1
18.30	0.860	(He)	0.868	-0.9		0.408	(HW)		1.1
23.60	1.700	(He)	1.688	0.7		0.390	(B,Wi)		-3.3
31.60	3.700	(He)	3.602	2.7		0.396	(O)		-1.8
44.70	8.780	(He)	8.730	0.6		0.396	(Bi)		-1.8
67.60	22.500	(He)	23.880	-5.8		0.459	(An)		13.8
83.40	36.600	(He)	38.060	-3.8		0.344	(J)		-14.7
114.00	67.000	(He)	73.140	-8.4		0.330	(A)		-18.2
					9.89	0.650	(He)	0.639	1.7
Ne (10)						0.643	(HW)		0.6
8.34	2.780	(He)	2.752	1.0		0.630	(B)		-1.5
	2.760	(B)		0.3		0.630	(O)		-1.5
	2.750	(W)		-0.1		0.553	(J)		-13.5
9.89	4.310	(He)	4.292	0.4		0.632	(Wi)		-1.1
	4.320	(B)		0.7		0.632	(Bi)		-1.1
	4.310	(W)		0.4	12.30	1.150	(B)	1.152	-0.2
12.30	7.700	(B)	7.463	3.2	(continued)				

APPENDIX I

Line (A)	Exp.	Table	% Dev.	Line (A)	Exp.	Table	% Dev.
Al (13)	(continued)			Cl (17)	(continued)		
12.30	1.150 (Wi)	1.152	-0.2		0.962 (W)		-6.0
13.30	1.468 (He)	1.453	1.0	9.89	1.610 (He)	1.618	-0.5
	1.440 (B)		-0.9		1.570 (W)		-3.0
	1.410 (O)		-3.0	11.90	2.500 (A)	2.667	-6.3
	1.440 (Wi)		-0.9	13.30	3.300 (He)	3.596	-8.2
14.60	1.840 (B)	1.848	-0.4	Ar (18)			
	1.830 (Wi)		-1.0	8.34	1.180 (He)	1.158	1.9
	2.290 (Hi)		23.9		1.150 (B)		-0.7
	1.740 (Ba)		-5.8		1.157 (W)		-0.1
16.00	2.330 (Ba)	2.379	-2.1	9.89	1.850 (He)	1.839	0.6
17.60	3.520 (Hi)	3.081	14.2		1.770 (B)		-3.8
	3.070 (Ba)		-0.4		1.865 (W)		1.4
19.40	4.050 (Ba)	4.040	0.2	12.30	3.200 (B)	3.247	-1.4
21.60	5.430 (Ba)	5.368	1.2	13.30	4.070 (He)	4.049	0.5
23.60	7.700 (F)	6.715	14.7		4.040 (B)		-0.2
	7.900 (LS)		17.6	14.60	4.600 (B)	5.070	-9.3
	7.330 (Hi)		9.2	16.00	6.390 (He)	6.453	-1.0
	6.920 (Ba)		3.1	18.30	8.940 (He)	8.924	0.2
31.60	16.000 (F)	13.830	15.7	23.60	15.900 (He)	16.160	-1.6
	15.700 (LS)		13.5		15.800 (De)		-2.2
44.70	24.400 (KS)	30.160	-19.1	31.60	30.200 (He)	29.510	2.3
	35.000 (F)		16.0	44.70	45.600 (He)	45.580	0.0
	32.000 (LS)		6.1		45.700 (D)		0.3
64.40	68.900 (F)	59.790	15.2		46.700 (De)		2.5
67.60	70.000 (LS)	65.170	7.4	64.40	8.200 (De)	8.630	-5.0
72.20	80.800 (F)	71.770	12.6	67.60	9.170 (He)	9.369	-2.1
82.10	97.400 (F)	87.170	11.7		8.900 (De)		-5.0
93.40	111.000 (F)	101.900	8.9	72.20	10.000 (De)	10.310	-3.0
109.00	128.000 (F)	102.800	24.5	82.10	11.800 (De)	12.610	-6.4
114.00	120.000 (LS)	107.000	12.1	83.40	12.700 (He)	12.850	-1.2
Si (14)				114.00	19.500 (He)	18.430	5.8
44.70	36.900 (KS)	36.980	-0.2	Ca (20)			
S (16)				44.70	6.800 (KS)	6.838	-0.6
8.34	0.868 (He)	0.870	-0.2	Sc (21)			
	0.793 (W**)		-8.8	8.34	2.000 (CR)	1.915	4.4
9.89	1.390 (He)	1.377	0.9	Ti (22)			
	1.387 (W)		0.7	8.34	2.280 (HW)	2.137	6.7
11.90	2.100 (A)	2.277	-7.8		2.236 (CR)		4.6
13.30	3.090 (He)	3.079	0.4		2.240 (Mc)		4.8
16.00	4.940 (He)	4.950	-0.2	9.89	3.090 (Mc)	3.300	-6.4
18.30	6.990 (He)	7.006	-0.2	44.70	6.900 (KS)	8.094	-14.8
23.60	13.000 (He)	13.010	-0.1	V (23)			
31.60	25.400 (He)	24.930	1.9	8.34	2.388 (CR)	2.379	0.4
44.70	47.500 (He)	47.940	-0.9	44.70	7.500 (KS)	8.840	-15.2
67.60	69.000 (He)	74.180	-7.0				
Cl (17)							
8.34	1.010 (He)	1.023	-1.3				

APPENDIX I

Line (Å)	Exp.		Table	% Dev.	Line (Å)	Exp.		Table	% Dev.
Cr (24)					Sr (38)				
44.70	7.700	(KS)	10.590	-27.3	44.70	49.000	(KS)	35.310	38.8
Fe (26)					Zr (40)				
8.34	3.430	(CR)	3.438	-0.2	8.34	1.749	(HW)	1.671	4.7
44.70	10.800	(KS)	13.300	-18.8	44.70	20.600	(KS)	31.130	-33.8
Co (27)					Nb (41)				
8.34	3.992	(CR)	3.738	6.8	44.70	15.800	(KS)	*33.990	-53.5
44.70	16.700	(KS)	14.730	13.4	Mo (42)				
Ni (28)					44.70	12.500	(KS)	*32.420	-61.4
8.34	4.557	(CR)	4.287	6.3	Ag (47)				
	3.140	(J)		-26.8	8.34	3.070	(An)	2.887	6.3
44.70	49.800	(KS)	17.270	188.4		3.145	(Bi)		8.9
Cu (29)						1.770	(J)		-38.7
8.34	4.680	(HW)	4.554	2.8	9.89	4.711	(Bi)	4.254	10.7
	3.450	(J)		-24.2		2.610	(J)		-38.6
16.00	2.940	(C)	2.537	15.9	16.00	11.500	(C)	12.100	-5.0
Zn (30)					23.60	18.857	(LS)	19.260	-2.1
8.34	3.971	(CR)	5.019	-20.9	31.60	5.676	(LS)	*3.900	45.5
44.70	68.400	(KS)	21.710	215.1	44.70	13.300	(KS)	5.507	141.5
Se (34)						6.333	(LS)		15.0
8.34	4.858	(Bi)	*5.618	-13.5	67.60	6.410	(LS)	6.462	-0.8
9.89	1.774	(Bi)	1.358	30.6	114.00	13.143	(LS)	5.880	123.5
Kr (36)					Cd (48)				
8.34	1.090	(He)	1.132	-3.7	8.34	3.576	(Bi)	2.974	20.2
9.89	1.740	(He)	1.709	0.1	9.89	5.352	(Bi)	4.421	21.1
13.30	3.550	(He)	3.452	2.8	Sn (50)				
16.00	5.450	(He)	5.191	5.0	8.34	4.329	(Bi)	3.331	30.0
18.30	7.270	(He)	6.960	4.5	9.89	6.261	(Bi)	4.926	27.1
23.60	12.300	(He)	11.890	3.4	Te (52)				
	9.510	(L)		-20.0	8.34	4.964	(Bi)	3.635	36.6
31.60	21.500	(He)	20.690	3.9	9.89	7.303	(Bi)	5.339	36.8
	14.385	(L)		-30.5	Xe (54)				
44.70	31.400	(He)	33.280	-5.6	8.34	4.500	(He)	4.201	7.1
	23.065	(L)		-30.7		3.839	(LG)		-8.6
	31.800	(D)		-4.4	9.89	6.200	(He)	6.178	0.4
64.40	26.172	(L)	40.490	-35.4		5.368	(LG)		-13.1
67.60	35.800	(He)	40.200	-10.9	13.30	8.800	(He)	8.921	-1.4
72.20	24.993	(L)	39.870	-37.3		8.918	(LG)		0.0
82.10	23.868	(L)	39.180	-39.1	14.60	9.546	(LG)	10.790	-11.5
83.40	32.400	(He)	38.160	-15.1	16.00	10.900	(He)	13.370	-18.5
109.00	17.894	(L)	13.760	30.0		11.347	(LG)		-15.1
114.00	13.000	(He)	11.600	12.1	17.60	20.893	(LG)	22.760	-8.2
					(continued)				

APPENDIX I

Line (Å)	Exp.		Table	% Dev.	Line (Å)	Exp.		Table	% Dev.
Xe (54)	(continued)				Ta (73)				
18.30	3.580	(He)	*2.848	25.7	44.70	8.500	(KS)	18.390	-53.8
	3.584	(LG)		25.8	W (74)				
19.40	2.956	(LG)	*3.152	-6.2	8.34	2.016	(LC)	1.776	13.5
21.60	4.433	(LG)	3.704	19.7	44.70	10.000	(KS)	18.750	-46.7
22.30	2.956	(LG)	3.868	-23.6	Pt (78)				
23.60	4.250	(He)	4.200	1.2	8.34	1.530	(J)	2.174	-29.6
	4.077	(LG)		-2.9	9.89	2.440	(J)	3.078	-20.7
27.40	5.368	(LG)	5.088	5.5	Au (79)				
31.40	6.030	(LG)	5.954	1.3	8.34	2.221	(LC)	2.305	-3.6
	5.130	(LZ)		-13.8		2.450	(An)		6.3
31.60	6.200	(He)	5.990	3.5	23.60	9.679	(LS)	11.760	-17.7
44.70	7.130	(He)	7.403	-3.7	31.60	12.008	(LS)	15.440	-22.2
	6.455	(LG)		-12.8	44.70	11.646	(LS)	15.210	-23.4
	6.319	(LZ)		-14.6		12.500	(D)		-17.8
	6.740	(D)		-9.0	67.60	6.729	(LS)	8.952	-24.8
58.40	5.130	(LZ)	7.437	-31.0	114.00	13.458	(LS)	5.812	131.6
64.40	3.975	(LZ)	6.144	-35.3	Pb (82)				
67.60	4.000	(He)	5.895	-32.1	8.34	3.499	(H+)	2.609	34.1
72.20	4.569	(LZ)	5.618	-18.7	9.89	5.723	(H+)	3.648	56.9
82.10	8.510	(LZ)	5.213	63.2	14.60	10.680	(H+)	7.122	50.0
83.40	10.200	(He)	5.157	97.8	16.00	12.900	(H+)	8.292	55.6
109.00	122.983	(LZ)	9.663	1170.0	18.30	15.390	(H+)	9.039	70.3
114.00	87.000	(He)	12.130	617.2	23.60	20.150	(H+)	12.470	61.6
Ba (56)					31.60	20.000	(H+)	14.020	42.7
44.70	6.300	(KS)	7.772	-18.9	44.70	13.590	(H+)	11.350	19.7
Nd (60)					67.60	7.866	(H+)	4.265	84.4
8.34	5.670	(OE)	5.467	3.7					
Sm (62)									
8.34	5.980	(OE)	5.897	1.4					
9.89	7.140	(OE)	7.031	1.6					
Gd (64)					** Value corrected for impurities				
8.34	5.250	(OE)	5.006	4.9	* Value near absorption edge				
9.89	6.860	(OE)	*7.480	-8.3					
Ho (67)					H+ This paper				
8.34	5.370	(OE)	5.742	-6.5					
9.89	1.810	(OE)	1.781	1.6					
Yb (70)									
8.34	1.360	(OE)	1.467	-7.3					
9.89	2.160	(OE)	2.079	3.9					
Hf (72)									
44.70	7.600	(KS)	18.030	-57.8					

REFERENCES
For characteristic line source measurements
listed in Appendix I

A   S. J. M. Allen, reported in Compton and Allison, <u>X-Rays in Theory and Experiment</u>, (Van Nostrand, 1935), p. 800.
An   C. L. Andrews, <u>Phys. Rev. 54</u>, 994 (1938).
B   A. J. Bearden, <u>J. Appl. Phys. 37</u> (4), 1681 (1966).
Ba   G. B. Bandopadhyaya and A. T. Maitra, <u>Proc. Roy. Soc. 21</u>, 869 (1936).
Bi   H. H. Biermann, <u>Ann. Physik 26</u>, 740 (1936).
BU   V. E. Baurmann and K. Ulmer, <u>Z. Naturforsch. 12A</u>, 670 (1957).
C   B. A. Cooke and E. A. Stewardson, <u>Brit. J. Appl. Phys. 15</u>, 1315 (1964).
CR   R. W. Carter, et al., <u>Health Physics 13</u>, 593 (1967).
CW   M. J. Cole, et al., Final Report, U. S. Dept. of Army Contract DA-91-591-EUC-3094, October 1964.
D   E. Dershem and M. Schein, <u>Phys. Rev. 37</u>, 1238 (1931).
De   D. R. Denne, <u>Brit. J. Appl. Phys. 3</u>, 1392 (1970).
F   V. A. Fomichev and A. P. Lukirskii, <u>Opt. and Spectrosc. 22</u>, 432 (1967).
H   B. L. Henke et al., Technical Report, AFOSR 67-1254, June 1967; and <u>Advances in X-Ray Analysis</u>, Vol. 13 (Plenum Press, New York, 1970), p. 639 (with R. L. Elgin).
Hi   R. D. Hill, <u>Proc. Roy. Soc.</u> (London) A161, 284 (1937).
HW   G. D. Hughes, et al., <u>Brit. J. Appl. Phys. 1</u>, 695 (1968).
J   E. Jonsson, Dissertation, Uppsala (1928).
KS   E. Kohlhaas and F. Scheiding, <u>Proc. Vth. Int'l Congress on X-Ray Optics and Microanalysis</u>, Tubingen, 1968 (Springer-Verlag, 1969), p. 193.
LC   P. Lublin, et al., Advances in X-Ray Analysis, Vol. 13 (Plenum Press, New York, 1970), p. 632.
LG   A. P. Lukirskii, et al., <u>Opt. and Spectrosc. 20</u>, 203 (1966).
LS   A. P. Lukirskii, et al., <u>Opt. and Spectrosc. 16</u>, 168 (1964).
LZ   A. P. Lukirskii, et al., <u>Opt. and Spectrosc. 17</u>, 234 (1964).
M   R. H. Messner, <u>Z. Physik 85</u>, 727 (1933).
O   W. T. Ogier, et al., <u>Appl. Phys. Lett. 5</u>, 146 (1964).
OE   B. Ortner, et al., <u>Mikrochim. Acta Suppl. 4</u>, 270 (1970).
Se   G. Senemaud, <u>J. Phys. 30</u>, 811 (1969).
W   B. Woernle, <u>Ann. Physik 5</u>, 475 (1930).
We   W. Weisweiler, <u>Proc. Vth Int'l Congress on X-Ray Optics and Microanalysis</u>, Tubingen, 1968 (Springer-Verlag, 1969), p. 198.
Wi   P. R. Wise, John Hopkins University Thesis NP-12661 (1961).
H+   B. L. Henke and E. S. Ebisu, this paper.

REFERENCES -

For continuous radiation source measurements, for indirect measurements, and for absorption data presented only as graphs. These works are listed below along with references to the elements and wavelengths covered.

- A. A. Aboud, et al., J. Opt. Soc. Am. 45, 767 (1955). [122-860 A; O]
- B. A. Cooke and E. A. Stewardson, Brit. J. Appl. Phys. 15, 1315 (1964). [7-17 A; Be, Mg, Al, Cu, Ag]
- Pierre Dhez, Thesis, University of Paris, Orsay, Series A, No. 747, (March 1971). [50-310 A; Bi]
- D. L. Ederer and D. H. Tomboulian, Phys. Rev. 133, A1525 (1964). [80-600 A; Ne]
- O. A. Ershov and A. P. Lukirskii, Sov. Phys. Solid State 8, 1699 (1967). [60-140 A; Si, SiO]
- V. A. Fomichev and I. I. Zhukova, Opt. and Spectrosc. 24, 147 (1968). [17.6-250 A; C]
- Christian Gahwiller and Frederick C. Brown, Phys. Rev. B 2, 1918 (1970). [60-175 A; Al, Si, SiO]
- R. Haensel, et al., Applied Optics 7, 301 (1968). [50-340 A; Cu, Ag, Sn, Au, Bi]
- R. Haensel, et al., Solid State Comm. 7, 1495 (1969). [20-410 A; Ta, W, Re, Pt]
- P. Jaegle, et al., Phys. Rev. 188, 30 (1969). [20-130 A; Ta, Pt, Au, Bi]
- P. Jaegle, et al., Physics Letters 26A, 364 (1968). [20-140 A; Ta, Pt]
- P. Jaegle and G. Missoni, C. R. Acad. Sc. Paris, Series B, 262, 71 (1966). [26-120 A; Au]
- P. Jaegle, et al., Phys. Rev. Lett. 18, 887 (1967). [25-85 A; Bi, Pb]
- D. F. Kyser, in G. Shinoda et al., Editors, Proceedings of the Sixth International Conference on X-Ray Optics and Microanalysis, University of Tokyo Press (1972). [12-28 A; Ti, V, Cr, Mn, Fe, Co, Ni, Cu, Zn]
- A. P. Lukirskii, et al., Sov. Phys. Solid State 8, 1525 (1966). [25-250 A; Te, Sn, Pb, PbTe, SnTe]
- B. Sonntag and R. Haensel, Solid State Comm. 7, 597 (1969). [40-310 A; Ti, V, Cr, Mn, Fe, Co, Ni]
- D. H. Tomboulian and D. E. Bedo, Phys. Rev. 104, 590 (1956). [70-200 A; Si, Ge]
- W. S. Watson, J. Phys. B: Atom. Molec. Phys. 5, 2292 (1972). [58-200 A; He, Ne, Ar]
- R. W. Woodruff and M. P. Givens, Phys. Rev. 97, 52 (1955). [100-400 A; Te]
- F. Wuilleumier, C. R. Acad. Sc. Paris, 270B, 272 (1970). [8-15 A; Ne, Ar]

- T. M. Zimkina, et al., Sov. Phys. Solid State 9, 1128 (1967). [25-250 A; Sn, Te, Xe, La, Ce, Pr, Nd, Sm, Eu, Gd, Ho, Er, Tu, Yb, Lu]
- T. M.. Zimkina and A. P. Lukirskii, Sov. Phys. Solid State 7, 1170 (1965). [23.6-190.3 A; KCl, KI, RbCl, RbBr, RbI, CsCl, CsBr, CsI]

APPENDIX II

MASS ABSORPTION COEFFICIENTS
μ (units of 10^3 cm^2/gram)

λ(Å)	H†(1)	He (2)	Li (3)	Be (4)	E (eV)
8.34	0.0018	0.0168	0.0647	0.1787	1487
9.89	0.0033	0.0286	0.1107	0.3005	1254
11.9	0.0061	0.0518	0.1969	0.5300	1041
12.3	0.0067	0.0568	0.2149	0.5783	1012
13.3	0.0089	0.0739	0.2796	0.7450	930
14.6	0.0119	0.0973	0.3663	0.9633	852
16.0	0.0160	0.1311	0.4876	1.2680	776
17.6	0.0227	0.1779	0.6553	1.6840	705
18.3	0.0258	0.2007	0.7353	1.8780	679
19.4	0.0316	0.2463	0.8929	2.2570	637
21.6	0.0434	0.3450	1.2330	3.0640	573
22.3	0.0478	0.3795	1.3480	3.3360	556
23.6	0.0582	0.4550	1.6020	3.9220	525
27.4	0.0967	0.7279	2.5090	5.9670	452
31.4	0.1475	1.1140	3.7260	8.6930	395
31.6	0.1513	1.1410	3.8120	8.8750	392
44.7	0.4727	3.3380	10.3100	22.0000	277
58.4	1.1110	7.4850	21.4400	42.7200	212
64.4	1.5020	9.8100	27.5400	53.1700	193
67.6	1.7790	11.4500	31.5900	60.5600	183
72.2	2.1670	13.7500	37.0900	69.9400	172
82.1	3.2530	19.9200	51.2300	93.5400	151
83.4	3.3920	20.6800	52.9500	96.3400	149
93.4	4.8400	28.4500	70.6900	122.9000	133
109.	7.8530	43.3100	101.9000	*169.0000	114
114.	9.0180	48.8800	112.4000	5.2300	109

†H-1 Interpolated from Veigele et al 1972
* - Values extrapolated near absorption edges

MASS ABSORPTION COEFFICIENTS
μ(units of 10^3 cm^2/gram)

λ(Å)	B (5)	C (6)	N (7)	O (8)	E (eV)
8.34	0.3828	0.7184	1.1070	1.5970	1487
9.89	0.6332	1.1740	1.7890	2.5330	1254
11.9	1.0930	1.9840	2.9700	4.1540	1041
12.3	1.1880	2.1470	3.2050	4.4760	1012
13.3	1.5100	2.7140	4.0220	5.6010	930
14.6	1.9470	3.4570	5.0690	6.9610	852
16.0	2.5310	4.4510	6.4960	8.8770	776
17.6	3.3140	5.7860	8.3440	11.3000	705
18.3	3.6770	6.3660	9.1610	12.3900	679
19.4	4.3930	7.4950	10.7400	14.5000	637
21.6	5.8680	9.9020	14.0000	18.7400	573
22.3	6.3320	10.6900	15.0700	20.1400	556
23.6	7.4160	12.3800	17.3100	*1.2000	525
27.4	11.0000	17.9600	24.8300	1.7560	452
31.4	15.5400	25.0300	1.6040	2.4780	395
31.6	15.8100	25.4900	1.6370	2.5270	392
44.7	37.0200	2.3730	3.9030	6.0440	277
58.4	68.5100	4.5960	7.4300	11.6200	212
64.4	83.5600	5.7550	9.3340	14.5500	193
67.6	3.3530	6.4560	10.5700	16.5300	183
72.2	3.8690	7.4940	12.2300	19.1600	172
82.1	5.4160	10.0500	16.5500	25.9800	151
83.4	5.5770	10.3500	17.0700	26.8000	149
93.4	7.2310	13.4200	22.1700	34.6600	133
109.	10.1900	18.8600	31.5000	48.8600	114
114.	11.2300	20.8100	34.8100	53.9500	109

MASS ABSORPTION COEFFICIENTS
μ(units of 10^3 cm^2/gram)

λ(Å)	F (9)	Ne (10)	Na (11)	Mg (12)	E (eV)
8.34	2.0370	2.7520	3.3560	4.2840	1487
9.89	3.1980	4.2920	5.2030	0.4889	1254
11.9	5.2160	6.9310	*0.5613	0.8147	1041
12.3	5.6200	7.4630	*0.6088	0.8806	1012
13.3	6.9410	9.1900	0.7751	1.1100	930
14.6	8.6470	*0.7154	0.9838	1.4110	852
16.0	10.9100	0.9178	1.2690	1.8180	776
17.6	13.8600	1.1780	1.6470	2.3640	705
18.3	*0.8677	1.3010	1.8230	2.6150	679
19.4	1.0210	1.5410	2.1660	3.1050	637
21.6	1.3440	2.0440	2.8840	4.1210	573
22.3	1.4570	2.2120	3.1290	4.4630	556
23.6	1.6880	2.5820	3.6450	5.1740	525
27.4	2.4900	3.8450	5.4290	7.6430	452
31.4	3.5320	5.4950	7.6440	10.7400	395
31.6	3.6020	5.6170	7.7960	10.9500	392
44.7	8.7300	13.5700	18.4600	24.9000	277
58.4	16.8900	25.7300	33.8800	43.8500	212
64.4	21.1100	31.8000	41.2900	52.7800	193
67.6	23.8800	35.7600	46.0100	58.1700	183
72.2	27.5600	40.8700	51.7700	64.7700	172
82.1	36.9600	53.3800	66.3800	80.1700	151
83.4	38.0600	54.7700	67.9600	81.9600	149
93.4	48.6100	67.9300	82.9400	96.9500	133
109.	67.1400	90.3300	106.3000	119.2000	114
114.	73.1400	97.5200	113.3000	125.3000	109

MASS ABSORPTION COEFFICIENTS
μ (units of 10^3 cm^2/gram)

λ(Å)	Al (13)	Si (14)	P (15)	S (16)	E (eV)
8.34	0.4034	0.5428	0.6751	0.8698	1487
9.89	0.6393	0.8644	1.0710	1.3770	1254
11.9	1.0680	1.4420	1.7760	2.2770	1041
12.3	1.1520	1.5580	1.9170	2.4590	1012
13.3	1.4530	1.9590	2.4050	3.0790	930
14.6	1.8480	2.4840	3.0410	3.8880	852
16.0	2.3790	3.1880	3.9020	4.9500	776
17.6	3.0810	4.1120	5.0070	6.3770	705
18.3	3.4070	4.5430	5.5260	7.0060	679
19.4	4.0400	5.3790	6.5270	8.1980	637
21.6	5.3680	7.0340	8.4680	10.6000	573
22.3	5.7950	7.6330	9.1520	11.3900	556
23.6	6.7150	8.7900	10.5400	13.0100	525
27.4	9.7200	12.7000	14.9700	18.3300	452
31.4	13.5900	17.3900	20.2400	24.5400	395
31.6	13.8300	17.6900	20.5700	24.9300	392
44.7	30.1600	36.9800	41.2800	47.9400	277
58.4	51.0300	59.9500	63.9500	61.1600	212
64.4	59.7900	69.1800	*73.2600	69.7900	193
67.6	65.1700	74.1800	66.1700	74.1800	183
72.2	71.7700	80.7800	71.7700	79.7800	172
82.1	87.1700	*94.5500	83.3800	++7.9170	151
83.4	88.7800	82.5800	84.5800	8.0920	149
93.4	101.9000	94.1800	++55.0300	9.6390	133
109.	102.8000	107.8000	8.4580	12.0300	114
114.	107.0000	111.6000	9.0450	12.7700	109

++ - Values interpolated from adjusted-Veigele for wavelengths longer than L_{III} edge value for P, S, Cl and Ar.

MASS ABSORPTION COEFFICIENTS
μ (units of 10^3 cm^2/gram)

λ(Å)	Cl (17)	Ar (18)	K (19)	Ca (20)	E (eV)
8.34	1.0230	1.1580	1.4770	1.7560	1487
9.89	1.6180	1.8390	2.3240	2.7130	1254
11.9	2.6670	3.0150	3.8120	4.3590	1041
12.3	2.8760	3.2470	4.1090	4.6850	1012
13.3	3.5960	4.0490	4.8870	5.7710	930
14.6	4.5220	5.0700	6.0910	7.1630	852
16.0	5.7500	6.4530	7.6820	9.0070	776
17.6	7.3660	8.1550	9.7100	11.3500	705
18.3	8.0470	8.9240	10.6200	12.3700	679
19.4	9.3670	10.3700	12.3500	14.3100	637
21.6	12.1000	13.2700	15.8300	18.1600	573
22.3	12.9300	14.2300	16.9700	19.4100	556
23.6	14.7900	16.1600	19.3700	22.0300	525
27.4	20.5700	22.2700	26.8800	30.3400	452
31.4	27.1900	29.0900	35.5100	35.1100	395
31.6	27.6100	29.5100	35.9000	35.5900	392
44.7	*50.7600	45.5800	6.3150	6.8380	277
58.4	64.1600	++7.3930	9.1800	10.5500	212
64.4	++6.9040	8.6300	10.4700	12.1400	193
67.6	7.5760	9.3690	11.2800	13.0100	183
72.2	8.4130	10.3100	12.3000	14.1100	172
82.1	10.3700	12.6100	14.7500	16.7200	151
83.4	10.5900	12.8500	14.9900	16.9600	149
93.4	12.4300	14.9000	16.7500	18.7100	133
109.	15.1800	17.5900	19.4800	21.3600	114
114.	15.9900	18.4300	20.3600	22.2000	109

MASS ABSORPTION COEFFICIENTS
μ(units of 10^3 cm^2/gram)

λ(Å)	Sc (21)	Ti (22)	V (23)	Cr (24)	E (eV)
8.34	1.9150	2.1370	2.3790	2.7650	1487
9.89	2.9570	3.3000	3.6570	4.2490	1254
11.9	4.7210	5.2470	5.7910	6.7310	1041
12.3	5.0680	5.6270	6.2060	7.2150	1012
13.3	6.2500	6.9300	7.6170	8.8360	930
14.6	7.7540	8.5570	9.3700	10.8400	852
16.0	9.7240	10.6700	11.6300	13.4300	776
17.6	12.2000	13.3700	14.4700	16.7300	705
18.3	13.3000	14.5400	15.7500	15.9100	679
19.4	15.3800	16.7400	18.1600	18.6700	637
21.6	19.5000	21.1900	19.9200	2.6120	573
22.3	20.8200	19.5300	21.3200	2.7840	556
23.6	23.5900	22.1400	24.2500	3.1430	525
27.4	28.4200	3.3180	3.6460	4.2560	452
31.4	3.9320	4.3040	4.7240	5.5500	395
31.6	3.9880	4.3640	4.7900	5.6300	392
44.7	7.4610	8.0940	8.8400	10.5900	277
58.4	11.4800	12.3900	13.5200	16.5100	212
64.4	13.2200	14.2500	15.5600	19.1500	193
67.6	14.1900	15.2800	16.7100	20.6700	183
72.2	15.4000	16.5900	18.1600	22.6000	172
82.1	18.2900	19.7000	21.6300	27.2600	151
83.4	18.5700	20.0100	21.9800	27.7400	149
93.4	20.6700	22.4600	24.8100	31.7400	133
109.	23.9100	26.2600	29.2500	38.1200	114
114.	24.9500	27.4900	30.6800	40.2000	109

MASS ABSORPTION COEFFICIENTS
μ (units of 10^3 cm^2/gram)

λ(Å)	Mn (25)	Fe (26)	Co (27)	Ni (28)	E (eV)
8.34	3.0270	3.4380	3.7380	4.2870	1487
9.89	4.6200	5.2450	5.6720	6.5110	1254
11.9	7.2560	8.2040	8.8290	10.1900	1041
12.3	7.7670	8.7750	9.4360	10.9000	1012
13.3	9.4730	10.6900	11.6600	11.5100	930
14.6	11.5600	13.2100	12.3400	1.8010	852
16.0	14.4400	14.1400	1.9240	2.2680	776
17.6	15.6100	2.1490	2.3710	2.7450	705
18.3	17.0200	2.3280	2.5680	2.9890	679
19.4	2.3170	2.6780	2.9390	3.4680	637
21.6	2.8970	3.3320	3.6720	4.3830	573
22.3	3.0820	3.5480	3.9100	4.6620	556
23.6	3.4680	4.0010	4.4070	5.2450	525
27.4	4.6730	5.4000	5.9610	7.0360	452
31.4	6.0750	7.0220	7.7780	9.1060	395
31.6	6.1600	7.1210	7.8880	9.2350	392
44.7	11.4700	13.3000	14.7300	17.2700	277
58.4	17.7800	20.6700	22.8300	26.6600	212
64.4	20.5700	23.9300	26.3700	30.7700	193
67.6	22.1600	25.7800	28.3400	33.0900	183
72.2	24.1700	28.1300	30.8200	36.0200	172
82.1	29.0200	33.7700	36.7800	43.0400	151
83.4	29.5100	34.3400	37.3700	43.7100	149
93.4	33.4800	38.8800	42.0800	48.8200	133
109.	39.7200	46.0200	49.4400	56.7200	114
114.	41.7500	48.3300	51.8100	55.4200	109

MASS ABSORPTION COEFFICIENTS
μ(units of 10^3 cm^2/gram)

λ(Å)	Cu (29)	Zn (30)	Ga (31)	Ge (32)	E (eV)
8.34	4.5540	5.0190	5.2950	5.7230	1487
9.89	6.9650	7.6000	6.9920	*7.6050	1254
11.9	9.5590	*10.5100	1.5280	1.7120	1041
12.3	10.3700	1.5200	1.6290	1.8280	1012
13.3	1.5860	1.8160	1.9890	2.2320	930
14.6	1.9680	2.2370	2.4500	2.7370	852
16.0	2.5370	2.7880	3.0430	3.3850	776
17.6	2.8830	3.4240	3.7560	4.1820	705
18.3	3.1590	3.7170	4.0720	4.5390	679
19.4	3.7520	4.2770	4.6670	5.2160	637
21.6	4.8800	5.3990	5.8610	6.5450	573
22.3	5.2160	5.7700	6.2570	6.9730	556
23.6	5.9200	6.5480	7.0860	7.8670	525
27.4	7.8250	8.8820	9.6170	10.6400	452
31.4	9.8570	11.5400	12.5300	13.8500	395
31.6	9.9970	11.7000	12.7100	14.0400	392
44.7	18.6900	21.7100	23.3300	25.5200	277
58.4	28.7800	32.9500	34.8000	37.4500	212
64.4	33.3300	37.6600	39.5500	42.3900	193
67.6	36.0600	40.1400	42.0300	45.0500	183
72.2	39.5400	43.2300	45.1200	45.2700	172
82.1	47.9600	50.5300	48.8500	50.9800	151
83.4	48.6500	51.2600	49.4300	51.5200	149
93.4	52.3900	53.7700	54.1100	55.6500	133
109.	56.4300	61.5600	61.1800	56.9200	114
114.	59.4300	64.0400	63.4100	57.8800	109

MASS ABSORPTION COEFFICIENTS
μ(units of 10^3 cm^2/gram)

λ(Å)	As (33)	Se (34)	Br (35)	Kr (36)	E (eV)
8.34	5.4000	*5.6180	1.0210	1.1320	1487
9.89	1.2430	1.3580	1.5370	1.7090	1254
11.9	1.9250	2.1140	2.3900	2.6540	1041
12.3	2.0570	2.2610	2.5560	2.8370	1012
13.3	2.5060	2.7590	3.1010	3.4520	930
14.6	3.0670	3.3730	3.7840	4.2120	852
16.0	3.7880	4.1610	4.6710	5.1910	776
17.6	4.6810	5.1460	5.7740	6.4150	705
18.3	5.0810	5.5850	6.2590	6.9600	679
19.4	5.8380	6.4150	7.1730	7.9890	637
21.6	7.3250	8.0360	8.9550	9.9730	573
22.3	7.8040	8.5540	9.5250	10.6000	556
23.6	8.8050	9.6330	10.7100	11.8900	525
27.4	11.8800	12.9600	14.3600	15.8800	452
31.4	15.4200	16.7800	18.5300	20.4300	395
31.6	15.6300	17.0000	18.7600	20.6900	392
44.7	27.9000	30.2100	32.5500	33.2800	277
58.4	40.2600	40.4100	41.8300	38.8000	212
64.4	42.5000	44.0400	44.9200	40.4900	193
67.6	44.4500	45.6000	42.7000	40.2000	183
72.2	46.8300	47.4900	43.7400	39.8700	172
82.1	52.2600	48.4300	46.0100	39.1800	151
83.4	52.6800	48.4900	45.7400	38.1600	149
93.4	51.1800	46.3300	39.5700	24.7600	133
109.	52.2200	43.5400	32.5200	13.7600	114
114.	52.5300	42.7700	30.7100	11.6000	109

MASS ABSORPTION COEFFICIENTS
μ(units of 10^3 cm^2/gram)

λ(Å)	Rb(37)	Sr (38)	Y (39)	Zr (40)	E (eV)
8.34	1.2500	1.3740	1.5310	1.6710	1487
9.89	1.8860	2.0700	2.2960	2.5050	1254
11.9	2.9090	3.1890	3.5390	3.8510	1041
12.3	3.1050	3.4030	3.7780	4.1080	1012
13.3	3.7760	4.1300	4.5790	4.9680	930
14.6	4.6100	5.0250	5.5650	6.0350	852
16.0	5.6720	6.1700	6.8210	7.3900	776
17.6	6.9780	7.5940	8.3710	9.0240	705
18.3	7.5510	8.2100	9.0490	9.7430	679
19.4	8.6270	9.3610	10.3200	11.0900	637
21.6	10.7300	11.5900	12.7700	13.6700	573
22.3	11.4100	12.2900	13.5400	14.4800	556
23.6	12.8200	13.7600	15.1400	16.1400	525
27.4	17.0200	18.1400	19.8400	21.2100	452
31.4	21.7000	23.0300	25.0700	24.8100	395
31.6	21.9600	23.3100	*23.8400	25.0300	392
44.7	35.5200	35.3100	31.3800	31.1300	277
58.4	37.7600	34.9800	31.9500	*35.6500	212
64.4	37.7700	*32.2200	*30.6900	*37.3900	193
67.6	35.8800	*27.3400	*30.0000	*38.4100	183
72.2	33.8000	*22.5700	*29.2200	3.7900	172
82.1	29.8000	*15.1000	3.6220	4.1620	151
83.4	*29.4200	*13.9700	3.6540	4.1920	149
93.4	*26.3600	*5.2630	3.8820	4.3750	133
109.	*21.8400	3.6060	4.2140	4.6350	114
114.	3.1990	3.7080	4.3160	4.7140	109

MASS ABSORPTION COEFFICIENTS
μ(units of 10^3 cm^2/gram)

λ(Å)	Nb (41)	Mo (42)	Tc (43)	Ru (44)	E (eV)
8.34	1.8490	1.9800	2.1080	2.3130	1487
9.89	2.7700	2.9600	3.1210	3.4230	1254
11.9	4.2620	4.5440	4.7910	5.2530	1041
12.3	4.5480	4.8470	5.1130	5.6060	1012
13.3	5.4990	5.8480	6.1730	6.7390	930
14.6	6.6760	7.0870	7.4480	8.1300	852
16.0	8.1690	8.6550	9.0470	9.8970	776
17.6	9.9680	10.5400	11.0100	12.0400	705
18.3	10.7700	11.3600	11.8500	12.9600	679
19.4	12.2800	12.9000	13.4200	14.6700	637
21.6	15.1400	15.8400	16.4800	16.8600	573
22.3	16.0300	16.7600	17.4700	17.7800	556
23.6	17.8500	18.6600	18.1200	19.6800	525
27.4	21.8700	22.1500	22.0400	20.6300	452
31.4	26.9800	23.0500	23.2000	25.0100	395
31.6	27.1900	23.2200	23.3100	25.1800	392
44.7	*33.9900	*32.4200	*28.9600	4.2400	277
58.4	*45.2400	4.3000	4.5930	5.2220	212
64.4	4.2200	4.5940	4.9160	5.5120	193
67.6	4.4170	4.7170	5.0530	5.5830	183
72.2	4.6590	4.8640	5.2160	5.6680	172
82.1	5.2110	5.1880	5.5780	5.8500	151
83.4	4.8970	5.2040	5.5940	5.8470	149
93.4	4.9700	5.1750	5.5560	5.6410	133
109.	5.0700	5.1360	5.5040	5.3720	114
114.	5.1000	5.1240	5.4890	5.2970	109

MASS ABSORPTION COEFFICIENTS
μ(units of 10^3 cm²/gram)

λ(Å)	Rh (45)	Pd (46)	Ag (47)	Cd (48)	E (eV)
8.34	2.5030	2.6800	2.8870	2.9740	1487
9.89	3.6970	3.9720	4.2540	4.4210	1254
11.9	5.6620	6.1040	6.4970	6.8190	1041
12.3	6.0410	6.5150	6.9280	7.2820	1012
13.3	7.2770	7.8100	8.3280	8.7220	930
14.6	8.7790	9.4000	10.0100	10.4400	852
16.0	10.6600	11.4100	12.1000	12.7100	776
17.6	12.9400	13.8300	13.7600	14.2900	705
18.3	13.9200	14.8000	14.7300	14.8700	679
19.4	15.7300	15.8100	16.4900	15.7200	637
21.6	17.6000	18.6400	16.4600	17.1900	573
22.3	18.2800	19.3100	17.3800	18.1700	556
23.6	19.6700	18.3300	19.2600	20.1900	525
27.4	21.5500	22.3400	*25.1800	*26.5800	452
31.4	*26.2000	*25.9300	*3.8640	4.1170	395
31.6	*26.4900	*26.1500	*3.9000	4.1520	392
44.7	4.6340	5.2500	5.5070	5.7930	277
58.4	5.6720	6.1380	6.3940	6.5890	212
64.4	5.9470	6.2910	6.5480	6.6920	193
67.6	5.9770	6.2040	6.4620	6.5670	183
72.2	6.0110	6.1040	6.3630	6.4250	172
82.1	6.0850	5.8990	6.1600	6.1350	151
83.4	6.0730	5.8920	6.1440	6.1150	149
93.4	5.8240	5.9560	6.0470	6.0190	133
109.	5.5010	6.0440	5.9170	5.8900	114
114.	5.4100	6.0700	5.8800	5.8530	109

MASS ABSORPTION COEFFICIENTS
µ(units of 10^3 cm^2/gram)

λ(Å)	In (49)	Sn (50)	Sb (51)	Te (52)	E (eV)
8.34	3.1700	3.3310	3.5170	3.6350	1487
9.89	4.6970	4.9260	5.1840	5.3390	1254
11.9	7.2170	7.5510	7.9200	8.1250	1041
12.3	7.7030	8.0570	8.4470	8.6590	1012
13.3	9.1990	9.6230	9.5090	9.7220	930
14.6	11.0700	10.8500	11.3000	11.4300	852
16.0	12.4600	12.7100	11.0300	11.3900	776
17.6	14.8900	12.7500	13.4700	13.9000	705
18.3	*15.9700	13.7600	14.6800	16.4000	679
19.4	14.8500	15.6400	17.0300	22.4900	637
21.6	18.4200	*19.3600	*21.7900	3.1400	573
22.3	19.6500	*20.5700	*23.3800	3.2810	556
23.6	22.2200	*23.0900	3.4180	3.5660	525
27.4	*30.6100	3.9420	4.1890	4.3410	452
31.4	4.4400	4.7250	4.9790	5.1130	395
31.6	4.4760	4.7600	5.0150	5.1480	392
44.7	6.1350	6.3320	6.6010	6.6660	277
58.4	6.8430	6.9480	7.1020	7.0800	212
64.4	6.8960	6.9610	7.0660	7.0030	193
67.6	6.7320	6.7710	6.8480	6.7560	183
72.2	6.5460	6.5570	6.6030	6.4800	172
82.1	6.1710	6.1280	*4.7850	*4.7260	151
83.4	6.1340	6.0790	4.8740	4.7780	149
93.4	5.8280	4.5230	5.6960	5.2540	133
109.	4.6310	5.8580	7.0390	5.9750	114
114.	5.0880	6.3150	7.4860	4.2060	109

MASS ABSORPTION COEFFICIENTS
μ(units of 10^3 cm^2/gram)

λ(Å)	I (53)	Xe (54)	Cs (55)	Ba (56)	E (eV)
8.34	3.9650	4.2010	4.4750	4.6660	1487
9.89	5.8220	6.1780	6.5740	6.3640	1254
11.9	8.3740	8.7770	9.8510	7.6150	1041
12.3	8.8790	9.3100	7.8380	8.1030	1012
13.3	10.7100	8.9210	9.4220	9.7440	930
14.6	10.1700	10.7900	11.5000	11.6000	852
16.0	12.4200	13.3700	15.4100	2.5290	776
17.6	15.6300	22.7600	2.8540	2.9620	705
18.3	*17.2400	*2.8480	3.0300	3.1400	679
19.4	*20.3500	*3.1520	3.3490	3.4640	637
21.6	3.4410	3.7040	3.9260	4.0450	573
22.3	3.5950	3.8680	4.0970	4.2150	556
23.6	3.9060	4.2000	4.4430	4.5600	525
27.4	4.7470	5.0880	5.3440	5.4420	452
31.4	5.5770	5.9540	6.2030	6.2720	395
31.6	5.6130	5.9900	6.2380	6.3060	392
44.7	7.0940	7.4030	7.6500	7.7720	277
58.4	7.3500	7.4370	6.5200	6.4480	212
64.4	7.1970	6.1440	6.1950	6.1470	193
67.6	*5.8800	5.8950	5.9960	3.6770	183
72.2	*5.5740	5.6180	5.7720	3.4230	172
82.1	4.9670	5.2130	2.9690	3.2160	151
83.4	4.9490	5.1570	3.0550	3.2920	149
93.4	5.1440	4.4270	5.2890	5.1460	133
109.	4.8510	9.6630	11.1400	9.4330	114
114.	6.3490	12.1300	13.8400	11.2500	109

MASS ABSORPTION COEFFICIENTS
μ(units of 10^3 cm^2/gram)

λ(Å)	La (57)	Ce (58)	Pr (59)	Nd (60)	E (eV)
8.34	4.9230	5.3420	5.3110	5.4670	1487
9.89	6.5460	7.1580	6.0980	6.0760	1254
11.9	7.9300	8.7960	9.4090	8.2840	1041
12.3	8.4420	9.3850	10.0500	8.6830	1012
13.3	10.1600	*11.4900	2.2190	2.2940	930
14.6	*12.3400	2.4550	2.5920	2.6790	852
16.0	2.6490	2.8820	3.0440	3.1440	776
17.6	3.0950	3.3790	3.5670	3.6870	705
18.3	3.2770	3.5840	3.7830	3.9090	679
19.4	3.6040	3.9540	4.1740	4.3120	637
21.6	4.1840	4.6180	4.8770	5.0390	573
22.3	4.3520	4.8130	5.0840	5.2550	556
23.6	4.6900	5.2070	5.5030	5.6910	525
27.4	5.5590	6.2330	6.6010	6.8420	452
31.4	6.3770	7.2130	7.6670	7.9700	395
31.6	6.4100	7.2540	7.7170	8.0260	392
44.7	7.8940	8.1020	8.9090	9.5430	277
58.4	6.4520	*8.4140	7.6160	10.9700	212
64.4	*3.9890	6.4540	8.1530	9.6570	193
67.6	3.8260	6.8110	8.7040	10.3900	183
72.2	3.6440	7.2520	9.3940	11.3000	172
82.1	3.2900	8.2740	11.0300	13.5000	151
83.4	3.3070	8.4130	11.2100	13.7100	149
93.4	3.9460	9.9670	12.8600	15.2900	133
109.	5.0150	12.5500	15.5000	13.6900	114
114.	5.3770	8.2630	11.6300	14.1400	109

MASS ABSORPTION COEFFICIENTS
μ(units of 10^3 cm^2/gram)

λ(Å)	Pm (61)	Sm (62)	Eu (63)	Gd (64)	E (eV)
8.34	5.7140	5.8970	4.9330	5.0060	1487
9.89	6.6790	7.0310	7.4190	*7.4800	1254
11.9	1.9440	2.0320	2.1340	2.1710	1041
12.3	2.0480	2.1450	2.2520	2.2910	1012
13.3	2.3920	2.5080	2.6330	2.6730	930
14.6	2.7960	2.9340	3.0800	3.1200	852
16.0	3.2830	3.4460	3.6170	3.6600	776
17.6	3.8490	4.0380	4.2350	4.2860	705
18.3	4.0810	4.2840	4.4950	4.5460	679
19.4	4.5000	4.7300	4.9680	5.0200	637
21.6	5.2650	5.5520	5.8350	5.8750	573
22.3	5.4960	5.8040	6.0960	6.1280	556
23.6	5.9640	6.3140	6.6270	6.6410	525
27.4	7.2170	7.6730	8.0610	8.0420	452
31.4	8.4660	9.0280	9.5050	9.4710	395
31.6	8.5320	9.1010	9.5830	9.5530	392
44.7	10.5300	11.4900	12.1900	*11.9700	277
58.4	12.5700	12.0800	13.0100	12.8500	212
64.4	11.5500	13.4200	14.4400	14.0300	193
67.6	12.4300	14.5600	15.6200	14.8900	183
72.2	13.5500	16.0100	17.1100	15.9600	172
82.1	16.2300	19.5400	20.7300	18.4600	151
83.4	16.6100	19.9300	21.0100	18.6000	149
93.4	17.9900	*23.5900	*18.0400	16.6600	133
109.	16.5900	18.8200	19.4100	17.5800	114
114.	17.0600	19.3100	19.8300	17.8500	109

MASS ABSORPTION COEFFICIENTS
μ(units of 10^3 cm²/gram)

λ(Å)	Tb (65)	Dy (66)	Ho (67)	Er (68)	E (eV)
8.34	5.4360	5.6650	5.7420	*6.3620	1487
9.89	*1.6280	1.7000	1.7810	1.8830	1254
11.9	2.3410	2.4520	2.5790	2.7380	1041
12.3	2.4740	2.5920	2.7280	2.8980	1012
13.3	2.8960	3.0400	3.2010	3.4040	930
14.6	3.3930	3.5720	3.7580	3.9990	852
16.0	3.9960	4.2200	4.4420	4.7260	776
17.6	4.6980	4.9670	5.2600	5.5870	705
18.3	4.9960	5.2840	5.5930	5.9500	679
19.4	5.5420	5.8630	6.1940	6.6170	637
21.6	6.5390	6.9200	7.3080	7.8270	573
22.3	6.8380	7.2360	7.6480	8.1860	556
23.6	7.4440	7.8770	8.3400	8.9150	525
27.4	9.0990	9.7000	10.3000	11.0200	452
31.4	9.9930	11.3200	11.3700	12.1800	395
31.6	10.0700	11.3900	11.4600	12.2800	392
44.7	12.4100	13.5900	14.5300	15.7200	277
58.4	16.1700	17.9000	19.2100	21.0800	212
64.4	17.7600	19.6100	20.9700	22.9200	193
67.6	18.7400	20.5400	21.8400	23.6000	183
72.2	19.9500	21.6800	22.8900	24.4200	172
82.1	22.7700	21.6600	22.5000	23.6500	151
83.4	22.8700	21.7800	22.6100	23.7500	149
93.4	21.1300	22.5500	23.1600	24.1800	133
109.	22.3400	23.6400	23.9400	24.7600	114
114.	22.7100	23.9700	24.1700	24.9300	109

MASS ABSORPTION COEFFICIENTS
μ(units of 10^3 cm^2/gram)

λ(Å)	Tm (69)	Yb (70)	Lu (71)	Hf (72)	E (eV)
8.34	*1.4110	1.4670	1.5300	1.6180	1487
9.89	2.0040	2.0790	2.1640	2.2850	1254
11.9	2.9200	3.0440	3.1610	3.3310	1041
12.3	3.0920	3.2250	3.3480	3.5270	1012
13.3	3.6260	3.7990	3.9290	4.1420	930
14.6	4.2690	4.4820	4.6240	4.8590	852
16.0	5.0640	5.3140	5.4810	5.7380	776
17.6	5.9970	6.2760	6.4840	6.7970	705
18.3	6.3830	6.6820	6.8970	7.2270	679
19.4	7.0880	7.4270	7.6480	8.0050	637
21.6	8.3800	8.7810	9.0200	9.4560	573
22.3	8.7690	9.1840	9.4310	9.9040	556
23.6	9.5610	10.0000	10.2700	10.1300	525
27.4	11.0200	11.5500	11.7200	12.1500	452
31.4	13.0200	13.6500	13.7800	12.8100	395
31.6	13.1300	13.7700	13.8800	12.9100	392
44.7	16.9300	17.7100	17.4100	18.0300	277
58.4	22.8000	23.4100	21.9100	20.8000	212
64.4	24.5100	24.8000	21.5500	21.4800	193
67.6	24.7900	23.1800	21.8900	21.6400	183
72.2	23.2700	23.7600	22.2900	21.8400	172
82.1	24.6200	25.0300	23.1600	22.2500	151
83.4	24.7000	25.1000	23.1800	22.2100	149
93.4	24.9400	25.1100	22.7700	21.2100	133
109.	25.2500	25.1200	22.2200	19.9200	114
114.	25.3500	25.1200	22.0700	19.5600	109

MASS ABSORPTION COEFFICIENTS
μ(units of 10^3 cm^2/gram)

λ(Å)	Ta (73)	W (74)	Re (75)	Os (76)	E (eV)
8.34	1.6880	1.7760	1.8600	1.9360	1487
9.89	2.4150	2.5380	2.6540	2.7410	1254
11.9	3.5080	3.6720	3.8380	3.9660	1041
12.3	3.7120	3.8830	4.0590	4.1940	1012
13.3	4.3680	4.5560	4.7510	4.9200	930
14.6	5.1280	5.3580	5.5680	5.7570	852
16.0	6.0480	6.3350	6.5650	6.7570	776
17.6	7.1420	7.4610	7.7240	7.9150	705
18.3	7.5880	7.9280	8.2270	8.4390	679
19.4	8.3970	8.7780	9.1560	8.7860	637
21.6	9.9270	9.7370	10.0100	10.1600	573
22.3	9.7500	10.1500	10.4400	10.5500	556
23.6	10.5600	10.9900	11.2900	11.3300	525
27.4	12.6100	13.0200	12.1400	12.1700	452
31.4	13.3300	13.7900	14.0300	14.0500	395
31.6	13.4200	13.8800	14.1200	14.1300	392
44.7	18.3900	18.7500	18.7100	17.0600	277
58.4	20.4400	19.8800	18.7800	17.3200	212
64.4	20.8700	19.9900	18.5700	16.8300	193
67.6	20.8200	19.6600	17.9800	16.0600	183
72.2	20.7500	19.2900	17.3300	15.2000	172
82.1	20.6100	18.5200	16.0300	13.5500	151
83.4	20.5100	18.3400	15.8200	13.3300	149
93.4	18.9500	16.2200	13.5300	11.2600	133
109.	17.0300	13.7200	10.9400	8.9890	114
114.	16.5100	13.0600	10.2900	8.4200	109

MASS ABSORPTION COEFFICIENTS
μ(units of 10^3 cm^2/gram)

λ(Å)	Ir (77)	Pt (78)	Au (79)	Hg (80)	E (eV)
8.34	2.0570	2.1740	2.3050	2.4150	1487
9.89	2.9150	3.0780	3.2530	3.4010	1254
11.9	4.2190	4.4490	4.6990	4.9430	1041
12.3	4.4630	4.7050	4.9670	5.2320	1012
13.3	5.2310	5.5110	5.7930	6.1150	930
14.6	6.1090	6.4520	6.7620	7.1840	852
16.0	7.1610	7.5920	7.9540	7.9430	776
17.6	8.4000	8.4030	8.7890	9.2920	705
18.3	8.4330	8.8930	9.2890	9.7320	679
19.4	9.2700	9.7720	10.1900	10.4600	637
21.6	10.6500	11.1500	10.5200	11.1300	573
22.3	11.0000	11.4700	10.9300	11.5700	556
23.6	11.7000	11.3400	11.7600	12.4400	525
27.4	12.7400	13.3200	13.6900	14.3200	452
31.4	14.6000	15.1400	15.3800	15.8600	395
31.6	14.6800	15.2100	15.4400	15.9200	392
44.7	16.4400	16.0500	15.2100	14.4300	277
58.4	15.9500	13.9800	11.8000	9.0250	212
64.4	15.0800	12.6000	10.2200	7.0250	193
67.6	14.0000	11.2700	8.9520	6.1800	183
72.2	12.8400	9.8970	7.6730	5.3230	172
82.1	10.7100	7.5350	5.5500	3.8890	151
83.4	8.1170	7.3380	5.4450	3.8590	149
93.4	8.0100	5.9180	5.2190	4.4260	133
109.	7.3560	5.9560	5.3330	4.4170	114
114.	7.1520	6.0580	5.8120	4.1850	109

MASS ABSORPTION COEFFICIENTS
μ(units of 10^3 cm^2/gram)

λ(Å)	Tl (81)	Pb (82)	Bi (83)	Po (84)	E (eV)
8.34	2.4970	2.6090	2.7270	2.8830	1487
9.89	3.5300	3.6480	3.8010	4.0060	1254
11.9	5.1530	5.2610	5.4640	5.7390	1041
12.3	5.4570	5.5610	5.7730	6.0610	1012
13.3	6.2760	6.4660	6.3750	6.6830	930
14.6	7.3040	7.1220	7.3800	7.7380	852
16.0	8.0130	8.2920	8.5330	8.9580	776
17.6	9.3330	9.5920	8.8100	9.2590	705
18.3	9.6900	9.0390	9.2850	9.7490	679
19.4	9.6200	9.8780	10.1300	10.6200	637
21.6	11.0300	11.3000	11.5500	12.1400	573
22.3	11.4300	11.6900	11.9300	12.5700	556
23.6	12.2100	12.4700	12.6900	13.4500	525
27.4	13.7100	13.7400	13.4300	13.0100	452
31.4	14.9400	14.0300	13.8600	13.7800	395
31.6	14.2600	14.0200	13.8100	13.6800	392
44.7	13.0300	11.3500	9.5330	8.2430	277
58.4	7.6460	5.5430	4.0690	3.6940	212
64.4	*5.7440	*4.3910	*3.0130	3.2040	193
67.6	*4.8850	*4.2650	*3.1360	2.5800	183
72.2	*3.0640	*4.1230	3.2850	2.1090	172
82.1	*3.7490	*3.8390	2.3930	2.7170	151
83.4	3.8270	3.8840	2.4810	2.8100	149
93.4	3.7550	3.1920	3.5670	3.9800	133
109.	4.7480	5.4740	5.8400	6.3840	114
114.	5.6160	6.4040	6.7400	4.7920	109

MASS ABSORPTION COEFFICIENTS
μ(units of 10^3 cm²/gram)

λ(Å)	At (85)	Rn (86)	Fr (87)	Ra (88)	E (eV)
8.34	3.0640	3.0670	3.1950	3.3180	1487
9.89	4.2490	4.2600	4.4670	4.6700	1254
11.9	5.7950	5.7830	6.0400	6.2360	1041
12.3	6.0980	6.0910	6.3610	6.5670	1012
13.3	7.0720	7.0590	7.3500	6.7280	930
14.6	8.1870	8.3160	7.5820	7.7790	852
16.0	8.4450	8.4540	8.7870	8.9990	776
17.6	9.7810	9.7480	10.1100	10.3300	705
18.3	10.2900	10.2400	10.6300	10.5600	679
19.4	11.1800	11.1200	11.5500	10.8200	637
21.6	12.2400	11.7900	10.9800	11.0400	573
22.3	12.3400	11.6600	11.2800	11.3200	556
23.6	12.5300	11.7600	11.8900	11.8700	525
27.4	13.1300	12.4700	12.1600	11.3800	452
31.4	13.5100	12.4300	11.3800	9.6010	395
31.6	13.3100	12.2600	11.1100	9.2010	392
44.7	6.5830	4.1730	*3.0580	2.4440	277
58.4	3.4440	1.9340	2.0910	2.2140	212
64.4	2.0260	2.1910	2.3570	2.4500	193
67.6	2.2380	2.4290	2.6170	2.6670	183
72.2	2.5140	2.7390	2.9540	2.9440	172
82.1	3.2070	3.5240	2.0510	2.1650	151
83.4	3.3110	3.6270	2.1520	2.2610	149
93.4	4.6190	3.0650	3.3410	3.4240	133
109.	7.2550	5.7780	6.0700	6.0110	114
114.	8.2730	6.9490	7.2220	7.0810	109

MASS ABSORPTION COEFFICIENTS
μ(units of 10^3 cm^2/gram)

λ(Å)	Ac (89)	Th (90)	Pa (91)	U (92)	E (eV)
8.34	3.4420	3.5280	3.7630	3.8080	1487
9.89	4.5300	4.5690	4.8580	4.9630	1254
11.9	6.3420	6.3120	6.0350	6.0920	1041
12.3	6.6740	5.9600	6.3500	6.3960	1012
13.3	6.8700	6.8680	7.3250	7.3590	930
14.6	7.9100	7.8860	8.3990	8.4200	852
16.0	9.1190	8.9780	9.2190	8.9040	776
17.6	10.4200	9.9800	9.1120	9.0940	705
18.3	10.3200	8.9420	9.5330	9.4920	679
19.4	9.8860	9.6430	10.2700	10.1600	637
21.6	10.9700	10.6700	11.1800	11.1200	573
22.3	11.2100	10.8800	11.2900	11.3100	556
23.6	11.6800	11.3000	11.5100	11.0600	525
27.4	10.5300	9.3340	6.9250	5.7840	452
31.4	8.3230	6.5930	3.5380	*2.5330	395
31.6	7.9890	6.2930	3.4610	*2.4170	392
44.7	2.6300	2.1200	2.3910	2.3170	277
58.4	2.4450	2.6060	1.4670	1.4830	212
64.4	2.6970	1.6440	1.7790	1.8620	193
67.6	*1.5880	1.8790	2.1040	2.2470	183
72.2	1.8810	2.1960	2.5570	2.7990	172
82.1	2.6840	3.0460	3.8540	4.4360	151
83.4	2.8000	3.1670	4.0270	4.6230	149
93.4	4.2260	4.6300	5.9520	6.2420	133
109.	7.3890	7.7530	10.1200	9.3810	114
114.	8.6930	9.0080	11.8000	10.5600	109

MASS ABSORPTION COEFFICIENTS
μ(units of 10^3 cm^2/gram)

λ(Å)	Np (93)	Pu (94)	Formvar $(C_5H_7O_2)X$	Parlodion $(C_{12}H_{11}O_{22}N_6)X$	X_E (eV)
8.34	3.7660	3.9220	0.9511	1.2830	1487
9.89	5.1420	5.3930	1.5290	2.0490	1254
11.9	6.3980	6.6930	2.5440	3.3790	1041
12.3	6.7250	7.0350	2.7470	3.6440	1012
13.3	7.7030	8.0700	3.4540	4.5680	930
14.6	8.6580	*8.3980	4.3430	5.7080	852
16.0	8.2820	8.7080	5.5640	7.2930	776
17.6	9.4450	9.9310	7.1560	9.3240	705
18.3	9.8360	10.2700	7.8600	10.2300	679
19.4	10.4900	10.7900	9.2250	11.9900	637
21.6	*10.6700	*9.2780	12.0500	15.5600	573
22.3	*10.3700	*8.1560	12.9800	16.7400	556
23.6	*9.8250	*6.3810	7.8930	6.1940	525
27.4	*5.0870	*3.2360	11.4600	8.9550	452
31.4	2.6400	2.7590	15.9800	7.8080	395
31.6	2.6540	2.7700	16.2700	7.9540	392
44.7	2.4310	2.4160	3.4230	4.7400	277
58.4	1.5910	1.6810	6.6160	9.1150	212
64.4	2.0100	2.2850	8.2920	11.4200	193
67.6	2.4780	2.9360	9.3760	12.9500	183
72.2	3.1620	3.9310	10.8800	15.0100	172
82.1	5.2770	7.2590	14.7100	20.3300	151
83.4	5.5250	7.6400	15.1700	20.9700	149
93.4	7.7230	10.7200	19.6700	27.1500	133
109.	12.1700	16.9800	27.7600	38.3100	114
114.	13.8900	*15.8000	30.6700	42.3100	109

MASS ABSORPTION COEFFICIENTS
μ(units of 10^3 cm^2/gram)

λ(Å)	Polypropylene $(CH_2)X$	Mylar $(C_{10}H_8O_4)X$	Kimfoil $(C_{16}H_{14}O_3)X$	Aluminum Oxide Al_2O_3	E (eV)
8.34	0.6154	0.9809	0.8445	0.9653	1487
9.89	1.0060	1.5770	1.3660	1.5310	1254
11.9	1.7000	2.6240	2.2840	2.5210	1041
12.3	1.8390	2.8330	2.4680	2.7170	1012
13.3	2.3250	3.5620	3.1090	3.4060	930
14.6	2.9620	4.4790	3.9270	4.2550	852
16.0	3.8140	5.7390	5.0400	5.4380	776
17.6	4.9580	7.3800	6.5070	6.9500	705
18.3	5.4550	8.1060	7.1510	7.6360	679
19.4	6.4220	9.5150	8.4030	8.9640	637
21.6	8.4850	12.4300	11.0200	11.6600	573
22.3	9.1610	13.3900	11.8800	12.5500	556
23.6	10.6100	8.1400	9.5860	4.1190	525
27.4	15.3900	11.8100	13.9100	5.9710	452
31.4	21.4500	16.4800	19.3900	8.3590	395
31.6	21.8500	16.7800	19.7500	8.5090	392
44.7	2.1000	3.5160	2.9600	18.8100	277
58.4	4.0950	6.7890	5.7280	32.4800	212
64.4	5.1440	8.5050	7.1790	38.4900	193
67.6	5.7840	9.6150	8.0980	42.2700	183
72.2	6.7280	11.1600	9.4000	47.0000	172
82.1	9.0730	15.0700	12.6800	58.3600	151
83.4	9.3500	15.5400	13.0700	59.6000	149
93.4	12.1900	20.1300	16.9500	70.2500	133
109.	17.2800	28.3900	23.9100	77.4100	114
114.	19.1200	31.3500	26.4100	82.0300	109

MASS ABSORPTION COEFFICIENTS
μ (units of 10^3 cm^2/gram)

λ(Å)	Quartz SiO$_2$	P 10 (CH$_4$)10%,Ar90%	Methane CH$_4$	Propane C$_3$H$_8$	E (eV)
8.34	1.1040	1.0960	0.5383	0.5874	1487
9.89	1.7530	1.7430	0.8798	0.9599	1254
11.9	2.8860	2.8620	1.4870	1.6220	1041
12.3	3.1120	3.0830	1.6090	1.7560	1012
13.3	3.8990	3.8480	2.0340	2.2190	930
14.6	4.8680	4.8220	2.5910	2.8270	852
16.0	6.2180	6.1410	3.3360	3.6400	776
17.6	7.9400	7.7730	4.3380	4.7320	705
18.3	8.7220	8.5090	4.7730	5.2070	679
19.4	10.2400	9.8950	5.6190	6.1300	637
21.6	13.2700	12.6900	7.4240	8.0990	573
22.3	14.2900	13.6100	8.0150	8.7440	556
23.6	4.7480	15.4700	9.2830	10.1300	525
27.4	6.8720	21.3900	13.4700	14.6900	452
31.4	9.4480	28.0600	18.7800	20.4800	395
31.6	9.6150	28.4700	19.1200	20.8600	392
44.7	20.5000	41.2100	1.8950	2.0260	277
58.4	34.2100	7.0260	3.7200	3.9590	212
64.4	40.0900	8.2360	4.6860	4.9770	193
67.6	43.4800	8.9600	5.2810	5.6010	183
72.2	47.9600	9.8950	6.1550	6.5200	172
82.1	58.0300	12.1800	8.3420	8.8070	151
83.4	59.0900	12.4300	8.6010	9.0780	149
93.4	62.4800	14.5400	11.2600	11.8500	133
109.	76.4100	17.4400	16.0900	16.8500	114
114.	80.9000	18.3700	17.8500	18.6500	109

SELF-ABSORPTION COEFFICIENTS FOR THE LOW ENERGY X-RAY REGION, $\mu_{K\alpha 1,2}$ and $\mu_{L\alpha 1,2}$. Also included are the associated <u>extrapolated</u> mass absorption coefficients μ_- and μ_+, on the short and long wavelength sides of the K and the L_I edges and the corresponding absorption jump ratios. The extrapolation to the L_I edge on the long wavelength side is of the absorption curve outside the L_{II} and L_{III} edges.

Element	λ_K (Å)	μ_-	μ_+	r_{KL}	$\mu_{K\alpha 1,2}$	$\lambda_{K\alpha 1,2}$
4 Be	111.71	178622.9	4974.0	35.91	5297.3	114.29
5 B	65.96	87685.0	3126.8	28.04	3353.0	67.65
6 C	43.69	53364.6	2233.4	23.89	2373.1	44.77
7 N	30.88	32927.5	1533.3	21.47	1632.6	31.60
8 O	23.31	22385.3	1160.2	19.29	1200.8	23.62
9 F	18.09	14850.0	847.1	17.53	867.7	18.32
10 Ne	14.30	11013.3	682.2	16.14	715.4	14.61
11 Na	11.57	7813.0	515.8	15.15	561.4	11.91
12 Mg	9.50	5986.1	439.5	13.62	489.3	9.89
13 Al	7.95	4497.4	354.5	12.69	403.6	8.34
14 Si	6.74	3648.5	306.8	11.89	356.0	7.13

Element	λ_K (Å)	μ_-	μ_+	r_{LM}	$\mu_{L\alpha 1,2}$	$\lambda_{L\alpha 1,2}$
22 Ti	22.00	21985.9	2161.6	10.17	3318.0	27.42
23 V	19.74	18734.0	1885.7	9.93	2847.6	24.25
24 Cr	17.85	17164.3	1738.9	9.87	2612.0	21.65
25 Mn	16.12	14792.6	1559.8	9.48	2317.0	19.45
26 Fe	14.66	13442.0	1453.6	9.25	2149.4	17.59
27 Co	13.40	11802.1	1310.0	9.01	1923.5	15.98
28 Ni	12.30	11004.1	1144.2	9.62	1804.2	14.56
29 Cu	11.31	9416.2	837.2	11.25	1586.0	13.34
30 Zn	10.39	5382.2	1098.2	4.90	1520.4	12.26
31 Ga	9.56	7503.7	925.9	8.10	1353.9	11.29
32 Ge	8.77	6470.2	849.9	7.61	1264.2	10.44

X-RAY FLUORESCENCE ANALYSIS OF PORTLAND CEMENT THROUGH THE USE OF EXPERIMENTALLY DETERMINED CORRECTION FACTORS

C. H. Anderson*, J. E. Mander**, and J. W. Leitner*

*Applied Research Laboratories, **Southwestern Portland Cement

ABSTRACT

Correction factors, termed α-factors, similar to those defined by LaChance and Traill have been generated by the addition of variable, known amounts of individual oxides, or other compounds, to a base cement sample and measuring the x-ray intensities of the elements of interest. The effects of all common constituents of cement on the determination of CaO, SiO_2 and Al_2O_3 were found. Factors for rhodium and chromium primary radiation were determined and, in general, showed small but significant differences. The factors for rhodium at 50kV and 30kV were substantially identical. The correction factors were tested through the use of the NBS 1011-1016 cements as reference standards to analyze the new proposed NBS cement series. The correction factors not only furnished improved calibration curves, but also allowed the determination of CaO, Al_2O_3 and SiO_2 with an average deviation of less than 0.2% (absolute) from the provisional values furnished with the standard samples.

INTRODUCTION

X-ray fluorescence methods have been applied to the analysis of Portland Cement for a number of years. In the more recent past, with improved x-ray tubes and higher available power, excitation and detection of all of the elements of interest in cement can now be determined by x-ray fluorescence. Such determinations can now be made with remarkably good precision. However when well analyzed cements are examined by x-ray fluorescence, the plot of intensity vs concentration for calcium, silicon and aluminum gives a plotting pattern whose points show an unacceptable deviation from the best straight line. The minor constituents do not show sufficient deviations to affect the practical analysis of cements. Poor sample grinding and lack of uniformity in crystalline structure may contribute to those deviations. Since interelement effects, arising from differential absorption of the primary and secondary x-rays, are known to take place it seemed reasonable to assume that these interelement effects were significant in cement analysis.

The objective of this work was to define quantitatively the interelement effects in the determination of calcium, silicon and aluminum and to demonstrate their applicability in the analysis of finished cements when NBS certified cements were used as standard samples. A further objective was to demonstrate the broad applicability of the interelement corrections to different instruments having the same x-ray optics. Since the type of exciting radiation is known to be a factor in generating interelement effects, these effects were determined using chromium and rhodium x-ray target tubes.

The interelement effects in x-ray fluorescence analysis and their physical basis are well known and a variety of approaches have been proposed. These range from the theoretical calculations as typified by Criss and Birks (1) to multiple regression correlations as proposed by Lucas-Tooth (2). Neither of these approaches appeared to be applicable to cement analysis because of the uncertainties of absorption coefficients for the light elements and the difficulty of determining multiple correlation coefficients in the relatively small concentration ranges in cements together with the uncertainties in the chemical analysis.

Our approach was based on the concept of the so-called α-factors, as proposed by LaChance and Traill (3) and further applied by Jenkins (4). In the field of electron probe analysis this concept has been widely and successfully applied, notably by Albee and Ray (5), who have determined correction factors for most elements of geological interest. The success of this approach in electron probe analysis is due largely to the ability of the focussed electron beam to analyze a small, homogeneous grain. On the other hand, in x-ray fluorescence (unless fusion techniques are employed) the individual particles are generally not homogeneous. Nevertheless, in finished cements, the minerals have been blended and fired at high temperature and approach some degree of homogeneity. It was felt worthwhile to investigate the so-called α-factor method for making interelement corrections in cement.

BASIS OF THE METHOD

Following LaChance and Traill (3) and Jenkins (4), the general form of an empirical correction can be expressed as,

$$I_A = \frac{K \cdot C_A}{1 + \Sigma \ (\alpha_{AB} \cdot C_B + \alpha_{AC} \cdot C_C + \ldots)} \quad (1)$$

Where

I_A = net (background and dead-time corrected) intensity of element A

K = constant for a given set of instrumental conditions

C_A = concentration of element A

α_{AB} = correction factor for the effect of element B on element A

C_B = concentration of element B

The above equation reflects the proportionality between x-ray intensity and concentration, appropriately adjusted by the correction factor α_{AB}. If two samples, one of which is considered to be a standard, the other an unknown, are analyzed, the relationship becomes,

$$C_A^x = \frac{I_A^x}{I_A^s} \cdot C_A^s \ (1 + \Sigma \ \alpha_{AB} \ (C_B^x - C_B^s) + \alpha_{AC} \ (C_C^x - C_C^s) + \ldots \quad (2)$$

Where the superscripts x, s refer to the unknown and standard respectively.

If we restrict our system to the effect of one element on another and rearrange equation (2) the following expression is obtained:

$$\frac{C_A^x \cdot I_A^s}{I_A^x \cdot C_A^s} = 1 + \alpha_{AB} \cdot (C_B^x - C_B^s) \quad (3)$$

A plot of the left hand side of the equation against $(C_B^x - C_B^s)$ then should give a straight line with the slope yielding α_{AB} directly and the intercept at unity.

EXPERIMENTAL DETERMINATION OF CORRECTION FACTORS

Equation 3 above was used as the basis for determining the interelement correction factors for CaO, Al_2O_3 and SiO_2 in cement. A series of samples of cement containing individual interfering oxides added to contain variable amounts of the interfering oxide were prepared. The interfering elements were added one at a time such that a set of four samples (base material plus three samples with incremental amounts added) were prepared for each interfering oxide. In the case of MgO, SiO_2, Al_2O_3 and Fe_2O_3, these oxides were added to the cement directly. To determine the effect of SO_3, Na_2O, K_2O, P_2O_5 and CO_2 it was necessary to add these components as compounds. Table 1 shows the compounds added.

The samples were thoroughly mixed and ground in a Bleuler Mill through the use of 5cc of absolute ethanol as a grinding aid. The samples were dried and pressed for thirty seconds at 50,000 psi. The net (i.e. background and dead-time corrected) x-ray intensities of all of the elements were obtained for each set of samples containing the individual added elements. The data were obtained on three different x-ray units representing two different models of similar geometry. Two different units of Applied Research Laboratories Model VXQ-25000, utilizing an OEG-60 Cr tube, and one Applied Research Laboratories Model VXQ-72000 with an OEG-75 Rh tube were used. Additional details of the experimental conditions are shown in Table 2. It will be noted that with the VXQ-72000 data were obtained at 50kV and 30kV. Data for the OEG-60 Cr tube were obtained on two VXQ-25000 x-ray Quantometers in order to determine whether the correction factors were materially different for units of the same geometry.

The treatment of the data followed the previous considerations. The expected intensity to be obtained through dilution only and assuming no interelement effects, was calculated and termed $I_{Theor.}$ The ratio of the I_{Theor} to the actual intensity found (I_{Exp}) was calculated and plotted against the percent added constituent. In the case of the pure oxide additions the α factor is found directly from the slope of the line. Figure 1 shows an illustrative example of the experimental data and plot for determining the effect of MgO on CaO.

Table 1. List of Added Compounds

Constituent	Added As	Constituent	Added As
CaO	$CaCO_3$	P_2O_5	$(NH_4)H_2PO_4$
K_2O	K_2SO_4	SO_3	Si_2SO_4
Na_2O	Na_2CO_3	L.O.I.	Li_2CO_3 $CaCO_3$

Table 2. Experimental Conditions.

Instrument	VXQ-72000 $\phi_1 = 90°, \phi_2 = 30°, 40°$	VXQ-25000 $\phi_1 = 90°, \phi_2 = 35°, 45°$
Tube Target	Rh	Cr
Tube Power	50kV, 40mA 30kV, 40mA	50kV, 25mA
Crystals and Detectors		
Silicon	EDdT, Ne Exatron, Be Window	
Aluminum	ADP, Ne Exatron, Al Window	
Calcium	LiF, Ar Multitron	

ϕ_1 = Angle of Incidence, ϕ_2 = Angle of Emergence

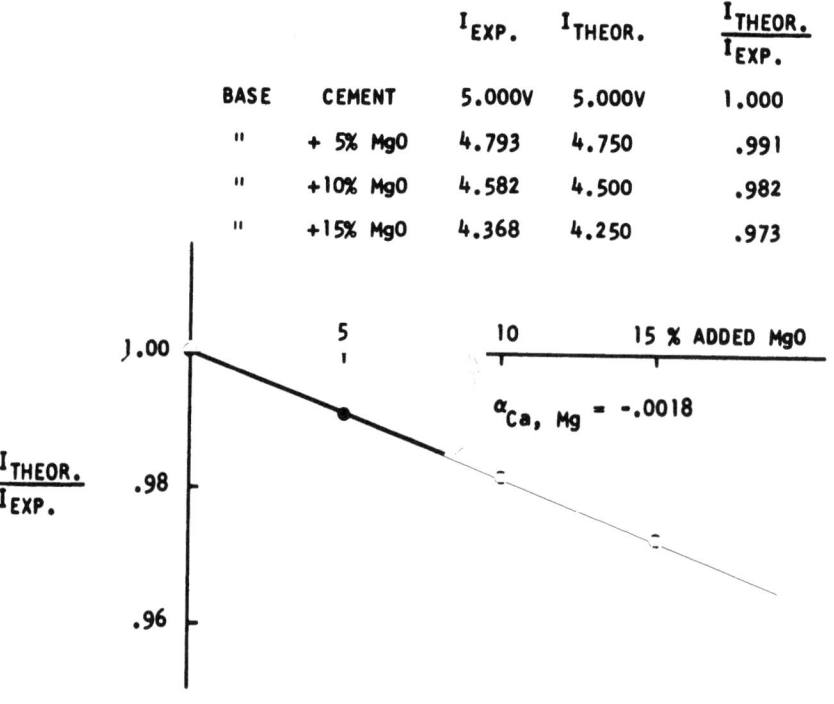

Figure 1. Determination of $\alpha_{Ca, Mg}$, Rh Tube.

It was found early in this investigation that the amount of loss on ignition (L.O.I.), i.e. CO_2 and H_2O had a significant effect on the x-ray analytical results. The α-factor for L.O.I. was determined for calcium in two separate experiments: (1) through the addition of $CaCO_3$ and calculating the effect as that due to CO_2 only and (2) through the addition of Li_2CO_3 and treating the total Li_2CO_3 as representing the L.O.I. effect. The results of these experiments indicated that the effect of CO_2 and Li_2CO_3 were, within experimental error, identical. It appeared therefore that Li_2CO_3 could be used to estimate the L.O.I. α factor on other elements.

In the case of the alkalis, SO_3 and P_2O_5, requiring the addition of a compound, it was assumed that the α factor for the compound was a linear combination of the separate oxides. The α factor of interest was then obtained by difference.

The results of these experiments, using both Rh and Cr radiation and summarizing the α factors found in this work, are shown in Table 3.

An investigation of the effect of 50kV vs 30kV for the OEG-75 Rh tube on the experimentally determined α-factors for Ca showed no significant differences.

One of the objectives of these experiments was to obtain factors which would be valid as universal correction factors for instruments of the same x-ray geometry. Experimental determination of the α-factors on two different x-ray units of the same design furnished α-factors for Cr excitation that, within experimental error, were identical. Subsequent data on four other x-ray instruments, three VXQ-72000s and one VXQ-74000 of the same geometry and using Rh excitation have shown the Rh correction factors listed in Table 3 to be valid for each instrument.

SIGNIFICANCE OF CORRECTION FACTORS

The factors listed in Table 3 can be assessed more readily in terms of their analytical significance if it is considered that the total effect is proportional to the amount of the element being determined and proportional to the amount of the interfering elements. The factors as listed show the effect of 1% absolute change of interfering element on the determination of 1% of the element. To estimate the total effect for example of 1% change in K_2O on the determination of CaO at 65% we obtain,

$$65\% \times .0228 \times 1\% = 1.48\% \text{ error (absolute)}$$

Table 3. Interelement Correction (α) Factors for Cement.

Analyte	CaO		SiO_2		Al_2O_3	
Interfering Element	Rh	Cr	Rh	Cr	Rh	Cr
Na_2O	-.0013	-.0007	+.0075	+.0110	+.0079	+.0096
MgO	-.0018	-.0013	+.0065	+.0100	+.0110	+.0120
Al_2O_3	-.00053	+.0002	+.0098	+.0130	------	------
SiO_2	+.0003	+.0013	------	------	-.0037	-.0050
SO_3	+.0020	+.0030	+.0027	+.0030	+.0078	+.002
K_2O	+.0228	+.0240	-.0001	+.0019	+.0020	+.0049
P_2O_5	-.00016	-.0015	-.0003	+.0008	-.0003	-.0007
CaO	------	------	+.0010	-.0015	-.0007	+.0043
Fe_2O_3	-.0028	-.0019	+.0071	+.0086	+.0093	+.0088
L.O.I.	-.0067	-.0060	-.0013	-.0013	-.0006	+.0028

Since potassium absorbs CaK_α radiation the result would be to cause a negative error (thus the correction factor is positive in sign).

Similarly, to again emphasize the error caused by L.O.I., at 65% CaO, and 1% difference in L.O.I.,

$$65\% \times .0067 \times 1\% = .44\%$$

In this case, since the correction factor is negative, the actual results would be too high by .44%.

APPLICATION OF CORRECTION FACTORS

The correction factors obtained in these experiments were used to correct the x-ray intensities, expressed as volts collected on integrating capacitors, of the NBS series 1011-1016 cement standard samples. Thus,

$$V^A_{corr} = V^A_{uncorr}(1 + \alpha_{AB} \cdot \%B + \alpha_{AC} \cdot \%C + \ldots)$$

The corrected voltage was then plotted against concentration. Figure 2 shows the improvement obtained when the corrections for a Rh target tube are used to generate the working curve for CaO in the NBS standard cements. In this figure the uncorrected voltages are also indicated, showing the characteristic scatter of points when uncorrected x-ray intensities are plotted with these standard samples. Identical results were obtained using a chromium tube for excitation. It should be pointed out that only the net CaO concentration was used, i.e. after the SrO concentration has been subtracted from the certificate value of (CaO + SrO).

In order to further test the correction factor approach to analysis, the new NBS cement standard samples, Nos. 633-639, were analyzed using the 1011-1016 series as standard samples. Figure 3 shows the data for CaO obtained for the new series when plotted on the curve obtained for the NBS 1011-1016 series. Similar improvement is obtained in the silica curve and somewhat less dramatic improvement is attained for alumina.

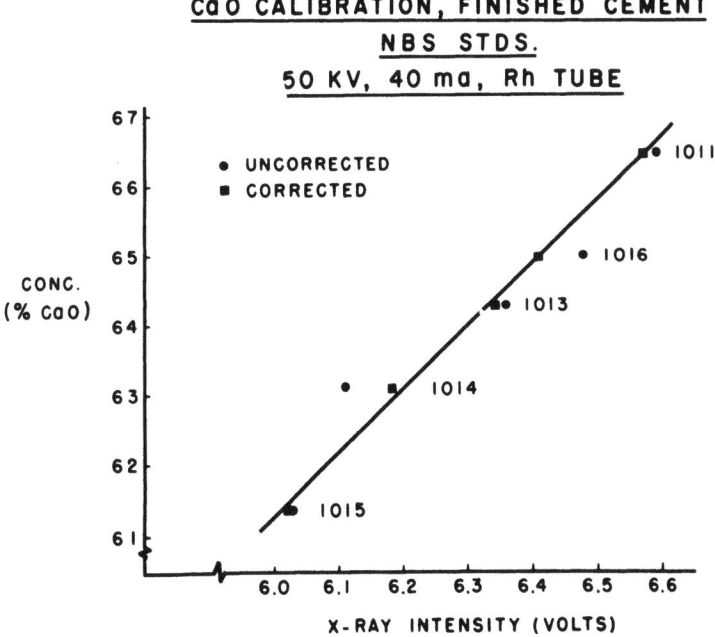

Figure 2. CaO Calibration NBS 1011-1016 Cement.

Table 5. Comparison of α-Corrected Results on NBS Nos. 633-639 Standard Cements.

Sample No.	CaO		SiO_2		Al_2O_3	
	Cert.	X-Ray	Cert.	X-Ray	Cert.	X-Ray
633	64.5%	64.6%	21.9%	21.9%	3.74%	3.69%
634	62.6	62.9	20.7	20.5	5.2	5.1
635	59.8	59.8	18.5	18.2	6.2	6.5
636	63.5	63.6	23.2	22.9	3.1	3.3
637	66.0	66.4	23.1	23.1	3.3	3.3
638	62.1	62.0	21.4	21.4	4.5	4.5
639	65.8	66.0	21.6	21.4	4.3	4.3

The analysis of the NBS series 633-639 for CaO, SiO_2 and Al_2O_3 by x-ray fluorescence using the correction factor approach is shown in Table 5. It will be noted that the average deviations of the x-ray results are approximately ±.2% of the provisional certificate value.

While the primary purpose of this work was to employ the correction factor approach to the analysis of finished cement, it is of practical importance to analyze raw mix by x-ray fluorescence. The correction factors should be universally valid for any oxide material and preliminary experiments have shown an improvement in analysis of raw mix by x-ray fluorescence. For example, Figure 4 shows the uncorrected and the α-corrected working curve for CaO in raw mix materials. It will be noted that the scatter of points in the uncorrected data is markedly improved through the use of the α-factor approach.

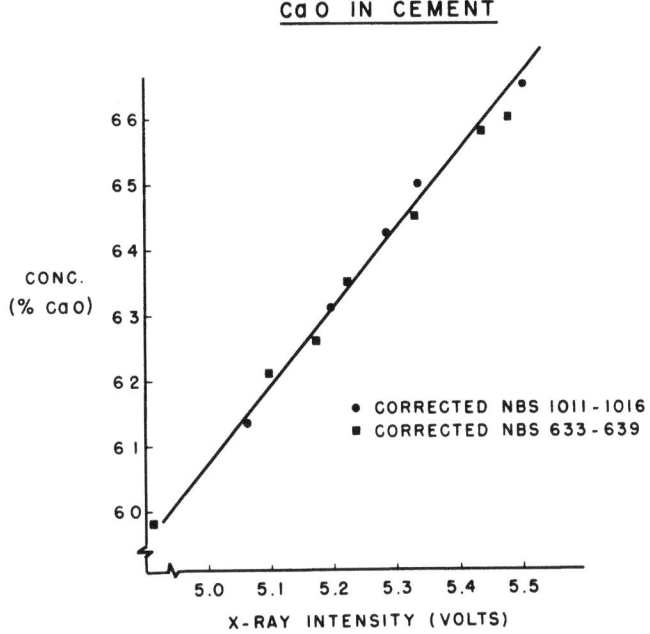

Figure 3. CaO Calibration NBS 1011-1016 and NBS 633-639.

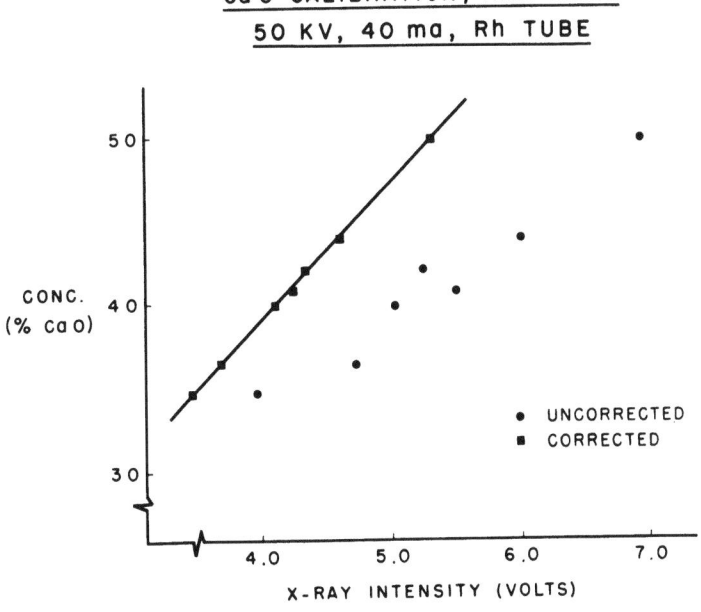

Figure 4. CaO in Raw Mix Cement.

Because of the volume of calculation required, the α-correction approach is best handled through simple computer programs. It has been found that the programmable calculators currently available are satisfactory.

SUMMARY AND CONCLUSIONS

This study has shown that correction factors for CaO, SiO_2 and Al_2O_3 in cement can be experimentally determined and that they are valid for different instruments of the same geometry. As a result improved analytical results are achieved by x-ray fluorescence spectrometry and establish x-ray fluorescence as a valid analytical procedure for cement analysis.

REFERENCES

1. J. W. Criss and L. S. Birks, "Calculation Methods for Fluorescent X-Ray Spectrometry," Analytical Chemistry, 40, 1080-1086 (1968).

2. J. Lucas-Tooth and C. Pyne, "The Accurate Determination of Major Constituents by X-Ray Fluorescence Analysis in the Presence of Large Interelement Effects," Advances in X-Ray Analysis, Vol. 7, Plenum Press New York, 526-554 (1964).

3. G. R. LaChance and R. J. Traill, "A Practical Solution to the Matrix Problem in X-Ray Analysis", Canadian Spectroscopy, 11, 43-48 (1966).

4. R. Jenkins and A. Campbell-Whitelaw, "Determination of Interelement Correction Factors for Matrix Correction Procedures in X-Ray Fluorescence Spectrometry," Canadian Spectroscopy, 15, 32-38 (1970).

5. A. L. Albee and Lily Ray, "Correction Factors for Electron Probe Microanalysis of Silicates, Oxides, Carbonates, Phosphates and Sulfates," Analytical Chemistry, 42, 1408-1414 (1970).

SPECIMEN STANDARDS FOR X-RAY SPECTROMETRIC ANALYSIS OF

ATMOSPHERIC AEROSOLS

Bruce E. Artz and Henry Chessin

State University of New York at Albany

Albany, New York 12222

ABSTRACT

Specimen standards have been prepared for elemental quantitative analysis of atmospheric aerosols by means of x-ray spectrometry. Efforts were made so that the standards were similar to the samples actually collected in the atmosphere. Soluable salts were placed in a completely enclosed, self-contained bubbler system so as to ensure uniform rate of flow through the filters on which elements were collected. The standards were deposited for various time intervals in the bubbler system and then weighed on a microbalance. Aerodynamic flow ensured that the specimens were uniformly distributed throughout the filter. Very satisfactory calibration curves were constructed of intensity versus mass of element. For low concentrations of the soluable salts and relatively small amounts of total material collected straight line relationships can be obtained for calibration curves. These amounts are mostly well within those actually collected in the field. The problem of impurity levels in collection filters which can far exceed the levels detected in the aerosols has been given careful consideration in the selection of filters used. Absorption-enhancement effects appear to be unimportant at the low concentration levels delt with. Interference problems with lead, arsenic, and bromine are discussed and appropriate calibration curves shown. A set of atmospheric specimens are analysed for several elements placing atmospheric aerosol analysis on a quantitative basis.

INTRODUCTION

This investigation was part of a program to develop a technique to monitor the elemental content of atmospheric aerosols. New York State has established stations throughout the state with equipment that will pass air through a filter of approximately 400 square centimeters in area. Chessin and McLaren (1) did preliminary work using x-ray spectrometry in the analysis of these filters and this study represents a continuation of these efforts. The particular aspect of the problem that will be discussed in detail is the preparation of suitable standards for quantitative analysis by x-ray fluorescent spectrometry (XFS) of the aerosol and particulate matter collected on the filters. The technique employed is that of calibration standardization (2).

APPARATUS AND SELECTION OF ELEMENTS

The potentially large number of collection sites mandates a rapid turn-around time in the analysis of the samples. To this end, it was thought that the method of calibration standardization would, with suitable standards, provide the quickest analysis with a minimum amount of specimen preparation. Once calibration curves are obtained for the various elements of interest, it is a relatively straight forward task to cut a specimen of proper size and perform XFS analysis on a computer controlled spectrometer.

The apparatus for the preparation of standards was described in detail by Chessin and McLaren (1) but some modification has occured. Figure 1 represents the present state of the bubbler apparatus. The most important change has been to enclose all but the bubbler in an isothermal environment. The reason that this was done was because it was observed that condensation was occuring in the relatively low pressure area preceeding the collector. This condensation was wetting the filter paper in the collector and it was feared that this could have a detrimental effect on the uniform deposition of salts on the filter. Enclosing this region and the collector itself, in the isothermal environment at about 45°C prevented this condensation and ensures an even deposition of analyte salt on the filterpaper. An additional change to the original apparatus was the removal of a liquid trap from between the final filter and the bubbler as this was found to be unnecessary.

The selection of a filter material for use in the bubbler apparatus was based on two criteria. The first criteria was that it be identical, or as near as possible, to the filters being used for sample collection and the second criteria was ease of handling and weighing. Analysis by Dams (3) and Chessin and McLaren (1) had indicated that Whatman 41 filter paper has the lowest background contamination of any filter available. For this reason New York State has been collecting samples on sheets of Whatman 41 filter

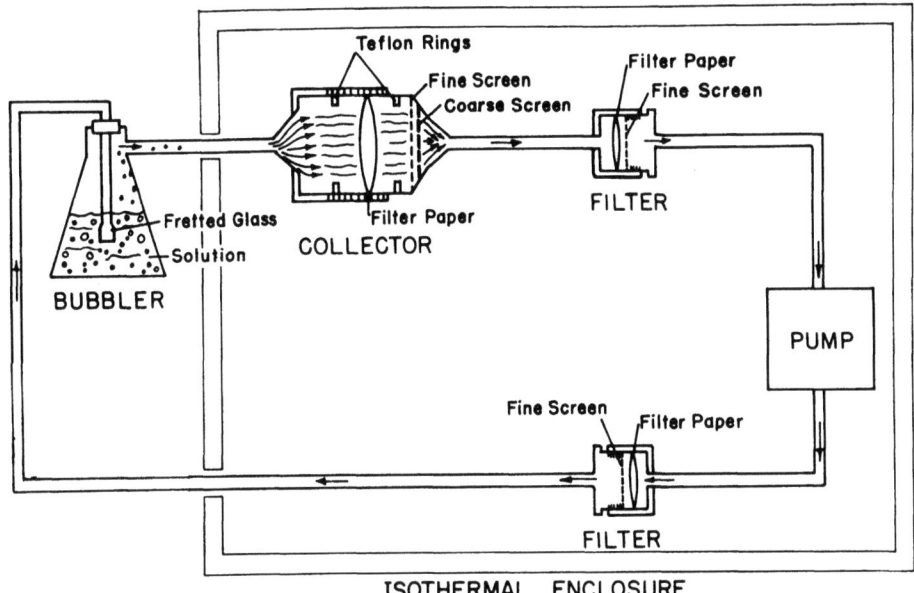

Figure 1. Apparatus for the Collection of Aerosol Standards

paper and consequently it seemed desireable to use this paper in the bubbler apparatus. Considerable effort was made to utilize the Whatman 41 paper but the apparent hydroscopic properties made consistent weighing of these papers impossible. Two other filter materials were readily available that were similar to Whatman 41, except that they were not of the paper type. MSA (fiberglass) and Microsorban (polystyrene) were examined to determine their suitability for accurate weighing. Both filters appeared not be hydroscopic to any great extent and appeared to reach equilibrium with room conditions within a few minutes. The major difference between the filters was the physical properties as the MSA filter was extremely weak and easily torn or damaged which made processing the filter very difficult. The Microsorban filter was better in this regard, although not as good as Whatman 41, and consequently Microsorban was chosen to be used in the bubbler apparatus. Background 2θ scans were made of both Whatman 41 paper and the Microsorban filter to see what levels of impurities were present and these are shown in Figure 2 along with a similar scan of New

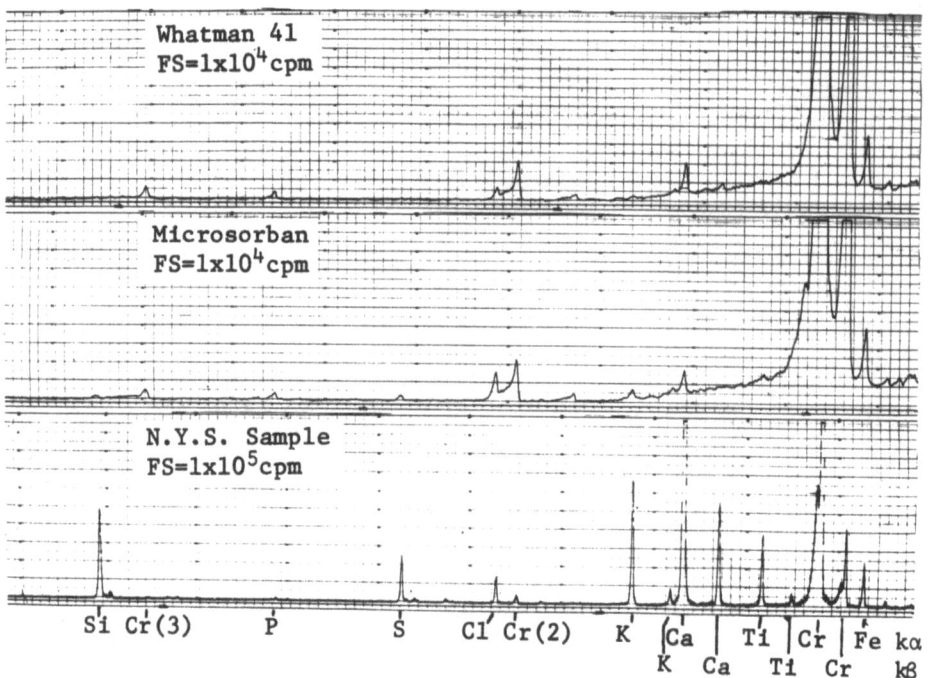

Figure 2a. Long wavelength scans- PET, Prop C, 0.15° slit, 50kV 34mA

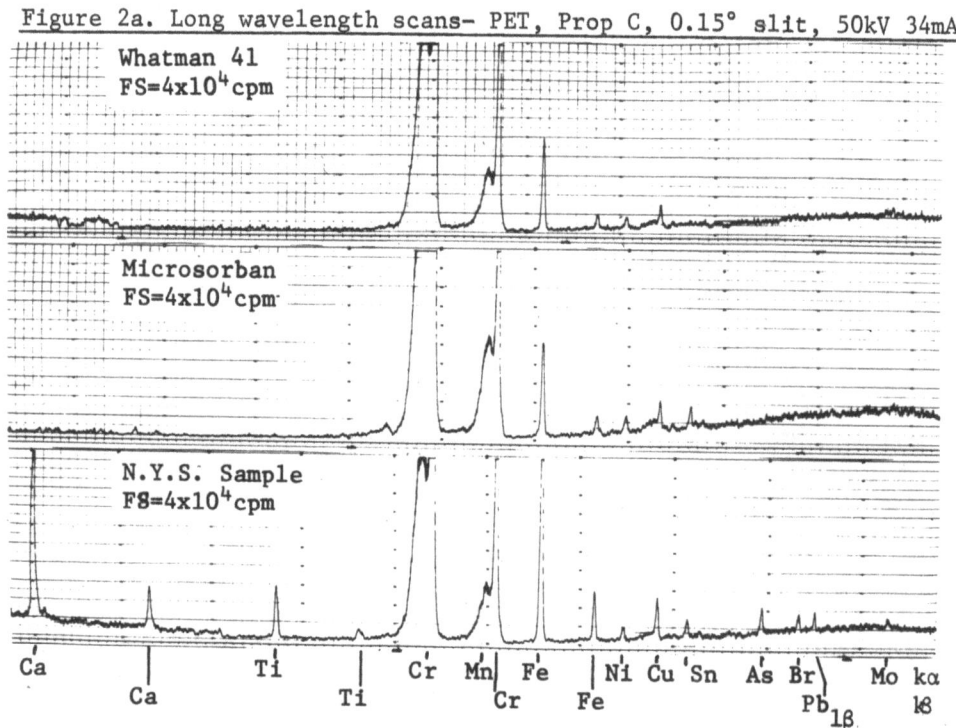

Figure 2b. Short wavelength scans- LiF, Sc Det. 0.15° slit, 50kV 34mA

York State specimen.

The initial choice of what calibration elements to use was determined from two considerations. First it was considered desireable to obtain calibrations for elements that existed in the samples collected and second it would be necessary to choose several elements so as to provide a wide range of conditions to check the system. Initially, approximately 30 N. Y. State samples were scanned in a semi-quantitative manner to determine the principal elements and approximate amounts. Figure 2 shows the typical scans made for a particular N. Y. State specimen.

The large amount of zinc present in the samples made this element a necessary choice. In addition, several non-hydroscopic salts were available and a comparison of the results would determine if the choice of salt had any effect on the calibration curve. The available salts were $Zn(NO_3)_2 \cdot 6H_2O$, $Zn(SO_4) \cdot 7H_2O$ and $ZnBr_2$. These salts also provided sulphur as well as bromine calibration curves. The abundance of calcium was somewhat unexpected and analysis of this element seemed desireable. The use of the salt $CaBr_2 \cdot 2H_2O$ would check the bromine calibration curve already obtained. The choice of arsenic and lead provided a system where spectral interference was particularly important. Since most specimens contain both arsenic and lead it will be necessary to have calibration curves other than the usual $As_{k\alpha}$ and $Pb_{\ell\alpha}$ lines which interfere. The $As_{k\beta}$ and $Pb_{m\alpha}$ lines will be measured in addition and used as calibration standards when necessary. The respective sensitivities of these lines can be compared to the preferable stronger lines also.

PROCEDURE AND PREPARATION OF STANDARDS

Without doubt the most difficult and sensitive part of the calibration procedure is the determination of the specimen weights. The balance used was a Mettler Gramatic balance with a precision of one microgram. The balance proved to be extremely sensitive to room temperature fluctuations in terms of zero-drift and consequently it was usually necessary to maintain the balance room to ±0.25°F while using the balance. This was accomplished primarily by monitoring the temperature and using the balance only during times of extreme temperature stability. Initially several 1-7/8" diameter Microsorban filters were pre-weighed, processed through the bubbler apparatus, which contained only distilled water and weighed again. The purpose of this was to determine the effects of this process on the weights of the filters. The filters were exposed for from five to fifteen hours and then dried in a vacuum dessicator for several hours before being reweighed. The results of this test indicated that one could expect a ±20 µg precision in the filter weights as a result of handling. These filters were also scanned by the spectrometer before and after treatment and no difference was noted.

A series of pre-weighed Microsorban filters were processed through the bubbler system containing an approximately 0.1 molar solution of a salt containing the analyte being calibrated. The filters were exposed for from one to about 20 hours on a random schedule-that is, the exposure times were random in length from one filter to the next. This minimized any cumulative time effects that might occur. The filters were dried and reweighed to determine the amount of salt deposited, from which the weight of analyte could be determined. The specimens were then measured in the X-ray spectrometer to determine the net intensity for the particular analyte, or analytes, desired. A three point measurement scheme was used which consisted of a peak measurement and a background measurement on each side of the peak. The net intensity, peak minus average background, was necessary because of the rather large fluctuations in background intensity encountered from filter to filter. This was caused primarily by different thicknesses of the Microsorban filters. It was also necessary to use net intensities since the sample analysis would be done on Whatman 41 paper which will have an entirely different background. The specimen holder exposed a slightly larger area of the filter than was exposed in the bubbler. This was necessary to make certain that all of the analyte was being exposed to the x-ray beam. The slight error introduced by the difference in sample exposure size, limited by the specimen holder, and the standard exposure size, limited by the teflon rings in the bubbler apparatus was very small since the additional area being exposed on the unknown was at the extreme of the specimen holder and well into the tail of the incident beam profile.

CALIBRATION CURVES AND RESULTS

Figures 3 through 6 show some of the calibration curves obtained for the aforementioned elements using the previously discussed technique. The importance of the various hydrated forms for a given salt is illustrated best in the zinc calibration data. Linke (4) gives the hydrated forms of the various zinc compounds used here as a function of temperature and solubility. It is not entirely clear what determines which hydrated form will be deposited in the collector which is at about 45°C in the isothermal enclosure. $ZnSO_4$ shows a transition from six waters of hydration to one as the temperature increases from 48°C to 75°C. $Zn(NO_3)_2$ shows a transition from $4H_2O$ to $2H_2O$ at 45°C to $1H_2O$ at 54°C, while $ZnBr_2$ exists in the $2H_2O$ condition below 35°C with no hydration above that temperature. As can be seen from the calibration data most salts deposited in their most hydrated form, the only exception being $Zn(NO_3)_2$ which showed no hydration. The proper hydration was determined by adjusting the amount of hydration on each compound until the best agreement was found. It would seem reasonable that the salts would exist in their most hydrated forms since they are being evaporated from an aqueous solution at room

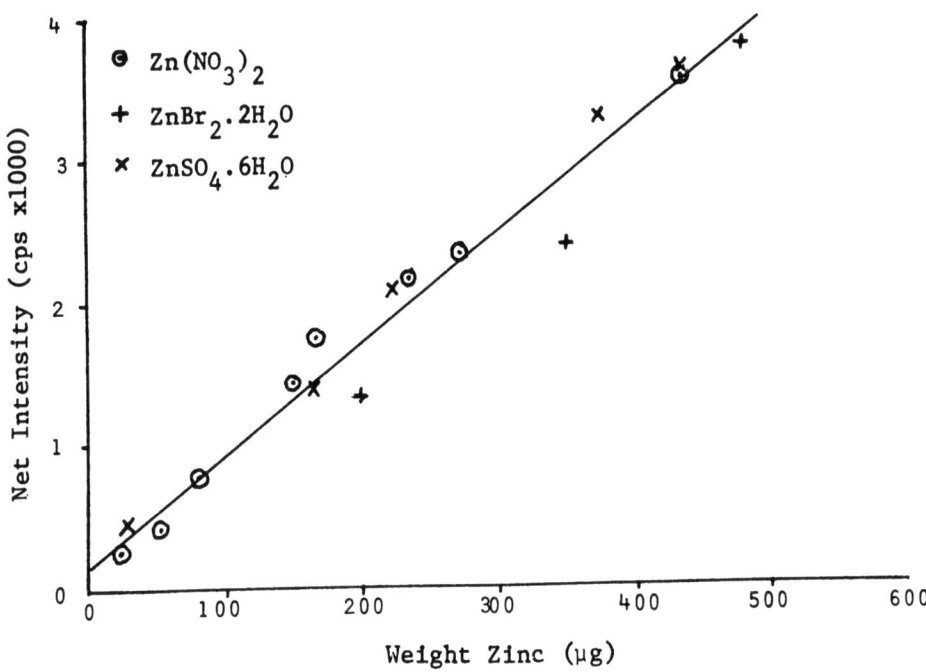

Figure 3. Zinc calibration curve. 50kV 34mA, Cr tube, 0.15° slit.

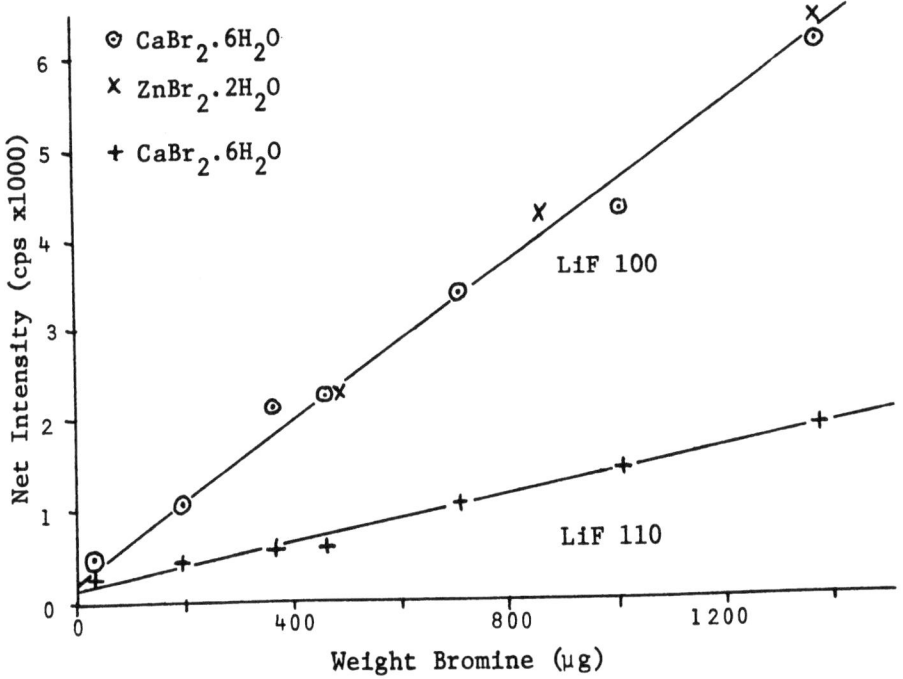

Figure 4. Bromine calibration curves. 50kV 34mA, Cr tube, 0.15° slit

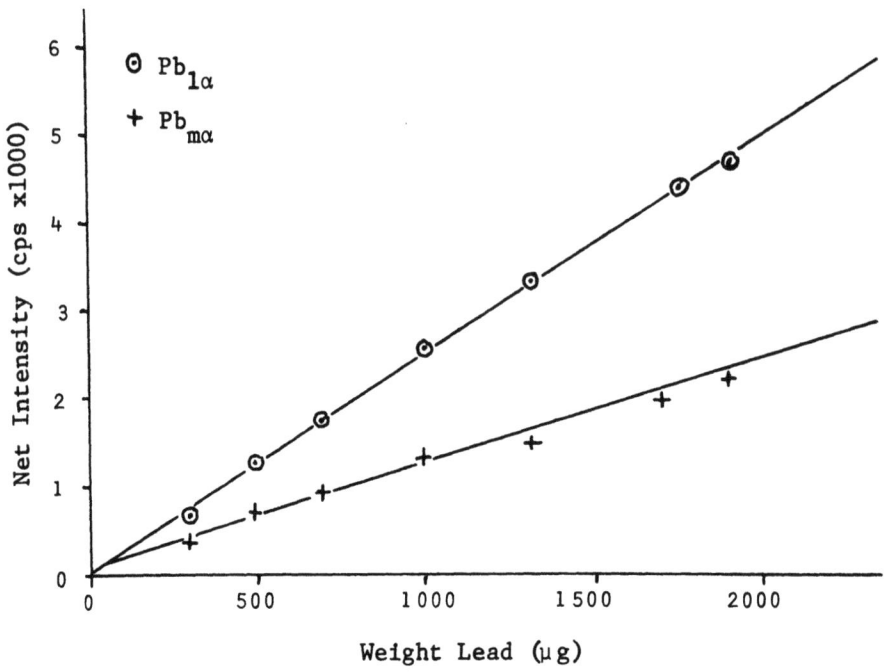

Figure 5. Lead calibration curves. 50kV 34mA, Cr tube, 0.15° slit.

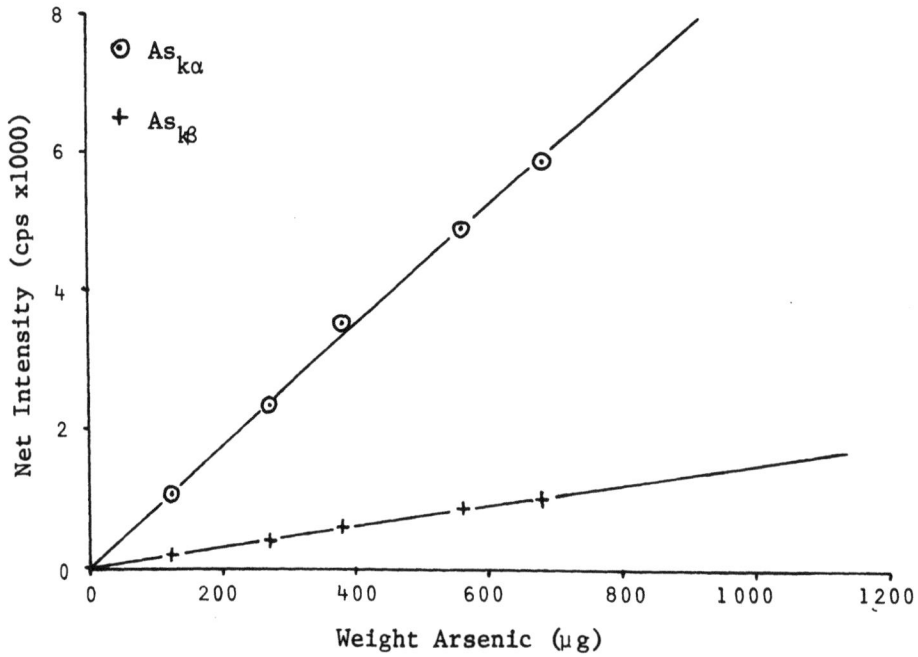

Figure 6. Arsenic calibration curves. 50kV 34mA, Cr tube, 0.15° slit

temperature and a possible explanation for the $Zn(NO_3)_2$ case being that the hydrated waters were removed in the collector at 45°C or during the drying process in the vacuum dessicator. The obvious problem in dealing with hydrated salts can be eliminated by using compounds that are water soluble but do not have any hydrated forms. This technique was used in determining the arsenic and lead calibration curves. As As_2O_3 and $PbNO_3$ do not show any hydrated forms (4). In the event that this type of salt cannot be obtained one method is to collect data for two or more salts with the same element and adjust the results for consistency, or another method is to use a salt that contains an element for which a reliable calibration curve exists and use this element to determine the proper number of hydrated waters. In either case knowledge of the possible hydrated forms and their transition temperatures is necessary to aid in the adjustment of the data.

Table 1 summarizes the sensitivities obtained from the calibration curves and the calculated lower limit of detection (l.l.d.) for each element on typical samples of the two types of filter material being used. The l.l.d. is calculated according to the definition put forth by Bertin (5) and is that concentration which gives three times the standard deviation σ of the background. For net intensities,

$$\sigma_{Net} \simeq (N_P + N_B)^{1/2}$$

where N_P and N_B are the total number of counts collected as the peak and background respectively. Defining R_P and R_B as the

---

Table 1. Summary of sensitivities and lower limits of detection for chromium x-ray tube at 50kV and 34 mA with a 2 cm. radius sample holder. Detection limits calculated for three point measurements of one minute each.

Analyte	Sensitivity (cps/µg)	l.l.d. Microsorban total µg	µg/cm²	l.l.d Whatman 41 total µg	µg/cm²
S kα PET	5.0±0.6	0.19	0.015	0.19	0.015
Ca kα LiF100	58.0±8	0.056	0.004	0.074	0.006
Zn kα LiF100	7.7±1.2	0.64	0.051	0.41	0.033
As kα LiF100	8.6±0.8	0.59	0.047	0.47	0.038
kα LiF100	1.5±0.2	3.30	0.26	2.90	0.23
Br kα LiF100	4.3±0.5	1.10	0.087	0.88	0.070
kα LiF100	1.2±0.1	2.10	0.17	1.73	0.14
Pb lα LiF100	2.5±0.2	1.67	0.13	1.36	0.11
mα PET	1.1±0.15	0.62	0.049	0.57	0.045

counting rates at the peak and background and T as the time spent at either peak or background gives,

$$l.l.d. = \frac{3(R_P T + 1/2 R_B T)^{1/2}}{mT}$$

where m is the sensitivity. The sensitivity is calculated from a computer generated least squares fit to the dead time corrected intensities as a function of weight of analyte. These results are shown in figures 3 through 6.

Table 1 shows the effect of using the Cr tube in that the sensitivity is much greater for calcium than any of the other elements considered. The $Cr_{k\alpha}$ is most effective for elements of atomic number less than chromium and this is reflected in the calcium sensitivity. The loss of sensitivity for sulphur is expected due to the decrease in excitation efficiency and absorption losses.

Table 2 shows preliminary results of analyses of atmospheric samples. All samples were collected at the same time in different locations for a period of one day. The concentrations are based on passing 2500 m^3 of air through Whatman 41 filters with a surface area of about 400 cm^2. The results appear to be consistent in that Binghamton and Elmira are fairly close to each other and have similar concentrations. The small concentration of arsenic made the use of the $As_{k\beta}$ line difficult and consequently a correction was made to the $As_{k\alpha}$ line and this was used. The correction con-

---

Table 2. Elemental concentrations in atmospheric aerosols collected simultaneously at several stations for one day.

City	Elemental Concentration (ng/m^3)					
	S	Ca	Zn	As**	Br	Pb*
Binghamton	600	3700	80	40	200	300
Elmira	670	3300	80	50	200	360
Grand Island	700	2600	86	20	35	90
N. Tonawanda	870	3600	130	16	35	90
Plattsburg	400	830	40	8	22	73
Troy	430	1100	40	11	45	80
Tupper Lake	630	300	16	15	80	125

*Lead determined from $Pb_{m\alpha}$ intensity.
**Arsenic determined from $As_{k\alpha}$ intensity after subtracting $Pb_{\ell\alpha}$ intensity since $As_{k\beta}$ intensity was not above l.l.d.

sisted of subtracting off the expected intensity from the amount of lead predicted by the $Pb_{m\alpha}$ line at the $As_{k\alpha}$ line position. The intensity remaining was used to determine the arsenic concentration. The concentrations given are expected to be accurate to about 10% except for arsenic which may be somewhat larger due to the technique employed.

ACKNOWLEDGMENT

This research was supported financially by Grant No. GA-31903 from the Atmospheric Sciences Section, National Science Foundation.

REFERENCES

1. H. Chessin and E. H. McLaren, "X-ray Spectrometric Determination of Atmospheric Aerosols," L. S. Birks, Editor, Advances in X-ray Analysis, Vol. 16, p.165-176, Plenum Press (1972).

2. E. P. Bertin, Principles and Practices of X-ray Spectrometric Analysis, p.387, Plenum Press (1972).

3. R. Dams, K.A. Rahn, J.W. Winchester, "Evaluation of Filter Material and Impaction Surfaces for Nondestructive Neutron Activation Analysis of Aerosols," Environmental Science and Technology, Vol.6, p.441-448, (1972).

4. W.F. Linke, Solubilities-Inorganic and Metal-Organic Compounds, Am. Chem. Soc.,(1965).

5. E.P. Bertin, Principles and Practices of X-ray Spectrometric Analysis, p.347-348, Plenum Press (1972).

A VERSATILE X-RAY FLUORESCENCE METHOD FOR THE ANALYSIS OF SULFUR IN GEOLOGIC MATERIALS

H. N. Elsheimer and B. P. Fabbi

U. S. Geological Survey

Menlo Park, California 94025

ABSTRACT

Nonproportionality of X-ray fluorescence intensity vs. concentration of sulfur occurs in geologic materials prepared as ground samples. Samples containing sulfur as sulfate yield higher intensities than an equivalent amount of sulfur as the sulfide. Although the intensity of free sulfur decreases markedly during a short exposure to X-rays in a vacuum, and sulfide intensities decrease over a much longer time period, sulfate intensities show no deterioration with time.

A fusion procedure utilizing a $LiBO_2-Ce(NH_4)_2(NO_3)_6$ flux has been developed for geologic materials to oxidize all sulfur in a multiplicity of oxidation states to a single oxidation state, namely, the sulfate. The procedure is applicable over a wide range of concentrations (0.5 - 28 wt. % total sulfur) with a detection limit of 0.01%. A suite of chemically analyzed rocks containing both sulfides and sulfates is used for standardization.

INTRODUCTION

Almost forty years ago, Parratt (1) showed conclusively that wavelengths of the K_α doublet for sulfur are not identical for the sulfate and sulfide. White et al. (2) reported a 0.10° difference in 2θ between the K_β peaks for sulfide and sulfate sulfur. Such wavelength shifts are generally small, but may cause considerable error in a quantitative analysis, particularly if both species are present. Faessler (3, 4) attributed the wavelength shifts primarily to the oxidation state of the sulfur, but both he and others (5, 6) recognized that other factors were involved relating to bonding characteristics, such as inductive effects, electronegativities of ligands, the nature of adjacent cations, and variable lattice structure.

Faessler (3) demonstrated that differences in oxidation state could also cause marked increases in emission line intensities owing to overlap of spectral lines. In this laboratory, analyses of ground samples of sulfur-containing geologic materials exhibited a nonproportionality of X-ray intensity with respect to concentration which was directly related to the oxidation state(s) of sulfur. Considerable analytical error may result from these nonproportional sulfur intensities.

EXPERIMENTAL

Gravimetric Determinations

Weigh 100 to 500 mg of hand-mixed sample into a platinum crucible and intimately mix with 3 g of Na_2CO_3 and 0.15 g of ground $NaNO_3$. First sinter, then fuse the mixture in an oxidizing gas flame. Leach the cooled fusion bead with water, transfer to a beaker for digestion on a steam bath for 2 hrs, and adjust to a final volume of about 80 ml. After overnight cooling, filter off the R_2O_3 group, precipitate the sulfate ions from the acidified filtrate with $BaCl_2$, and determine gravimetrically as $BaSO_4$. Perform all determinations in duplicate.

Sample preparation for X-ray fluorescence determination

Mix samples by the hand-rolling technique just prior to analysis because high density sulfide and sulfate minerals may segregate during mechanical mixing, causing as much as a 15% error. For samples known to contain 10% or less of sulfur as sulfide or sulfate, take 100 mg for each analysis. Weigh 50-mg portions for samples having a sulfur content significantly greater than 10%, known to contain large amounts of ferrous iron, or thought to contain elemental sulfur. Mix each sample intimately on weighing paper with 200 mg of $Ce(NH_4)_2(NO_3)_6$ (99.97% pure, G. F. Smith Chemical Co.), prepared for use by grinding to 100 mesh and drying for 1 hr at <85°C. Lightly mix into this mixture 150 or 200 mg of pure quartz, depending on the sample size (100 or 50 mg). Transfer the resulting mixture to a 30-ml, black-glazed porcelain crucible containing 1.7500 g of dried $LiBO_2$ (7).

Thoroughly mix the contents in the crucible and carefully transfer them to a 30-ml, preignited graphite crucible with a hemispherical bottom (7). Fuse the sample mixture for 15 min in a furnace at 900°C and cool the resulting bead in air.

Remove the bead from the crucible, weigh, and place in a large steel ball mill (Spex Industries, Inc., Model No. 8001) containing two 1/2-inch-diameter steel balls. Grind for 4 min and transfer the resulting powder to a boron carbide mortar containing a weighed portion of chromatographic cellulose (Whatman CF-11) sufficient to bring the combined weight of bead and cellulose to 2.4000 g. After mixing to a uniform consistency, return the mixture to the ball mill and grind for an additional 5 min. To insure uniform particle size and homogeneity, transfer the ground mixture to the mortar again and hand grind for a few min. Split the finely ground mixture into approximately equal parts on weighing paper and prepare 2 pellets for XRF determination (8). Analyze each sample or standard in duplicate.

Apparatus and Experimental Conditions

A General Electric XRD-6 vacuum spectrograph equipped with a 4-crystal changer and a Cr-W dual target X-ray tube was used in the analyses. The experimental operating conditions were as shown in Table I. Second order titanium K_{α_1} radiation was effectively discriminated against by the pulse height selector. The intensity for each sample pellet was compared to that of a 26 wt. % sulfur standard pellet which had an intensity of 1950 cps and recorded as an intensity ratio. This technique neutralized instrumental fluctuations.

RESULTS AND DISCUSSION

Figure 1 illustrates the nonproportionality of X-ray intensity with concentration for a series of sulfur-containing geologic materials that had been mixed with cellulose binder in a 1:1 ratio and pelletized (9). For the suite of samples containing sulfur as the sulfide, a good calibration curve was obtained. Samples containing only sulfates, or mixtures of sulfates and sulfides, gave results as much as 28% high when compared with the sulfide calibration curve, although each type indicates a linear relationship when compared with samples of the same kind. Samples containing principally elemental sulfur would be expected to yield lower results than the sulfides because of elemental sulfur volatilization under conditions of analysis.

Table I. Operating conditions for X-ray fluorescence determination of sulfur in geologic materials

Target:	Cr	Detector:	Flow Proportional Counter
Crystal:	PET	Voltage:	1400 V
Voltage:	50 KV	Base:	0.40 V
Amperage:	50 ma	Window:	1.00 V
Collimator:	5 mil	Vacuum:	<200µ
Tunnel:	20 mil	Gain:	16 x 1
2 θ:	75.76°	Time:	2 x 50 sec.

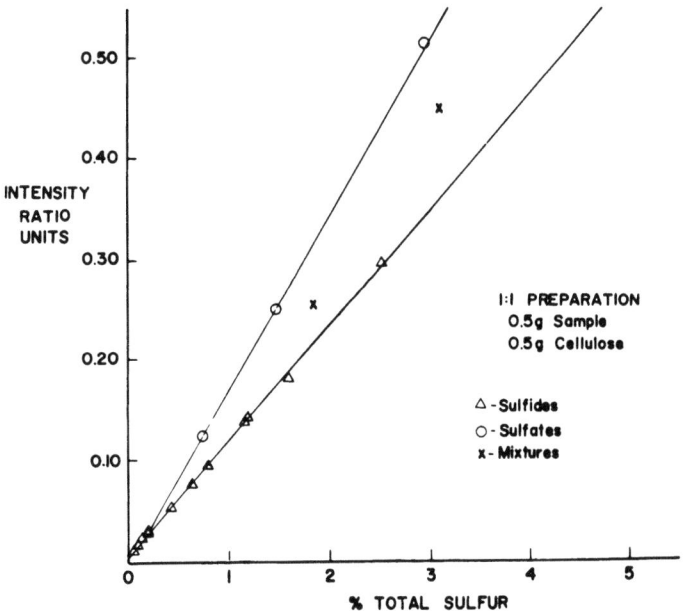

Figure 1. Nonproportionality of sulfide and sulfate sulfur X-ray intensities.

The disparity of results between the various sample types has a two-fold explanation. The first is that variations in sulfur oxidation state cause spectral line shifts, which may result in intensity differences through spectral line overlap, and/or a shift in 2θ. Spectral line overlap was considered to be more important, since only a $0.04°$ 2θ shift was observed. Second, intensity decreases with time for both elemental and sulfide sulfur under X-ray irradiation in vacuum. A pellet of a synthetic sample containing 5% elemental sulfur mixed with USGS standard BCR-1 decreased approximately 35% in intensity over a span of 5 successive counting periods of 100 sec each, with counts being taken at 10-sec intervals. Pellets of sulfide samples were found to lose 5-10% intensity over a period of several months. By contrast, pellets of sulfate samples show no change in intensity with time.

To eliminate nonproportional sulfur intensities due to variations in oxidation state, all sulfur species were converted to the sulfate prior to X-ray irradiation. Several oxidants were investigated--$NaNO_3$, $KBrO_3$, $KMnO_4$, CeO_2, and $Ce(NH_4)_2(NO_3)_6$. Sodium nitrate produced beads of poor quality that were not easily removed from the graphite crucibles. Ceric oxide beads were inferior to those of the ceric ammonium nitrate salt. Fusions with potassium bromate yielded significantly lower intensities than those with ceric salt above 8% sulfur and gave erratic results above 20%. The permanganate salt yielded results comparable to the ceric salt for samples with a greater than 20% sulfur content, but these were very inconsistent. Manganese would also interfere spectrally with the determination of other elements on these same pellets.

The oxidant found to give the most satisfactory results was the ceric ammonium nitrate salt (CAN). Two hundred mg of CAN added to 1.7500 g of $LiBO_2$ flux, plus 150 or 200 mg of pure quartz to improve the bead quality, was adequate to oxidize sulfur in 100- and 50-mg portions of rocks and minerals. The smaller sample portions were necessary for samples containing more than 10% sulfur or large amounts of ferrous iron, which competes for the available oxidant. The final sample mixtures taken for pelletizing represented a 1:23 dilution for the 100-mg sample portions and a 1:47 dilution for the 50-mg portions. If 50-mg portions of all samples were taken for analysis, the sensitivity for low concentrations of sulfur would decrease, as well as that of other elements of interest. In addition, small portions could lead to deleterious sampling effects. Use of larger sample portions, while preserving the same 4:1 ratio of flux to sample, was not feasible because of the size limitations of the graphite crucibles available.

Figure 2 illustrates the complete oxidation of sulfur by CAN for a suite of 10 rock and mineral standards used for calibration whose sulfur contents were determined gravimetrically. Only the sample with a sulfur content of approximately 24% exhibited a significant spread of values which was probably caused by heterogeneity through sample depletion. These standards and their XRF vs. chemical values are listed in Table II. The average relative error for the XRF vs. the gravimetric values was 1.2%.

Table II. Sulfur content of standard samples

Sample	XRF(% S)[a]	Gravimetric (% S)[1]
457 Limestone	0.54 ± 0.01	0.54
255 Jasper	1.69 ± .01	1.65
888 Metallized hornfels	2.97 ± .24	3.02
355 Metallized quartz-mica schist	5.01 ± .11	5.10
386 Andesite (altered)	5.36 ± .22	5.37
369 Metallized quartz-mica schist	7.70 ± .40	7.87
359 Metallized quartz-mica schist	10.24 ± .04	10.36
365 Metallized quartz-mica schist	20.58 ± .38	20.23
354 Metallized quartz-mica schist	24.27 ± 1.05	24.32
379 Metallized quartz-mica schist	25.64 ± .01	25.91

[a] Precision is relative to duplicate samples.

[1] H. N. Elsheimer

Average relative error: 1.2%

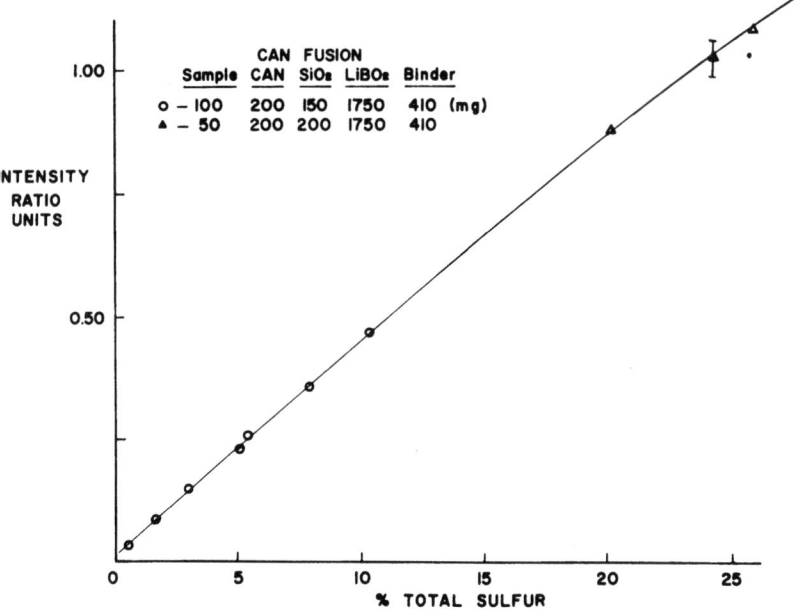

Figure 2. Calibration curve for sulfide-sulfate standards using $Ce(NH_4)_2(NO_3)_6$ as oxidant.

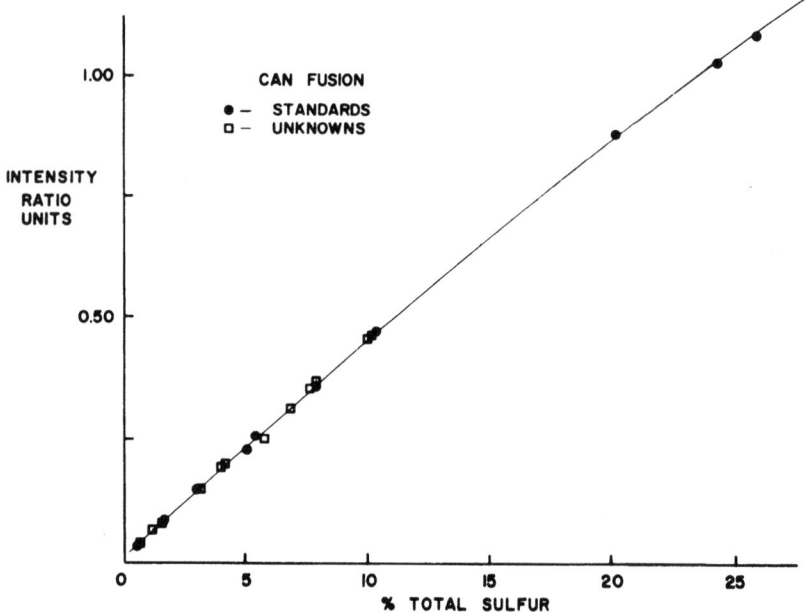

Figure 3. "Unknown" sulfide-sulfate samples vs. a sulfide-sulfate standard curve.

Table III. Sulfur content of "unknown" samples

Sample		XRF(% S)[a]	Gravimetric(% S)
373	Andesite (altered	10.20 ± 0.15[a]	10.16^{1}, 9.87^{2}
378	" "	$7.87 \pm .02$	7.54^{1}, 7.56^{2}
387	" "	$9.72 \pm .15$	10.08^{1}
383	" "	$4.30 \pm .01$	4.19^{1}, 3.53^{2}
803	Mineralized diabase	$1.53 \pm .03$	1.66^{1}
807	" "	$6.68 \pm .04$	6.56^{1}
826	Quartz diorite	$0.67 \pm .00$	0.67^{1}
842	Sericitic alteration	$8.04 \pm .02$	7.87^{1}
260	Greenstone (altered)	$4.12 \pm .07$	3.99^{1}
I 4088	Basalt (altered)	$5.23 \pm .06$	5.83^{1}, 5.86^{3}
I 4090	Tuff	$1.30 \pm .03$	1.20^{1}, 1.16^{3}

[a] Precision is relative to duplicate samples.

[1] H. N. Elsheimer

[2] S. T. Neil

[3] L. C. Peck

XRF average relative error: 4.3%

Gravimetric average relative error (between values reported): 4.2%

Duplicate portions of another suite of gravimetrically analyzed samples were fused and pelletized for XRF analysis to serve as "unknowns." The resultant curve for the "unknowns" vs. the standards is shown in Figure 3. "Unknowns" 373, 378, 387, and I 4088 are mixed sulfide-sulfate samples, as was standard 386. The numerical, least squares-derived results for the XRF determinations are compared with the gravimetric values in Table III. The average relative error for the "unknown" XRF determinations vs. their gravimetric values was 4.3%, which compares favorably with the average relative error between analysts for the gravimetric results. The somewhat low value for sample I 4088 undoubtedly resulted from its high ferrous iron content (21 wt. %), which suggests that a 50-mg sample portion should have been taken.

Two analyses not included in Table III and Figure 3 were those of a highly pure stibnite (Sb_2S_3) mineral and the SSC sulfide ore SO-1. The XRF value obtained for the stibnite mineral was 28.33% sulfur, which is in agreement with the theoretical value of 28.32% sulfur. Only 7.2% and 8.5% sulfur were indicated, respectively, using 50 and 25 mg sample portions, for SO-1 in contrast to the reported value of 12.10% sulfur. The large ferrous iron content (31%) is certainly responsible for the low results.

For evaluation of the CAN method for samples containing elemental sulfur, high purity sulfur (99.999+%) was ground to 100 mesh, and appropriate amounts to make 2.5, 5.0, 7.5, and 10.0 wt. % standards were mixed in a mortar with the USGS standard BCR-1. Because of excessive volatility, 50 mg sample portions were prepared. The resultant analytical curve for elemental sulfur samples vs. the sulfide-sulfate calibration standards is shown in Figure 4. The spread between duplicate samples was greatest at the 2.5% level, which may be due to heterogeneity and associated sampling difficulties. The higher point at the 10% level represents a separate, prepared standard. As sulfur-containing geologic materials usually contain elemental sulfur in quantities markedly less than 10%, the proposed method should be generally applicable to such samples.

The overall results indicate the general applicability of the method to geologic materials containing elemental sulfur, sulfides, sulfates, and mixtures of sulfides and sulfates. The detection limit for the technique is 0.01% sulfur.

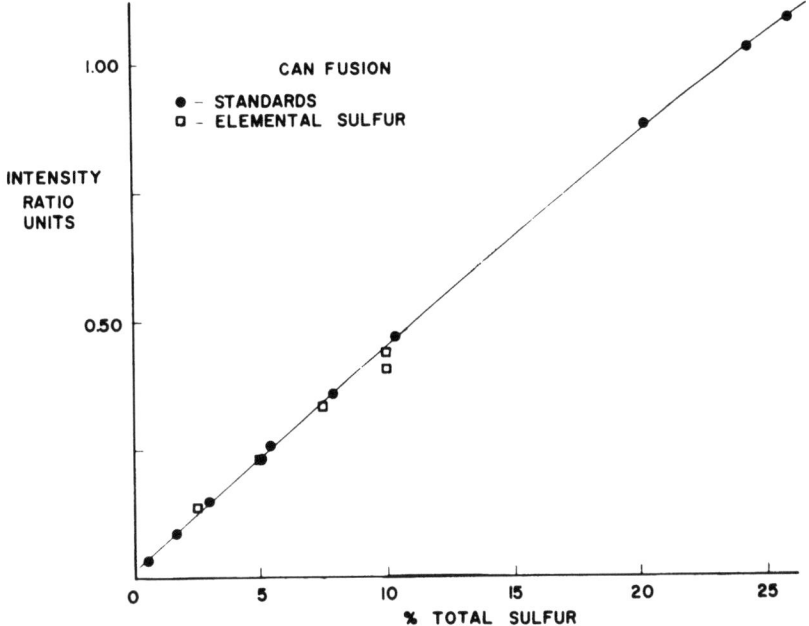

Figure 4. Synthetic mixtures of elemental sulfur vs. a sulfide-sulfate standard curve.

CONCLUSIONS

Fusion of sulfur-containing geologic materials with a $LiBO_2$-$Ce(NH_4)_2(NO_3)_6$ flux is at least twice as rapid as the gravimetric method and yields precise and accurate analytical results for a variety of samples regardless of the sulfur species present. Caution must be exercised for individual samples where excessively large amounts of highly reducing elements such as ferrous iron may be present, but the relative amounts of sample and oxidant may be adjusted accordingly in such cases. The highly absorbing character of cerium should favor the use of pellets prepared by this method for the analytical determination of other elements by XRF.

ACKNOWLEDGMENTS

We thank L. F. Espos for his assistance in sample preparation, S. T. Neil for her gravimetric determinations on certain of the standards, and G. K. Czamanske and H. J. Rose, Jr. for their perceptive reviews.

REFERENCES

1. L. G. Parratt, "Effect of Chemical Binding on the X-ray $K_{\alpha 1, 2}$ Doublet Lines of Sulfur Studied with a Two Crystal Spectrometer," Phys. Rev., 49, 14-16 (1936).

2. E. W. White, H. A. McKinstry and T. F. Bates, "Crystal Studies by X-ray Fluorescence," Advances in X-Ray Analysis, Vol. 2, Plenum Press, 239-45 (1960).

3. A. Faessler and M. Goehring, "X-Ray Spectrum and Bond Characteristics; K_α Fluorescence Radiation of Sulfur." Naturwiss., 39, 169-177 (1952).

4. A. Faessler and P. Mecke, "Bonding in Sulfides and Polysulfides," Z. Electrochem., 64, 587-591 (1960).

5. F. A. Gianturco and C. A. Coulson, "Influence of Chemical Combination on K Fluorescence Lines of Compounds of Aluminum, Silica, and Sulfur," Inorg. Chim. Acta, 3(4), 607-611 (1969).

6. D. Conti and H. Estour, "Chemical Effect on Wavelength of Characteristic X-Rays. Application Test," J. Microsc. (Paris), 15(3), 297-303 (1972).

7. B. P. Fabbi, "A Refined Fusion X-Ray Fluorescence Technique and Determination of Major and Minor Elements in Silicate Standards," Amer. Mineral., 57, 237-245 (1972).

8. B. P. Fabbi, "A Die for Pelletizing Samples for X-Ray Fluorescence Analysis," U.S. Geol. Surv. Prof. Paper 700B, B187-B189 (1970).

9. B. P. Fabbi and W. J. Moore, "Rapid X-Ray Fluorescence Determination of Sulfur in Mineralized Rocks from the Bingham Mining District, Utah," Appl. Spectrosc., 24, 426-428 (1970).

SAMPLING AND STANDARDS IN A RECYCLED WORLD

H. E. Marr III, W. J. Campbell, and D. L. Neylan

U. S. Bureau of Mines

College Park, Maryland   20740

ABSTRACT

A number of procedures for fluorescent X-ray analysis have been introduced to accommodate the samples that are produced from research on the recovery of values from secondary metal sources. For some applications, standards are conveniently available such as those that can be purchased from the National Bureau of Standards. For other applications, secondary standards must be prepared and analyzed by independent methods. The sample preparation procedures vary considerably. For monitoring process efficiency, sample preparation is often kept at a minimum such as simply pouring loose powders into disposable cups. For the most accurate analyses, sample preparation requires casting the alloys and finishing the surfaces. Matrix correction procedures are employed where concentrations of major constituents vary over wide ranges.

INTRODUCTION

At the College Park Metallurgy Research Center, several of the research programs are directed toward improving conservation of the Nation's resources by increased recycling and utilization of secondary materials (1). By applying traditional methods of beneficiating and processing ores to waste products and scrap materials, the Interior Department's Bureau of Mines is reclaiming values from these "urban ores". Although the original materials are effectively unsuitable for sampling, the product fractions have been characterized at appropriate stages of the processing operations.

The Bureau has developed and operated a pilot plant to recover metal and mineral materials from municipal incinerator residues. A large effort is now focused on the operation of a new type of recycling plant that will reclaim metals, minerals, and combustibles from raw, unburned refuse (2). Projects such as the recovery of aluminum from drosses and smelter slags are designed to increase reutilization of industrial wastes. Other efforts are directed toward improving the efficiency of recycling by the introduction of new technology to the recycling industries such as the application of portable X-ray analyzers for the rapid identification of alloys.

The techniques in fluorescent X-ray spectrography that have been applied to this research on secondary resource recovery reflect a broad range of needs for analytical determinations. For example, samples are analyzed throughout all stages of research to evaluate the efficiency of the recovery processes. In addition, products of recycling research that might be marketable are compared to commercial grades of competitive materials. In the area of new technology, X-ray methods have been evaluated for the sorting of alloys in scrapyards.

The diversity of materials encountered in recycling research complicates the development of appropriate routine standards. In the later stages of the pilot plant operations, the materials are separated into ferrous, nonferrous, aluminum, glass, paper, and plastic fractions, where a wide variation in composition may be required in the metal and mineral standards. Although the mixtures separated from the raw refuse are often unique to recycling operations and therefore not available as NBS-type standards, the latter standards are suitable for a surprising number of applications after the fractions are beneficiated.

The suitability of the standards for recycled materials is checked by confirmation analyses using other analytical methods and, where available, statistical comparisons between the results obtained for two or more sets of NBS-type standards. For the latter, convenient timesharing computer programs are used to process the data and evaluate the agreement between sets of standards.

APPLICATIONS

Steels From Ferrous Scrap

One of the objectives of the Bureau of Mines resource recovery program is to establish a correlation between the composition of steels made from secondary sources and the physical and chemical properties, to determine where secondary steels can be competitive with present commercial-grade steels. For this research it was necessary to reliably determine the concentration of eight minor elements--aluminum, copper, chromium, manganese, nickel, phosphorus, silicon, and tin (3). Certified standards from NBS and United States Steel were available as disks that covered the concentration ranges of interest in these low-alloy steels.

In our studies we compared three sets of standards, two of which are commercially available. The new NBS series 1261 to 1265 standards and five United States Steel standards were compared to an old set of NBS 1160 series steels. Surface preparation is a critical step, especially for the determination of low atomic number elements. The ASTM method and other published methods for low-alloy steels are based on ground and polished surfaces. Lathe-finished surfaces were used in our studies on both samples and standards to eliminate surface contamination from residual amounts of polishing agents such as silicon carbide or alumina. No significant differences between the two finishing procedures were noted in the determinations (3).

The reliability of the X-ray procedure for these low-alloy steels was evaluated by two methods--statistical analyses of results obtained on sets of standards treated as unknowns, and the use of other analytical techniques such as spectrophotometry, atomic absorption, and optical emission spectrography. By using a computer to calibrate results and to determine concentrations, personal bias in the treatment of calibration curves was reduced. All of the calculations were made on a timesharing computer using a program, XRF1, written in BASIC. The calibration equation was a linear least square fit of intensities and concentrations. No corrections were applied for interelement effects or overlapping spectral lines. The reliability of the calibration equation was indicated by statistical information supplied by XRF1. In our laboratories we have learned to be wary of overconfidence in computer-calculated data. Therefore, we make an added effort to include extra safeguards in the routine programs that are utilized by personnel less familiar with the actual mathematics and statistics. One of the features of XRF1 is that the program tests the data for improvement in the least square fit when each standard, in turn, is not included in the fit. In this way, outlier data points are brought to the attention of the analyst by noting the improvements in the numbers representing the standard deviation and the "goodness-of-fit."

In the evaluation studies and periodically during the research, confirmatory checks on the X-ray results were run by other chemical and spectroscopic methods. Samples of steels prepared from ferrous scrap were submitted for X-ray and optical emission analysis in the form of 1-1/4-inch-diameter disks. Ten to fifteen grams of turnings from the disks were used for atomic absorption and spectrophotometric analysis. Prior to X-ray analysis all of the standard and sample disks were machined with a tungsten carbide cutter to provide a lathe-finished surface. This lathe-finished surface was machined on the opposite side from the polished surfaces prepared by the National Bureau of Standards. After preparation, samples and standards were kept in closed containers to minimize surface contamination. A comparison of results for three representative samples is shown in table 1. For tin it was possible to compare the results obtained using the tin $k\alpha$ and the tin $l\alpha$ lines.

TABLE 1 - Comparison of results obtained by X-ray spectrography and other analytical methods

Determination	Sample Number		
	5291	5391	5664
Aluminum			
X-ray	0.073	0.073	0.094
Atomic absorption	.067	.070	.090
Copper			
X-ray	1.91	.33	.64
Atomic absorption	1.95	.34	.67
Chromium			
X-ray	.034	.027	.024
Atomic absorption	.027	.021	.019
Manganese			
X-ray	.42	.28	.22
Atomic absorption	.39	.28	.21
Optical emission	.41	.28	.22
Nickel			
X-ray	1.07	.17	.12
Atomic absorption	1.03	.17	.12
Phosphorus			
X-ray	.043	.029	.014
Spectrophotometry	.042	.034	.019
Silicon			
X-ray	.35	.21	.26
Optical emission	.35	.21	.24
Tin			
K X-ray	.166	.109	.166
L X-ray	.172	.112	.169
Atomic absorption	.17	.11	.18

Copper and Tin in Steel Scrap

Copper and tin are two contaminants that make certain types of ferrous scrap unattractive for manufacture of new steels. Rapid analytical turnaround times were desired for a project on the removal of tin from ferrous scrap by a chlorination process. In the initial tests, tin at the 0.1 to 1 weight-percent level in carbon-saturated molten iron was treated with various solid sources of chlorine. Samples of the treated molten iron were taken periodically and, after cooling, pulverized in a shatterbox. These powdered iron samples were compared to previously analyzed samples used as secondary standards. Sample preparation for X-ray analysis was limited to pouring the sample into a disposable plastic cup covered with Mylar film and tapping the sample holder approximately five times against the bench top to "pack" the powders. Typical calibration data shown in table 2 illustrate that this rather crude approach provided surprisingly good agreement between atomic absorption values and X-ray measurements. Subsequent tests at the 0.02 to 0.20 initial weight-percent tin level gave similar agreement. In practice, the X-ray data is only reported to two significant figures; however, three figures were used for statistical evaluations.

One promising approach to removal of copper from molten scrap is the use of an alkali slag. The process involves the addition of sodium sulfate to a molten bath of high-carbon iron scrap to form copper sulfide which preferentially reports to the slag. X-ray spectrography was used to obtain rapid analyses of samples for measuring the efficiency of the copper removal process. Typical calibration data from a wide selection of samples used as secondary standards is shown in table 3. The atomic absorption-analyzed values are compared to the X-ray data on pulverized samples. Again, the only sample preparation after pulverizing was tapping the sample holder a few times against the bench top.

TABLE 2 - Agreement of calibration data for tin in treated steel

Tin, weight-percent	
AA	X-ray
1.08	1.074
.87	.874
.62	.620
.41	.427
.38	.377
.20	.193
.15	.151

TABLE 3 - <u>Agreement in calibration data for copper in treated steel</u>

Copper, weight-percent	
AA	X-ray
1.06	1.064
1.00	1.002
.95	.954
.94	.911
.56	.556
.42	.422
.35	.36
.31	.308
.29	.282

One unexpected problem that was encountered recently in some determinations of copper in an iron matrix was brought out by some unaccountably high values of copper determined by atomic absorption techniques. The erratic results showed up at copper levels of 4 to 6 weight-percent, levels at which copper was expected to be miscible in this Fe-Cu-C-X system. Abnormally high copper $K\alpha$ intensities would be expected if some of the copper was not dissolved in the iron-carbon lattice, because the mass absorption of copper X-rays by free copper is less than the mass absorption by iron. Electron probe microanalysis confirmed that the 4 and 6 weight-percent copper samples had free metallic copper in addition to the copper in solid solution with iron and carbon. The limit of solubility of copper for these samples proved to be approximately 3 percent. In this instance, the X-ray techniques fortuitously provided information that would not have been obtained from the wet chemical procedures.

Analysis of Nonferrous Alloys

The nonferrous fraction recovered from the processing of urban refuse and incinerator residues is only a small portion of the total material. This small fraction, however, represents the greatest dollar value per ton of residue, and is therefore the focus of some interesting research on recovering marketable products.

The nonferrous metals fraction recovered from incinerator residues cover a wide range of compositions: Aluminum 40-80 weight-percent, copper 3-25, lead 0-20, tin 0-2, zinc 0-35, iron 0-5, and nickel, manganese, and magnesium 0-1 each. Standards for these analyses were not available commercially and ingots are now being cast that will cover the composition ranges of interest. It will be particularly important to maintain consistency in the preparation of these samples and standards which vary so widely in composition to assure ourselves that differences in X-ray results are due to composition and not variations in metallurgical history. For this reason samples will be periodically examined on the electron probe for evidence of significant segregation problems.

Iron in Glass

Glass is one of the economically recyclable constituents in municipal incinerator residues. The glass fraction is separated from the other metallic and nonmetallic components by various physical methods such as screening, magnetic separation, and flotation. Iron is the principal impurity and is present in two forms--as particles of metallic iron or iron oxides and as iron dissolved in the silicate structure. Knowledge of the concentration of each form of iron is necessary to evaluate the potential for upgrading the quality of the glass. The discrete iron-bearing particles can be removed by physical methods.

Depending on the processing stage, the total iron concentration in the glass fraction ranged up to 5 to 10 weight-percent, whereas the dissolved iron was only 0.1 to 0.3 weight-percent maximum. Using -200 mesh samples poured into disposable cups, excellent agreement was obtained between atomic absorption and X-ray spectrography for the feed samples as demonstrated by the data in table 4. These analyses are more than adequate for the plant operator to determine the amount of the beneficiation necwssary to reach the desired grade of glass. The next step was to determine the amount of iron dissolved in the silicate structure.

Our approach to this analytical problem was to determine the removable iron by leaching the glass with hydrochloric acid. The iron remaining in the leached glass was then determined by X-ray spectrography. Samples of the leached powdered glass were poured into disposable cups and compared to other leached samples

TABLE 4 - Agreement of calibration data for iron in glass from municipal incinerator residues

Iron, weight-percent	
Chemical	X-ray
0.28	0.285
.37	.378
.50	.477
.59	.567
.70	.665
1.21	1.24
1.65	1.69
1.94	1.90

previously analyzed for iron by a dichromate titration after fusion of the sample. By using X-ray procedures it was possible to avoid the lengthy fusion process required for the chemical analysis.

Aluminum From Dross and Municipal Waste

A non-polluting metallurgical procedure being developed at College Park is designed to recover a product containing more than 95 percent metallic aluminum from the dross generated by secondary aluminum processes. Commercially available standards ranging in concentration from 85 to 100 percent aluminum are being used to determine the concentrations of silicon, iron, manganese, and copper in order to monitor product purity. For this research problem, 2-1/2-inch-diameter by 1-inch-thick disks were purchased from the Aluminum Company of Canada (Alcan). The disks were machined to 1-1/4-inch diameter, and the surface was turned down on a lathe for X-ray examination.

The samples are cast in disposable ceramic molds, and the surface is lathe-finished. One recent test was designed to determine if polishing the surfaces yielded results different from the data obtained on lathe-finished surfaces. Several repeated runs on determinations of iron and silicon in aluminum showed that the more tedious polishing process did not significantly improve the reliability of the determinations.

Another source of aluminum samples is an aluminum-rich fraction recovered from the nonferrous portion of the incinerator residues. Alcan standards were used for the determination of chromium, copper, iron, manganese, nickel, silicon, tin, titanium, and zinc. High levels of copper and zinc were encountered in some of the samples. Standards containing up to 6.85 percent copper and 7.65 percent zinc were used. Therefore, it was expedient to check for interelement corrections in some of the other determinations.

Fitted to the intensity data from seven standards was the equation:

$$I_i = A_0 + A_1 C_i + A_2 C_{Cu} + A_3 C_{Zn}$$

where I_i is the observed intensity,
  A_0 represents the background,
  A_1 is the coefficient of the term representing the element being determined, and
  A_2, A_3 are coefficients for concentrations of copper and zinc.

The standard deviations (σ) listed in table 5 indicated that the corrections for copper and zinc brought the matrix errors within acceptable limits.

TABLE 5 - Regression equations from typical calibration curves for determination of iron, nickel, tin, and chromium with standard deviation from curves

	A_0	A_1	A_2	A_3	σ, Wt-pct
Iron	9 +	1658*Fe -	10*Cu +	6*Zn	0.007
Nickel	52 +	9831*Ni -	4*Cu -	4*Zn	.001
Tin	3428 +	20887*Sn -	216*Cu -	221*Zn	.005
Chromium	78 +	1093*Cr +	2*Cu +	2*Zn	.002

Rapid Identification of Copper-Base Alloys

Recent investigations at the College Park Metallurgy Research Center concerned the evaluation of X-ray instrumentation for application in scrapyards and foundries. Three types of X-ray analytical devices were initially compared--a conventional X-ray spectrograph, a portable X-ray spectrograph, and a radioisotopic X-ray analyzer using balanced filters (4). Energy dispersion X-ray analysis using radioisotopic sources and a semiconductor detector was also investigated to anticipate the use of these systems in on-line sorting operations (5).

For these studies on the sorting of copper-base alloys, 20-pound ingots of 23 brass and bronze alloys, purchased from R. Lavin & Sons, were used as standards. The ingots were cut into approximately 1/2-inch-thick slices and the surface of each slice for X-ray analysis was turned smooth on a lathe. One set of samples was cut into 1-1/4-inch disks to fit into the sample holders of the portable and conventional X-ray spectrographs. A set of slices from the end of the ingot was used to provide both smooth and rough surfaces. A certificate of lot analyses was provided for each of the alloys. For confirmatory analyses of individual ingots, the turnings and shavings were saved from the cutting of the slices.

The copper-bearing alloys cover a wide range of compositions: Copper 55-100 weight-percent, tin 0-10, lead 0-25, zinc 0-40, aluminum 0-12, nickel 0-26, iron 0-5, manganese 0-5, and silicon 0.5. One of the objectives of this research was to determine the magnitude of the analytical error when using calibration curves that included standards of widely varying compositions. Simple calibration curves with no interelement corrections were suitable for some applications; other applications required extensive matrix corrections, as would be expected. Two general applications were considered: Rapid identification for sorting scrap metals and quantitative analyses to provide guidance in the melting and casting of alloys.

These investigations proved that each instrument can be useful for certain problems encountered in the scrapyard or in the foundry. The radioisotopic X-ray analyzer can distinguish between certain classes of alloys with analytical times of 5-10 seconds. The conventional X-ray spectrograph and the portable X-ray spectrograph will provide quantitative determination of the significant components in copper-base alloys with high precision and accuracy. All the above instrumentation is commercially available and suitable for use by skilled nonprofessional personnel. Tests to date show that, in just a few seconds of analytical time, energy dispersive X-ray analysis using semiconductor detector systems and computer processing of the data can determine the concentrations of copper, tin, lead, zinc, manganese, iron, and nickel. To make the technique useful for application in the scrapyards or foundries, suitable methods of preparing the surfaces and presenting the samples to the detectors will have to be devised.

SUMMARY

X-ray methods utilizing both commercial and in-house standards have been applied to the analysis of recycled materials to monitor research progress in developing recovery processes, to compare commercially competitive materials to products recovered from solid waste, and to introduce new applications of technology to the secondary metals industries. Examples of rapid control-type analyses are the monitoring of the removal of tin and copper from ferrous scrap and monitoring iron in the glass separated from municipal incinerator residues. Examples of high-precision determinations are the analysis of minor elements in steels prepared from ferrous scrap and minor elements in refined aluminum reclaimed from aluminum dross. Other research includes the evaluation of energy dispersive X-ray analysis and other X-ray techniques that have been applied to the rapid identification of copper-base alloys. A variety of interesting problems are therefore encountered in recycling research in the Bureau of Mines, and X-ray methods of analysis have been applied to this research in several ways.

REFERENCES

1. C. B. Kenahan, R. S. Kaplan, J. T. Dunham, and D. G. Linnehan. Bureau of Mines Research Programs on Recycling and Disposal of Mineral-, Metal-, and Energy-Based Wastes. U. S. Bureau of Mines Information Circular 8595 (1973).

2. P. M. Sullivan, M. H. Stanczyk, and M. J. Spendlove. Resource Recovery From Raw Urban Refuse. U. S. Bureau of Mines Report of Investigations 7760 (1973).

3. W. J. Campbell and D. L. Neylan. Determination of Eight Minor Elements in Low Alloy Steels by Fluorescent X-Ray Spectrography. U. S. Bureau of Mines Report of Investigations 7773 (1973).

4. W. J. Campbell and H. E. Marr III. Identification and Analyses of Copper Base Alloys by Fluorescent X-Ray Spectrography. U. S. Bureau of Mines Report of Investigations 7635 (1972).

5. H. E. Marr III. Rapid Identification of Copper Base Alloys by Energy Dispersion X-Ray Analysis. U. S. Bureau of Mines Report of Investigations   (1973).

X-RAY CROSS-SECTIONS IN DESIGN AND ANALYSIS OF NON-DISPERSIVE SYSTEMS

Benton C. Clark

Martin Marietta Aerospace

Denver, Colorado 80201

ABSTRACT

In designing a radioisotope, energy-dispersive (proportional counter) x-ray fluorescence spectrometer for elemental analysis of geologic specimens, a theoretical framework has been developed to model instrument response. This model is based upon the fundamental physics of x-ray excitation, absorption, and scattering, and employs the most modern available values of the applicable physical constants. The model includes matrix absorption and enhancement effects. By explicitly including scattering in the model and the measurements, element concentrations can be calculated from the shape alone of the x-ray spectrum and the presence of elements having non-observable fluorescences can be detected.

INTRODUCTION

The model was originally developed for specific use in ultra-miniature x-ray fluorescence spectrometers employing ^{55}Fe and ^{109}Cd radioisotope sources and thin-window, sealed proportional counter detectors. As a result, the energy range of one to 22 keV was of prime interest. Data presented below will cover only this range, but the formulae given are applicable over a much greater range and the additional data required are available in the cited references. The reader may also apply the equations to other cases, e.g. x-ray tube excitation (requires knowledge of the tube emission spectrum and integration of the equations over energy) and detectors such as scintillators or cryogenic solid state detectors. Streamlined notation is employed to facilitate inspection of the equations for the influence of the various parameters.

THE MODEL

Fundamental Data

Cross-sections. The cross-section for a given type of interaction has a value such, that when multiplied times the number of target atoms per unit area in a thin target gives the probability that an incident particle will undergo the interaction during its passage through the target. Thus, the probability, P_i, of interaction with atoms of type i is

$$P_i = \sigma_i\, n_i\, dx \tag{1}$$

where σ_i = cross-section of atoms of type i (in cm^2/atom), n_i is the number of target atoms per cm^3, and dx is the target thickness (in cm). In x-ray work, it is convenient to define a quantity, μ, which is proportional to cross-section:

$$\mu_i = \frac{N_o\, \sigma_i}{A_i} \quad cm^2/g \tag{2}$$

where N_o = Avogadro's number and A_i is the atomic weight of atom i. This new quantity, conventionally called the x-ray attenuation coefficient, retains the properties of a cross-section. It is readily shown that equation (1) can be written

$$P_i = \mu_i\, w_i\, dt \tag{3}$$

where w_i is the weight fraction of type i atoms in the target material, and dt is the target thickness measured in g/cm^2 (i.e., $dt = \rho dx$, where ρ is the material density in g/cm^3).

Problems in x-ray fluorescence involve three types of interaction of practical importance: (1) photoelectric effect, (2) coherent scattering, and (3) incoherent scattering. The short-form notation to be used herein for these interaction coefficients (cross-sections!) for x-ray photons of energy, E, is as follows,

$\mu_i(E)$ = photoelectric coefficient of atoms of type i

$\mu(E) = \sum_i w_i\, \mu_i$ = photoelectric coefficient of a material with composition w_1, w_2, ...

$\mu_{SC}(E)$ = coherent scattering coefficient of total material

$\mu_{SI}(E)$ = incoherent scattering coefficient of total material

In this paper I have employed the very extensive set of compiled and smoothed values of x-ray coefficients prepared by Veigele, et al (1,2). Selected data are given in Table I.

TABLE I - Fluorescence and Coefficient Data for Selected Elements

Ele	Z	K_K	ω_K	Y_i	μ_i	^{55}Fe μ_{SC}	μ_{SI}	μ_i	^{109}Cd μ_{SC}	μ_{SI}
C	6	.971	.001	.001	10.8	.30	.115	0.2	.057	.162
O	8	.959	.002	.002	28.7	.50	.100	0.4	.086	.156
Na	11	.933	.019	.018	71.8	.83	.077	1.3	.143	.142
Mg	12	.928	.027	.025	95.9	.93	.079	1.7	.167	.145
Al	13	.924	.036	.033	121.8	.96	.077	2.2	.181	.139
Si	14	.920	.047	.043	151.3	1.05	.080	2.8	.205	.143
S	16	.912	.076	.069	220.9	1.17	.080	4.4	.245	.140
K	19	.901	.138	.124	340.4	1.36	.075	7.4	.305	.132
Ca	20	.897	.163	.146	401.4	1.47	.077	8.9	.332	.135
Ti	22	.891	.219	.195	469.1	1.54	.068	11.2	.350	.122
Fe	26	.880	.347	.305	89.0	2.07	.060	18.3	.450	.119
Zn	30	.870	.479	.417	136.2	2.64	.052	27.5	.560	.113
Sr	38	.853	.691	.589	259.8	3.13	.048	48.5	.750	.100

Fluorescence Probability. Emission of a K fluorescence x-ray depends upon (1) the probability, K_K, of a K-shell vacancy rather than L, M, or higher shell vacancy being created during a photoelectric interaction, and (2) the branching ratio between fluorescence emission and de-excitation via Auger electron emission. The former can be calculated from the K absorption edge jump ratio, J_K, under the assumption that the fraction of K shell absorption in all shell absorptions is independent of photon energy above the K shell threshold:

$$K_K = (1 - 1/J_K) \tag{4}$$

The latter is described by the fluorescence yield, ω_K, which is the probability of photon emission following a K shell vacancy. The total K fluorescence probability, Y, given that a photoelectric interaction has occurred is

$$Y = K_K \omega_K \tag{5}$$

Values of K_K, ω_K, and Y are given in Table I. K_K were calculated from eq. (4) using the jump ratios of Veigele, et al (1). The ω_K were calculated using eq. (3-11) of Bambynek, et al (3). It should be cautioned that values of ω_K for Z < 13 are significantly dependent upon the chemical state of the atom (4).

Energy Shifts. Fluorescence x-ray energies may be found in many handbooks as well as review journals (5). Coherently scattered radiation has exactly the same energy as the incident radiation. The energy of an incoherently scattered photon is given by the equation for the Compton shift (6) which for low energies

($E \ll 511$ keV) can be written as

$$E' \simeq E - (1 - \cos \theta) (E^2/m_o c^2) \tag{6}$$

where E and E' are the initial and final energies of the photon, θ is the angle between the direction of the scattered photon and the original direction, and $m_o c^2$ is the rest energy of the electron (511 keV). Fig. 1 plots eq. (6) for two special cases of interest: the 5.90 keV Mn K_α primary emission of ^{55}Fe, and the 22.2 keV Ag K_α primary emission of ^{109}Cd radioisotope.

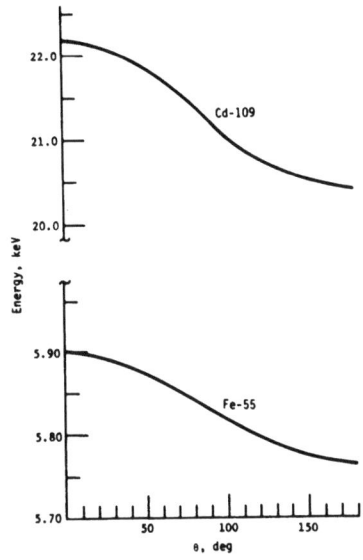

Figure 1 - Compton Energies

Angular Dependence of Scattering. Whereas fluorescence x-ray emission has no preferred direction, all x-ray scattering is anisotropic. To describe this, the cross-section may be written as differential with respect to solid angle, $(d\sigma/d\Omega)$. Coherent scattering by an atom whose electrons are loosely bound is given by the Thomson formula (6)

$$\frac{d\sigma_{Th}}{d\Omega} = \frac{Z r_o^2}{2} (1 + \cos^2 \theta) \tag{7}$$

where r_o is the classical electron radius (0.28×10^{-12} cm). Incoherent scattering by free electrons is described with the aid of the scattering equation for unpolarized radiation developed by Klein and Nishina (6), which for $E \ll 511$ keV becomes

$$\frac{d\sigma_{KN}}{d\Omega} \simeq \left(\frac{d\sigma_{Th}}{d\Omega}\right) \left[1 - \frac{2E}{m_o c^2} (1 - \cos \theta)\right] \tag{8}$$

Neither of these equations is strictly correct for the x-ray energies of interest here because of the non-negligible electron binding

TABLE II - Angular Functions for Scattering in Oxygen, Silicon*, and Iron

θ (deg)	^{55}Fe $g_c(\theta)$		^{55}Fe $g_I(\theta)$		^{109}Cd $g_c(\theta)$			^{109}Cd $g_I(\theta)$		
	O	Fe	O	Fe	O	Si	Fe	O	Si	Fe
0	3.3	2.6	.00	.00	8.5	7.7	7.0	.00	.00	.00
10	3.0	2.5	.09	.14	5.5	4.6	4.3	.50	.49	.42
20	2.6	2.2	.20	.36	2.3	2.6	2.5	.98	.81	.69
30	2.2	1.8	.45	.51	.72	1.5	1.3	1.2	.97	.91
40	1.7	1.4	.59	.64	.38	.85	.75	1.2	1.1	1.0
60	.78	.87	.78	.73	.21	.23	.30	.98	.97	.96
80	.41	.54	.82	.75	.11	.10	.18	.83	.83	.83
90	.34	.46	.85	.80	.09	.08	.15	.82	.81	.81
100	.29	.44	.93	.87	.08	.07	.13	.84	.86	.85
110	.27	.44	1.1	1.0	.08	.06	.13	.90	.95	.93
120	.26	.45	1.2	1.2	.08	.06	.12	.99	1.1	1.0
140	.27	.50	1.5	1.6	.08	.07	.12	1.3	1.3	1.3
160	.29	.55	1.9	1.9	.08	.07	.13	1.5	1.5	1.6
180	.30	.55	2.0	2.1	.08	.08	.13	1.6	1.6	1.7

*For ^{55}Fe, silicon values are very close to iron.

energies. Binding prohibits incoherent energy transfers insufficient to excite or ionize the atom, so that the incoherent cross-section sharply decreases at low scattering angles and the coherent cross-section increases for the same reason (7). Corrections are made by invoking multiplicative factors called the atomic form factor, f, and the incoherent scattering factor, S:

$$\frac{d\sigma_{SC}}{d\Omega} \equiv \left(\frac{f^2}{Z}\right)\left(\frac{d\sigma_{Th}}{d\Omega}\right) \quad \text{and} \quad \frac{d\sigma_{SI}}{d\Omega} \equiv S\left(\frac{d\sigma_{KN}}{d\Omega}\right) \quad (9, 10)$$

where f and S are functions of Z and (sin θ/λ), tabulated in Ref. 1. From these equations, the angular dependence of scattering of ^{55}Fe and ^{109}Cd radiation have been calculated and normalized such that they represent the amount of radiation scattered into solid angle dΩ at any θ relative to a value of 1.0 for omnidirectional scattering. These angular dependence factors, $g_c(\theta)$ and $g_I(\theta)$, for coherent and incoherent scattering, respectively, are given in Table II and Fig. 2.

Formulae

Detector Efficiency. Although inferior to cryogenic silicon detectors in resolution, proportional counters do possess the attributes of satisfactory operation at high temperature and the capability of performing with a variety of working gases. This permits selection of gas composition and pressure as a means of enhanc-

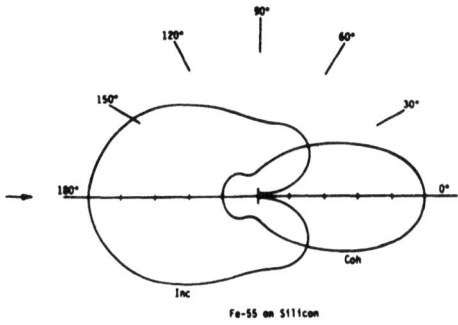

Figure 2 - Scattering Distribution

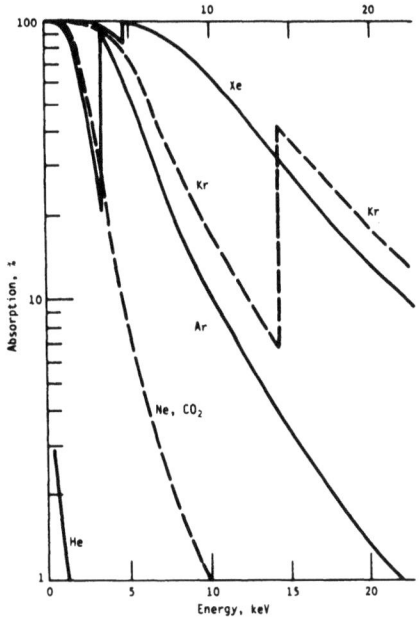

Figure 3 - Absorption in 1 cm Gas at STP

ing sensitivity to various portions of the energy spectrum. Detector efficiency, ϵ, is the product of transmission, τ, through the window aperture and absorption, α, in the counter gas.

$$\epsilon_i = \alpha_i \tau_i, \quad \tau_i = \exp\left[-\mu(E_i) T\right], \quad \alpha_i = 1 - \tau'_i \quad (11)$$

where μT is for the window in the case of τ_i and for the gas in the case of α_i. Values of α_i for a one cm path length in several gases are given in Fig. 3. Examples of tailored proportional counter detection efficiencies have been given in Ref. 8.

Scattering Power of a Material. Coherent scattering cross-sections increase approximately as Z^2 and incoherent scattering as Z. Why, then, does the scattering from a bulk specimen decrease, in general, with Z? The answer is tied to the fact that the photoelectric cross-section increases as Z to the 4 to 4.5 power in this energy range. Consider scattering from an incremental layer, dt, located at depth, t, in a material of total thickness, T, Fig. 4. (Let pr. = probability.) The probability of scattered radiation appearing at the observer is

$$F_S(T) = \int_o^T (\text{pr. of photon reaching t}) (\text{pr. scattering}) (\text{pr. of photon reaching surface})$$

which can readily be shown to be

$$F_{SC}(T) = F_{SC}(\infty) \left[1 - \tau(\theta_1) \tau(\theta_2)\right] \quad (12)$$

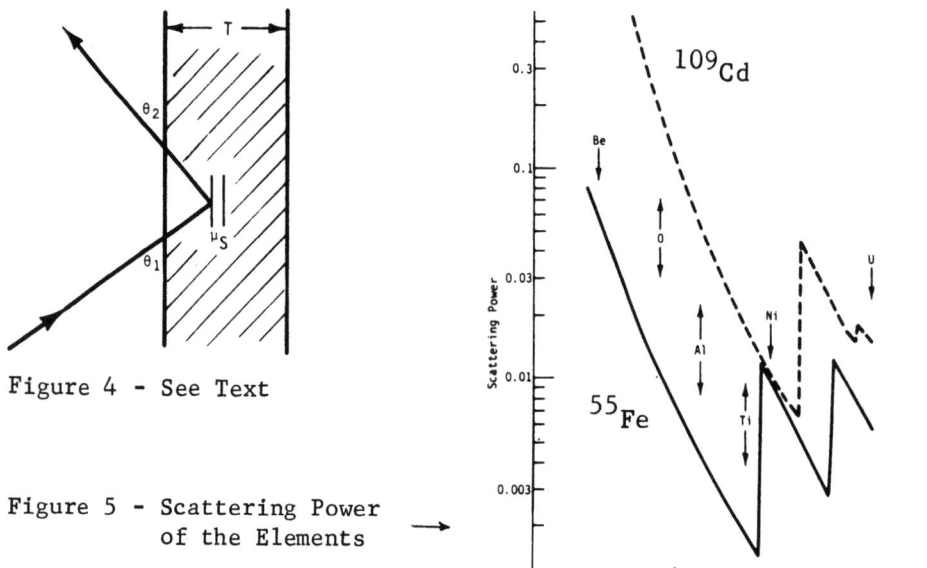

Figure 4 - See Text

Figure 5 - Scattering Power of the Elements

where $\quad F_{SC}(\infty) = \dfrac{\mu_{SC}\, g_c\,(\theta)}{\mu(\csc\theta_1 + \csc\theta_2)} \quad$ and $\theta = \theta_1 + \theta_2 \qquad (13)$

and analogously for F_{SI}.

Neglecting for the time being the g's, and assuming $\theta_1 = \theta_2 = 90°$, we obtain

$$\text{Scattering Power} = \frac{(\mu_{SC} + \mu_{SI})}{2\mu} \qquad (14)$$

which shows that the scattering from a semi-infinitely thick material depends upon the ratio of scatter to photoelectric coefficients. This quantity is plotted in Fig. 5 for ^{55}Fe and ^{109}Cd radiation. For a short distance in the periodic table, elements such as Ni are effective equally in scattering both energies. Note that titanium is an excellent 'sink' for ^{55}Fe radiation. It would not, however, be a good material for anti-scatter assemblies because of the resulting very high titanium K_α emission. Multi-layer materials offer the best relief of this problem. For example, a 0.1 mil coating of nickel over aluminum base makes a good absorber of ^{55}Fe radiation, with the aluminum fluorescence absorbed by the nickel.

Matrix Absorption. Consider x-rays of energy E incident upon a semi-infinite slab, producing fluorescence of element i at a depth t in the slab. The intensity, I_i, of fluorescence photons (energy E_i) which emerges at angle θ_2 to the slab is proportional to

$\qquad I_i \propto$ (attenuation of incident photons to depth t)
$\qquad\qquad$ (cross-section for photoelectric absorption by

$$\text{or,} \quad I_i \propto \left[e^{-\mu(E)t} \right] \underset{\text{atom i)}}{\left[-\mu_i(E) \, w_i \, dt \right]} \underset{\text{(pr. of K fluorescence)}}{\left[\delta_i \, Y_i \right]} \underset{\text{(attenuation of fluorescence photons)}}{\left[e^{-\mu(E_i)t} \right]}$$

Integrating from $t = 0$ to ∞,

$$I_i \propto w_i \, Y_i \, P_i \tag{15}$$

where
$$P_i \equiv \frac{\delta_i(E) \, \mu_i(E)}{\mu(E) \sec \theta_1 + \mu(E_i) \sec \theta_2} \tag{16}$$

and $\delta_i(E) = 1$ if E is above the absorption edge of i, and zero if below. The μ's given in the denominator above are written as the photoelectric coefficients, but more rigorously are the total attenuation coefficients and should include scattering. The author chooses to neglect scattering on the assumption that each scatter-out is balanced by an equivalent scatter-in, and the net scatter effect is satisfactorily close to zero. This assumption is clearly justified for ^{55}Fe radiation where the scattering coefficient is less than 3% of that of attenuation for all elements above carbon. It is less clearly warranted for ^{109}Cd in, say, geologic specimens where scattering is as high as 25% of the total attenuation coefficient. The above model also assumes that the photons are observed at a distance large with respect to the photon mean-free-path so that $1/R^2$ effects inside the specimen can be neglected.

P_i is matrix dependent and is a result solely of matrix absorption. For this reason it is often called the absorption factor. Note that when $\mu(E_i) \sec \theta_2$ is considerably more than $\mu(E) \sec \theta_1$, the effect of variation of θ_1 is small. This is clearly the case in the example of Fig. 6.

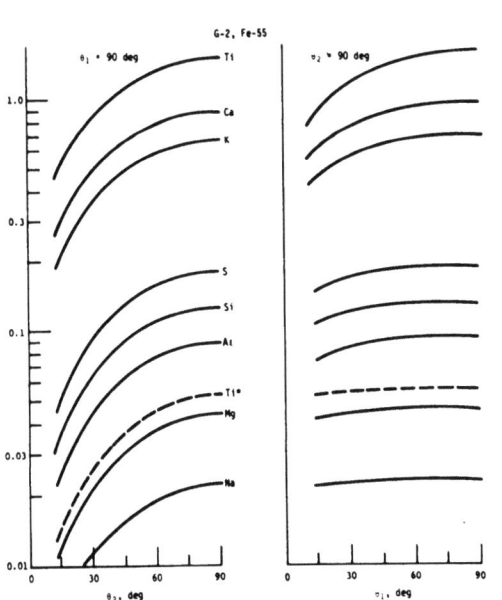

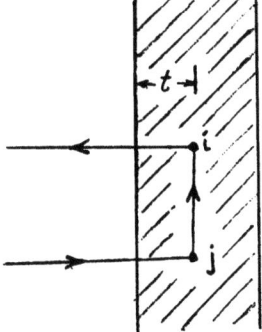

Figure 7 - Enhancement

Figure 6 - P_i for Specimen G-2 and ^{55}Fe. Dotted line is ^{109}Cd Excitation

Matrix Enhancement. Suppose that instead of interacting directly with element i, the incident energy photon fluorescences element j, whose resulting fluorescent x-ray then interacts with element i to produce the observed E_i photon. The exact derivation of this enhancement effect is difficult, but it is instructive to consider the special case of normally incident excitation and detection, with emission by atom j restricted to the plane parallel to the slab face, Fig. 7. The chronology is thus, (1) pr. photon E reaches depth t, (2) pr. E interacts with element j, (3) pr. j fluoresces, (4) pr. photon E_j penetrates distance s, (5) pr. E_j interacts with element i, (6) pr. i fluoresces, and (7) pr. photon E_i reaches the surface. In this case, the intensity observed is proportional to

$$I_i' \propto \sum_j \left[e^{-\mu(E)t}\right]\left[-w_j \mu_j(E)\, dt\right]\left[Y_j\, \delta_j(E)\right]\left[e^{-\mu(E_j)s}\right]$$
$$\left[-w_i \mu_i(E_j)\, ds\right]\left[Y_i\, \delta_i(E_j)\right]\left[e^{-\mu(E_i)t}\right] \quad (17)$$

Upon integrating over ds and dt from 0 to ∞, and dividing by eq. (15), one obtains

$$\frac{I_i'}{I_i} = \sum_j [w_j][Y_j]\left[\frac{\delta_i(E_j)\,\delta_j(E)}{\delta_i(E)}\right]\left[\frac{\mu_i(E_j)}{\mu(E_j)}\right]\left[\frac{\mu_j(E)}{\mu_i(E)}\right] \quad (18)$$

In this equation, all five factors are less than one, so that the product is always much less than one, and this secondary fluorescence effect is small compared to the primary factor.

The complete equation, allowing for variable source and detector angles, as well as omnidirectional fluorescence emission and a continuum source of x-ray energies has been reported by Shiraiwa and Fujino (9). For a discrete energy source, and the notation employed here, eq. (18) becomes

$$\frac{S_i}{P_i} = \sum_j w_j Y_j \left[\frac{\delta_i(E_j)\,\delta_j(E)}{\delta_i(E)}\right]\left[\frac{\mu_j(E)}{2\mu_i(E)}\right]\left[\mu_i(E_j)\,(A+B)\right] \quad (19a)$$

where
$$A \equiv \frac{1}{\mu(E)\,\csc\theta_1} \ln\left[1 + \frac{\mu(E)\,\csc\theta_1}{\mu(E_j)}\right] \quad (19b)$$

$$B \equiv \frac{1}{\mu(E_i)\,\csc\theta_2} \ln\left[1 + \frac{\mu(E_i)\,\csc\theta_2}{\mu(E_j)}\right] \quad (19c)$$

With aid of the approximation $\ln(1+x) \simeq x$ for small x, it is readily shown that eq. (19) reduces to eq. (18) under the conditions

$$\mu(E)\csc\theta_1 \ll \mu(E_j) \text{ and } \mu(E_i)\csc\theta_2 \ll \mu(E_j)$$

The value of $S_i/(P_i + S_i)$ is given for some cases of interest in Table III.

TABLE III - Enhancement Using ^{55}Fe Excitation

Ele	Granite (G-2)		Basalt (BCR-1)		Dolomite	
	100 w_i	$S_i/(P_i+S_i)$	100 w_i	$S_i/(P_i+S_i)$	100 w_i	$S_i/(P_i+S_i)$
O	48.4	-	45.7	-	52.4	-
Mg	0.47	7%	2.00	7%	10.9	13%
Al	8.10	6%	7.25	6%	0.85	13%
Si	32.4	2%	25.5	3.5%	1.31	14%
K	3.74	1%	1.39	3.5%	0.17	17%
Ca	1.42	0.2%	4.97	1%	21.0	0.0%
Ti	0.32	-	1.34	-	0.05	-

Total Matrix Effect. The total intensity is therefore

$$I_i'' \propto w_i Y_i \left[P_i + S_i + T_i \right] \quad (20)$$

where T_i is a tertiary factor to account for triple fluorescences. In general, T_i is negligibly small from a practical standpoint (9, 10).

Final Model Equations. Let N_i be the number of events/sec in the detector due to radiation from element i. Let $\mathcal{A}$ be the combination of influences which scale all events by an equal factor: e.g., source strength, geometric factors, and sample density. Then

$$N_i = \mathcal{A} \epsilon_i w_i Y_i \left[P_i + S_i + T_i \right] \quad (21)$$

$$N_{SC} = \mathcal{A} \epsilon_{SC} F_{SC} (\infty) \quad (22)$$

$$N_{SI} = \mathcal{A} \epsilon_{SI} F_{SI} (\infty) \quad (23)$$

where N_{SC} and N_{SI} are the detected coherent and incoherent scattering events/sec. This provides (i + 2) equations in only (i + 1) unknowns and so is more than sufficient information to determine the element concentrations, w_i, with no direct measurement of $\mathcal{A}$. Even if a number of elements are present which do not produce observable fluorescent emissions (e.g., C, N, O, etc. in this case), they will contribute to the N_{SC} and N_{SI}, and their presence will be detected, albeit as a group. Precise measurement of the N_{SC}/N_{SI} ratio could indicate the average atomic number of this light element group.

CONCLUDING REMARKS

The model embodied in eq. (21) - (23) has been tested (see Ref. 8) with good results using empirical correction factors for the Y_i. The required corrections are partly due to uncertainties in Y_i, ϵ_i, and perhaps other parameters. Nonetheless, the composition of a specimen is in principle exactly calculable from the shape alone of the resulting fluorescent x-ray spectrum when the scattered components are included.

REFERENCES

1. W. J. Veigele, E. Briggs, L. Bates, E. M. Henry, and B. Bracewell, "X-ray Cross-section Compilation from 0.1 keV to 1 MeV," Defense Nuclear Agency Report DNA-2433F, Vol. I and II, Rev. I (1971).

2. B. L. Bracewell and W. J. Veigele, "Elemental X-ray Cross-sections at Selected Wavelengths," Adv. X-ray Anal., Vol. 15, p. 352-364, Plenum Press (1972).

3. W. Bambynek, B. Crasemann, R. W. Fink, H.-U. Freund, H. Mark, C. D. Swift, R. E. Price, and P. V. Rao, "X-ray Fluorescence Yields, Auger, and Coster-Kronig Transition Probabilities," Rev. Mod. Phys., Vol. 44, p. 716-813 (1972).

4. R. W. Fink, private communication, 1973.

5. J. A. Bearden, "X-ray Wavelengths," Rev. Mod. Phys., Vol. 39, p. 78-101 (1967).

6. R. D. Evans, The Atomic Nucleus, McGraw-Hill Book Co., 1955.

7. W. J. Veigele and P. T. Tracy, "Compton Effect and Electron Binding," Am. J. Phys., Vol. 34, p. 1116-1121 (1966).

8. B. C. Clark and A. K. Baird, "Ultraminiature X-ray Fluorescence Spectrometer for in-situ Geochemical Analysis on Mars," Earth and Planet. Sci. Letters, in press (1973).

9. T. Shiraiwa and N. Fujino, "Theoretical Calculations of Fluorescent X-ray Intensities in Fluorescent X-ray Spectrochemical Analysis," Japan J. Appl. Phys., Vol. 5, p. 886-899 (1966).

10. J. W. Criss and L. S. Birks, "Calculation Methods for Fluorescent X-ray Spectrometry," Anal. Chem., Vol. 40, p. 1080-1086 (1968).

USE OF MULTIPLE STANDARDS FOR ABSORPTION CORRECTION AND

QUANTITATION WITH FRIEDA

P. S. Ong, E. L. Cheng and G. Sroka

The University of Texas System Cancer Center, M. D.

Anderson Hospital and Tumor Institute, Houston, Texas

ABSTRACT

The computerized fluorescence radiation induced energy dispersive analyzer (FRIEDA) (1) described earlier uses an x-ray beam with a well defined energy for the excitation of fluorescence radiation, and an Si(Li) detector to measure the total x-ray spectra emitted. Such a system can also simultaneously provide supplemental data for the determination of the dry mass and the sample mass absorption which is necessary for accurate quantitation of the results. This instrumental capability has been utilized in the measurement of the trace elements iron, copper, and zinc in serum.

Known amounts of two elements are thoroughly mixed with the sample. One element has a 'high energy' K line, the other a 'low energy' K line. The ratio of these intensities, in the absence of absorption, is a known constant and dependent only on the relative amounts of the respective elements, and on the energy of the exciting radiation. Whenever absorption is present, the ratio will change in a manner directly related to the mass absorption of the sample for these radiations.

INTRODUCTION

One of the unique features of an energy dispersive analyzer is that it can quantitatively determine a large number of elements simultaneously. In a situation where the maximum energy of the x-ray line to be used for analysis is equal to E_{max}, and in which the detector has an energy resolution of E_R, the analyzer can be considered as a system with a number of discrete channels equal to

E_{max}/E_R. It is, however, a rare occasion that all of these channels are used simultaneously. Typically, the instrument is used for determining a limited number of elements. Consequently, a number of channels become available which may be utilized to measure or monitor various parameters related to the sample mass, mass absorption, x-ray intensity or other factors.

BIOLOGICAL APPLICATION

Of the many types of samples, for which multielement analysis is required, biological specimens are a special group for the following reason:

First of all, the bulk of the sample consists of the lightest elements, hydrogen, carbon, nitrogen, and oxygen. The quantities of these elements have no biological interest as they bear no relationship to the components in which they are present. X-ray

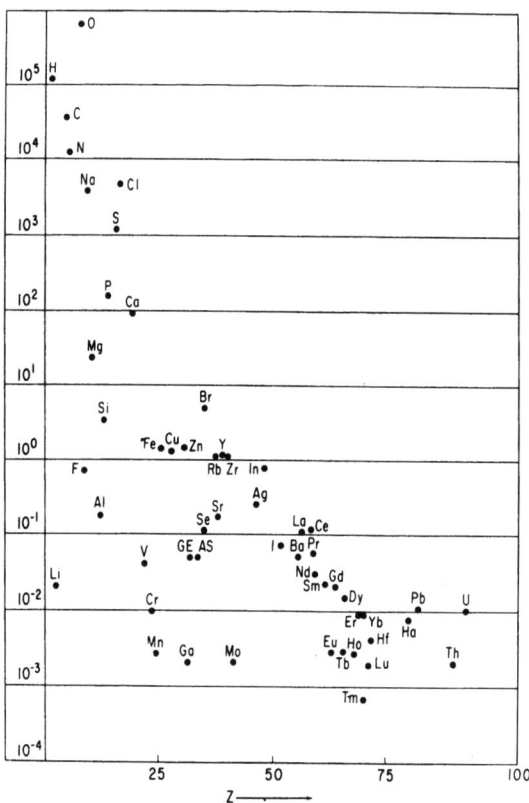

Fig. 1. Composition of human serum in microgram per milliliter sample, against atomic number Z, from ref. 3.

elemental analysis is therefore of value only in the determination of inorganic components. These may be divided into two categories, the major elements phosphorus, sulphur, chlorine, potassium, and calcium; and the trace elements which constitute the remaining components. The quantities at which these elements are present may cover a range of eight orders of magnitude as shown in Figure 1.

Thus, the energy dispersive system might not appear to be a suitable method for biological applications because 1) the high degree of Compton scattering would render the method unsuitable for determining trace elements, 2) the wide range of concentrations, and 3) elements which are present in large amounts are the light elements for which the detector is least suitable, and for which serious mass absorption effects occur. In practice, however, it was proved that the energy dispersive system may be an ideal system for certain biological samples if certain shortcomings can be resolved.

Despite the wide range of quantitative values for the elements present in biological specimens, these elements are present in groups of approximately similar amounts as shown in Figure 1. Compositional patterns may be similar in tissues; Figure 2 shows the 'spectral profile' obtained from two types of specimens and covers two groups of elements. The first group includes phosphorus, sulphur, chlorine, potassium, and calcium; the second group includes iron, copper, and zinc. Bromine and rubidium would also belong to this group but are not shown on these graphs. The typical patterns of biological samples show no strong spectral lines between calcium and iron and between iron and copper because the corresponding elements are normally present in much lower concentrations, often below the detectable limit. If determination of these elements is not pursued, the addition of one or more of these elements in known amounts gives useful supplemental data which will be further explored here.

MASS DETERMINATION

Compton scattering generally is undesirable because it increases the background and puts an additional load on the detector. However, because of the rather constant composition of the bulk material in biological samples, Compton scattering can be used to determine the mass of the sample being analyzed (2). Figure 3 shows a calibration curve for determination of mass using Compton scattered radiation obtained from two samples, lung tissue and paper. The lung tissue, being inhomogeneous, is expressed in absolute weight. With a sample having uniform area such as paper, the mass per unit area is used.

When the element to be determined is expressed in weight fraction, all the information is present in the spectrum and no additional information is required. The use of a calibration

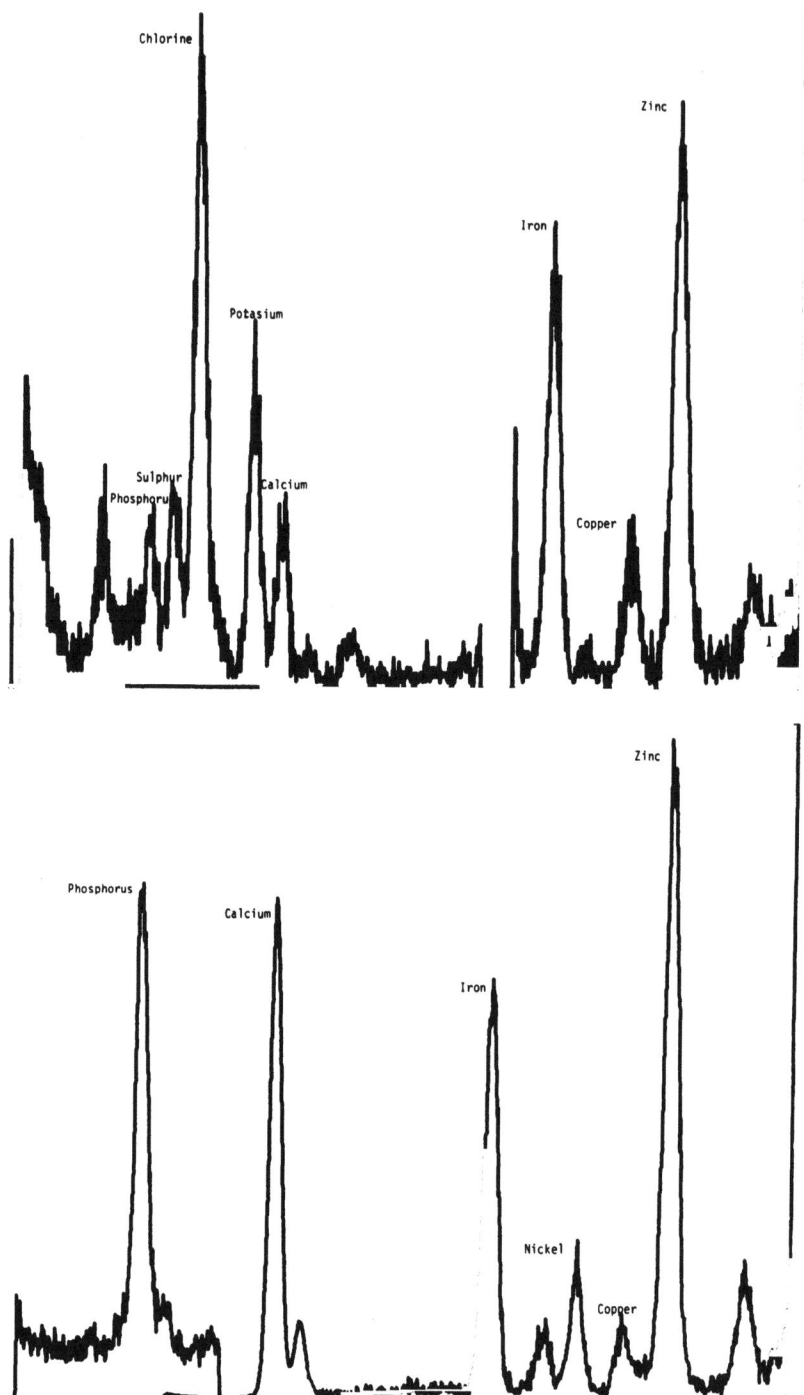

Fig. 2. Spectral profiles obtained from a liver needle biopsy (top) and from a bone biopsy (bottom).

curve, such as shown in Figure 3, however, requires that time integrated intensities be the same for the analysis of the calibration standard and sample. Instead of keeping both time and x-ray intensity constant, the use of a reference standard is proposed as follows.

The sample and calibration standard are always measured covered with a thin film on which a metal layer has been vacuum deposited. The line intensity of this metal could serve as a reference for the total integrated intensity. Thin and strong films of stretched polypropylene could easily be produced to serve as a sample support and cover, on which the thin film is deposited.

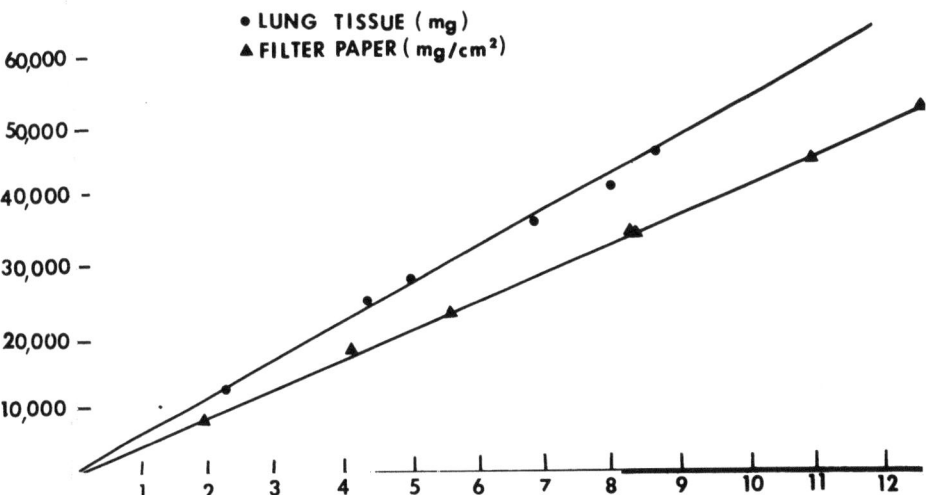

Fig. 3. Calibration curves for mass determination of lung tissues (in mg) and Whatman filter paper (in mg/cm^2) as obtained from FRIEDA, 20 mins accumulation time.

USE OF INTERNAL STANDARD

In the characterization of body fluids (serum, cerebrospinal fluid), it is customary to use the expression µg per ml liquid. The dry mass may vary from sample to sample and therefore could not be used as a reference. Here the addition of an element such as nickel may serve as a means for quantitation. For example, routinely, 2 ppM nickel is added in serum for the analysis of iron, copper, and zinc. From the relative efficiencies, the elements iron, copper, and zinc may be expressed in ppM. In such a method, intensity ratios are used rather than the absolute values. Time dependent variances are automatically eliminated because all of the elements, including the reference element, are measured simultaneously. The amount of sample being analyzed need not be known. Therefore, sample loss during preparation time does not affect the results, so long as this loss occurs after the addition of the reference element.

Because of mass absorption, variance in thickness effects the precision and accuracy of the determination. For copper and zinc determinations on serum, effect of variance in thickness is negligible. For iron, however, absorption correction is necessary.

ABSORPTION CORRECTION

Absorption correction requires the knowledge of thickness t, mass absorption coefficients μ, and density ρ. In biological samples, the absorption is caused by the bulk of the sample for which μ and ρ are known and the thickness t is the variable. The thickness in a dried serum sample is not easy to measure, however. The effect of attenuation of the x-ray may be eliminated if a parameter related to the thickness could be simultaneously measured. For this purpose, another reference element is added to the sample. This element, vanadium, has been selected because of the high absorption of its radiation, its close proximity to iron, and because it is present in low levels in biological samples. Since all measurements are relative to the nickel reference, the parameter to be used as an index for absorption will be 'vanadium to nickel' intensity ratio. This ratio is dependent on the absorption and the 'effective' efficiency may now be defined, which is dependent on the V/Ni ratio. The effective efficiency of iron has been determined by the following method.

Figure 4 shows two graphs of two sets of samples. The lower graph was obtained from a number of samples prepared from pooled sera. The top graph was obtained from the same sera to which 2 ppM iron was added. Subtraction of these two curves yields a relation which shows the efficiency as related to the V/Ni value.

To test this correction procedure, a run of about 40 samples was prepared from the same pooled sera, and analyzed after they were dried. Subsequently, the organic matter was removed by dry ashing as described previously (1), which reduces the absorption. Figure 5 shows the histogram of these results after being corrected for absorption (left) and the raw data (right).

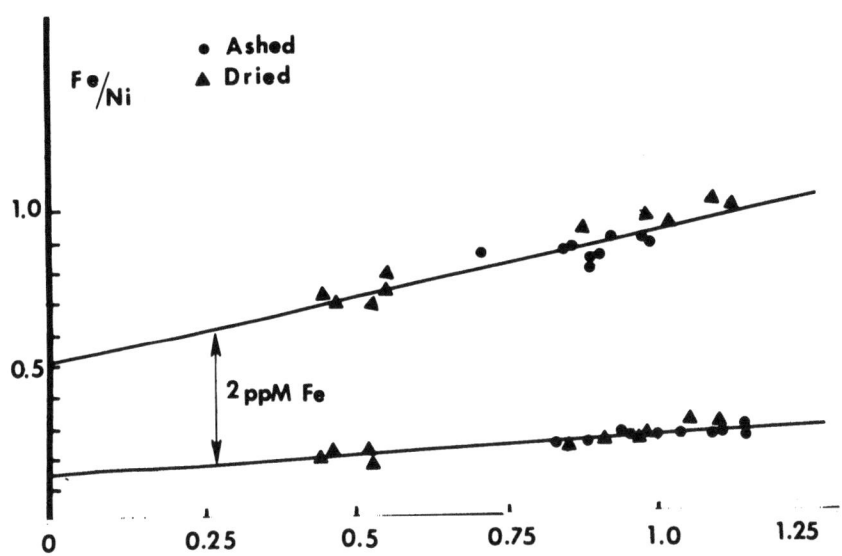

Fig. 4. The response of 2 ppM iron in serum in various mass absorption as reflected in the V/Ni ratio.

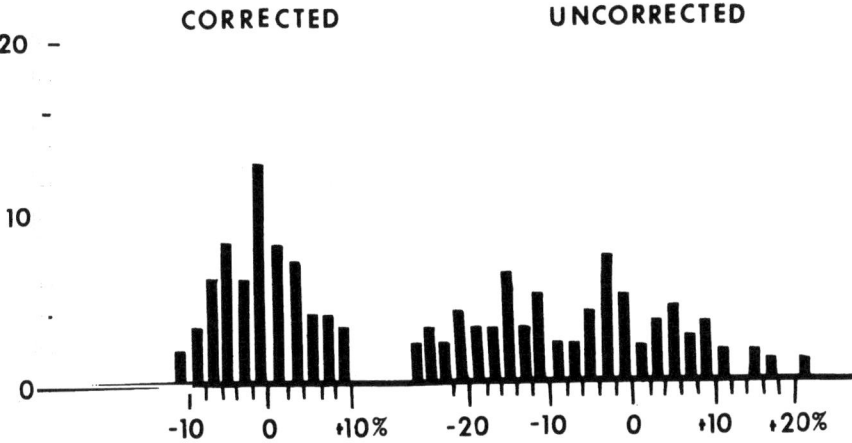

Fig. 5. Histogram of iron determination in pooled sera before (right) and after mass absorption correction using the graphs shown in Figure 4.

SIMPLIFIED SAMPLE PREPARATION

In serum analysis, as described previously (1), the sample is dried followed by dry ashing. This procedure is time consuming (4 hours), results in loss of the volatile material (Cl, Br) and, is a potential source of contamination.

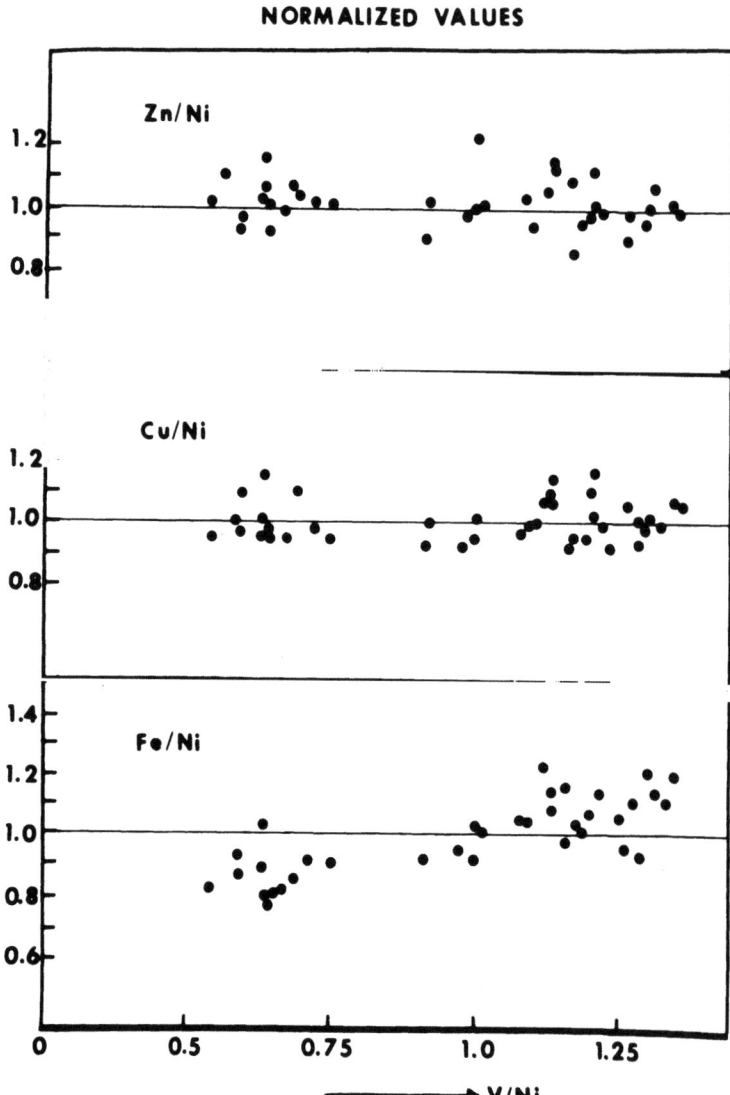

Fig. 6. Values of Zn/Ni, Cu/Ni, and Fe/Ni for various values of V/Ni. Only iron shows a systematic increase with increasing V/Ni ratio.

While ashing improves precision, background was found to be reduced only by a factor of 3. Depending on the required precision, such a procedure may or may not be worthwhile. Absorption variance in iron proved to be the largest source of error if unashed samples were used, as shown in Figure 6. With the described correction procedure, ashing may be eliminated if the slight loss in precision is acceptable.

The following measurements have been conducted. Forty samples were prepared from pooled sera, dried and analyzed for iron, copper, and zinc. After ashing, the measurements were repeated. The results are shown as a histogram in Figure 7. Iron values have been corrected for mass absorption.

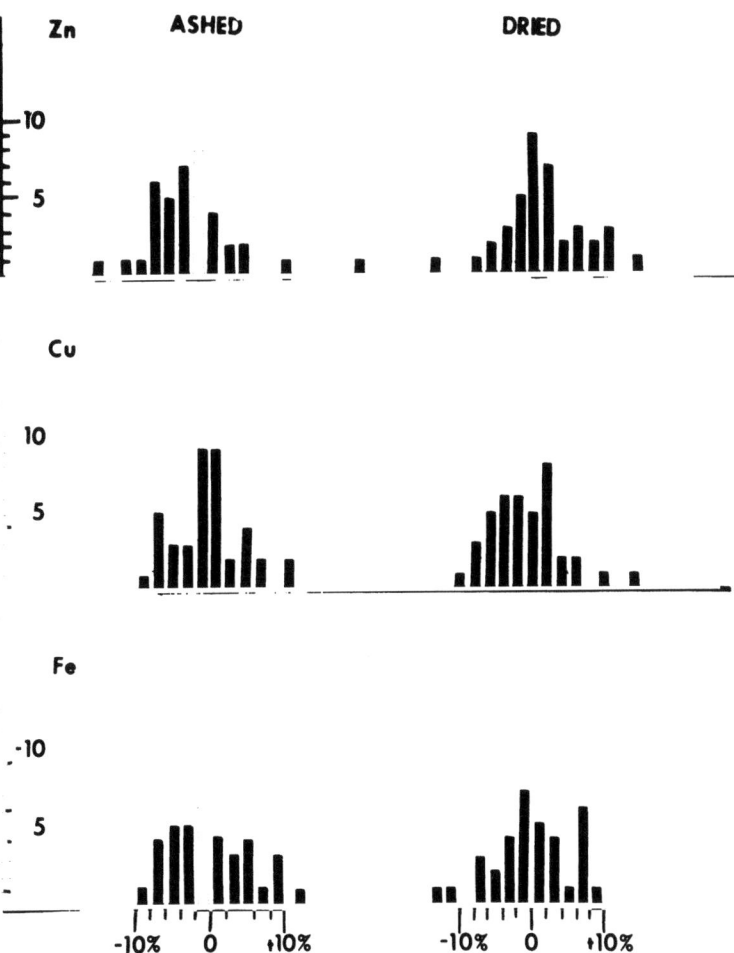

Fig. 7. Histogram of zinc, copper, and iron determination of dried sera samples (right) and after low temperature ashing. The values for iron were corrected for mass absorption in both cases.

ACKNOWLEDGMENT

The authors wish to acknowledge the dedicated assistance of Mary McCracken and Catherine Johns for the preparation of this manuscript and Ella Vallier for sample preparation.

This work was supported by an Institutional Research Grant No. RR05511-10.

REFERENCES

1. P. S. Ong, P. K. Lund, C. E. Litton, and B. A. Mitchell, An Energy Dispersive System for the Analysis of Trace Elements in Human Blood Serum. Advances in X-ray Analysis, Vol. 16, Proc. 21st Ann. Conf. on Application of X-ray Analysis, Denver, 1972.

2. L. Zeits, X-ray Emission Analysis in Biological Specimen. In Progress in Analytical Chemistry, Vol. 3 (X-ray and Electron Probe Analysis in Biomedical Research). K. M. Earle and A. J. Tousimis, Eds. Plenum Press, N.Y., 1969.

3. H. J. M. Bowen, The Elementary Composition of Mammalian Blood, United Kingdom Atomic Energy Authority Research Group, Report, 1963.

RESIN-LOADED PAPERS - A VERSATILE MEDIUM FOR SAMPLING AND STANDARDIZATION

Stephen L. Law and William J. Campbell

U. S. Bureau of Mines

College Park, Maryland 20740

ABSTRACT

Resin-loaded papers composed of approximately 50% cellulose and 50% ion-exchange or chelating resin provide an ideal matrix for many X-ray spectrographic analyses. Standards are prepared by multiple filtration of solutions of known composition through the paper to achieve quantitative collection or by the use of a radiotracer as a monitor for nonquantitative collection. Solutions prepared from unknown samples are processed in the same manner as the standards.

Advantages of the resin-loaded papers are: reduction of interelement effects because standards and unknowns are present in a similar low X-ray absorbing matrix; physical parameters such as metallurgical history, grain size, and surface preparation are eliminated; and sampling errors are significantly reduced and sensitivity greatly increased by concentrating trace elements separated from large samples.

Application of these papers to a variety of metallurgical, geological, and water samples will be summarized. The possible use of resin-loaded papers as standards for air pollution monitoring will be examined.

INTRODUCTION

The analytical requirements of an increasingly trace-element-conscious world cannot be satisfied by conventional direct X-ray spectrographic analysis because of inadequate sensitivity, interelement effects, and lack of suitable standards. However, the capabilities of X-ray spectrography can be significantly extended by using simple chemical separating and concentrating procedures to circumvent these limitations.

Separating the elements of interest from a large volume into the almost ideal matrix and volume of a resin-loaded paper disk has proven to be a versatile and effective approach to trace metals determination by X-ray spectrography (1,2). These papers are composed of approximately 50% cellulose and 50% powdered ion-exchange or chelating resin, providing a matrix of low-atomic-number elements (C, H, O, N, S) that have little effect on the X-ray determination of most elements. Incorporation of the resin in a thin paper disk provides a convenient media for handling small quantities of resin and for supporting the resin in the X-ray spectrograph or energy dispersive system. Standards and unknowns are prepared on similar resin-loaded papers, providing a match that is often impractical to achieve in direct X-ray analysis.

The general analytical procedure consists of the following steps:

1) Dissolution of the sample, or selective dissolution of the element or elements of interest.

2) Adjustment of pH, addition of complexing or masking agents, or other chemical treatment that may be necessary to achieve the selectivity desired in the ion-exchange process.

3) Collection of the desired elements on a resin-loaded paper disk by filtration or by suspension of the disk in the solution.

4) X-ray determination of the elements on the dried resin-loaded disk, using disks containing known quantities of the elements as standards.

The purpose of this report is to provide a state-of-the-art review of ion-exchange resin-loaded papers including both chemical and X-ray parameters, to summarize published applications, and to briefly consider potential applications of environmental interest.

CHEMICAL CONSIDERATIONS

The commercially available resin-loaded papers are made using ion-exchange resins that are strong acid or weak acid cation exchangers or are strong base or weak base anion exchangers. These resins are relatively nonselective, collecting almost all cations or all anions from solution according to an order to preference based on the size, charge, or other physical properties of the ions. Table 1 lists the properties of four commercially available ion-exchange resin-loaded papers. A resin-loaded paper not commercially available has been made by using a mixture of cation-exchange and anion-exchange resins for an even wider range of ion collection (3). It is often desirable to concentrate only a single element or group of elements away from the major constituents. Such selectivity is possible using an appropriate chelating resin that combines complexation reactions with ion-exchange properties. Because resin-loaded papers containing chelating resins are not commercially available, two types of chelating resin-loaded papers were prepared for our research on a special order basis by H. Reeve Angel and Co.[1/] One paper identified as SRXL resin paper, was made using a Srafion NMRR-type chelating resin, which is selective for gold and platinum metals (4), and more recently was found to collect both methylmercuric and mercuric forms of mercury (5). The Chelex 100 paper was prepared using Bio-Rad Chelex 100 resin, which has a high affinity for Hg^{2+} and Cu^{2+} compared with other heavy metals, and almost no affinity for alkali metals (6).

Selectivity may also be achieved, even for the more general ion-exchange and chelating resins, by introducing the appropriate complexing agents into the solution prior to ion exchange. For example, mercury forms chloro complexes in dilute hydrochloric acid that are readily collected by anion resin-loaded paper. Using an anion paper, mercury can be separated from metals that do not form strong chloro complexes in dilute HCl (7,8).

Manipulation of the solution pH can also provide a certain amount of selectivity. The Chelex 100 resin, for example, has been observed to undergo a transition from an ion-exchange collection mechanism to a chelating mechanism around pH 2 to 4 in studies of alkaline earth (9), copper (10), and mercury collection (11). Figure 1 compares the X-ray energy-dispersion spectra of two Chelex 100 resin-loaded papers after passing dilute HCl solutions of identical metal concentrations but different pH through the two disks (11). At pH 0.6, the Chelex 100 paper is very selective for the mercuric chloride complex as compared to the low collection at pH 4. These spectra were obtained using an Americium-241 excitation source with a molybdenum fluorescer. A Si(Li) detector and a 400-channel analyzer were used to collect the 15-min. count.

[1/] Trade names and company names are used for identification only and do not imply endorsement by the Bureau of Mines.

Table 1 — <u>Some Commercially Available Resin-Loaded Papers</u>[1]

	Properties			
Trade name	SA-2	WA-2	SB-2	WB-2
Resin type	Strong acid	Weak acid	Strong base	Weak base
Resin form	Na^+	H^+	Cl^-	OH^-
Thickness, mil	14	14	14	14
Wet strength	Good	Good	Good	Good
Flow rate	Fast	Fast	Fast	Fast
Capacity, meq/g	2	5	1.5	5
pH range	1-14	5-14	0-12	0-9

<u>Typical reactions</u>

SA-2	$RzSO_3^-H^+ + Na^+Cl^- \rightleftarrows RzSO_3^-Na^+ + H^+Cl^-$
WA-2	$RzCOO^-H^+ + Na^+Cl^- \rightleftarrows RzCOO^-Na^+ + H^+Cl^-$
SB-2	$(RzNR_3)^+OH^- + Na^+Cl^- \rightleftarrows (RzNR_3)^+Cl^- + Na^+OH^-$
WB-2	$(RzNH_2R)^+OH^- + H^+Cl^- \rightleftarrows (RzNH_2R)^+Cl^- + H_2O$

[1] Supplier: H. Reeve Angel and Co., 9 Bridewell Place, Clifton, N.J. 07014

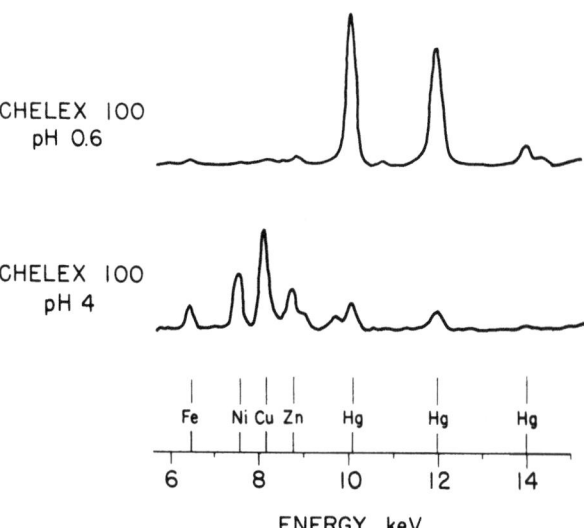

Figure 1 — X-ray spectra for comparison of selectivity for Hg, Fe, Ni, Cu, and Zn by Chelex 100 resin-loaded paper at pH 4 and pH 0.6 (11).

Of particular interest in metal determination is the form of mercury in solution. The Chelex 100 resin-loaded paper shows no tendency to collect the methylmercuric ion at pH 4 or below, but readily collects this toxic form of mercury at pH greater than 6 (11). Figure 2 shows a comparison of four different resin-loaded papers and their affinity for the two forms of mercury at different pH values, further emphasizing the importance of this parameter in achieving selectivity.

In some cases, it may be more convenient or even necessary to achieve selectivity by using a preliminary chemical separation technique such as solvent extraction, selective precipitation, or column ion exchange. For example, in determining trace impurities in molybdenum and tungsten it was necessary to separate the trace metals from the bulk molybdenum or tungsten in the dissolved sample by using a resin-column or solvent extraction before concentrating the trace metals in a cation resin-loaded paper for X-ray determination (12-14). Otherwise, the resin-loaded paper would have been filled to capacity with the molybdenum or tungsten and no advantage would have been realized.

The capacity of the different types of resin-loaded papers varies, but is usually around 1.5 to 5 milliequivalents per gram (meq/g) of paper as given in table 1. For example, a 3.5-cm-diam. SA-2 resin-loaded paper has a usable capacity of about 0.2 meq. This is equal to several thousand micrograms of the common metals; e.g., 4,000 µg Fe^{3+}.

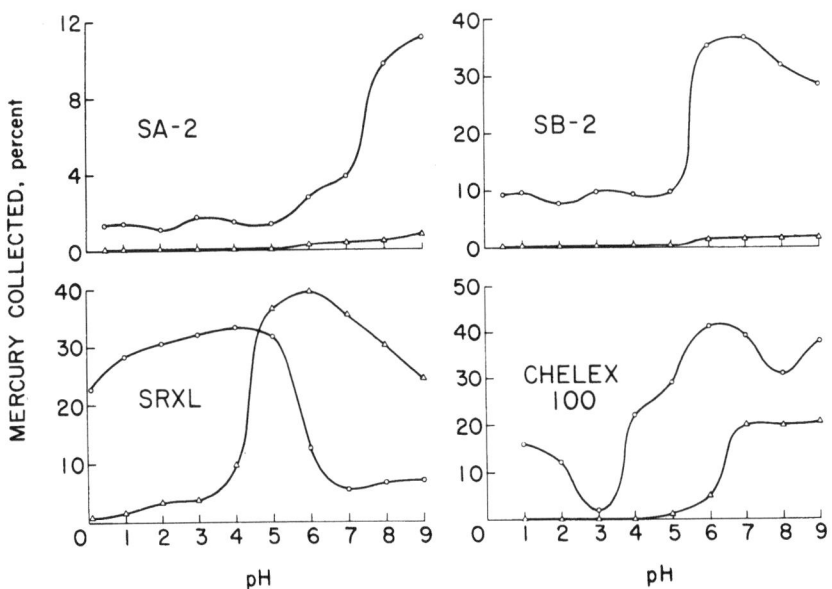

Figure 2 - Comparison of resin-loaded papers for pH effect on the collection of methyl mercuric (△) and mercuric (O) ions (11).

A resin-loaded disk is a very short ion-exchange column--the column length is equal to about the thickness of construction paper. Several filtrations, usually five to eight, are necessary to reach equilibrium between the solution and the resin particles. A variety of filtering apparatus have been reported for unidirectional filtering of the solution through the paper disk manually (12, 15-21) or using a recycling peristaltic pump (7). A simple device for filtering the solution through the disk in one direction and then back through in the reverse direction has also been reported (22). To insure quantitative recovery of an ion from solution, it is advisable to filter the solution through a second and third disk and examine all three by X-ray. Generally, the first disk will contain greater than 95% of the collectable ions. An alternate approach to determining collection efficiency is to use a radioactive isotope of the element of interest as a collection monitor. A carrier-free solution of known activity can be added prior to collection of the ions and the activity before and after collection compared to accurately establish the amount collected. Using such a monitor, it is not necessary to collect 100% of the ions of interest, thus permitting the analyst to simply suspend a resin-loaded paper disk in the solution overnight rather than filter (4). It is essential that the radioactive isotope be in the same ionic form as the ion in the sample in order to act as a true collection monitor.

X-RAY CHARACTERISTICS

The theory and advantages of using thin film techniques for determining microgram amounts of elements by X-ray spectrography have been documented (1, 23-25). A relative freedom from matrix effects is achieved using resin-loaded papers because they not only meet the thin film criteria but also, the organic portion of the disk (cellulose and resin) constitutes more than 95% of the weight, and is identical for both the standards and the unknowns. The exchange of the ions in solution for the original ions attached to the resin during the ion collection is the only change that occurs in the bulk composition of the disks other than degree of hydration. This small change generally has a minor effect on the resulting intensity, especially if an ion such as sodium is being replaced by an ion of similar X-ray absorption characteristics. The change of absorption characteristics is a function of the difference between the mass absorption coefficients of the exchanging ion and the exchanged ion (1). If the difference is small, the intensity to concentration relationship will be essentially constant. A low-energy X-ray photon such as 6.4 keV iron Kα shows only a 5% to 10% decrease in sensitivity over the range of 0 to 1,000 µg of iron. The linear response range is, of course, greater for elements having characteristic X-rays of higher energy. Standards of the same resin-loaded paper will show the same departure from linearity as the unknowns if both have similar concentrations of the significant elements. If a highly X-ray absorbing element, such as lead, is collected along with an element of interest that has a low-energy characteristic X-ray line, standard disks containing the highly absorbing element may need to be prepared or the measured intensities corrected. However, in most analytical applications encountered to date, the X-ray determination of an element has been only slightly affected by the presence of other elements on the same disk. The analyst can readily determine if there is significant difference in the sample and standard disks by measuring the transmission of the X-ray line of interest through each paper. Such analytical errors are generally smaller in magnitude than the initial sampling errors for trace-element determinations of environmental interest.

After collecting ions from a solution filtered through a resin-loaded disk, an exponential distribution of the ions might be expected, with the highest concentration at the top of the disk. In actual practice, intensity ratios between the top and bottom sides of a disk are close to unity, especially for high-energy X-rays. Disks containing approximately 1,000 µg of mercury show an average difference of 3% between the top and bottom intensities from the 11.8 keV Hg $L\beta_1$ line. The effect of silver distribution on photographic film showed that varying thickness or unequal coatings had a minor effect using the $AgK\alpha$ line (26).

The resin particles in a resin-loaded paper range in diameter from 10 to 40 µm, which is approximately one-tenth the thickness of the paper. The particles are fairly uniformly distributed with an equal weight of cellulose and a comparable volume of void space, as shown in figure 3. Each ion in solution, therefore, contacts only two to four resin particles per filtration, resulting in uniform distribution after several filtrations. Increasing concentration of competing ions, increasing flow rate, and ions of lower distribution coefficient all increase the certainty that the intensity ratios between sides of the disk will approach unity. Of course, if the resin-loaded paper disk is suspended in the solution containing the collectable ions rather than being used as a short ion-exchange column, both sides of the disk will show equivalent collection of ions.

Autoradiography of radioactive ions collected on resin-loaded paper and examination of random small sections of resin paper after ion collection have shown a fairly uniform distribution of ions over the entire area (27). Rotating the disk during X-ray excitation helps to insure reproducible intensities even if small inhomogeneities should be present. If rotation of the disk is not possible, the use of small disks (e.g., 1.5 cm diam or less) may result in greater reproducibility because essentially all of the disk is being viewed by the X-ray beam. However, in general, the small disk is used for increased sensitivity rather than improved reproducibility.

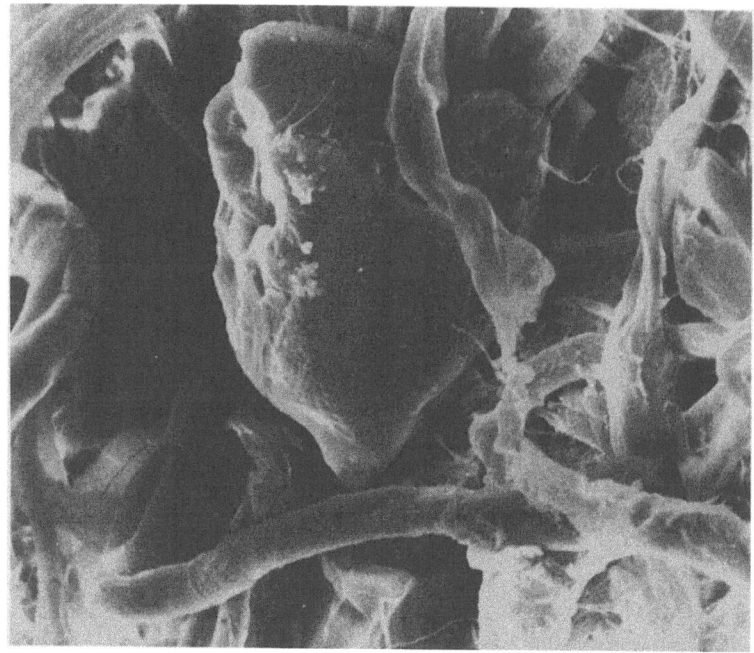

Figure 3 - A 1000X scanning electron microscope picture of a resin particle in SA-2 paper (21).

PREPARATION OF STANDARDS

Either the multiple disk filtration or the radioisotope monitor method may be used to establish the concentration of an element in a resin-loaded paper disk for use as a standard. In the multiple-disk filtration method, an accurately measured volume of a solution of known concentration of the element or elements of interest is passed six to eight times through a resin-loaded paper disk. The same solution is then filtered through a second and third disk for the same number of times. The X-ray intensity above background for each element of interest is obtained for each disk. The background is established by using a disk through which reagent blank solution only has been filtered. Usually the third disk will show no counts above background and may be discarded. The total X-ray intensity of all the disks represents the total amount of the element in the original standard solution. Comparison of the count rate from the first disk with the total counts gives the percent of the element present in the first disk, which is then used as the standard. For example, a solution containing 500 µg of mercury was filtered six times through each of three successive 3.5-cm-diam SRXL resin-loaded paper disks. At the particular X-ray instrument settings used, the first disk contained 1,019 counts per second (cps) above the reagent blank disk, the second disk contained 17cps and the third disk gave no detectable signal above background. The first disk, therefore, contained 98.4% of the total counts or 429 μg of mercury. A standard containing several different elements may be prepared in the same manner from a solution containing a mixture of known amounts of the elements. For example, a solution containing 500 μg each of Pd, Pt, and Au was passed six times through each successive disk of 3.5-cm-diam SRXL paper, resulting in a first-disk standard containing 500 μg Pd, 469 μg Pt, and 496 μg Au.

The radioisotopic monitor method of establishing the concentration of an element on a resin-loaded paper disk requires the use of only one disk, and quantitative recovery is not important. A standard solution is prepared containing a known amount of the element of interest and a known activity of an isotope of the element in the same chemical form. The solution is either filtered through the resin-loaded paper as before, or the paper disk may be simply suspended in the solution for a convenient period of time. All that is required is the collection of a sufficient concentration of the element to give a good X-ray signal and an adequate radioactivity count. Comparison of the activity added to the solution with the activity collected on the disk gives the fraction of the element that has been transferred to the resin-paper standard. For example, a 1.3-cm-diam SRXL resin-loaded paper disk was suspended for 40 hours in a solution containing a total of 100 μg of gold and 1,105 cps of Au-195 gamma activity. After drying, the disk contained 587 cps gamma activity, or 53.1% of the original activity. The standard disk thus contains 53.1 μg of gold. By using the same amount of radiotracer monitor in standards and unknowns, it is very convenient to use a calibration curve of the ratio of X-ray to gamma-ray intensities versus the solution concentration. The need to establish absolute amounts on each disk is thus avoided.

The radioactivity of the disk is measured on the scintillation detector of the X-ray spectrograph thereby avoiding the need for extra equipment. The gamma rays from the radiotracer monitor are shielded from the detector during the actual X-ray generation, and the use of carrier-free tracers insures that detectable amounts of the element are not added to the sample.

APPLICATIONS

Any sample that can be put into solution, or the elements of interest dissolved from the matrix, may be amenable to the advantages of analysis by resin-loaded paper X-ray spectrography. Although ion-exchange resin-loaded papers are especially adapted to determining trace constituents in large samples, the technique can also be used to determine major components where only small samples are available as, for example, in studies of artifacts of artistic or historical value. A summary of various applications of resin-loaded papers in X-ray spectrography is given in table 2.

Table 2 - Applications of Ion-Exchange Resin-Loaded Papers in X-Ray Spectrography

Sample type	Elements
Glass	Sn (28)
Metals	
Base metals	Os, Ru (29)
Copper	Ag, Au (30)
Molybdenum	Co, Cu, Fe, Mg, Mn, Ni, Pb, Zn (14)
Ni-Co-Cr alloys	Co, Cr, Fe, Mn, Ni (1)
Plutonium	Sc (18), U (16)
Tungsten	Co, Cu, Ni, Pb, Zn (12,13) Co, Ni, (15)
Ores and minerals	Au (4), Sc, Y, Rare Earths (31)
Petroleum	Ni, V (17)
Pigments	Hg (8)
Pottery	Pb (32)
Rat kidneys	Ca, Cd, Co, Cu, Fe, Mn, Ni, Zn (1, 33)
Terephthalic acid	Ca, Cr, Co, Fe, Mn, Mo, Ni, Ti (34)
Uranium mill products	Dy, Eu, Gd, Sm (35)
Waters	
Dilute acids	Cr, Hg, Pt (36), Cs(3) Hg (5), Y (37) Rare Earths (3)
Industrial waste waters	Cd (38)
Natural waters	Br (39), Cu (20), Hg (11), As, Cd, Cr, Cu, Fe, Mn, Pb, Se, Zn (44)
Reactor feedwater	Co, Cu, Fe, Ni, Zr (40)

One of the factors favorable to the use of resin-loaded papers in X-ray spectrography is the advantage of having both the sample and the standards in the same resin-loaded paper matrix. However, many of the samples of environmental interest are collected as discrete particles on filters of cellulose or glass fibers. Fortunately, samples of this type as well as the resin-loaded paper standards generally meet the thin-specimen criteria (1,23-25). Thus the advantages of ease of preparation and durability of the resin-loaded paper standards may be applicable to air filter samples. Disks from an air pollution monitor (41-42) could be routinely compared with a set of resin-loaded paper disk standards containing the elements of interest, using the multielement capability of energy-dispersion X-ray analysis. The resin-loaded paper standards contain the collected elements in discrete resin particles of 10 to 40 μm in diameter as shown in figure 3. Air pollution particulates usually fall in the range of 0.1 to 10 μm or somewhat larger. The major effect on the observed X-ray intensity will therefore arise not so much from size differences as from the fact that the air pollution particulates are mainly concentrated on one surface, whereas the resin particles are distributed throughout the resin-loaded paper.

Precipitated samples collected on the surface of a filter paper should be similar to air particulate samples thus serving to establish if there is any bias between surface collected samples and resin-loaded paper standards. Known quantities of mercury were coprecipitated with $(NH_4)_2S$ from a solution containing 11 other elements onto the surface of SB-2 paper disks. These disks were compared for $HgL\beta_1$ intensities to the same quantities of mercury collected by ion exchange on SB-2 paper from solutions containing mercuric chloride only. A Hg-203 tracer was used to monitor each collection and correct for differences in mercury collection. The data indicates there was no detectable difference in the 11.8 keV $HgL\beta_1$ line intensity from precipitated and ion-exchange-collected mercury. For low-energy X-rays a bias would be expected (1), especially if resin-loaded-paper standards are compared with air particulate samples collected on an X-ray absorbing filter.

A critical evaluation of ion-exchange resin-loaded papers as air pollution standards is therefore warranted. In addition to the precipitation-radiotracer method outlined previously, air particulate samples could be compared for intensities as received, and again after dissolution and recollection on ion-exchange resin-loaded paper.

Another support medium with high potential for environmental application is the reagent-impregnated filter paper. An ordinary filter paper is soaked in a specific reagent, or has a selective compound precipitated within its fibers. This paper is then used in the same way as an ion-exchange resin-loaded paper. As an example, the silver in silver sulfide precipitated in the fibers of a cellulose filter paper is replaced by mercury, which forms a very insoluble mercury sulfide (43).

CONCLUSIONS

The combination of ion-exchange separation and collection greatly extends the range of fluorescent X-ray spectrography so that the X-ray technique is applicable to many problems of environmental concern. The chemical preconcentration reduces or eliminates problems of matrix correction, variations in physical properties of the sample, and increases sensitivity by several orders of magnitude. In addition, reliable standards are easily prepared. With the commercial availability of computerized energy dispersive systems, ion-exchange resin-loaded papers will be of increasing importance in monitoring our waters and as standards for air monitors.

REFERENCES

1. Campbell, W. J., E. F. Spano and T. E. Green, Anal. Chem., 38 987 (1966).

2. Campbell, W. J., T. E. Green and S. L. Law, Amer. Lab., p. 28, (June 1970).

3. Hooton, K. A. H. and M. L. Parsons, Anal. Lett., 6, 461, (1973).

4. Green, T. E., S. L. Law and W. J. Campbell, Anal. Chem., 42, 1749 (1970).

5. Law, S. L., Science, 174, 285 (1971).

6. Bio.Rad Laboratories, Technical Bulletin 114 (1972).

7. Becknell, D. E., R. H. March and W. Allie, Jr., Anal. Chem., 43, 1230 (1971).

8. Link, W. B., K. S. Heine, J. H. Jones and P. Wattlington, J. Assoc. Offic. Agri. Chemists, 47, 391 (1964).

9. Luttrell, G. H., Jr., C. More and C. T. Kenner, Anal. Chem., 43, 1370 (1971).

10. Heitner-Wirguin, C., and G. Markovits, J. Phys. Chem., 67, 2263 (1963).

11. Law, S. L., Amer. Lab., p. 91 (July 1973).

12. Spano, E. F., T. E. Green and W. J. Campbell, BuMines Rept. of Invest., 6565 (1964).

13. Hubbard, G. L. and T. E. Green, Anal. Chem. 38, 428 (1966).

14. Spano, E. F. and T. E. Green, Anal. Chem. 38, 1341 (1966).

15. Spano, E. F., T. E. Green and W. J. Campbell, BuMines Report of Invest. 6308 (1963).

16. Hayden, J. A., Talanta, 14, 721 (1967).

17. Bergmann, J. G., C. H. Ehrhardt, J. Granatelli, and J. L. Janik, Anal. Chem., 39, 1258 (1967).

18. Hakkila, E. A., R. G. Hurley, and G. R. Waterburg, Anal. Chem., 41, 665 (1969).

19. Tackett, S. L., Anal. Chem., 43, 972 (1971).

20. Malissa, H. and I. L. Marr, Mikrochim. Acta, p.241, (1971).

21. Walton, R. D., Develop. Appl. Spectrosc., 9, 287 (1971).

22. Hooton, K. A. H. and M. L. Parsons, Anal. Chem., 45, 436 (1973).

23. Campbell, W. J., ASTM Special Technical Publication 349, 48, (1963).

24. Rhodes, J. R., A. Pradzynski, R. D. Sieberg and T. Furuta, "Applications of Low Energy X- and Gamma Rays", Ziegler, C. A. (ed.), New York; Gordon and Breach, Science Publishers, Inc. p. 317 (1971).

25. Giauque, R. D., F. S. Goulding, J. M. Jaklevic and R. H. Pehl, Anal. Chem., 45, 671 (1973).

26. Ehn, E., X-Ray Spectrometry, 2, 27 (1973).

27. Fukuda, K., K. Sugiyama and A. Mizuike, Radioisotopes (Tokyo), 19, 247 (1970).

28. Chamberlain, B. R. and R. J. Leech, Talanta, 14, 597 (1967).

29. Taylor, H. and F. E. Beamish, Talanta, 15, 497 (1968).

30. Fukasawa, T., T. Fujii and A. Mizuiki, Japan Analyst, 17, 713 (1968).

31. Eby, G.N., Anal. Chem., 44, 2137 (1972).

32. Tackett, S. L., G. H. Bender, T. R. Brunner, D. J. Duncan, M. G. Fedak, R. F. Gentile, J. F. Hiller, K. A. Hooker, A. J. McAuley, K. L. Rollick, J. F. Sandolfini, J. L. Smith, J. D. Vojtko, P. H. Pekala, and S. A. Williams, Anal. Lett., 6, 355 (1973).

33. Doolan, P. D., S. L. Schwartz, J. R. Hayes, J. C. Mullen, and N. B. Cummings, Toxiocology and Appl. Pharmacology, 10, 481 (1967).

34. Bergmann, J. G., C. H. Ehrhardt, L. Granatelli, and J. L. Janik, Anal. Chem., 39, 1331 (1967).

35. Latimer, J. N., W. E. Bush, L. J. Higgins, and R. S. Shay, USAEC RMO-3008, 271 (1970).

36. Minns, R. E., Proc. Conf. on Limit. of Detect. in Spec.-Chem. Anal., Univ. of Exeter, p. 45 (1964).

37. Klecka, J. F., AEC Report UCRL-17144 (1966).

38. Tanaka, H., T. Yamamoto, M. Akamatsu and G. Hashizume, Japan Analyst, 20, 784 (1971).

39. Radcliffe, D., Anal. Lett., 3, 573 (1970).

40. Montford, B., Canad. Spec., 13, 4 (1968).

41. Rhodes, J. R., A. H. Pradzynski, C. B. Hunter, J. S. Payne, and J. L. Lindgren, Environmental Sci. Tech., 6, 922 (1972).

42. Chessin, H. and E. H. McLaren, Adv. X-Ray Anal., 16, 165 (1972).

43. Krivan, V., Z. Analyt. Chem., 253, 192 (1971).

44. Burkhalter, P. G., Naval Research Laboratory Report 7637 (1973).

CHELATING ION EXCHANGE RESINS AND X-RAY FLUORESCENCE

Donald E. Leyden

Department of Chemistry, University of Georgia

Athens, Georgia 30602

ABSTRACT

Functional groups which form chelate compounds with metal ions have long been of analytical usefulness. Such compounds have served as titrants, precipitating reagents, extractants, and as ionogenic groups on ion exchange resins. These ion exchange resins often have high distribution coefficients which enable the quantitative extraction of metal ions from solution in a single batch treatment with the resin. This report describes the application of chelating resins to sampling and determination of certain trace metals using x-ray fluorescence. Two commercial resins will be discussed. Chelex-100 (Bio Rad) is a resin containing iminodiacetic acid functional groups and has chelation chemistry quite similar to EDTA. NMRR (Srafion) is a resin containing a functional site selective for gold, platinum metals, mercuric ions and methyl mercury.

INTRODUCTION

The concept of using ion exchange resins in combination with x-ray analysis is not new. Luke (1) reported the use of ion exchange impregnated paper in 1964. Campbell, Spano and Green (2) published a detailed study of the factors involved in the utilization of these papers in 1966. However, the use of chelating functional groups on the ion exchange resins is more recent. The use of ion exchange materials offers two major advantages in combination with x-ray fluorescence. First, a valuable preconcentration may be obtained. Trace ions in a liter or more of solution may be placed on a few milligrams of resin, or on a small resin impregnated paper disc. Second, matrix effect variations are essentially eliminated as the standards may be prepared in the same way as the samples. The resins are applicable to a wide variety of materials as long as the samples can be placed in solution form.

There are several factors which must be considered before using a chelating resin to extract metals from a solution and to prepare the treated resin for x-ray fluorescence. The chemistry of each resin must be considered in detail, however a few points may be discussed in a general way. One of the first factors to consider is the distribution ratio of the metal ion between the solution and resin phases. Most chelating resins offer an advantage in this regard. In many cases, the desired metal ion may be removed in a single batch extraction. However, because many of these functional groups are weak acids, the rate of equilibration may be slow, especially if the resin is in an acid form rather than as a salt. The extraction may require several hours to complete. A batch extraction offers the advantage of a homogeneous distribution of the ion in the resin. A weighed amount of the resin may be stirred in the solution, filtered and pressed into a pellet which is conveniently analyzed by x-ray or neutron activation. Most resins may be pelletized without the addition of a binder, therefore there is no dilution of the sample or the risk of contamination. Batch equilibration and pelletization techniques have been reported in several papers (3-5).

In cases in which the distribution ratio for the metal of interest is not sufficiently high for batch extraction, the solution may be passed through a column prepared from the resin. In this case, the ions are not dispersed homogeneously on the resin and some form of mechanical mixing may be required after the resin is removed from the column. Collin (7) has reported the use of a column for the ion exchange separation of a strontium from calcium acetate, with subsequent x-ray analysis of segments taken from the column.

Already mentioned is the use of ion exchange impregnated paper. Applications of this method have been further reviewed by Campbell, Green and Law (8). Law (9) has recently discussed resin impregnated papers utilizing Chelex-100 and NMRR resins which will be discussed here. The resin impregnated papers have an advantage in ease in handling. However, there are few varieties of these papers commercially available. A further disadvantage is that ion exchange chromatography can occur as the solution is passed through the paper. This leads to an inhomogeneous sample and requires correction factors for the x-ray counting data when long wavelength x-rays are used.

Finally, one must consider the chemistry of the resin to the advantage of a particular determination. Few highly specific chelating groups are available. However, several methods may be employed to gain selectivity. Masking agents may be used which prevent the uptake of undesirable ions by the resin. The use of pH control can frequently aid in selectivity, as can the control of the oxidation state of the ions. Before presenting examples of successful use of these resins, some of the sampling problems and related applications will be discussed.

SAMPLING WITH CHELATING RESINS

If a representative sample of a material to be analyzed is taken, a solution must be prepared to extract ions with any resin material. The advantage of the use of chelating ion exchange resins lies in their ability to absorb trace levels of metal ions from large volumes of solutions, i.e. preconcentration. If however, this solution contains large quantities of ions which also form chelates with the resin, the available functional sites may be loaded to the exclusion of the trace elements. Frequently this problem may be solved by simple chemical manipulation. For example, Chelex-100 may be used to extract Bi^{+3} in nitric acid at a pH of about 1.5 with only a partial extraction of copper and mercury. Figure 1 shows the percent metal ion extracted by Chelex-100 as a function of pH for several ions. Ferric iron, a common interference in silicate rocks can be masked by the addition of ascorbic acid which reduces the iron to the ferrous ion (6). In other applications the iron may be removed by passing the solution through an ion exchange column of Dowex-50 (6).

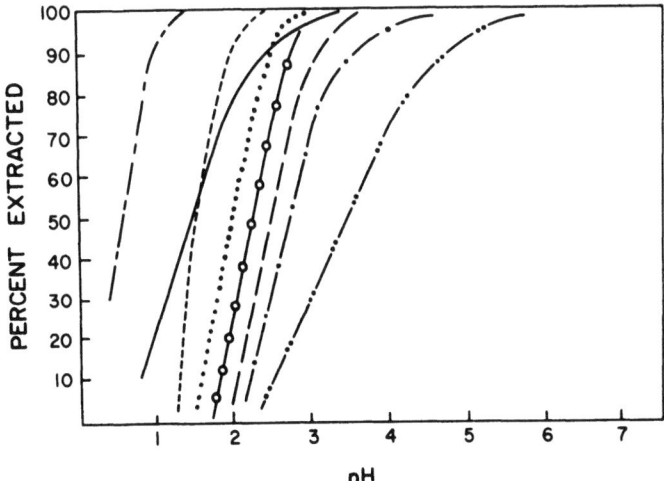

Figure 1. Per cent metal ion extracted by Chelex-100 as a function of pH

Bi, Tl — - — -   Cd - O - O -
Cu, W, Mo - - - -   Zn, Co — — — —
Fe^{3+} ———   Sr, Ba —·—·—·
La, Pb, Ni . . . . Fe^{2+} —··—··—

(Reprinted by permission of the American Chemical Society.)

The nature of many chelating functional groups makes complete equilibration of the resin with the solution slow (10). There is evidence that the exchange process with Chelex-100 may occur in two steps (9,11). The first step is a simple ion exchange followed by a much slower chelation reaction. We have found that Bi^{+3} at pH 1.5 is 90% extracted by Chelex-100 in about ten minutes. However, there is a slow increase to 100% recovery after ten hours (12). Surprisingly, Cr(III) which is known to have slow coordination rates is quantitatively recovered on Chelex-100 in less than thirty minutes (5). This further suggests that the first step in the exchange is other than chelation.

If resin beads are used rather than impregnated papers, the effect of particle size upon the x-ray intensity must be considered. Allen and Rose (13) have investigated this point using copper loaded loose beads of Chelex-100. They report a maximum intensity of particles smaller than 56 microns, and a size independent region between 150 and 500 microns. This intensity effect decreases with resin loading. However, if the beads are pressed into fused pellets, the particle size effect becomes unimportant.

In summary, chelating ion exchange resins offer a valuable method of quantitative recovery of trace elements for x-ray determination of metals. Problems of selectivity, interference (chemical and spectroscopic) must be treated on an individual sample basis. In many cases multielement extraction and determination may be performed on a single pellet.

Determination of Cobalt and Nickel in Silicate Rocks

The determination of cobalt and nickel in four USGS standard silicate rock samples was made (6). Accurately weighed portions of each sample were dissolved in a mixture of hydrofluoric and nitric acids. Because these samples contained up to 10% iron, and the cobalt K_α and iron K_β lines are quite close, the iron was removed by passing the solution through a column of Dowex 50-X8. The effluent was evaporated to dryness, the residue was dissolved in 100 ml of water and extracted with 200 mg of Chelex-100. The resin was filtered from the solution, air dried and pressed into a pellet. The results are reported in Table I which also shows the mean and range of values found by others using x-ray methods (14).

Determination of Bismuth

Bismuth was determined in rock samples supplied and previously analyzed by Kennecott Exploration Inc. (6). The major interfering elements in these samples are iron, aluminum and copper. The rock samples were dissolved with hydrofluoric acid, or hydrofluoric/nitric acid mixtures. The iron was easily masked by the addition of ascorbic acid. The aluminum reacts slowly with the resin and only presented problems by precipitation in the HF solutions. Attempts to mask the aluminum with tris(hydroxymethyl)aminoethane

Table I. Means and Ranges (ppm) for Published Values of Cobalt and Nickel in AGV-1, BRC-1, PCC-1 and DTS-1 (14)

USGS Rock Types	Cobalt		Nickel	
	X	Range	X	Range
PCC-1	112	80-300	2430	1750-3400
	(107 ± 2)[a]		(2450 ± 3)	
DTS-1	132	96-200	2330	1770-3300
	(186.6 ± 0.8)		(2306 ± 2)	
BCR-1	35.5	29-60	15.0	8-30
	(47 ± 2)		(19.3 ± 1.6)	
AGV-1	15.5	10-30	17.8	11-27
	(17.1 ± 0.7)		(23.8 ± 0.3)	

[a]Values in parentheses from this work

(THAM) resulted in low bismuth recovery. Results of bismuth determinations are presented in Table II. The analyses are based on repeated measurements of aliquots taken from the same rock sample and on replicate rock samples. The reproducibility of measurements from the same rock sample is better than those between separate samples. Results of standard addition samples are essentially the same as those obtained by direct analysis. Solutions obtained by dissolving the sample using HF and nitric acid agree quite well with those obtained using nitric acid alone. The data for nitric acid digestion indicates that a HF attack may not be necessary. The variation in results for all samples is larger than expected. However, this could be due to variation in the bismuth content within the samples.

Determination of Cr(VI) and/or Cr(III)

The determination of chromic ion using Chelex-100 is an easy task. However, the speed with which this ion is recovered was surprising (5). As a first approximation, it would seem that the coordination properties of chromium toward Chelex-100 should be similar to those of chromium toward EDTA because the active functional group in each case is the iminodiacetate group. The quantitative chelation of Cr(III) with EDTA has been known to be slow for many years. The coordination kinetics of Cr(III) are accelerated by increases in both pH and temperature; however, at room temperature total reaction is achieved only after approximately 50 hours. At room temperature, quantitative uptake of Cr(III) by Chelex-100 is realized within 30 minutes. In a mixture of chromic and chromate ions, only chromic is extracted; unless a reducing agent is added to convert the anionic chromate to chromic. Aikens and Reilley (15) showed that chromium may be reacted more rapidly with EDTA by reducing chromate in the presence of EDTA, than by direct reaction with pre-reduced chromate. The determination of Cr(III) in a mixture of chromic

Table II. Results of Determination of Bismuth in Geochemical Rock Samples

Sample No.	Digestion Procedure	Average Bi, ppm	KEI value, ppm[a]
GRLD 102	HF-HNO$_3$	15.08 ± 2.2	12
102	HNO$_3$	16.0 ± 1	
102	HF-HNO$_3$ + Std. Add.	16.5 ± 0.9 14.0 ± 1.5	
105	HF-HNO$_3$	5.60 ± 0.21	7.8
105	HNO$_3$	4.9 ± 0.5	
105	HF-HNO$_3$ + Std. Add.	5.1 ± 1	
114	HF-HNO$_3$	0.51 ± 0.05	1
114	HNO$_3$	0.42 ± 0.05	
115	HF-HNO$_3$	0.77 ± 0.1	1
115	HNO$_3$	0.88 ± 0.1	
100	HF-HNO$_3$	0.05 ± 0.01$_5$	0.5
100	HNO$_3$	0.18 ± 0.06	

[a] Kennecott Exploration Inc. determination by emission spectrography, personal communication.

and chromate is rapidly done by direct extraction from an aliquot. A second aliquot is extracted in the presence of sodium bisulfite, which reduces chromate to chromic, thereby permitting the determination of total chromium. The difference between the two gives the chromate (5).

Determination of Gold in Silicate Rocks

NMRR resin (formerly known by the designation SRXL) has been used by Green, et.al. (16), for the analysis of gold in silicate rocks using both x-ray fluorescence and nuetron activation to determine the gold content of resin loaded filter papers. Gold recovery was much less than complete so that a small quantity of the radioactive isotope Au^{195} was added to each sample to monitor recovery. Ions of other elements are not strongly absorbed by this resin and the minor amount present was washed out of the filter paper with 2.4 N HCl after the gold had been taken up.

The use of Chelex-100 was briefly tested for gold analysis. Recovery was good over a pH range from 1 to 3. The gold-Chelex-100 complex produced a purple coloration. The potential use of Chelex-100 for gold analysis was not fully examined, because of the lack of selectivity between gold and other transition or heavy metal ions. It is almost certain that ferric iron would interfere even at a pH near 1 because of the high iron content of solutions derived from the digestion of rock samples in aqua regia. The use of ascorbic acid as employed in the determination of bismuth could not be applied because gold would be reduced also. At present, it appears that NMRR resin is to be preferred because of its selectivity. A few results are shown in Table III.

Table III. Results of Analysis of Some Geochemical Standard Samples

Sample Kennecott Exploration Inc. Geochemical Standards	Sample Size (grams)	Results ppm	Reported Value ppm
GRLD 102	5	1.44 ± .09	1.42 (a)
GRLD 105	10	0.44 ± .04	0.46 (a)
GRLD 115	20	0.04 ± .01	0.04 (a)
GRLD 116	20	0.08 ± .02	0.11 (a)

[a] Kennecott Exploration Inc. values determined by Atomic Absorption.

USE OF ION EXCHANGE RESINS AS STANDARDS

Ion exchange resins represent a potential convenient form of a standard for trace analysis. However, there are many problems involved. Commercial resins vary in properties from batch to batch. Variations in bead size and density are of concern. A group at the National Bureau of Standards has studied the use of resins as microstandards (17). A provisional certification for a standard for calcium has been issued. These standards are taken as individual beads which contain nanogram quantities of the ion. The resin is carefully prepared and one may calculate the quantity of ion present by measuring the diameter of the bead with a micrometer and using factors provided with the resin.

Secondary standards for x-ray may be prepared by treating chelating resins with standard solutions. However, if the resins are pressed into pellets, careful storage is required as they tend to curl with excess humidity and crack if kept excessively dry.

SUMMARY

The use of chelating ion-exchange resins is a convenient method for extracting and preconcentrating metal ions in solution. These resins may be analyzed by x-ray, neutron activation or other techniques. The resin may be left in loose bead form, pressed into pellets or impregnated in paper. Batch extraction or column techniques may be used with beads. Impregnated papers may be suspended in the solution, or held as a filtration device. Either experimental technique is simple. Multielement determinations are conveniently made. Matrix effects are small as standards may be prepared in the same manner as the sample. When the original sample matrix effects recovery of the desired ion, masking or standard additions techniques may be used.

ACKNOWLEDGMENTS

The author acknowledges the aid of C. W. Blount and the students who have worked on this research. This research was supported in part by Research Grant GP-24311 from the National Science Foundation.

REFERENCES

1. C. L. Luke, "Ultratrace Analysis of Metals with a Curved Crystal X-ray Milliprobe", Anal. Chem., 36, 318-322 (1964).

2. W. J. Campbell, E. F. Spano and T. E. Green, "Micro and Trace Analysis by a Combination of Ion Exchange Resin-Loaded Papers and X-Ray Spectrography", Anal. Chem., 38, 987-996 (1966).

3. J. N. Van Nickerk, J. F. DeWet and F. T. Wybenga, "Trace Analysis by Combination of Ion Exchange and X-Ray Fluorescence", Anal. Chem., 33, 213-215 (1961).

4. C. W. Blount, W. R. Morgan and D. E. Leyden, "An Improved Technique for the Preparation of Pellets for X-Ray Spectrographic Analysis of Ion Exchange Resins", Anal. Chim. Acta, 56, 456-458 (1971).

5. D. E. Leyden, R. E. Channell and C. W. Blount, "Determination of Microgram Quantities of Chromium(VI) and for Chromium(III) by X-Ray Fluorescence", Anal. Chem., 44, 607-610 (1972).

6. C. W. Blount, D. E. Leyden, T. L. Thomas and S. M. Guill, "Application of Chelating Ion Exchange Resins for Trace Element Analysis of Geological Samples Using X-Ray Fluorescence", Anal. Chem., 45, 1045-1050 (1973).

7. R. L. Collin, "X-Ray Fluorescence Analysis Using Ion Exchange Resin for Sample Support", Anal. Chem., 33, 605-607 (1961).

8. W. T. Campbell, T. E. Green and S. L. Law, "Ion Exchange Papers in X-Ray Spectroscopy", Amer. Lab., 2, 28-34 (1970).

9. S. L. Law, "Resin-Loaded Papers for Methyl Mercury and Inorganic Mercury Determination", Amer. Lab., 5, 91-97 (1973).

10. D. E. Leyden and A. L. Underwood, "Equilibrium Studies with the Chelating Resin Dowex A-1", J. Phys. Chem., 68, 2093-2097 (1964).

11. G. H. Luttrell, Jr., C. More and C. T. Kenner, "Effect of pH and Ionic Strength on Ion Exchange and Chelating Properties of an Iminodiacetate Ion Exchange Resin with Alkaline Earth Ions", Anal. Chem., 43, 1370-1375 (1971).

12. D. E. Leyden and C. W. Blount, unpublished results.

13. A. L. Allen and V. C. Rose, "Effect of Ion Exchange Resin Particle Size on X-Ray Fluorescent Analysis", in Advances in X-Ray Analysis, K. F. J. Heinrich, Editor, Vol. 15, p. 534-538, Plenum Press (1972).

14. F. Flanagan, "U. S. Geological Survey Standards-II", Geochim. Cosmochim. Acta, 33, 81-120 (1969).

15. D. A. Aikens and C. N. Reilley, "Rapid Chelometric Determination of Chromate as Cr(III)-EDTA by Reduction in the Presence of EDTA", Anal. Chem., 34, 1707-1709 (1962).

16. T. E. Green, S. L. Law and W. J. Campbell, "Use of Selective Ion Exchange Paper in X-Ray Spectrography and Neutron Activation", Anal. Chem., 42, 1749-1753 (1970).

17. D. H. Freeman, L. A. Currie, E. C. Kuehner, H. D. Dixon and R. A. Paulson, "Development and Characterization of Ion-Exchange Bead Microstandards", Anal. Chem., 42, 203-209 (1970).

CAN REGRESSION EQUATIONS BE OPTIMIZED BY FINAGLING X-RAY INTENSITIES?

M. Fatemi and L. S. Birks

Naval Research Laboratory

Washington, D. C. 20375

ABSTRACT

The application of the null-determinant technique to eliminate the inconsistency of x-ray empirical equations is investigated. Experiments with a series of stainless steel standards show that, in general, calculated compositions obtained by the null-determinant technique agree more closely with chemical analysis than do those obtained by simple averaging. Since it is usually errors in measured intensities which lead to inconsistent estimates of concentrations, the advantages of the null-determinant technique in adjusting the intensities become evident.

INTRODUCTION

The purpose of this work is to demonstrate the application of a recent technique to resolve the inconsistency problem associated with the use of regression equations. These equations have been widely used in the analysis of multicomponent systems to account for interelement absorption and secondary fluorescence. In a recently published paper (1) it was pointed out that the solution of regression equations for the composition of alloys almost invariably leads to inconsistencies in the estimated compositions. A simple test was, therefore, presented, through which the degree of inconsistency in these equations could be established. A procedure was also outlined that would eliminate the inconsistency by adjusting the measured intensities in a reasonable fashion.

REVIEW OF REGRESSION EQUATIONS

In Fig. 1 the relative intensity of Fe Kα radiation excited from a variety of stainless steel alloys is shown.

As is well known, there is no simple exact relationship between the measured intensities and concentrations of iron in the samples. We know that the measured intensity from a certain element in the alloy is dependent upon the presence of other elements. In the empirical coefficient technique this dependence is written as a <u>linear</u> combination of all weight fractions:

$$W_i = R_i \sum_j \alpha_{ij} W_j, \tag{1}$$

where α_{ij} is an empirically determined coefficient that signifies the effect of element j on the measured intensity of element i. Equation (1a) shows the explicit form for one of the equations.

$$W_A = R_A (\alpha_{AA} W_A + \alpha_{AB} W_B + \alpha_{AC} W_C + \ldots) \tag{1a}$$

Note that the measured intensity of only one of the components, R_A, appears in this equation, but that all W's are present.

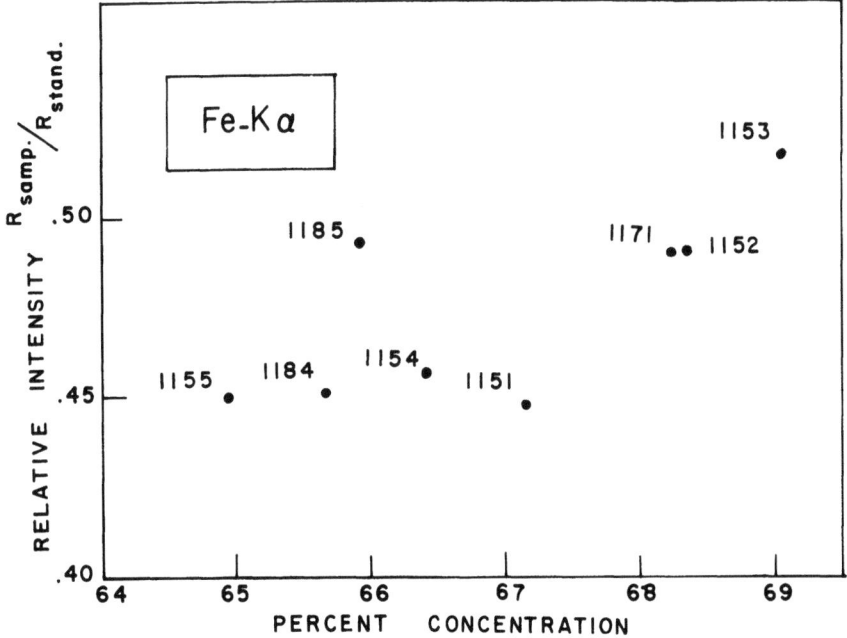

Fig. 1. Variation of measured intensity of Fe Kα with concentration of iron in NBS standard samples.

Equation (1a) may be rewritten to collect all W's together.

$$(R_A \alpha_{AA} - 1) W_A + R_A \alpha_{AB} W_B + R_A \alpha_{AC} W_C + \ldots = 0$$

$$R_B \alpha_{BA} W_A + (R_B \alpha_{BB} - 1) W_B + R_B \alpha_{BC} W_C + \ldots = 0 \quad (2)$$

Etc.

The lack of any non-zero constant on the right shows that the equations are homogeneous. The significant point is that there are as many equations as there are unknowns to be found. Thus if a sample containing five elements is considered, the resulting five equations could only give the <u>ratios</u> of the concentrations and not their individual values. It would, therefore, be necessary to solve any four of equations (2) together with a supplementary (non-homogeneous) equation such as equation 3, which states that the sum of the weight fractions should be a constant:

$$W_A + W_B + W_C + \ldots = \text{Const.} \quad (3)$$

The value of the constant would be unity if all the constituents have been accounted for, but would be less than unity when impurities and/or other elements are present.

In practice one finds that the calculated compositions which should be the same, regardless of which four equations are chosen for solution, are in fact not the same; that is, the equations are inconsistent. It was the need to obtain consistent results that led to the formulation of the data adjustment technique presented in Ref. (1). The test for the consistency of empirical equations is simply that the determinant of the multipliers of W_i in equation 2 be equal to zero:

$$\begin{vmatrix} R_A \alpha_{AA} - 1 & R_A \alpha_{AB} & R_A \alpha_{AC} & \ldots \\ R_B \alpha_{BA} & R_B \alpha_{BB} - 1 & \ldots & \\ R_C \alpha_{CA} & R_C \alpha_{CB} & \ldots & \\ \ldots & & & \end{vmatrix} = 0 \quad (4)$$

If the determinant (which in practice is almost never zero) is very small compared to its own members, one may safely assume that the calculations will lead to consistent results. If not, one may often be able to adjust the intensities, in a systematic fashion, to reduce the determinant to a null-value. As can be seen, there are two distinct quantities (R_i and α_{ij}) in the determinant that could, in principle, be adjusted. Hence, a few comments on the sources of inconsistency would be in order. These may be classified in the following way:

1 - Random Errors in Measured Intensities From Unknown Samples

2 - Systematic Variations in Experimental Conditions Which Affect Unknown Intensities

3 - Errors in Measurement of Standard Intensities Which Affect Computation of α_{ij}

Of these, the most important are the first two, since the quantities α_{ij} are usually very exactly determined through repeated measurements. This at least would be a matter of convenience, as one would only be concerned, in later measurements, with the adjustment of intensities from <u>unknown compositions</u>.

Recently, measurements were performed in our laboratory on a set of NBS stainless steel standards. These were 1151, 1152, 1153, 1154, 1155, 1171, 1184, and 1185. Of these, 1152, 1154 and 1185 were chosen as "unknown samples" and the remaining five as standards of "known samples" from which α_{ij} could be calculated. $K\alpha$ line intensities were recorded from Cr, Ni, Fe, Mn and Mo constituents using a vacuum single crystal spectrometer, Ag-FAQ 60 x-ray tube and an abraded LiF (200) single crystal.

Two examples of the type of answers obtained from these measurements are shown in Table I. Because of the limited number of samples, we chose to adjust the intensities by the method of least squares, namely, that the squares of the fractional corrections on various measured intensities would have a minimum sum:

$$\sum_i \left(\frac{\Delta R_i}{R_i} \right)^2 = \min. \qquad (5)$$

where ΔR_i are the actual adjustments on the R_i. In the computer program, called FINAGLE, and written in the conversational mode, the compositions and the recorded intensities from five standards are entered and the α_{ij} are calculated. The program then asks for the measured intensities of an unknown sample, and prints five sets of calculated W's and the determinant of coefficients of W's to demonstrate their inconsistency. It then proceeds to correct the measured intensities by the prescribed method, and again prints the new determinant of the coefficients as well as five consistent sets of calculated W's, together with the percent correction of each intensity required to produce this consistency.

TABLE I.

1185

	Cr	Ni	Fe	Mn	Mo
	By N-1 Equations, Sum W_i = 98.6%				
Set 1	17.51	13.07	65.70	1.23	1.88
Set 2	16.44	14.08	65.74	1.23	1.92
Set 3	17.31	13.01	65.86	1.23	1.99
Set 4	17.43	13.03	66.71	.22	2.01
Set 5	16.74	13.12	67.03	1.23	1.29
Ave	17.09	13.26	66.21	1.03	1.82
	By Null-Determinant Method Sum W_i = 98.6%				
$\dfrac{\Delta R_i}{R_i}$ (%)	+1.68	+1.34	−2.68	− .15	− .35
Consistent Set	17.22	13.19	65.81	1.23	1.95
	Chemical Analysis				
	17.09	13.18	65.90	1.22	2.01

1152

	Cr	Ni	Fe	Mn	Mo
	By N-1 Equations, Sum W_i = 98.6%				
Set 1	18.35	10.37	68.31	1.20	.362
Set 2	18.99	9.79	68.26	1.20	.358
Set 3	18.51	10.39	68.14	1.20	.352
Set 4	18.46	10.38	67.74	1.67	.351
Set 5	18.80	10.35	67.54	1.20	.710
Ave	18.62	10.26	68.00	1.29	.427
	By Null-Determinant Method Sum W_i = 98.6%				
$\dfrac{\Delta R_i}{R_i}$ (%)	− .88	− .53	+1.48	+0.08	+0.03
Consistent Set	18.51	10.33	68.19	1.19	.36
	Chemical Analysis				
	18.49	10.21	68.36	1.19	.36

In the case of (1185), it is interesting to note that the Cr concentration by averaging technique agrees exactly with chemical analysis, but the remaining four components give better agreement by null-determinant calculation.

It may be argued that especially in the case of Mn and Mo, two of the numbers are so poor that they should be discarded from the average, and that the agreement with chemical analysis would then be exact also. The reason for their inclusion in the calculation of averages is two-fold. First, the fact that only one number in the group is very poor for a given concentration may be due to rather careful measurements. If these measurements had been performed with less certainty (as sometimes happens in actual practice), a much more pronounced discrepancy would have been observed among the results, thus making it difficult to decide which result is poor and which is not. Secondly, the reader may note that in our example it is the diagonal numbers that are clearly wrong. The reason for this effect is that in each set of computed results, (N-1) empirical equations are used, and the missing equation corresponds to the particularly wrong answer. For example, in set 4, the fourth empirical equation corresponding to R_{Mn} is missing and this causes the calculated W_{Mn} to be greatly influenced by the errors in the remaining intensities that are used in the calculation. However, if the diagonal terms are eliminated from the averaging, the results for Cr, Ni and Fe are worse than before, i.e., Cr_{Ave} would be 16.98, Ni_{Ave} would be 13.06 and Fe would be 66.29. Further, it should be noted (see set 2 of 1185) that a poor answer in one component (Ni) might affect only one concentration (Cr) without significantly affecting the others.

A question now arises: certainly the inconsistency of equations (2) is not new. How, then, has it been handled so far by various experimenters? We suspect that either following the observation of various solutions to the compositions one "reasonable" set has been chosen hopefully to represent the correct solution, or an average has been found from among the more reasonable data. A study of several experimental results will clearly show, however that the estimated compositions will be in better agreement in the case of null-determinant method than with straight-averaging. The two methods obviously would give similar results, when all the measurements are accurately performed. However, it often happens that with only a few percent error in the intensities some sets of calculated concentrations become extremely "unreasonable." It is precisely this problem that the null-determinant technique can overcome, whereas the averaging technique cannot.

In conclusion, it has been shown that the null-determinant technique is applicable to real data with very good agreement. In the light of this new recognition in resolving the inconsistencies of empirical equations, we plan in a future work to reconsider a comparison between the experimental and computational methods of x-ray analysis, namely, the empirical coefficient and the fundamental parameter techniques.

ACKNOWLEDGEMENTS

The authors would like to thank S. D. Rasberry and K. F. J. Heinrich of the National Bureau of Standards for their kindness in supplying the NBS stainless steel samples. We thank J. W. Criss for preparing the computer program, and J. V. Gilfrich for his comments on the manuscript.

REFERENCE

1. M. Fatemi and L. S. Birks, "On Obtaining Consistent Solutions of Empirical Equations in X-Ray Fluorescence," Anal. Chem. 45, 1443-1447 (1973).

X-RAY FLUORESCENCE ANALYSIS OF HIGH-TEMPERATURE SUPER-ALLOYS — CALIBRATION AND STANDARDS

K. F. J. Heinrich and S. D. Rasberry

Institute for Materials Research

National Bureau of Standards

Washington, D. C. 20234

ABSTRACT

The current experimental work extends our calibration concept of separating the effects of absorption and fluorescence to the high-temperature superalloys. The new calibration procedure produces calibration equations which are valid over wide ranges of composition — a feature which is useful in the analysis of high-temperature superalloys. For the specimens considered, the elements iron, nickel, chromium, cobalt and molybdenum can be present at levels greater than 10%; while tantalum, aluminum, titanium, manganese, silicon and vanadium may be present at levels between 1 and 6%. The calibration for this group of alloys has required, in the past, a large number of standards; the number is reduced by judicious application of the given correction equations. Analytical errors can be limited to 1 to 2% in the use of this method.

INTRODUCTION

In 1971, we presented a new empirical method for the calibration of x-ray fluorescence analysis in the presence of interelement effects (1). The equation proposed by us offered the advantage of improved accuracy by using separate terms for the effects of absorption and fluorescence:

$$C_i/R_i = 1 + \sum_{\substack{k=1 \\ k \neq i}}^{n} A_{ik} C_k + \sum_{\substack{k=1 \\ k \neq i}}^{n} \frac{B_{ik} C_k}{1 + C_i} \qquad (1)$$

In this equation,

C_i, C_k are the concentrations of the ith and kth elements,

R_i is the measured relative x-ray intensity, $I_i/I_{pure\ element}$,

A_{ik} is a coefficient used when the significant effect of element k on the analyte i is absorption, and

B_{ik} is a coefficient used when the significant effect of element k on the analyte i is fluorescent (tertiary) excitation.

This method is discussed further in a recent article (2), where the proposed approach is compared to those given in sixteen other publications. The Fe-Ni-Cr system, over a range of compositions for all three elements from 0 to 100%, and the calcium carbonate-silica system are used as demonstration cases.

In the present extension of this work we have determined the coefficients necessary for the analysis of high-temperature superalloys, especially those which contain major quantities of molybdenum, nickel, cobalt, iron, chromium and titanium, but do not contain major amounts of tungsten or niobium. Comparative results are given for analyses of such alloys over very wide ranges of composition.

A new adaptation of the equation (the Delta Adaptation) is described herein. It permits analyses of high accuracy over relatively narrow ranges of composition, and with relaxed requirements for accuracy of coefficients, by using a single type-standard as a pivotal point for calibration.

HIGH TEMPERATURE SUPERALLOYS

X-ray intensities were measured at NBS for the elements molybdenum, nickel, cobalt, iron, chromium and titanium on the pure elements, and in 37 specimens of high temperature alloys, maraging steel, and electronic and magnetic alloys (most are NBS-SRM's[1]). For each of the 37 specimens, the compositions had been previously certified by methods other than x-ray fluorescence analysis. Instrumental conditions for the x-ray measurements are given in Table I.

TABLE I. INSTRUMENT SETTINGS

PARAMETER	ELEMENT					
	Mo	Ni	Co	Fe	Cr	Ti
Tube Voltage (kV)	45	45	45	45	45	40[1]
Tube Current (mA)	10	5	5	5	15	20
Detector Voltage (V)	1000	1150	1150	1150	1150	1200
Amplifier Gain	80	80	80	80	80	640
PHA Baseline (V)	1.2	1.2	1.2	1.2	1.2	1.2
PHA Window (V)	9.0	10.0	10.0	10.0	10.0	5.0
Measurement Time (s)	100	30	100	100	100	100
2ΘLiF (°)	20.33	48.66	52.79	57.52	69.35	86.13

[1] Chromium-anode x-ray tube used for titanium, tungsten-anode used for other elements.

---

[1] Available from Office of Standard Reference Materials, National Bureau of Standards, Washington, D. C. 20234.

The compositions and intensities from 10 of the specimens were used to determine interelement coefficients for the six elements according to the second method of Reference (2). The other 27 specimens were treated as unknowns for the purpose of testing the accuracy of calibration. The interelement coefficients determined in this work are given in Table II. Six of them — those for the Fe-Ni-Cr system — were also determined two years ago. For comparison, the values obtained in the earlier work are shown in Table III. The close agreement in the values is gratifying; the two measurement sets were taken two years apart, by two different operators, and the first was calibrated with ternary specimens of the system Fe-Ni-Cr while the later set was obtained from high temperature alloys. The instrumental conditions were essentially the same except that two different tungsten-anode x-ray tubes were used.

TABLE II. INTERELEMENT COEFFICIENTS DETERMINED 1973

ANALYZED ELEMENT	INTERACTING ELEMENT					
	Mo	Ni	Co	Fe	Cr	Ti
Mo	***	A 0.09	A 0.05	A -0.02	A -0.34	A -0.84
Ni	B 0.75	***	A -0.15	A 1.70	A 1.20	A 0.20
Co	B 0.60	A 0.05	***	A -0.05	A 1.66	A 0.60
Fe	B 0.50	B -0.49	A 0.49	***	A 2.10	A 1.20
Cr	B 0.35	B -0.23	B -0.34	B -0.46	***	A 2.20
Ti	B 0.22	B -0.30	B -0.35	B -0.42	B -0.52	***

TABLE III. INTERELEMENT COEFFICIENTS, DETERMINED 1971
[Taken from Reference (2)]

ANALYZED ELEMENT	INTERACTING ELEMENT					
	Mo	Ni	Co	Fe	Cr	Ti
Mo	***					
Ni		***		A 1.71	A 1.23	
Co			***			
Fe		B -0.47		***	A 2.10	
Cr		B -0.23		B -0.42	***	
Ti						***

Results for the analysis of five specimens are given in Table IV; the basis of calibration was Equation (1) and the coefficients given in Table II. The first three specimens are high temperature superalloys; the fourth is a maraging steel, and the fifth is an electronic and magnetic alloy. The specimens in this table were chosen so as to cover the widest possible range of compositions. The results for them are typical of those obtained on the 27 specimens treated as unknowns. The difference between the certified and determined values for chromium in 1185 is one of the largest deviations in the entire test. The deviations for the other entries in Table IV are within the limits of experimental error using Equation (1).

DELTA ADAPTATION

In some cases, especially in industry, analyses are required over smaller ranges of composition than those shown in Table IV. Even in such cases, interelement effects may cause inaccuracies unless the slopes of the calibration curves in the region of interest are known. We have made an adaptation to the method given in Reference (2) which will correct for interelement effects, but which is less sensitive to errors in interelement coefficients than is the general procedure.

TABLE IV. RESULTS USING RASBERRY-HEINRICH EQUATION

Concentration (weight percent)

	Fe	Ni	Cr	Co	Mo
NBS 1185					
Certified	65.9	13.2	17.1	0	2.01
Determined	65.1	13.4	16.2	0	2.04
NBS 1190					
Certified	0.60	51.9	17.0	19.1	3.80
Determined	0.61	52.1	17.2	19.1	3.69
NBS 1192					
Certified	1.58	57.3	17.9	11.4	7.33
Determined	1.63	57.5	17.1	11.2	7.11
NBS 1156					
Certified	69.7	19.0	0.20	7.30	3.10
Determined	70.3	19.2	0.21	7.16	3.22
NBS 1160					
Certified	14.3	80.3	<0.1	<0.1	4.30
Determined	13.5	79.6	<0.1	<0.1	4.22

The change, which we call the Delta Adaptation, consists of pivoting the calibration around a single type-standard. All analyses are done relative to the type-standard so that errors in coefficients have shorter ranges over which to propagate. Equation (2) defines the delta value of the C_i to R_i ratio:

$$\Delta(C_i/R_i) = (C_i/R_i)_s - (C_i/R_i)_u \qquad (2)$$

where the subscripts s and u denote type-standard and unknown. Equation (1) is substituted for the s and u terms. After subtraction and collection of similar terms, we obtain Equation (3):

$$\Delta(C_i/R_i) = \sum_{\substack{k=1 \\ k \neq i}}^{n} A_{ik}\Delta(C_k) + \sum_{\substack{k=1 \\ k \neq i}}^{n} B_{ik}\Delta\left(\frac{C_k}{1+C_i}\right) \qquad (3)$$

where the deltas in the summation terms indicate differences between type-standard and unknown values. It should be noted that, throughout the derivation, no change is made in the definition or values of the A and B coefficients.

The iterative technique described in Reference (2) is employed to obtain $\Delta(C_i/R_i)$ values, which permit the solution for C_i's in the unknown.

A practical test of the Delta Adaptation has been made by determining molybdenum, nickel, cobalt, iron, chromium and titanium in six high temperature superalloys. Results are given in Table V for nickel; the calibration is based on the coefficients given in Table II used with Equation (1) [Rasberry-Heinrich Method] and used with Equation (3) [Delta Adaptation]. In this case, little difference is discernable between the results from the two equations. However, if the interelement coefficients are less accurately known, the Delta Adaptation is less sensitive to error, as is demonstrated in Table VI, in which the Ni coefficients have been altered on purpose.

TABLE V. DETERMINATIONS USING THE SAME SET OF INTERELEMENT COEFFICIENTS

Specimen	Ni Concentration (weight percent)		
	Given	Basic Rasberry-Heinrich Method	Delta Adaptation
IN-132	54.3	54.4	54.3
133	55.0	55.0	54.9
134	55.0	54.3	54.3
135	54.6	55.1	Std.
136	55.2	54.9	54.9
137	54.1	54.6	54.6

TABLE VI. DETERMINATIONS USING Ni COEFFICIENTS WHICH HAVE BEEN INCREASED BY 20% OVER THE BEST ESTIMATE

Specimen	Ni Concentration (weight percent) Relative Errors in Parenthesis		
	Given	Basic Rasberry-Heinrich Method	Delta Adaptation
IN-132	54.3	58.1 (7.0)	55.2 (1.7)
133	55.0	58.2 (5.8)	55.3 (0.5)
134	55.0	57.4 (4.4)	54.5 (0.9)
135	54.6	58.3 (6.8)	Standard
136	55.2	58.1 (5.3)	55.2 (0.0)
137	54.1	57.7 (6.7)	54.8 (1.3)

SUMMARY

The Rasberry-Heinrich Method has been extended to calibration of high temperature superalloys, over wide ranges of composition, with errors not usually exceeding 1 to 2%. A new adaptation is described which offers similar accuracy even though calibration is based on a single type-standard and approximate estimates of coefficients. The coefficients may be available or approximated by reference to previous analyses, published values for similar conditions, or interpolation between values accepted for bracketing elements. Extension to tool steels and other alloy types seems reasonably certain of success as soon as the additional coefficients required by tungsten and niobium have been determined.

ACKNOWLEDGMENTS

The authors gratefully acknowledge the careful assistance given by Ms. Lesley Zinger in the measurement and tabulation of all the x-ray intensities needed for this work. Our thanks also goes to Ms. Sarah Degenkolb, of Pratt and Whitney Aircraft, who provided the IN-100 series superalloys that we measured and mentioned in Tables V and VI.

REFERENCES

1. S. D. Rasberry and K. F. J. Heinrich, Colloquium Spectroscopicum Internationale XVI, Heidelberg, Germany, (Oct. 1971).

2. S. D. Rasberry and K. F. J. Heinrich, "Calibration for Interelement Effects in X-ray Fluorescence Analysis," Anal. Chem. (Expected publication date: Jan. 1974.)

QUANTIFICATION OF SUB-MICROGRAM ELEMENTAL CONCENTRATIONS USING MICRO-DOT SAMPLES

James M. Mathiesen

Finnigan Corporation

Sunnyvale, California 94086

ABSTRACT

The following study describes the experimental results obtained by energy dispersive x-ray analysis on a series of representative samples, concentrated by precipitation, to spots 1.5 mm in diameter on Millipore $^{(R)}$ membranes.

Using modified detector collimation, the contributions of air scatter and membrane (Substrate) scatter to the background under a peak, can be minimized to achieve limits of detection approaching a nanogram of the element present in the precipitate.

The equipment used here is composed of an 80 mm^2 Si(Li) detector connected to a computer based data handling unit. Excitation was achieved with a Rhodium target x-ray tube having a maximum output of 50 kV at 5.0 ma. Even with the high detection efficiency of the Si(Li) detector and the capability of minimizing the source to sample and sample to detector path lengths, it was found with these samples that at least 3.0 ma of tube current is needed in the excitation source.

A mechanism has been developed for this work that allows reproducible sample positioning for standards and unknowns. This mechanism is used to produce element profiles on larger precipitated standards to measure radial uniformity.

INTRODUCTION

The great concern today over trace elements in the environment has presented a challenge to all scientists and equipment manufacturers to improve the sensitivity of their equipment and to develop simple, rapid and reproducible sampling procedures.

When one tries to improve the sensitivity of an x-ray system it is usually by adding more components to the beam path (eg. secondary radiators) which in turn make the optics more complicated, add new impurities to the spectrum, and create errors due to re-positioning of these components. In short, the system needs an experienced user in order to achieve its best performance.

Since the development of the energy dispersive detector, many have marvelled at its simplicity, and the x-ray field has opened its doors to new colleagues from vastly different disciplines, forensic scientists, botanists, and medical doctors to name a few. These users, lacking the years of experience in x-ray analysis, should be able to use equipment now and trust their results now.

On the following pages is described the preliminary results obtained from a small modification of the detector collimator of a standard Finnigan OM 900 x-ray analysis system and a simple sample concentration procedure ideally suited to energy dispersive analysis that was recently described by J. W. Mitchell[1] of Bell Laboratories.

X-RAY SPECTROMETER SYSTEM

Figure 1 is a block diagram of the x-ray energy spectroscopy system used to provide the data in this paper.

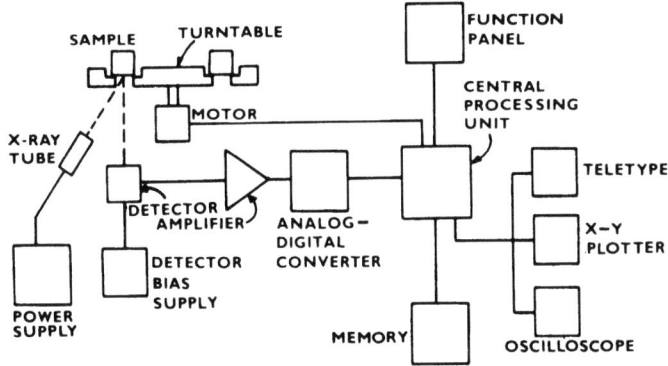

Figure 1 Block Diagram of the Automated Materials Analysis System

The system consists of low power (50 kV @ 5.0 ma) x-ray power supply and tube, a 12-position sample changer, a 16 bit central processing unit, an oscilloscope display and associated interface circuitry.

The above standard Finnigan QM-900 system was fitted with a 80 mm^2 Si(Li) detector having a resolution of 163 ev FWHM at 1000 counts per second on the manganese K alpha line. The sample changer wheel was removed and a tin mini-collimator with a 1/16 inch diameter opening was installed. Figure 2 shows the mini-collimator in position with an 80 microgram flake of automobile paint adjacent to the detector hole.

Figure 2   Mini-Collimator in Position

The uniform exciting beam comes in at a 45 degree angle with respect to the sample surface and the detector looks only at that part of the sample directly above the hole.

A microscope stage has been added to the collimator to allow the positioning of opaque samples by removing the entire assembly and centering the sample while viewing it through the hole, then re-inserting the assembly in the system.

A motor has been added to the microscope stage to drive it along one axis at a uniform rate. When combined with a multi channel scaling program in the computer, the system has been used to measure the concentration of an element versus position on large diameter precipitates.

ANALYSIS OF PRECIPITATES

Standards and unknowns were prepared using an apparatus as described by Mitchell, et. al.[1]. The samples were precipitated both by the coprex technique and as sulfides. Standards were produced as large depositions on filter paper and their uniformity measured.

Figure 3 is a spectrum of a standard containing 1.2 micrograms each of manganese, copper and arsenic. The sample was run at 40 kV @ 4.0 ma for 100 seconds.

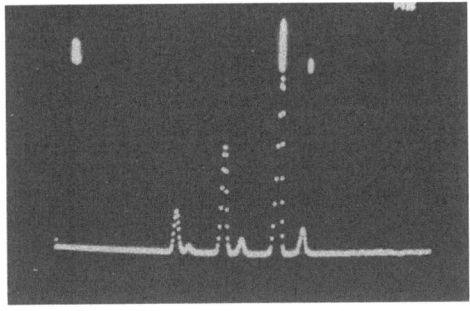

Figure 3   1.2 Micrograms Manganese, Copper and Arsenic

Limits of detection for these and other elements were calculated using the following formula:

$$\text{Minimum detectable limit} = \frac{3 \, (\text{Background})^{1/2} \times \text{Concentration}}{\text{Peak Net Area}}$$

Table 1 shows the limits for these elements based on a 1000 second counting time.

Table 1

Limits of Detection

Element	Limit in Nanograms
Mn	2.1
Fe	1.3
Cu	1.0
As	0.8
Pb	1.1

Figure 4 is a spectrum of the residue from the firing of a 38 caliber revolver over a control taken before firing. The sample preparation was after firing, a cube of sugar was moistened with dilute nitric acid and rubbed on the top surface of the firing hand. The cube was then dissolved in 20 cc of deionized distilled water and the entire solution precipitated down to a 70 mil dot. The dot was then inserted into the collimator assembly, centered and run in the system for 200 seconds at 40 kV @ 4.0 ma.

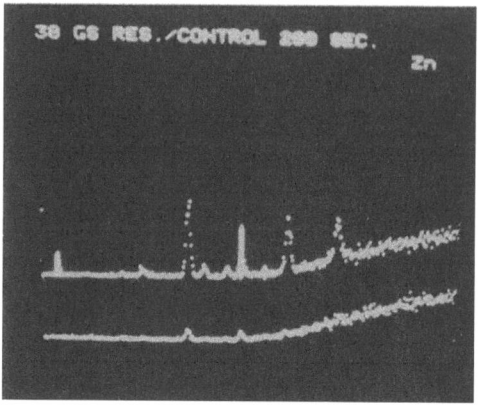

Figure 4   Gun Shot Residue

The main peaks of interest were quantitated by comparing to known standards. The results are iron = 0.161 micrograms, copper = 0.036 micrograms and lead = 0.072 micrograms.

Figure 5 is a 1000 second run on 20 cc of tap water from Sunnyvale, California, precipitated in approximately two minutes to a micro dot sample.

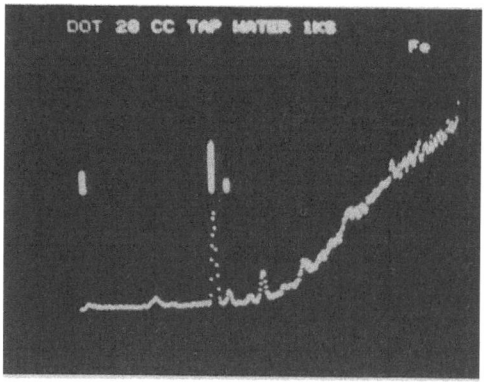

Figure 5   20 cc Tap Water

The preceding figure shows the presence of iron 0.040 micrograms, copper 0.002 micrograms, zinc 0.007 micrograms and lead 0.007 micrograms. With respect to the original 20 cc sample the lead is equivalent to 0.3 parts per billion. Controls were run with deionized distilled water and no lead was found.

Figure 6 is a spectrum of a sample of dissolved sediment from Baltimore harbor. It was found to contain manganese 0.042 micrograms, iron 1.48 micrograms, copper 0.005 micrograms and lead 0.009 micrograms.

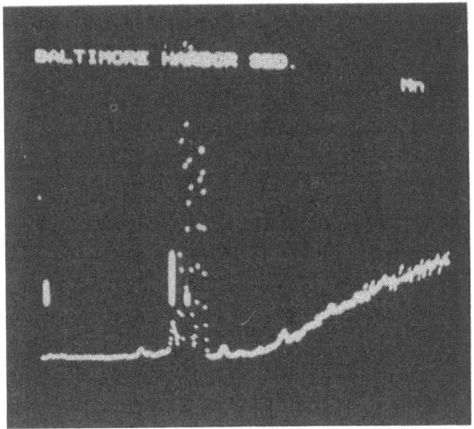

Figure 6   Baltimore Harbor Sediment

Figure 7 is a multi channel scaling of a coprex precipitate 25 mm in diameter. The precipitate was inserted into the collumator assembly and then driven across its diameter at 1.0 mm/2.5 minutes with a clock motor; each dot represents one mm of movement.

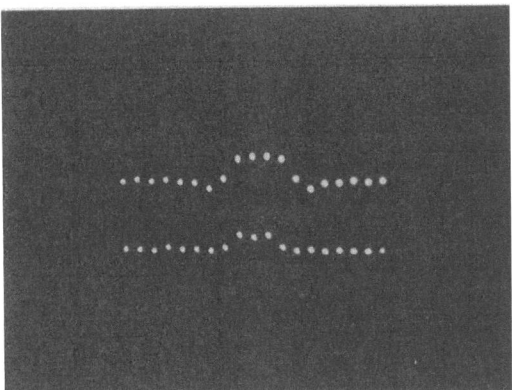

Figure 7   Precipitate Scan 1.0 mm/Point

The top curve is the copper distribution and the bottom curve is the distribution of zirconium. This technique is very useful in assuring that one's standards are indeed uniform.

CONCLUSION

It appears that the micro dot precipitation technique is an excellent method of achieving concentration of a sample in liquid form. An automated x-ray system frees the technician to do jobs he is best at doing, namely sample preparation. We have found that these samples can be prepared in very few minutes which is comparable to the x-ray analysis times.

REFERENCES

1. J. W. Mitchell, C. L. Luke, W. R. Northover, Analytical Chemistry, Vol. 45, No. 8, 1503 (1973).

A RAPID DIRECT X-RAY FLUORESCENCE METHOD FOR SIMULTANEOUSLY DETERMINING BRASS COMPOSITION AND PLATING WEIGHT FOR BRASS-PLATED-STEEL TIRE CORD WIRES

James Gianelos

The B.F. Goodrich Company Research and

Development Center, Brecksville, Ohio 44141

ABSTRACT

A rapid direct x-ray fluorescence (XRF) method has been developed for simultaneously determining brass composition and plating weight for brass-plated-steel tire cord wires. On a manual XRF instrument, a complete analysis can be done in ten minutes including calculations. A complete analysis can be done in only two or three minutes on a multi-channel Quantometer-type instrument. The method can easily be adapted to "on line" use.

Analysis is done directly on the "as received" wires. Sample preparation is simple and non-critical with no surface pre-treatment of any kind required. Sample weighing and dissolution of the brass plating are entirely eliminated. Wire samples may be either stranded or unstranded. Rather large variations in sample size can be tolerated because of the ratio-type calculation system used. Analytical precision is high.

Calibration curves were established by wet chemical, atomic absorption, and solution-XRF methods. No serious calibration problems were encountered because the plating is an alloy, or because count data were taken from round, non-flat samples. Low impurity levels of iron in the brass plating are shown to have an insignificant effect on the plating weight calibration curve.

INTRODUCTION

During the last ten years a number of new tire constructions have been introduced in the United States. One of the newest and best of these is the steel-belted radial tire. The steel wires used in these tires have a very thin brass plating. The actual thickness of the plating is generally in the range of 0.25 to 0.75 microns. The purpose of the plating is to promote good adhesion between wire and rubber, and to prevent rusting and corrosion.

It is important to know both the composition of the brass and the thickness of the plating for quality assurance reasons. In the past, solution methods have generally been used to analyze these wires. Once the plating has been dissolved off, the analysis can be completed using wet chemical, atomic absorption, or solution-XRF methods. However, these solution methods all require considerable sample preparation. This includes weighing out and unstranding the wire, and dissolving off the plating with subsequent boiling, aliquoting, etc. I decided to investigate the possibility of a direct XRF method, which hopefully would permit analyzing these wires in their "as received" form, and eliminate most or all of these time-consuming sample preparation steps.

SAMPLE PREPARATION

Figure 1 shows a typical mounted sample. After being cut into pieces about $1\frac{1}{2}$ inches long, the wires were fastened side by side onto a flat Plexiglas block about $1\frac{1}{2}$ x $1\frac{1}{2}$ x $\frac{1}{4}$ inches using Scotch "double stick" adhesive tape (Cat. No. 136). The wires receive no surface treatment of any kind before mounting. Therefore, sample preparation only takes a minute or two, and is very simple, straight-forward, and non-critical, except that it is important that no wire ends be in the x-ray beam. Wire ends leave the steel substrate exposed and this will distort the plating thickness determination.

During manufacture single wires are stranded or bundled together. In Figure 1 each wire shown is actually three single wires spiral-wrapped together. The single wires are all the same size. But the number of single wires stranded together can vary. Typical wire constructions use from two or three up to seven or eight single wires.

FIGURE 1. TYPICAL MOUNTED SAMPLE

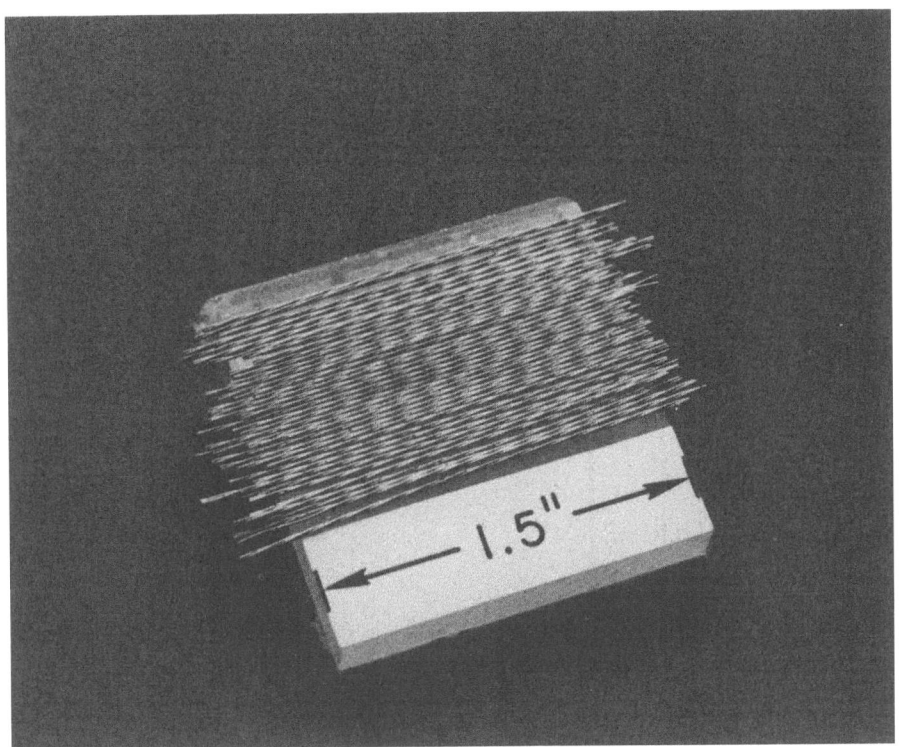

Mounted samples may consist of either stranded wires or unstranded single wires, or they may be mixed. It is not necessary to weigh out the wires, or to use the exact same number of wires for each analysis. The reasons for this will be discussed later.

INSTRUMENTAL PARAMETERS

Samples were analyzed in a General Electric XRD-6VS spectrometer, equipped with a 4 sample changer. Table I summarizes the instrumental parameters used.

BRASS COMPOSITION CALIBRATION

Figure 2 shows the x-ray spectrum obtained from a typical mounted sample. The brass plating is so thin that the most intense

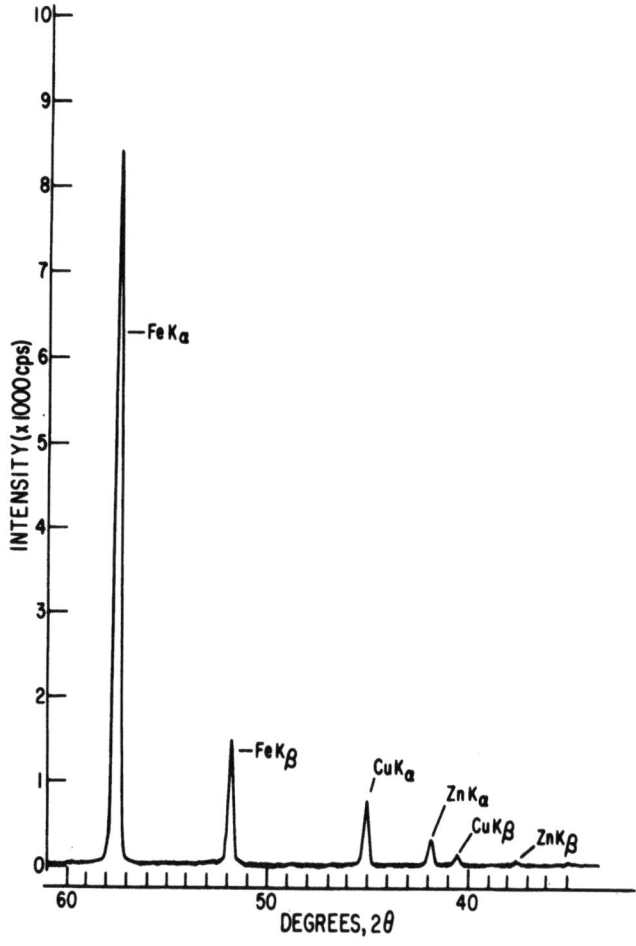

Figure 2. X-RAY SPECTRUM of BRASS-PLATED-STEEL TIRE CORD WIRES

SHOWING RELATIVE INTENSITY OF ALL ANALYTICAL LINES

0.25mm DIAMETER STRANDED WIRES

TABLE I

Chromium target (G.E. dual target tube)
Counting time: 100 seconds at each 2 θ angle
Secondary slit width: 5 mils
Sample Mask: rectangular 1 x 3/4 inches
Counter Tube: G.E. Model 9-A flow proportional counter at
    1550 volts D.C. for all elements, using P-10 gas
All counting in air path. No count rates exceeded 10,000 cps.
    Sample orientation was constant.
Pulse Heighth Selector: E_1 mode with E_1 = 2.5 volts
Analytical Lines Used:

°2 θ	Crystal	Line	Power
41.80	LiF (100)	Zn K_α	25 ma/50 kvp
43.50	LiF (100)	background	25 ma/50 kvp
45.02	LiF (100)	Cu K_α	25 ma/50 kvp
51.73	LiF (100)	Fe K_β	25 ma/50 kvp
52.23	EDDT	Fe K_γ n=2	25 ma/50 kvp
57.52	LiF (100)	Fe K_α	2.5 ma/50 kvp

lines in the spectrum are not from the brass, but from the steel substrate. Typically iron K_α is ten times more intense than either the copper or zinc K_α lines from the plating. And this is so even though the brass plating obviously must absorb and attenuate some iron K_α radiation.

At the present time there are no commercially available wire standards. Therefore, small spools of twelve previously analyzed wire samples were obtained from a vendor to serve as standards. These spanned both the brass composition and plating weight ranges of interest. The vendor's values for these standards were obtained using his unpublished solution-XRF method. To verify their accuracy, these wires were then re-analyzed using both atomic absorption spectroscopy and a wet chemical compleximetric procedure.

Figure 3 shows a typical calibration curve for copper in the brass plating. The ratio of net copper to zinc counts is plotted against the vendor's values for percent copper. Only Standard #5 falls off the smooth curve drawn. However, when re-analyzed this standard was found to fall on the curve, so there was no real discrepancy.

A calibration curve could be constructed in the same way to determine the zinc in the brass plating. However, since the brass used has very low levels of impurities, the percent zinc was usually obtained by difference (100%-%Cu).

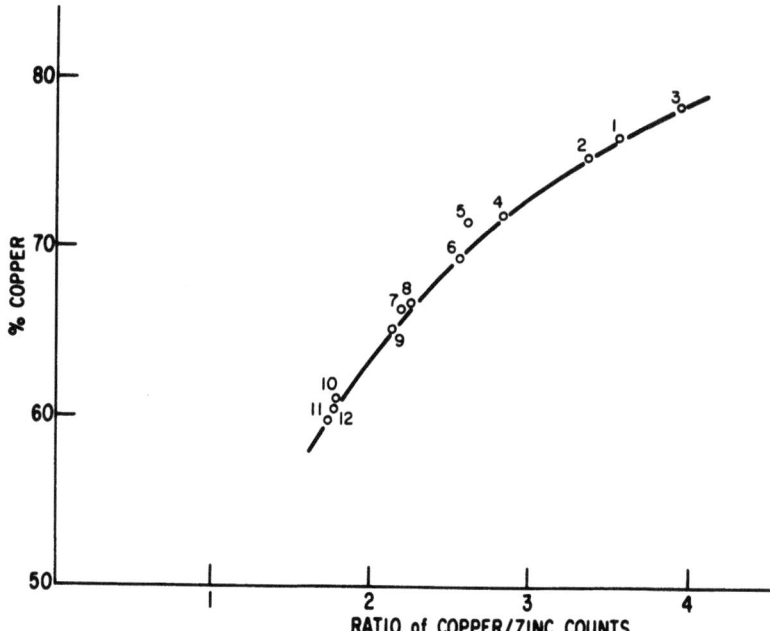

Figure 3. BRASS COMPOSITION
%COPPER vs. RATIO of COPPER/ZINC COUNTS
0.25mm DIAMETER STRANDED WIRES
VENDOR'S SOLUTION-XRF ASSAYS

PLATING WEIGHT CALIBRATION

Figure 4 shows the plating weight calibration curve obtained from these 12 standards. The term "plating weight" requires explaining. In the industry, plating thicknesses in microns are not generally used. Instead, thickness is usually expressed as "grams per kilogram" (g/kg), or plating weight. The term simply means the number of grams of brass plating on one kilogram of plated wire. By using an appropriate factor, it is easy to convert from microns to g/kg, and vice versa.

The plating weight values are plotted against the ratio of net brass to iron counts. The values for plating weight here were obtained by atomic absorption. There was excellent agreement with the vendor's solution-XRF values. With background correction, the curve was found to be linear over its entire length. Although there are no data for wires with less than 2 g/kg, the curve does extrapolate through zero. A similar broad ranged linear

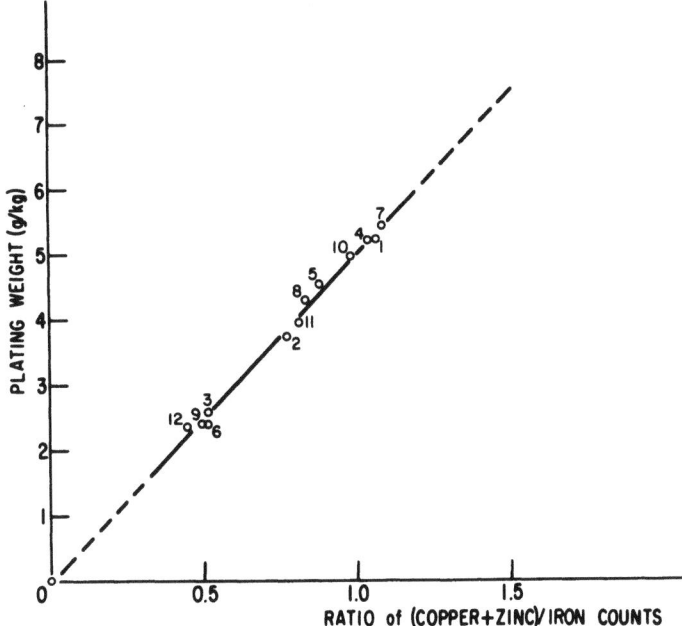

Figure 4. BRASS PLATING WEIGHT CALIBRATION CURVE
PLATING WEIGHT (g/kg) vs. RATIO of COPPER+ZINC/IRON COUNTS
SINGLE UNSTRANDED 0.25mm WIRES
CALIBRATED BY ATOMIC ABSORPTION

relationship was also obtained by Liebhafsky and Zemany (1, 3) when they studied tin-plated steel, and other thin plating systems. They pointed out that the dominant effect was simple attenuation of iron K_α by the plating, obeying Beer's law of absorption. My data show this linear relationship also applies where the plating is an alloy, and with no adverse effect introduced because the data were obtained from round, non-flat samples.

In obtaining the brass to iron ratio used in plotting Figure 4, the net copper and zinc counts were just added together as if they were equal and the same. No intensity corrections were necessary. There are a number of reasons why this can be done. Both x-ray lines are K_α lines and are measured only 3.22° apart on a LiF crystal. Their fluorescent yields are almost identical. So it is not surprising that pure copper and pure zinc should both count nearly the same under test conditions. The mass absorption coefficients for copper K_α and zinc K_α for both zinc and copper are low and nearly the same. Therefore, advantage can be taken of this fortunate combination of facts to permit the relatively simple calculation procedure used.

ANALYTICAL PRECISION

Table II shows that high analytical precision was obtained, both when the sample was undisturbed and when it was randomly rearranged. To permit rearrangement, the double-stick tape was not used, so the wires were loose.

TABLE II

ANALYTICAL PRECISION

A. Eleven sets of counts, sample undisturbed

Ratio Cu/Zn	Ratio (Cu + Zn)/Fe
$\bar{x}$ = 2.61880	$\bar{x}$ = 1.14908
σ = ±0.00738	σ = ±0.00302
C_v = ±0.2819%	C_v = ±0.2580%
C_v = ±0.063 copper	C_v = ±0.012 g/kg

B. Eleven sets of counts, sample removed and wires rolled and randomly rearranged between each set.

$\bar{x}$ = 2.62032	$\bar{x}$ = 1.14036
σ = ±0.00828	σ = ±0.01274
C_v = ±0.316%	C_v = ±1.12%
C_v = ±0.071% copper	C_v = ±0.052 g/kg

This sample was 69.8% copper and 30.2% zinc, and had a plating weight of 4.7 g/kg. In the worst case, the one sigma coefficients of variation, 0.071% copper and 0.052 g/kg plating weight, are both very satisfactory for routine production control purposes.

SAMPLE SIZE TOLERANCE

One thing that is not obvious from the precision data in Table II-B is that when the wires were rearranged, sample size varied by as much as 25%. This is shown by the count data in Table III.

The extremes are the seventh and tenth count sets. These data show that sample size is not critical, and indeed, rather large variations in sample size can be tolerated. The reason is that both the brass composition and plating weight calibration curves are based on _ratios_. When large count variations were obtained, all analytical lines varied by proportional amounts, so the _ratios_ remained unchanged.

TABLE III

COUNT DATA USED FOR TABLE II-B

Set	Zn K$_\alpha$	Bkgd	Cu K$_\alpha$	Fe Kα	Ratio Cu/Zn	Ratio (Cu+Zn)/Fe
1	233,620	14,220	591,640	703,990	2.6318	1.1319
2	233,730	11,810	592,620	705,950	2.6172	1.1371
3	238,640	14,400	597,480	687,070	2.6002	1.1750
4	253,820	13,540	644,370	754,990	2.6254	1.1538
5	239,020	11,630	606,600	726,770	2.6165	1.1315
6	199,900	11,270	506,020	599,650	2.6229	1.1396
7	261,530	11,910	663,760	793,480	2.6114	1.1361
8	240,940	12,120	612,390	730,040	2.6233	1.1357
9	207,970	14,200	522,520	617,220	2.6233	1.1375
10	197,220	12,170	498,340	595,860	2.6272	1.1265
11	227,960	12,870	577,330	684,210	2.6243	1.1393

I found that it made no difference whether the wires analyzed were stranded or unstranded, or were mixed. There was no significant difference in the ratios obtained. This was verified by unstranding all 12 standards and comparing them directly against the same wires in their stranded "as received" form. While count differences similar to those in Tables II and III were obtained because of differences in sample size, the count ratios remained the same, so the same calibration curves could be used interchangeably.

OTHER IMPORTANT FACTORS

All of the data presented so far have been for tire cords where the individual single wires were 0.25 mm diameter. While this is the most commonly used size, several other sizes are sometimes used. I found that separate plating weight calibration curves are needed for each wire size. This is shown in Figure 5. The 12 points which define the 0.25 mm diameter wire curve have been left out to minimize clutter. While many of these curves are defined by only a single point, it is clear that a family of curves has been generated, which differ only in slope. The reason for this is that the brass/iron count ratio used in the calibration curves is proportional to actual plating thickness expressed in microns. The count ratio is not proportional to plating weight except for a single specific wire size. Different diameter wires could have the same plating thickness, but have different plating weights. Thus, separate plating weight calibration curves are required for each wire size. However, I found that the same brass composition calibration curve could be used interchangeably for all the wire sizes shown.

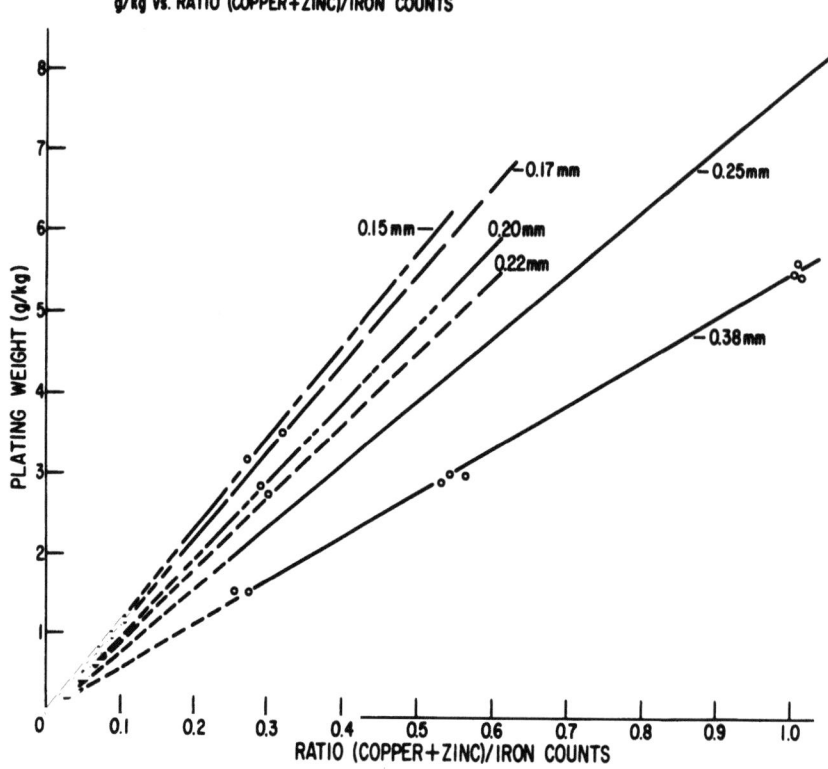

Figure 5. PLATING WEIGHT CALIBRATION CURVES for VARIOUS DIAMETER WIRES

g/kg vs. RATIO (COPPER+ZINC)/IRON COUNTS

In order to determine plating weight using this method, it is important that the brass plating should either contain no iron impurity, or should have an insignificant level. The thicker the plating the more important this requirement is. Otherwise, determining plating weight by this method would be impossible. Fortunately, these brass platings normally contain less than 0.1% iron, which is insignificant. Because the brass platings are so thin, it can be estimated that even a 1.0% iron impurity level is permissable, by using the Criss-Birks (2) Basic Parameters Computer program. A brass sample with a composition of 1.0% iron, 66.3% copper, and 32.7% zinc should give the following relative intensities:

Z	Wt. Fraction	Relative Intensity	Pure Z Count	Theoretical Sample Counts
Fe	0.01000	0.01397 x	1,771,491 =	24,748
Cu	0.66300	0.64025 x	2,486,027 =	1,591,679
Zn	0.32700	0.34249		

Since from Figure 2 it is known that the iron substrate is typically 10 times as intense as the copper K_α line, then:

$$\left(\frac{24,748}{1,591,679 \times 10}\right) \times 100 = 0.17\%$$

The 1.0% impurity iron in the brass would only alter the total iron count by 0.17%. This would cause a negligible error of -0.01 g/kg in determining plating weight.

The XRF spectrometer used in this investigation is manually operated, and reads one analytical line at a time. On this instrument, the total time per analysis is about 8 minutes, plus a minute or two for sample mounting. However, on a multi-channel computer controlled Quantometer-type instrument, this analysis could be done in a minute or two. Since signal levels are quite high, this method could readily be adapted to on-line use in manufacturing.

While XRF is well known as a non-destructive method, I found that after several months of continuous heavy use the standards showed some deterioration. With further continued use, more corrosion and some rusting-through occurred. I would urge anyone wishing to set up this method in their own laboratory to obtain large spools of wire, so that ample replacement standards are available.

CALCULATOR PROGRAM

A desk top calculator program was written which eliminates all routine manual calculations and curve plotting. By just entering the four x-ray intensities, the brass composition and plating weight are instantaneously calculated. This was done by obtaining equations for the calibration curves, and then programming them on a Wang 500 calculator. The program is much too long for inclusion as part of this paper, but is available on request.

ACKNOWLEDGMENT

I would like to express my appreciation to Mr. W. H. Parry and his staff for their help in obtaining the wire standards used, and for providing the many atomic absorption and compleximetric analyses which were required.

REFERENCES

1. H. A. Liebhafsky, H. G. Pfeiffer, E. H. Winslow, P. D. Zemany, "X-Rays, Electrons, and Analytical Chemistry," John Wiley & Sons, Inc. (1972).

2. J. W. Criss and L. S. Birks, "Calculation Methods for Fluorescent X-Ray Spectrometry," Analytical Chemistry 40, 1080-1086 (1968).

3. H. A. Liebhafsky and P. D. Zemany, "Film Thickness by X-Ray Emission Spectrography," Analytical Chemistry 28, 455-459 (1956).

QUANTITATIVE NONDISPERSIVE X-RAY FLUORESCENCE ANALYSIS OF HIGHLY
RADIOACTIVE SAMPLES FOR URANIUM AND PLUTONIUM CONCENTRATION[*]

W. L. Pickles and J. L. Cate, Jr.

Lawrence Livermore Laboratory, University of California

Livermore, California 94550

ABSTRACT

Fluorescent x-ray energy spectra were successfully acquired from uranium and plutonium in 400:1 ratios in samples containing 2 Ci/gram of mixed fission products. The analytical system consists of a silver transmission anode x-ray tube, a low-Z scattering chamber, a magnetic β-ray trap, a beam monitor probe, a commercial Si(Li) detector, a set of modified electronics to handle the large γ-ray overload rate, and a computer analyzer using a higher-level language to handle data reduction.

The computer programs used to obtain peak areas from the closely spaced uranium and plutonium (Lα 1+2) peaks were constructed to make use of knowledge of upper-edge tailing gained in this experiment. Programs are being developed to properly remove background under the uranium and plutonium peaks. Absorption effects in the larger samples have been measured using the ratio of uranium (Lℓ) to uranium (Lγ) peak area and are incorporated in the data-analysis schemes. A titanium monitor probe, consisting of a fixed titanium plate near the sample, introduces a constant-area titanium K x-ray line into the spectrum. The program uses the area of this peak to correct for effects of total exciting flux, geometry, and system dead-time losses. Standard samples of various types are used to generate calibration curves from which quantitative results are obtained.

Samples are taken from dissolved high-burnup power-reactor fuel rods. The liquid sample is acidic and has a radiation level at one foot of approximately 2 R/hr β and 300 mR/hr γ. Sample preparation involves only the evaporation of the liquid sample on a 1/2-mil polycarbonate substrate and subsequent sealing with another layer of polycarbonate film. The samples are then mounted in standard 35-mm slide-holders.

Preliminary testing on a limited number of prepared uranium and plutonium samples indicates a precision of about 1% and an accuracy of about 2%, over a range of 1 to 58 μg total mass. The samples have not yet been verified by independent chemical analysis. The system has been installed at the AEC Savannah River facility for extensive testing.

INTRODUCTION

The prototype nondispersive x-ray fluorescence system assembled and put into operation in a test and evaluation program at du Pont Savannah River Plant is intended to analyze "hot" dissolver solutions for elemental uranium and plutonium concentrations with a minimum of sample preparation and without the removal of the copious quantities of radioactive mixed fission products present. The solutions obtained by dissolving fuel rods from power reactors have uranium concentrations of about 50 g/liter, uranium to plutonium ratios ranging from 80 to 400, and radiation levels of about 2 Ci/g (β, γ).

SYSTEM DESCRIPTION

The system is made up of three parts: a movable unit (Fig. 1), referred to as the "X-ray data head," a computer analyzer, and a sample preparation stand.

The x-ray data head requires only 110 V AC power and a periodic liquid-nitrogen fill for the detector. It contains the tube-exciter system, the scattering chamber to hold the samples, a Si(Li) detector, and all the power supplies and electronics needed to produce the signal used for analysis. The following sections describe the subsystems that make up the X-ray data head. Figure 2 is a schematic drawing of the unit, and Figures 3, 4, and 5 are photographs for clarification of the description.

X-Ray Tube Exciter

The x-ray tube exciter is a pumpable, silver transmission anode of the air-cooled type, similar to that described by Jacklevic (1). This tube dissipates up to 50 W and delivers a flux of approximately 10^{10} photons/cm^2 per second at the sample (3-5 R/sec). The

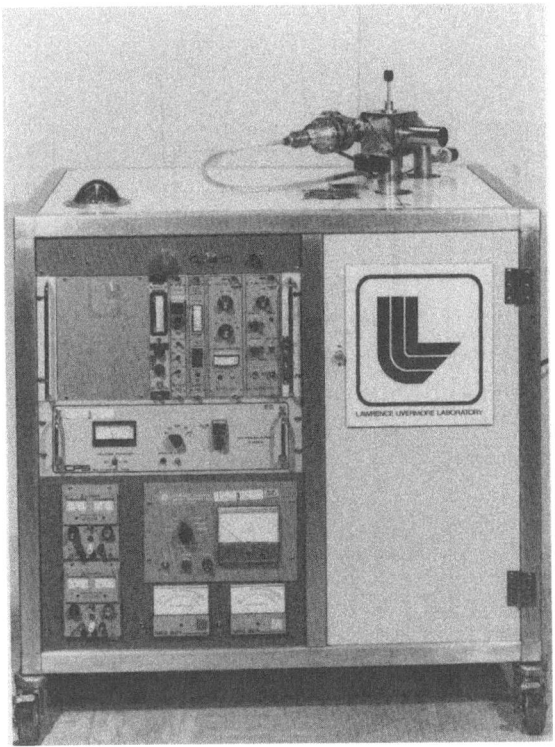

Figure 1. X-ray data head, showing complete movable unit.

energy spectrum of the photon beam is shown in Fig. 6. The quality of the beam is expressed by the percent of the total flux in each of four regions. Region II contributes the major part of the Compton-scattered background under the peak used for analysis (uranium (Lα)). Regions III and IV are available for excitation of the sample, and comprise about 95% of the spectrum. The tube can be cycled up to air, have the target changed to another material, and be producing a useful x-ray beam in about 30 minutes. The tube is evacuated by a combination of a cryogenic molecular-sieve roughing pump, and an 8-liter/sec ion pump; the operating pressure is approximately 10^{-7} torr.

The photon beam can be varied by both the filament current and the grid cap bias. The beam flux has been measured to be stable and resettable to 1% (1 sigma).

The construction of this tube is difficult, but we have found it to be very reliable in operation.

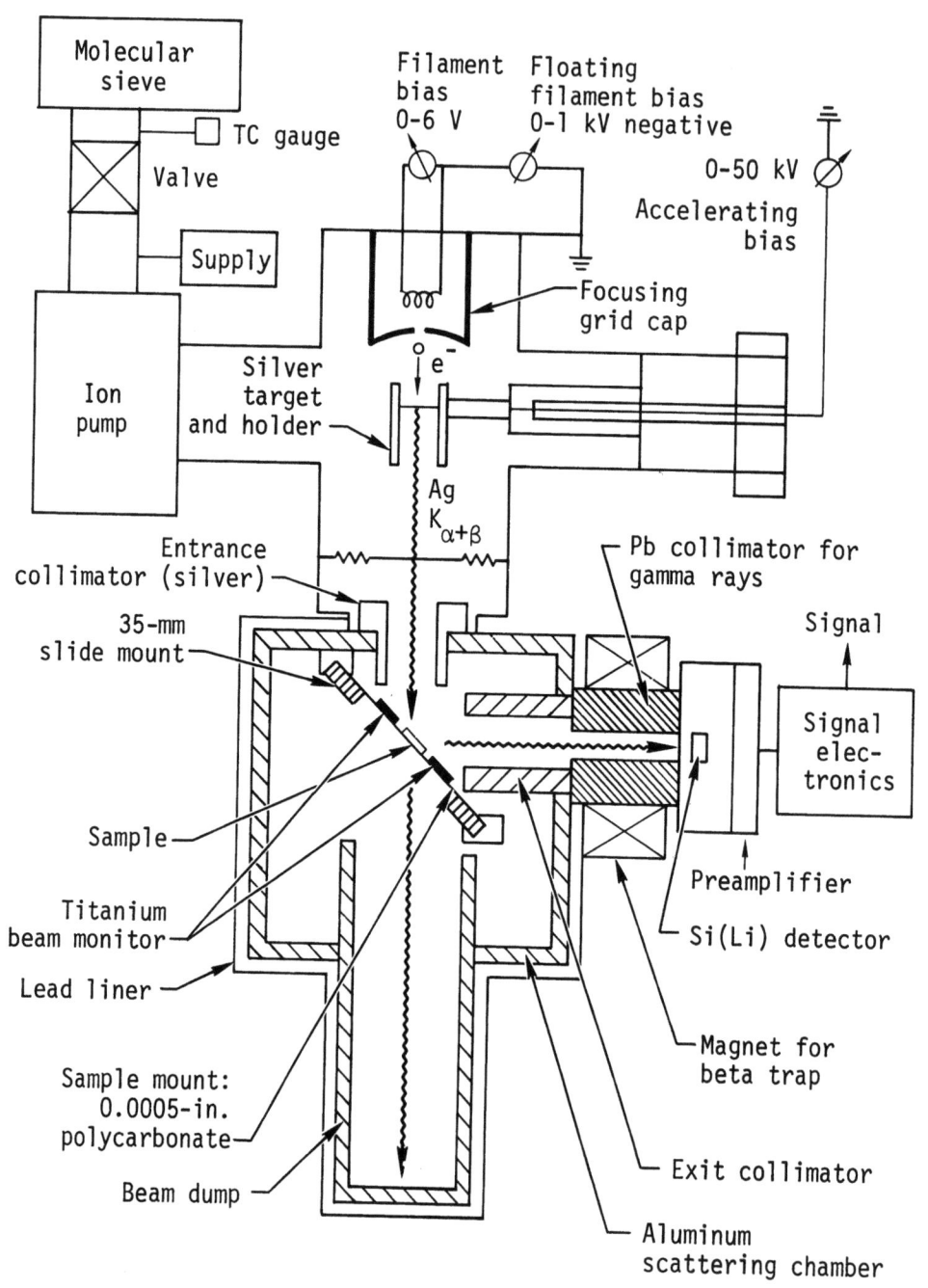

Figure 2. Schematic drawing of equipment contained in the "x-ray data head" unit.

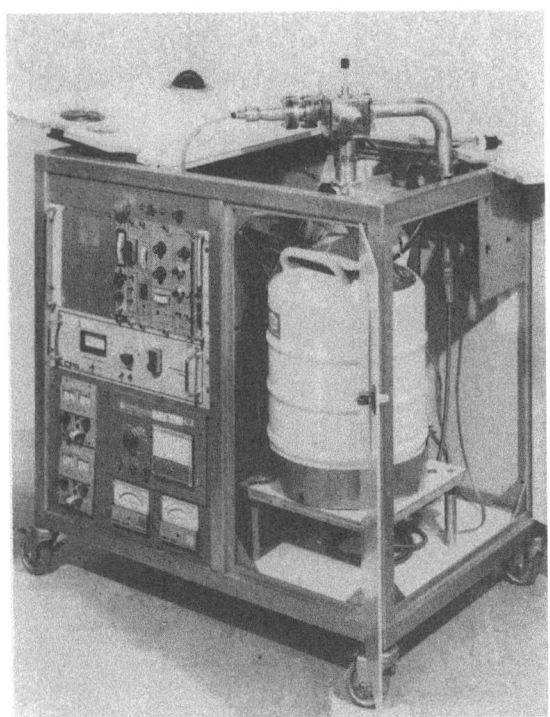

Figure 3. X-ray data head, front view.

Figure 4. X-ray data head, rear view.

Figure 5. X-ray data head, top view.

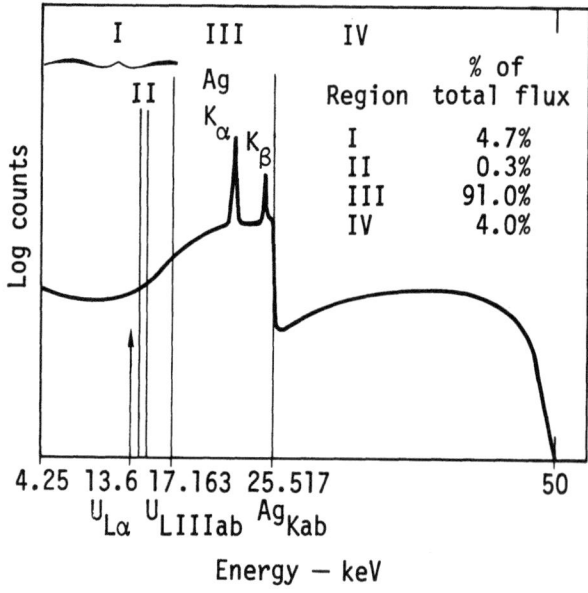

Figure 6. Energy spectrum of x-ray beam produced by the silver transmission anode tube; silver target 0.005 in. thick.

Scattering Chamber

The sample is held at 45 degrees to the exciting beam and to the detector by a plastic slide-holder. The chamber itself is made of aluminum, with lead shielding added to the outside for radiation protection of personnel.

Two sheets of titanium are placed on either side of the sample location so that the fluorescence from the titanium enters the detector directly. Figure 7 shows the scattering chamber, the plastic slide-holder, the titanium beam monitor sheets, and an actual sample, as seen through the beam dump port. An additional beam monitor can be inserted behind the sample to provide absorption information. The use of any beam monitor increases the Compton-scattering in the spectral region of interest, but its presence provides a normalization peak in each spectrum. The beam monitor is an "external standard" which completely removes the effect of fluctuations in beam fluence, all system dead-time corrections, pulse pile-up rejection losses, and changes in geometry.

The magnetic β-ray trap and the lead collimator (aluminum-lined) are located between the scattering chamber and the detector.

Figure 7. View of complete scattering chamber with sample and beam monitor as seen through the beam dump port.

Detector and Electronics

The system has been operated with Si(Li) detectors and electronic packages obtained from two different commercial sources. One detector is 10 mm^2 in area by 2 mm deep, while the other is 30 mm^2 by 3 mm deep. In both cases, the manufacturer developed special electronics packages to allow the systems to operate in the large flux of high-energy (100-keV to 3-MeV) γ-radiation from the sample. The γ-rays produce overload pulses in the detector. We find that the β-trap prevents any β-rays from reaching the detector, and the special electronics package allows the detector to operate in a γ-ray field of about 1 R/hr with only a 5% to 10% degradation in resolution. The detector resolution at 13.6 keV with the radioactive sample in the chamber is about 210 eV, FWHM.

The electronics package contains a preamplifier (pulsed-optical feedback), an amplifier with pulse pileup rejection and modifications for the overload gamma pulses, and high-voltage control and supply. A rate-meter was added to monitor the overall input count rate. Typical overload pulse rates are 3000 to 5000 cps, and the signal pulse rate is approximately the same.

Sample Preparation

The sample preparation equipment (Fig. 8) consists of a sintered stainless steel pellet which has a vacuum line attached. The vacuum line sucks a 0.5-mil polycarbonate sheet onto the slightly concave surface of the steel pellet. One 50-microliter drop of dissolver solution is placed on the film at the center and allowed to dry;

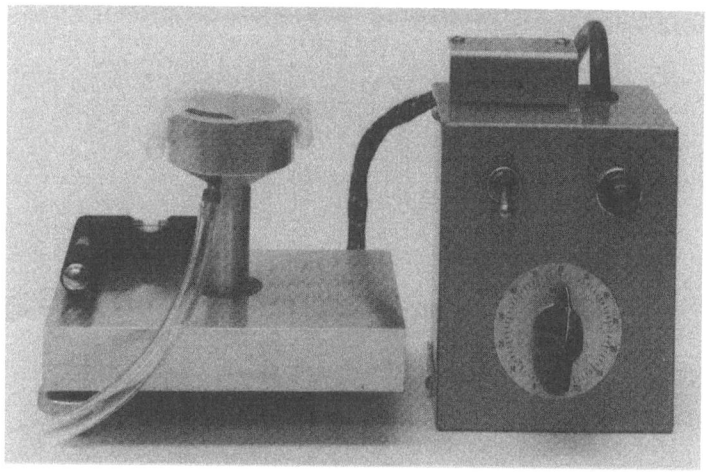

Figure 8. Sample preparation stand with heater control and polycarbonate film in place.

this takes a few minutes with the help of an infrared lamp or the heater element in the stand. After the sample is dry, adhesive is sprayed on the film and a second sheet of film is placed over the sample. This sandwich is then mounted in a standard 35-mm slide. Care must be taken to not allow large crystals to develop when the sample is drying, because they make an absorption correction necessary during analysis. We are currently exploring the use of freeze-drying techniques for an alternate method of sample preparation. Prefreezing of the sample before evaporation of the water prevents the formation of large single crystals.

Figure 9 shows some actual samples being inserted into the scattering chamber. The slide-holder is withdrawn for display. Note the radiation damage to the bottom of glass storage bottle in the foreground.

Analysis

Of the 104 ℓ x-ray lines of uranium and plutonium, only the uranium (Lα 1+2) and the plutonium (Lα 1+2) are free from interference. The uranium (Lα) doublet is at about 13.613 keV and the plutonium (Lα) is at 14.279 keV, a difference of 666 eV. Figure 10 shows this part of a spectrum (uranium (Lℓ), uranium (Lα), and plutonium (Lα) peaks) of an actual dissolver-solution sample. The uranium-plutonium ratio was 250:1 and the sample was highly radioactive. Figure 10 shows a similar spectrum from a sample containing

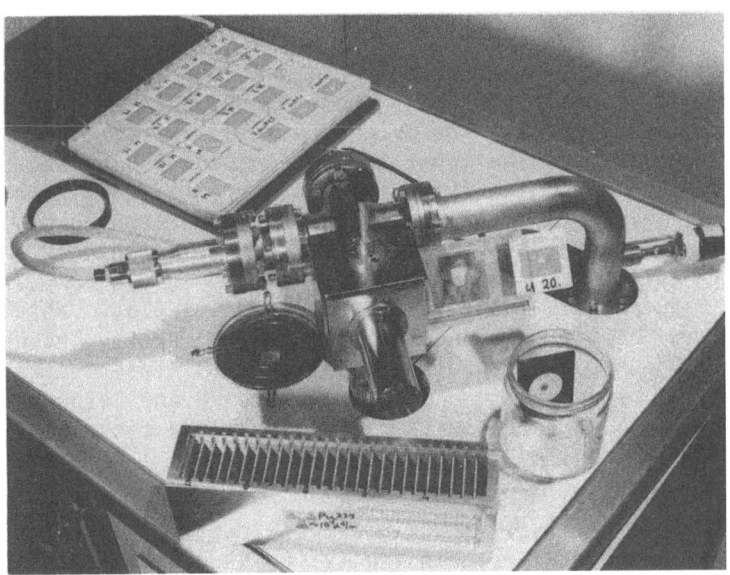

Figure 9. View of scattering chamber in use with various types of samples.

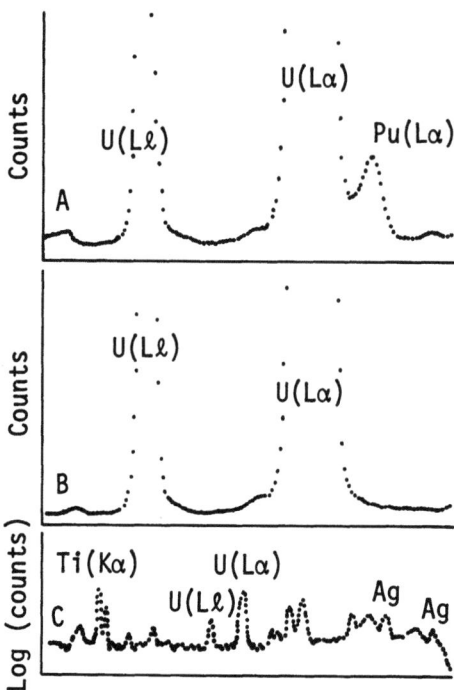

Figure 10. Actual energy spectra: (a) dissolver solution, U/Pu = 250; (b) uranium only, prepared standard; (c) log plot of complete spectrum of prepared standard (note titanium beam monitor peaks).

no plutonium or fission products; there is very little change in background or resolution. The primary analysis problem in using this data is to remove the upper tail of the large uranium (Lα) peak from beneath the plutonium (Lα). We are currently working on a technique which uses a double gaussian with upper and lower folded tails to represent the shape of the peaks in this region.

Figure 10 shows a typical spectrum in a log plot. Note the titanium beam monitor peak and the uranium (Lα) peak. The net peak areas of these two peaks are obtained with linear background subtraction routines that use clear areas above and below the peaks. The uranium (Lα) peak area is then divided by the titanium (K α+β) peak area and multiplied by a constant.

We have begun extensive calibration experiments both at LLL and du Pont Savannah River. Initially, we used prepared standards with known quantities of uranium ranging from 1 to 100 μg, with no fission products.

Our tests at LLL have resulted in a measured reproducibility for a particular sample of 0.2% (1 sigma) and a variance between samples of supposedly equal masses of 0.6% (1 sigma). We have not yet been able to check our standards by an independent method.

The data analysis takes approximately 30 seconds and the typical acquisition time is about 10 minutes. Our reproducibility checks showed no dependence on count rate, on sample position, or on uranium amounts, to within 0.2% (1 sigma). With the prepared standards we have measured the linearity of the system for uranium mass to be 1% (1 sigma) from 1 µg to 58.1 µg total uranium mass.

We are greatly encouraged by these results and are looking forward to using the sets of standards now being made at du Pont Savannah River. These standards will contain the complete inventory of fission products and will have various ratios of uranium to plutonium.

ACKNOWLEDGMENTS

This work was performed under the auspices of the U.S. Atomic Energy Commission. The authors gratefully acknowledge the helpful efforts of R. Taylor, T. O. Hoeger, R. Howell, and M. Coops in the development of this equipment. We also thank R. Jaroszeski of Nuclear Fuel Services Company, and the four power companies (Pacific Gas & Electric, Yankee Atomic Electric Co., South Carolina Electric & Gas Co., and Consumers Power Co.) who donated the dissolver solutions from their reactor fuels.

REFERENCE

1. J. M. Jaklevic, R. D. Giaugue, D. F. Malone, and W. L. Searles, "Small X-Ray Tubes for Energy Dispersive Analysis Using Semiconductor Spectrometers," in Kurt F. J. Heinrich, Editor, Advances in X Ray Analysis, Vol. 15, p. 266-275, Plenum Press (1972).

THE EFFECTS OF SELF-IRRADIATION ON THE LATTICE OF $^{238(80\%)}PuO_2$. III*

R. B. Roof, Jr.

Los Alamos Scientific Laboratory

Los Alamos, New Mexico 87544

ABSTRACT

The results of the fourth year of observations on the effects of self-irradiation on the lattice of $^{238(80\%)}PuO_2$ confirm the predictions derived from previous observations. Namely, the lattice constant continues to decrease with time but at a slower rate indicating an approach to equilibrium. Very small reductions are also noted in the half-width of the diffraction lines.

INTRODUCTION

In previous papers (1,2) the effects of self-irradiation on the lattice of $^{238(80\%)}PuO_2$ were reported for observations of X-ray line broadening that extended over a three year period. The current paper presents an extension of the observations through the fourth year of self-irradiation. Details of sample preparation, data collection, and methods for analysis are given in (1).

---

*Work performed under the auspices of the U. S. Atomic Energy Commission. The Los Alamos Scientific Laboratory is operated by the University of California.

ANALYSIS OF EXPERIMENTAL PEAKS

X-ray Line Broadening

In Table 1 are listed the Gaussian half-widths of six selected reflections from a diffraction examination of $^{238}(80\%)$PuO$_2$. The experimental intensity data from the six reflections 111, 200, 220, 311, 333(511), and 622 were, in general, averaged in 3-wk unit blocks and subjected to a least-squares fit of an equation consisting of the summation of Gaussian curves plus a polynominal background. The Gaussian half-widths were used to determine whether significant line broadening had occurred as a function of time. The data of Table 1 are shown in Fig. 1. It is apparent that a maximum in line broadening occurred approximately 2-1/2 years after the start of self-irradiation with the trend thereafter suggesting a decrease.

The constants of a polynominal fit,

$$y = P_1 + P_2 x + P_3 x^2$$

Table 1. Gaussian half-widths, in °2θ, of $^{238}(80\%)$PuO$_2$ reflections as a function of time after an initial anneal.

Time wks	Reflections					
	111	200	220	311	333(511)	622
4, 5, 6	0.093±1	0.092±4	0.113±3	0.122±3	0.151±17	0.283± 14
7, 8, 9	0.095±1	0.089±4	0.114±3	0.118±3	0.162±14	0.266± 16
10, 11, 12	0.094±1	0.090±4	0.112±3	0.120±3	0.165±15	0.272± 17
13, 14, 15	0.094±1	0.089±4	0.116±3	0.118±3	0.152±12	0.249± 18
16, 17, 18	0.092±1	0.096±4	0.109±3	0.118±3	0.165±14	0.277± 20
19, 20, 21	0.095±1	0.092±4	0.119±3	0.128±3	0.163±17	0.294± 27
22, 23, 24	0.094±1	0.088±4	0.117±3	0.121±3	0.171±21	0.326± 27
25, 26, 27	0.093±1	0.088±4	0.118±3	0.123±3	0.155±13	0.314± 39
28, 29, 30	0.093±1	0.091±4	0.112±3	0.127±3	0.184±24	0.293± 28
--, 32, 33	0.097±1	0.089±4	0.115±2	0.138±4	0.189±21	0.286± 22
95, 96, 97	0.107±1	0.098±3	0.142±3	0.155±3	0.280±22	0.659± 90
98, 99,100	0.107±1	0.096±3	0.140±3	0.154±3	0.266±22	0.660± 93
113,114,115	0.107±1	0.095±3	0.141±2	0.164±3	0.288±23	0.711±101
129,130,131	0.105±1	0.094±2	0.144±2	0.165±3	0.284±20	0.673± 90
142,143,144	0.104±1	0.093±2	0.140±2	0.159±3	0.304±20	0.648± 81
156,157,158	0.101±1	0.092±2	0.141±2	0.166±3	0.275±19	0.660± 89
169,170,171	0.099±1	0.091±1	0.140±2	0.165±3	0.281±19	0.711± 99
---,190,---	0.099±1	0.091±1	0.130±2	0.156±3	0.280±18	0.623± 81

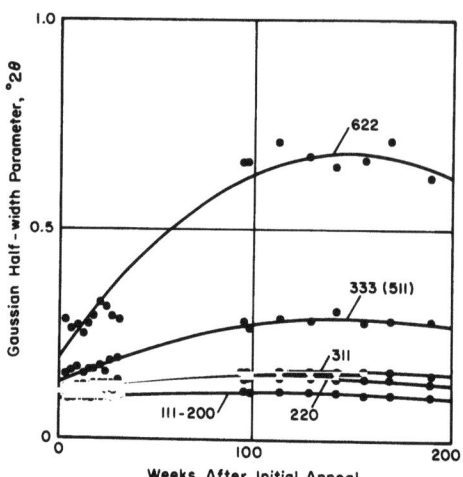

Fig. 1. Gaussian half-widths, in °2θ, of $^{238}(80\%)$PuO$_2$ reflections as a function of time after an initial anneal.

to the data of Fig. 1, where y = Gaussian half-width and x = time in weeks after an initial anneal are listed in Table 2. Examination of the coefficient P_3 indicates that a decrease has occurred for all diffraction lines.

Table 2. Constants of a polynominal equation fitted to the change in Gaussian half-width parameter σ of selected reflections from $^{238}(80\%)$PuO$_2$.

$$\sigma (\text{in °2}\theta) = P_1 + P_2 (\text{weeks}) + P_3 (\text{weeks})^2$$

Reflection	P_1	P_2	P_3
111	0.0847± 9	0.00028± 3	-0.0000012± 2
200	0.0886± 11	0.00012± 4	-0.0000006± 2
220	0.1056± 18	0.00055± 6	-0.0000021± 4
311	0.1117± 20	0.00068± 7	-0.0000022± 4
333,511	0.1308± 54	0.00210±19	-0.0000069±10
622	0.1740±210	0.00693±73	-0.0000234±40

Lattice Constant Values

Lattice constant values as a function of time after an initial anneal are listed in Table 3 and shown graphically in Fig. 2. The change in the lattice constant as a function of time appears to be best expressed in a polynominal

$$a_0 = 5.41360 \pm 24 + 0.00011 \pm 1(t)$$
$$- 0.0000011 \pm 2(t^2) + 0.0000000028 \pm 6(t^3)$$

where t is time in weeks.

Table 3. Lattice constants of $^{238}(80\%)PuO_2$ as a function of time after an initial anneal

Time wks	a_0, Å	Time wks	a_0, Å
5	5.41408±16	96	5.41631±33
8	5.41438±12	99	5.41597±20
11	5.41481±15	114	5.41636±38
14	5.41497±14	130	5.41537±35
17	5.41502±15	143	5.41417±30
20	5.41567±22	157	5.41364±23
23	5.41565±13	170	5.41407±20
26	5.41584±15	190	5.41336±32
29	5.41558±14		
32.5	5.41580±15		

After the main experiment had been terminated at 190 weeks and the sample removed from the diffractometer, the author examined a second sample of $^{238}(80\%)PuO_2$. This second sample was a batch-mate of the original sample and had been held at 900°C for 163 weeks while He generation and release were measured (3). These measurements indicated that approximately 20% of the He generated had been retained in the sample. The lattice constant for this sample is 5.41487±32 Å and this value is shown as the triangular symbol in Fig. 2. It is of interest to note that this value does not appear significantly different from the values obtained in the main experiment.

CONCLUSIONS

The predictions made previously (2) concerning the effects of self-irradiation on the lattice of $^{238}(80\%)PuO_2$ have been confirmed by an additional year of experimental observation. Namely, the lattice constant continues to decrease but shows a definite trend toward flattening out to an equilibrium value. The half-widths of the diffraction lines also continue to show a slight decrease. These features were explained (2) by proposing that

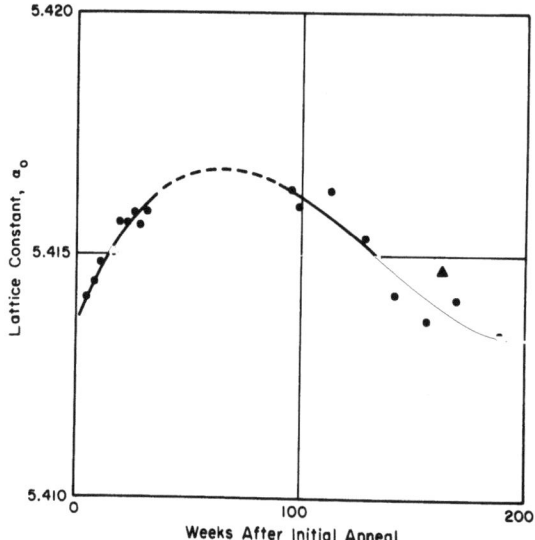

Fig. 2. Lattice constants of 238(80%)PuO_2 as a function of time after an initial anneal.

interstitials (including He atoms) and vacancies are formed and combine at different rates, and that the eventual loss of these entities from the lattice is manifested in a decrease in the lattice constant.

Superimposed on the general trends shown by the solid curves in Figs. 1-2 there appears to be a secondary effect. This effect is a small sinusoidal variation with time of the general effect and is particularly noticeable in the lattice constant and the half-width for the 622 reflection. While this may be no more than normal statistical fluctuations the possibility does exist that it is a true short time effect of gain and loss of interstitials and vacancies reflected in the variation of the parameter in question.

REFERENCES

1. R. B. Roof, Jr., "The Effects of Self-Irradiation on the Lattice of 238(80%)PuO_2" in K. F. J. Heinrich, C. S. Barrett, J. B. Newkirk, and C. O. Rudd, Editors, <u>Advances in X-ray Analysis</u>, Vol. 15, Plenum Press (1971).

2. R. B. Roof, Jr., "The Effects of Self-Irradiation on the Lattice of 238(80%)PuO$_2$. II" in L. S. Birks, C. S. Barrett, J. B. Newkirk, and R. O. Rudd, Editors, <u>Advances in X-ray Analysis</u>, Vol. 16, Plenum Press (1972).

3. B. Mueller, private communication (1973).

THE EFFECTS OF X-RAY OPTICS ON RESIDUAL STRESS MEASUREMENTS IN STEEL

Chester F. Jatczak and Harald H. Boehm
Research Division   formerly with
The Timken Company   The Timken Company
Canton, Ohio, U.S.A.

ABSTRACT

The effects of various combinations of divergence, receiving and Soller slits on x-ray measurements were investigated for Siemens-Halske and General Electric diffractometers. Influences of the following factors which also affect accuracy and precision of x-ray R.S. results were determined in addition: (a) parafocus versus stationary detector focusing geometry, (b) method of peak location, (c) LPA intensity correction, (d) diffractometer electronic stability and (e) elastic constants.

The optimum choices of beam optics and factors (a-e) were defined with regard to accuracy, precision and minimum time for stress determination, on sharp and broad line specimens of soft (annealed) and hardened steel and of annealed Cr-powder.

INTRODUCTION

Stress measurements by x-ray techniques involve rather precise measurements of lattice strains between specially oriented $(hk\ell)$ planes within the body of the specimen. To measure lattice strains we make use of an x-ray beam of suitable monochromatic wavelength λ and record the 2θ shift in position of a particular high angle diffraction line $(hk\ell)$ or its equivalent in d-values as a function of the specimen orientation angle ψ. These observed changes in lattice spacings with orientation angle ψ are then expressed in such strain units as $\Delta d/d$, $\Delta a/a$, $(\varepsilon_\psi - \varepsilon_\perp)$ or $\Delta 2\theta$ for conversion to stress values by use of appropriate stress-strain relationships and proper material elastic constants (1-7).

One of the following three equations will satisfactorily calculate stresses from lattice strains with any x-ray technique used internationally.

$$(\varepsilon_\psi - \varepsilon_\perp) = \frac{\Delta d}{d_\perp} = \frac{\Delta a}{a_\perp} = -\frac{\cot\theta}{2}(\Delta 2\theta) = \frac{1}{2} S_2 \sigma_\phi \sin^2\psi \qquad (1)$$

$$\sigma_\phi = K'_\psi (2\theta_\perp - 2\theta_\psi) = K'_\psi (\Delta 2\theta) \qquad (2)$$

$$\sigma_\phi = K* \frac{(d_\psi - d_\perp)}{d_\perp} = K* \frac{\Delta d}{d_\perp} \qquad (3)$$

Barring significant preferred orientation and/or plastic deformation in the specimen satisfactory stress determination can be made with independent observations of the specimen lattice spacings at just two ψ orientations, such as 0 degrees and 45 or 60°. However, multiple observations involving measurements of $2\theta_{hkl}$ or d_{hkl} shifts at more than two ψ orientations of the specimen surface are often used to minimize the random errors associated with individual measurements of strain. The former has been given the name the two exposure or $\Delta 2\theta$ technique and the latter the multi-exposure or $\sin^2\psi$ method because with this technique, the data are usually handled most conveniently by plotting the lattice strains; $\Delta d/d$, ε, etc. against $\sin^2\psi$, and then making use of the line slope with Equation 1 above to calculate the stress, σ_ϕ.

Lattice strain measurements have to be made with the greatest accuracy and precision or reproducibility, if the x-ray stress procedure is to be useful. Many factors are known to influence or affect these measurements. Some of the most important are listed in Table 1 where they are broadly classified into three categories.

Table I. Factors Which Influence Accuracy and Precision in X-Ray Stress Measurements

Equipment and Instrumental Factors	Technique and Geometrical Factors	Specimen and Material Factors
Alignment Technique	Selection of Radiation	Type of Material
Specimen Alignment Technique	Selection of (hkl) Plane	Crystal System
Availability of $\theta/2\theta$ Goniometer Rotation	Selection of Apertures, etc.	Specimen Preparation Technique
	Beam Focusing Technique	Specimen Size
Availability of Parafocusing	Method of Correction for LP and Absorption Effects	Specimen Geometry
Available Apertures		Elastic Constants
Available Soller Baffles	Method of Peak Location	Mass Absorption Coefficients
Detector Type	Counting Statistics	Grain Size
Availability Proper Radiation	Scanning Procedure	Effect of 2nd Phase
Noise Discrimination Circuit	Stress Calibration Technique	Type of Deformation Causing Stress
Power Stability	Deviation from Plane Stress	Stacking Faults
Electronic Stability	Corrections for Beam Penetration	Twinning
Availability High KW Tubes	Corrections for Material Removal	Texture

The general effects of most of these factors have been fairly well documented by individual workers in Germany (4, 6) in Japan (8-13) as well as in the U.S.A. (1-3, 5, 14-18). Yet despite all this work, it is still very difficult to compare x-ray stress results between any two laboratories much less between countries. The writer feels that much of this is due to the fact that each entity uses a somewhat different stress technique that involves beam optics, focusing geometry and other geometrical and instrumental conditions which are strange to the other. Hence, a direct evaluation of the proper application of each individual technique cannot be readily made. Clearly, a study is needed to define the influence of the various geometrical factors listed in Table I on lattice strain measurements when performed with instrumentation and techniques used internationally.

Such a study is described herein where both the $\Delta 2\theta$ and $\sin^2\psi$ methods were used to establish the errors in stress value produced by the following factors while making actual stress measurements on soft and hardened steel specimens and annealed chromium powder:

(1) influence of various divergence, receiving and Soller slits on the beam optics

(2) influence of parafocusing and stationary beam focusing methods

(3) influence of various 2θ peak location methods

(4) influence of correction of line intensities for Lorentz, polarization and x-ray absorption effects

(5) influence of electronic stability of the diffractometer

EXPERIMENTAL TECHNIQUE

Two different diffractometers, the Siemens-Halske Type F of European make and a modified General Electric XRD-3 unit (USA) with no diffracted beam Soller baffles were used in this program. The operating conditions for both instruments are given in Table II. The various combinations of divergence, receiving and Soller slits investigated, and other important aspects of beam optics and focusing conditions, are shown in Table III. Initial alignment of both diffraction units was made according to the manufacturer's recommended technique. The Cornu method was used for the S.H. unit and the beam dividing method for the G.E. unit. Final alignment of both instruments was checked with the SAE precision technique in which lattice parameter measurements of a standard tungsten powder are used to check the final adjustments of the radial slide track and diffractometer optics (14).

Two methods of beam focusing are presently used internationally in x-ray diffractometer stress measurements:

Table II. Measuring Apparatus and Conditions for X-Ray Diffraction Techniques

	Siemens-Halske Type F	General Electric XRD-3 Mod.
Radiation	Cr-Kα	Cr-Kα
Filter	.025 mm. V	.025 mm. V
Diffraction Plane	(211)	(211)
Divergence Slits	See Table III	See Table III
Receiver Slits	"	"
Tube Voltage	44 kV	42 kV
Tube Current	30 mA	38 mA
Detector Type	Scintillation	Argon Prop.
Detector Voltage	1380 V	1305 V
Pulse Ht. Discrimination (Steel)	20 V x 40 V	None
" " " (Cr-Powder)	5 V x 20 V	"
Goniometer Radius	17 cm	14.6 cm (5.73 in.)

Table III. Data on Beam Optics and Focusing Conditions

	Siemens-Halske Type F	General Electric XRD-3 Mod.
Divergence Slits	0.5, 1.0, 2.0, 4.0°	1°, 3°
Soller Baffles	Incident and Diffraction Beams	Incident Beam Only
	1.8°	2.9°
Vertical Divergence	0.5° - 23 mm.²	1° - 39 mm.²
	1.0° - 45 mm.²	3° - 115 mm.²
Irradiated Specimen Area	2.0° - 90 mm.²	
	4.0° - 180 mm.²	
Receiving Slits	.10 mm.	.2° - .50 mm.
	.13 mm.	.4° - 1.0 mm.
	.20 mm.	.5° - 1.25 mm.
	.25 mm.	
	.40 mm.	
	1.00 mm.	
Parafocus Distance (R) at 156°-2θ		
ψ = 0	17.0 cm	5.73 inches
" = 15	15.0 cm	5.1 "
" = 30	13.1 cm	4.4 "
" = 45	10.9 cm	3.7 "
" = 60	9.5 cm	2.6 "

Table IV. Bend Specimen Data

Size: 100 mm. x 16 mm. x 2.5 mm.

Composition: 46100

C	Mn	Si	Cr	Ni	Mo	Al
1.01	.56	.26	.36	1.75	.23	.045

Structure: 73% Tempered Martensite, 27% Retained Austenite
Hardness: Rockwell "C" 62

Heat Treatment: Prior Structure: Spheroidize Annealed
Austenitized at 845°C - 30 minutes, oil quenched, Tempered at 180°C - 1 hour

Table V. Steels Used for Stressed Specimens

Steel Type	C	Mn	Si	Cr	Ni	Mo	Specimen No.
TBA	.75	1.2	.30	1.1	1.5	1.3	32, 100, 101
Vega	.70	2.0	.30	1.0	-	1.4	103, 106
SAE 1010	.10	.45	.15	-	-	-	1010

Heat Treatment:
TBA & Vega Steels: Austenitized in Vacuum at 860°C/2 Hours, Cooled in Argon at .2 Atmosphere Pressure, Tempered at 190°C/4 Hours, to Rc 60/62
SAE 1010 Steel: Austenitized at 820°C/1 Hour, Slow Cooled in Vacuum, Hardness 130 Brinell

Table VI. Stress Levels in Experimental Specimens

Specimen	Condition	Stress kp/mm.²	Stress ksi	(211) Peak Profile
TBA 32	Hard. and Temp.	3.0	4.5	Broad
TBA 100	"	3.0	4.5	"
TBA 101	Annealed	0	0	Fine
Vega 103	H.T. & Shot Peened	-61	-85	Broad
Vega 106	H.T. & Glass "	-36	-51	"
SAE 1010	Slow Cooled	-6.0	-8.4	Fine

(1) the stationary method, where the receiving slit and detector remain on the goniometer circle

(2) the parafocus technique, where the receiving slit and/or detector are moved along a radius of the goniometer toward the specimen in order to fulfill the changing beam focusing conditions when the specimen orientation angle ψ is varied (7, 14).

There are two variations of the stationary method, the standard method as used in Europe and the parallel beam method as used in Japan. In the former, the specimen rotates through an angle θ when the goniometer rotates 2θ to keep the same planes in focus at all times, whereas in the latter the specimen is always fixed; i.e. does not rotate with 2θ and thus somewhat differently oriented planes are being exposed as 2θ is changed. The parafocus method which is most extensively used in the U.S.A. also contains $\theta/2\theta$ rotation of the specimen and is specifically used to increase diffracted intensity and/or to minimize times for individual stress determinations (14).

The parafocusing and the European stationary focusing techniques were compared at all variations of beam optics. The parallel beam technique was not studied per se since it has already been compared with the parafocus method by Fukura and Fujiwara (12, 13) and the effects of beam optics well established by Aoyama et al. (19).

Two methods for $2\theta_{hk\ell}$ peak location were explored:

(1) fitting a parabola or higher order curve to the upper (>85% I max.) portion of the peak, at 3 or 5 different 2θ positions

(2) bisection of chart recorder traces of peak profiles.

For the parabola technique, intensity measurements of the Fe(211) planes were made by recording times to achieve a preset count of 10^5 impulses at each of the 3 or 5 - 2θ points for ψ specimen rotations up to 30° and 2×10^5 counts for $\psi > 30°$. The 2θ interval between points was ± 1.5° for the broad or hardened steel peaks and 0.1 to 0.2° for fine or sharp line samples such as annealed steel and chromium powder.

Intensity corrections for LPGA (Lorentz, Polarization, Geometry and Absorption) effects were always made to each 2θ value before fitting a parabola, etc., except in those cases when the effects of omitting this correction were being assessed. Corrections for background were not made nor efforts made to separate broad peaks into $K\alpha_1$ and $K\alpha_2$ components. Parabola fits were made using three and five data points. For the five point method (15) second, third and fourth order curve approximations were also applied.

Recorder traces for the chart scan technique of $2\theta_{hkl}$ peak locations were made at scan rates of 1/4 to 1/2°/min. and at fast chart advances to obtain the best peak resolution. The peak position was located by two methods, center of gravity calculation and by drawing bisecting lines parallel to the background at approximately 50, 66 and 75% of total peak height. Because the latter method was found to be more expedient and equally as accurate as the c.g. technique, it was used most extensively.

As noted before, both the two exposure ($\Delta 2\theta$) and the multi-exposure or $\sin^2\psi$, techniques were employed for stress calculation. The following stress constants for CrK_α radiation and the Fe_{211} and Cr_{211} reflections were used with Equations 1-3:

	S_1	$1/2\, S_2$	$(1/2\, S_2)^{-1}$	K'_{45}
Steel:	-1.33×10^{-5} mm²/kp (-9.35×10^{-6} ksi^{-1})	6.10×10^{-5} mm²/kp (4.3×10^{5} ksi^{-1})	16.4×10^{3} kp/mm² (23.2×10^{3} ksi)	60.9 kp/mm² (86.3 ksi/1° 2θ)
Chromium:	--	--	19.6×10^{3} kp/mm² (27.8×10^{3} ksi)	82.7 kp/mm² (117.3 ksi/1° 2θ)

These elastic constants and coefficients were determined experimentally by comparing the lattice strains developed at various ψ specimen orientation in a bent beam of the particular material mechanically loaded to increasing stress levels short of yielding or brittle fracture.

Under uniaxial or bend loading Equation (1) above becomes

$$\varepsilon_{\phi,\psi} = \frac{1}{2} S_2\, \sigma_{xx} + S_1\, \sigma_{xx} = \sigma_m\, \frac{1}{2} S_2 \sin^2\psi + S_1 \quad (4)$$

and
$$\varepsilon_{\phi,\psi}/\sigma_m = \frac{1}{2} S_2 \sin^2\psi + S_1 \quad (5)$$

Thus plotting $\varepsilon_{\phi,\psi}/\sigma_m$ versus $\sin^2\psi$ will yield both of these elastic constants plus their equivalences K' and K* as used with Equations 2 and 3. The stress constants K' and K* were obtained in yet another and simpler fashion by plotting the strain $\Delta 2\theta$ or $\Delta d/d$ directly versus the mechanically applied stress σ_m. In both instances, a four point bending device employing four strain gauges (2-90° rosettes) mounted on either side of an electropolished area was used to determine the mechanical stress while x-ray strains were being measured on the polished area. An example of an actual determination of stress constant for a hardened and tempered 46100 type steel is given in Figure 10. The specimen dimensions, composition and heat treatment are listed in Table IV.

Comparative data on the reproducibility of residual stress measurements with each diffractometer were obtained by repetitive measurements over an 8 hour operating period using unaltered instrument conditions. For the Siemens unit, the operating conditions were selected to maximize the intensity while maintaining the detector stationary. For the General Electric unit a 3°/0.5° slit combination was explored along with the largest slit combination recommended for fine line resolution (3°/0.2°). The resulting data include all errors due to electronic drift, counting statistics and specimen positioning because the specimen was removed and replaced for each individual test run (see Table IX).

MATERIALS

Flat steel specimens with essentially zero stress, light tensile and medium to high compressive stresses were prepared from the three steels shown in Table V by heat treatment and by steel shot and glass bead peening. The approximate stress levels obtained and the types of (211) peak profiles generated are given in Table VI.

Before stress measurements, all specimens were electropolished in a 7% solution of perchloric acid in ethanol to a depth of .05 to .10 mm (.002/.004 inches). Chromium powder specimens of -325 mesh were also prepared by annealing 99.98% pure Cr at 1000°F/4 hours and furnace cooling in vacuum. The Cr powder was used as a standard for stress calibration and for rapidly checking the resolution of the (211) doublet when the various slit combinations were evaluated.

RESULTS AND DISCUSSION

Stress data obtained on the Siemens-Halske unit with variations of the four divergence slits and six receiving apertures with parafocus conditions and two Soller collimation are shown in Table VII. These results were obtained with the two exposure ($\Delta 2\theta$) technique, $\psi = 0, 45°$. Similar data using the $\sin^2\psi$ technique and stationary focusing conditions, but for only two receiving slit combinations, with the four divergence slits are given in Figure 1. Additional stress data comparing the parafocus and stationary detector techniques may be observed in Table VIII.

Table VII. Effects of Slit Combinations on R.S. Results (kp/mm.^{2}), ψ = 45° Parafocus Techniques and S.H. Unit

Sample	Div. Slit	Receiving Slit Aperture in mm.						Average
		0.10	0.13	0.20	0.25	0.40	1.00	
#100	0.5°	+2.5	+2.1	+4.3	+0.9	+2.7	0	+2.1
Stress-Free	1°	+3.1	+5.1	+1.7	+1.7	+4.9	+3.6	+3.3
Standard	2°	+1.8	+4.5	+2.0	+4.8	+2.9	+2.1	+2.4
	4°	+2.9	+1.1	0	+6.0	+2.6	+5.1	+3.0
#103	0.5°	-61.3	-63.3	-58.0	-63.1	-63.2	-60.2	<-61.5>
Shot-Blast	1°	-62.1	-59.5	-60.5	-62.1	-60.4	-64.1	<-61.5>
Specimen	2°	-58.6	-59.5	-60.8	-59.1	-57.0	-57.2	<-58.7>
	4°	-54.5	-51.4	-51.0	-54.3	-54.8	-49.0	<-52.5>

Table VIII. Comparison of Stress Results Between Parafocus (P) and Stationary Detector (S) Techniques for ψ = 45° Two Exposure and Sin$^2\psi$ Methods, S.H. Unit, kp/mm.^{2}

Sample	Rec. Aperture in mm.	Divergence Slit								Results From Slopes in Figure 1 Sin$^2\psi$ Method			
		0.5°		1°		2°		4°		0.5°	1°	2°	4°
		P	S	P	S	P	S	P	S	Stationary			
#100 Hard	0.40	+2.7	+2.7	+4.9	+3.0	+2.9	+4.9	+2.6	+5.3	+3.7	+3.3	+4.6	+4.8
Stress-Free Standard	1.00	0	+5.1	+3.6	+4.3	+2.1	+2.0	+5.1	+3.2	+4.6	+3.9	+2.5	+3.4
#103 Shot-Blast Standard	0.40	-63.3	-59.8	-60.4	-63.7	-59.8	-57.8	-54.8	-54.1	-66.2	-64.7	-59.2	-56.3
	1.00	-60.2	-55.6	-64.2	-63.2	-60.0	-55.7	-49.0	-51.7	-61.5	-64.4	-60.5	-54.9
#106 Glass-Peened Standard	0.40	-43.2	-44.0	-38.2	-36.1	-36.5	-34.8	-31.0	-27.1	-35.1	-37.4	-37.4	-33.0
	1.00	-42.2	-38.9	-40.1	-41.1	-37.4	-37.4	-30.0	-26.0	-43.3	-44.3	-39.5	-26.5

These three sets of data for broad line (hardened) steel specimens reveal the following: (1) as the divergence slit increases in size an increasingly more positive (tensile) stress value is obtained at any fixed receiving slit choice in all specimens, (2) this effect is observed for both the parafocus and stationary detector focusing conditions, (3) no clear effect of receiving slit size per se on the stress data is apparent, (4) for any given slit combination the parafocus and stationary focusing techniques produced essentially equivalent stress values.

The effects of the available slit combinations and focusing conditions on strain measurements made with the General Electric XRD-3 unit are shown in Figure 2. Strain measurements were made on all specimens listed in Table VI. However, specific data on only the fine line (1010) and one broad line specimen are shown in Figure 2, because the stress results and data trends for specimen Nos. 100, 103 and 106 were similar to those shown for these specimens in Table VIII when the 1° and 2° divergence slits were used on the S.H. unit. Figure 2 again shows that the stress (strain) data become more positive (tensile) with increasing divergence slit size and that the receiving slit shows no significant nor specific effect.

Figure 3 shows results obtained on the S.H. equipment when the diffracted beam Soller baffles were removed. For both

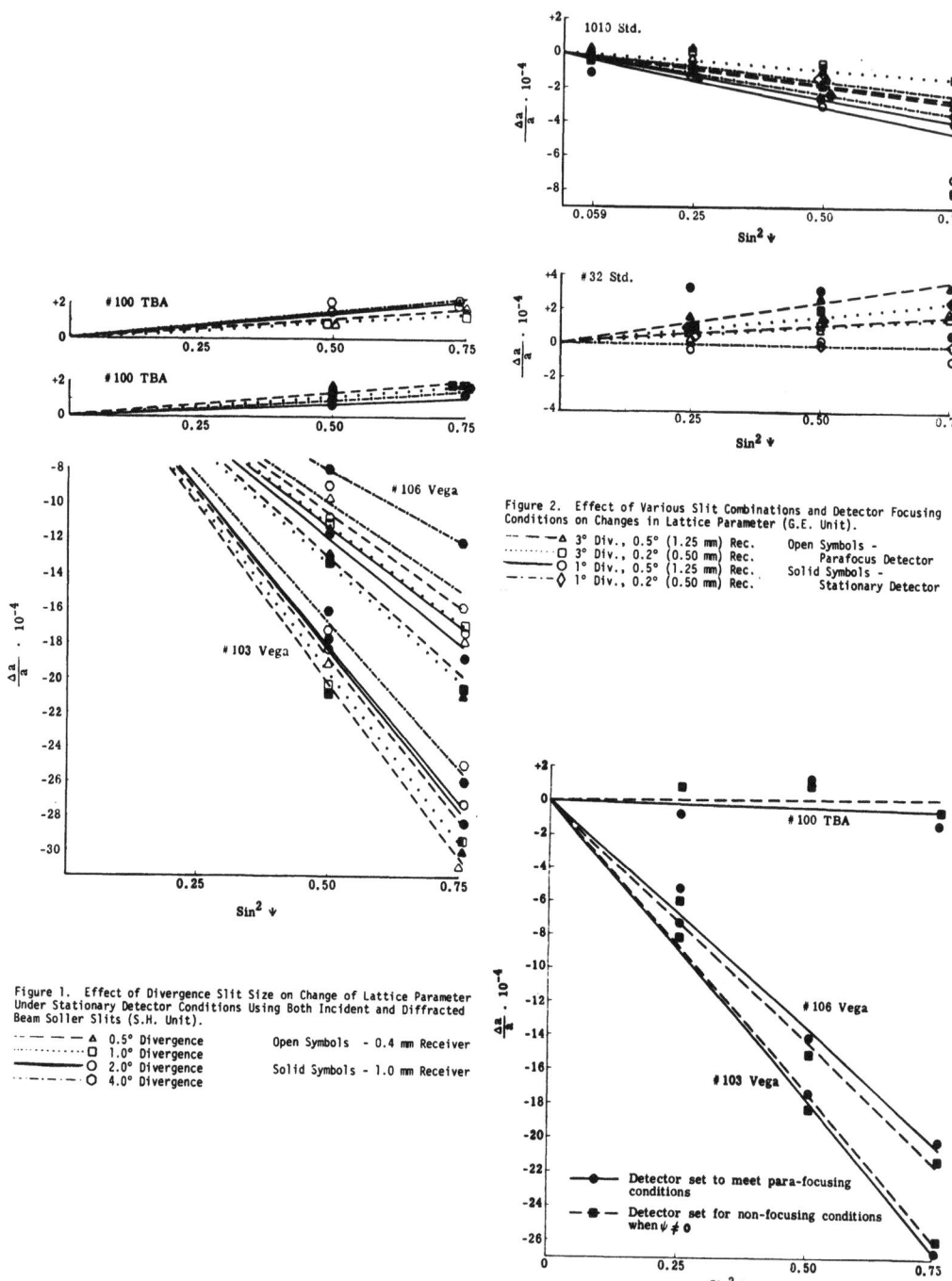

Figure 1. Effect of Divergence Slit Size on Change of Lattice Parameter Under Stationary Detector Conditions Using Both Incident and Diffracted Beam Soller Slits (S.H. Unit).

— — △ 0.5° Divergence
·······□ 1.0° Divergence
———○ 2.0° Divergence
—·—·○ 4.0° Divergence

Open Symbols - 0.4 mm Receiver
Solid Symbols - 1.0 mm Receiver

Figure 2. Effect of Various Slit Combinations and Detector Focusing Conditions on Changes in Lattice Parameter (G.E. Unit).

— — △ 3° Div., 0.5° (1.25 mm) Rec.
·······□ 3° Div., 0.2° (0.50 mm) Rec.
———○ 1° Div., 0.5° (1.25 mm) Rec.
—·—·◇ 1° Div., 0.2° (0.50 mm) Rec.

Open Symbols - Parafocus Detector
Solid Symbols - Stationary Detector

Figure 3. Effect of Removal of Diffracted Beam Soller Slit on Lattice Parameter Change (2° Div., 1.0 mm Rec., Incident Beam Soller Slit Only, S.H. Unit.

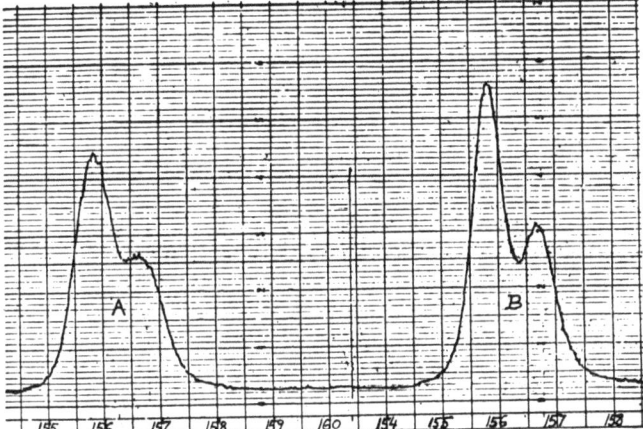

Figure 4. Resolution of (211) Doublet for Annealed TBA Steel for the Stationary Detector Condition When Using (A) 2° Divergence Slit and (B) 1° Divergence Slit and 1 mm Receiving Slit and 2 Sollers at $\psi = 45°$. (S.H. Unit).

focusing conditions there is little scatter in data points and no significant difference in comparative $\Delta a/a$ strain values from those obtained with two sets of Soller (incident and diffracted beam) baffles, see Figure 1.

The conclusion is then, that of all elements in the x-ray beam optics system, the divergence slit is the only one we need be concerned with, since its effect on measured strain and stress values can certainly be significant.

If we are to achieve reproducible and comparative results with the European stationary focusing method and the American parafocusing technique, the optimum choice of slit combinations must be identified for both geometric conditions. To achieve this end, we need to establish one other essential factor: the degree of $K\alpha_1 - K\alpha_2$ (211) doublet resolution which can be attained on fine or sharp line specimens with each slit combination at the various ψ tilt angles and focusing conditions.

Chart traces showing the type of peak profiles and the degree of $K\alpha$ doublet resolution of the Fe (211) reflection in annealed steel achieved on the S.H. unit with the various slit selections are reproduced in Figures 4 and 5. Satisfactory resolution of the (211) doublet is obtained with a 2°/0.4° (1.0 mm) slit system at $\psi = 45°$ when either the stationary or parafocus technique is used. However, when a 4°/1.0 mm combination is employed (Figure 5), doublet resolution at $\psi = 45°$ is not achieved with the stationary detector focusing method.

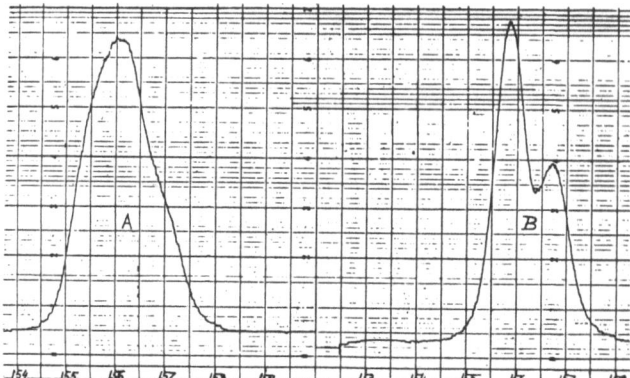

Figure 5. Resolution of (211) Doublet for Annealed TBA Steel When Using 4° Divergence Slit With 1.0 mm Receiver and 2 Soller Slits for (A) Stationary and (B) Parafocus Detector Conditions at $\psi = 45°$, (S.H. Unit).

The types of peak profiles and doublet resolution obtained with the available slit combinations on the G.E. equipment for Cr-powder are indicated in Figures 6 and 7. The line-resolving advantage of the parafocus technique is again demonstrated. With parafocusing, satisfactory resolution is achieved with a 3°/0.4° (1.0 mm) combination, but the largest slit combination permitted with the stationary detector focusing method appears to be a 1°/0.5° system. However, it is felt that had a 2°/0.4° slit system been available for the G.E. unit it, too, would show adequate resolution of the (211) doublet. The optimum divergence/receiver slit combinations for both diffractometer types and focusing techniques are given in the conclusions. These selections were based on obtaining satisfactory fineline peak resolution with x-ray intensities of at least 10^3 to 10^4 counts per second.

Although these optimum combinations assure proper peak resolution and thus freedom from wild deviations in strain and stress results should resolved and nonresolved peaks be inadvertently compared at the various ψ specimen orientations, they do not assure freedom from the systematic tendency of stress values becoming more tensile (positive) with increasing divergence slit size. This effect is due to beam divergence per se and is present with both stationary or parafocusing conditions. It was shown in (16, 17) that beam divergence always produces a shift in $2\theta_{hkl}$ peaks under ψ rotation of the specimen. This produces focusing aberrations that in turn produce systematic errors in stress which are always more tensile the wider the beam. Specimen surface geometry also produces focusing aberrations and systematic stress errors, but since flat specimens were always compared herein, specimen geometry was not a factor. The relative rise in stress and strain to more positive values with increasing divergence slit size in Tables VII and VIII and Figures 1 and 2 is about that predicted and measured in reference (17).

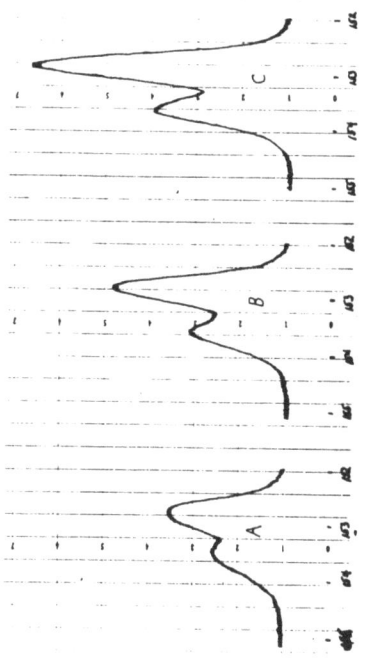

Figure 7. Cr-Powder (211) Doublet Resolution on G.E. Unit When Using a 1°/0.5° (1.25 mm) Slit Combination With Incident Soller Baffles Only and Detector Stationary. Note: A 1°/0.4° (1.0 mm) Combination Produced Satisfactory Resolution at ψ = 60°.

(A) ψ = 60°   (B) ψ = 45°   (C) ψ = 0°

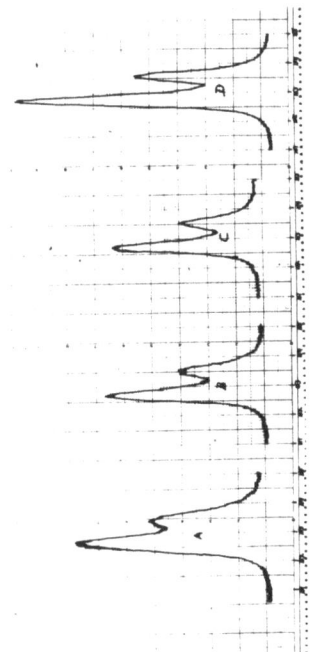
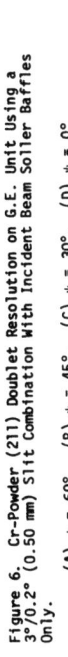

Figure 6. Cr-Powder (211) Doublet Resolution on G.E. Unit Using a 3°/0.2° (0.50 mm) Slit Combination With Incident Beam Soller Baffles Only.

(A) ψ = 60°, (B) ψ = 45°, (C) ψ = 30°, (D) ψ = 0°

Figures 8 and 9 compare the two different methods of 2θ peak location explored: (1) the three point parabola and (2) the bisection of recorder peak traces. Although no significant difference is indicated, the peak bisection technique appears to have less random scatter. Figures 8 and 9 also indicate that for sharp line specimens, corrections for Lorentz-Polarization-Absorption effects is not required since the data points are only 0.2° 2θ apart. LPA correction to line intensities must, however, be made on broad line data (10, 14).

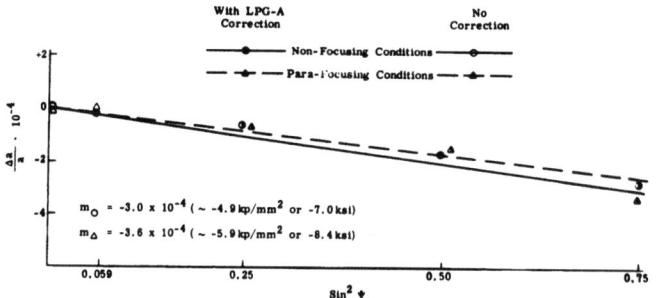

Figure 8. Change in Lattice Parameter for the SAE 1010 Steel Specimen Using Parabola Method at Parafocus and Stationary Detector Conditions (1° Div., 1.0 mm Rec., 2 Soller Slits, S.H. Unit).

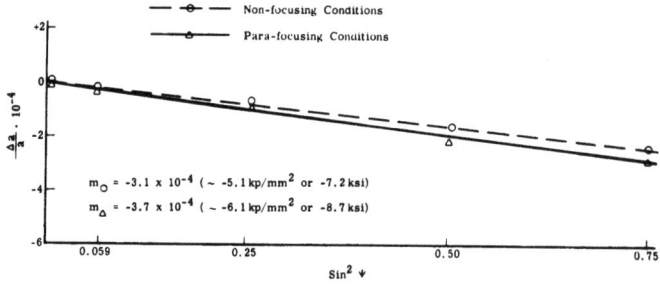

Figure 9. Change in Lattice Parameter for the SAE 1010 Steel Specimen Using Chart Peak Bisection Method at Parafocus and Stationary Detector Conditions (1° Div., 1.0 mm Rec., 2 Soller Slits, S.H. Unit).

Originally Ogilvie (15) proposed a parabola fit using a least square fit to 5 points. This method was applied to a broad line or hardened steel specimen #103 using a Univac computer. The results of fitting 2nd, 3rd and 4th order curves through five (5) data points are listed below:

Interval, $2\theta_{211}$	Order of Equation	Stress kp/mm², Using Δ2θ Method, ψ = 0 and		
		ψ = 30°	ψ = 45°	ψ = 60°
	3 Point Parabola			
1.5° 2θ Apart	-	-57.2	-53.4	-56.0
3.0° 2θ Apart	-	-54.7	-52.7	-51.5
	5 Point Parabola			
1.5° 2θ Apart	2nd	-57.8	-54.5	-53.7
1.5° 2θ Apart	3rd	-59.9	-53.2	-57.5
1.5° 2θ Apart	4th	-61.2	-56.2	-59.1

The second and third order curves produced good correspondence with the 3-point parabola data. Thus, it does not seem justified, to apply the more elaborate and time consuming five point method.

One set of results of the experimental determinations of elastic constants for hardened 46100 steel is given in Figure 10. The measured constants agree rather closely with those reported in (6-9) for high carbon steel and in fact also with bulk constants measured by standard mechanical tests. Of course, this would not have been so had plastic deformation been present.

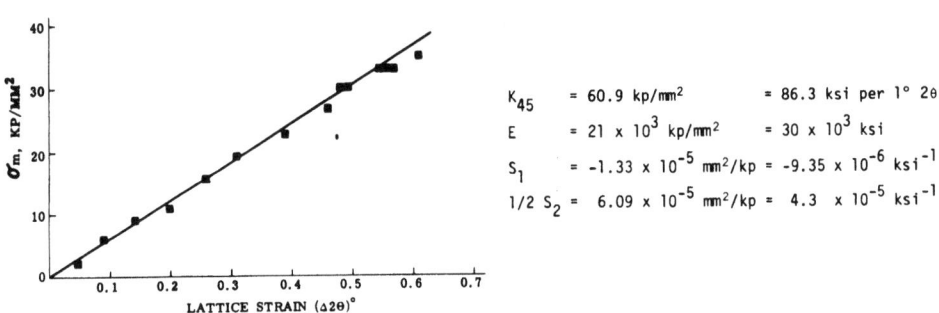

Figure 10. Determination of Stress Constant (K_{45}) and Elastic Constants, $1/2\ S_2$, S_1 and E for (211) Planes of Hardened and Tempered 46100 Steel (see Table IV).

K_{45} = 60.9 kp/mm² = 86.3 ksi per 1° 2θ

E = 21 x 10³ kp/mm² = 30 x 10³ ksi

S_1 = -1.33 x 10⁻⁵ mm²/kp = -9.35 x 10⁻⁶ ksi⁻¹

$1/2\ S_2$ = 6.09 x 10⁻⁵ mm²/kp = 4.3 x 10⁻⁵ ksi⁻¹

Data on the electronic stability of the x-ray instrumentation used are presented in Table IX. As noted before, the results include errors due to counting statistics and specimen positioning practices in addition to electronic drift. It is clear that stress measurements with present day equipment and techniques should easily approach a precision of ±2 kp/mm² (±3 ksi).

Table IX. Reproducibility of Residual Stress Values on Broad Line Specimens over 8 hour Period Using Largest Available Divergence and Receiving Slits*

	Siemens-Halske 4 deg./1.0 mm-Detector Stationary						General Electric 3 deg. Div.-With Parafocused Detector (parabola scanned in both directions)					
	Parabola Scanned in Both Directions			Parabola Scanned in One Direction Only			0.2 degrees Receiver			0.5 degrees Receiver**		
Test	2θ	$2\theta_{60}$	σ kp/mm²	2θ	$2\theta_{60}$	σ kp/mm²	2θ	$2\theta_{45}$	σ kp/mm²	2θ	$2\theta_{45}$	σ kp/mm²
1	154.525	155.867	-54.2	154.498	155.855	-54.9	155.285	155.326	-2.59	155.331	155.288	+2.61
2	154.523	155.822	-52.6	154.511	155.822	-53.1	155.299	155.290	+0.54	155.293	155.294	+0.25
3	154.511	155.852	-54.2	154.509	155.856	-54.6	155.319	155.311	+0.49	155.282	155.305	-1.40
4	154.470	155.867	-56.5	154.455	155.889	-58.1	155.293	155.297	-0.25	155.312	155.288	+1.46
5	154.498	155.859	-55.1	154.498	155.864	-55.3	155.292	155.311	-1.15	155.300	155.270	+1.82
6	154.498	155.845	-54.5	154.484	155.848	-55.2	155.300	155.287	+0.79	155.261	155.332	-1.31
7	154.511	155.871	-55.0	154.498	155.900	-56.7	155.308	155.291	+1.03	155.293	155.299	-0.37
8	154.498	155.853	-54.8	154.498	155.833	-54.1	155.312	155.298	+0.84	155.338	155.278	+3.64
9	154.498	155.872	-55.6	154.510	155.855	-54.5	155.307	155.307	±0	155.325	155.298	+1.64
10	154.523	155.846	-53.4	154.511	155.837	-53.7	155.272	155.307	-2.12	155.298	155.290	-0.12
Avg.	154.506	155.855	-54.6	154.497	155.856	-55.0	155.299	155.303	-0.24	155.304	155.294	+0.52
Range 3σ			±2.0			±2.5			±2.0			±4.0
∼σ, kp/mm			±0.7			±0.8			±0.7			±1.3
σ, ksi+			±1.0			±1.2			±1.0			±2.0

* Data include errors due to electronic drift, counting statistics, and specimen positioning errors (specimen removed and replaced for each test run).
** The 0.5 deg. receiver does not give sufficient resolution for this stress measurement, but data included nevertheless.
+ Conversion from kp/mm² to ksi, multiply kp/mm² by 1.422.

CONCLUSIONS

(1) The most important factor in x-ray stress measurements is that proper peak resolution be achieved of all ψ specimen orientations. To obtain satisfactory $K\alpha_1$-$K\alpha_2$ doublet resolution of fine line (211) peak profiles in steel and Cr powder at ψ specimen orientations greater than 45°, the largest slit combination cannot exceed 2°/0.4° (1.0 mm) for the European or stationary method and 3°/0.4° (1.0 mm) for the American parafocus technique. For broad line specimens larger slit choices can be made, but they are not recommended due to the inherently larger aberration errors (17).

(2) When proper conditions for peak resolution are employed, the parafocus and stationary focusing techniques produce essentially equivalent results. However, for the flat specimens studied a slight tendency toward more positive (tensile) stresses (approximately 1 kp/mm²) is indicated for the parafocus detector method. Reasons for this are given by the author in (17).

(3) An increase in divergence slit with double or single Soller collimation produces an increasingly more positive (tensile) stress value due to focusing aberrations when either parafocusing or stationary methods are used.

(4) The receiving aperture and Soller baffles do not appear to influence residual stress measurements significantly when proper peak resolution and counting statistics are observed.

(5) Of the two methods of peak location, the bisection of slowly scanned chart traces produces lower random errors than the parabola fit techniques. However, times for completion were longer by a factor of three to four.

(6) Although the confidence level of the five point parabola fits is slightly better than for the conventional three point method, its use does not appear justified except in special instances.

(7) Annealed Cr-powder of -325 mesh appears to be an excellent stress free standard material for calibration of residual stress measurements and for tests of peak resolution. The stationary and the parafocus methods both yielded essentially a zero stress value (± 1 kp/mm^2) for Cr-powder on the Siemens and the General Electric diffractometers.

(8) Elastic constants for hardened steel as determined in bending are in excellent agreement with the constants obtained from theoretical values for Cr-Kα radiation and (211) reflection.

(9) The electronic stability and sample positioning instrumentation of present diffractometer equipment allows a precision in stress measurement of better than ± 2 kp/mm^2 (± 3 ksi). The $\sin^2\psi$ method permits slightly better precision than the two exposure technique (ψ = 0, 45° or 60°), however, if high counting statistical conditions are maintained (10^5 - 2×10^5 counts/point) the two exposure method is satisfactorily accurate and much less time consuming.

REFERENCES

1. C. S. Barrett, Structure of Metals, (1952), p. 316, McGraw-Hill.

2. A. L. Christenson and E. S. Rowland, "X-Ray Measurements of Residual Stress in Hardened High Carbon Steel," Trans. ASM, 45, (1953), p. 638.

3. B. D. Cullity, Elements of Diffraction, (1959), 436, Addison-Wesley.

4. V. Hauk, "Present State of the Determination of Stresses by X-Ray Investigation," Archiv, Eisenhuttenwessen, 38, (1967), p. 233.

5. D. P. Koistinen and R. E. Marburger, "Calculating Peak Positions in X-Ray R.S. Measurements on Hardened Steel," Trans. ASM, 51, (1959), p. 537.

6. E. Macherauch, "X-Ray Stress Analysis," Proc. Soc. Exp. Mechanics, 23, (1966), p. 140.

7. TR-182, "Measurement of Stress by X-Rays," Soc. Automotive Engineers, Div. 4, (1952), SAE, New York.

8. S. Taira, J. Arima, and T. Shiroyama, Journal Soc. Mat. Sci., Japan, 12, No. 123, (1963), p. 865, "Measurement of Elastic Constants by X-Ray Methods".

9. K. Hayashi, "X-Ray Inv. on Lattice Strain and Deformation of Polycrystalline Metals," Doctor Engr. Thesis, Kyoto Univ., (1970).

10. S. Taira and Y. Yoshioka, "X-Ray Investigation of the Residual Stress of Metallic Materials," Journal Soc. Mat. Sci., Japan, 13, No. 135, (1964), p. 949.

11. T. Shiraiwa and Y. Sakamoto, "X-Ray Stress Measurement and Its Application to Steel," Sumitomo Search No. 7, May, 1972.

12. J. Fukura and H. Fujiwara, "Evaluation of Various Techniques of X-Ray Stress Measurement With the Diffractometer," Journal Soc. Mat. Sci., Japan, 15, No. 159, (1966), p. 825.

13. J. Fukura and H. Fujiwara, "Developments in X-Ray Stress Measurements," Bulletin Faculty Engr., Tokushima Univ., 7, No. 1, (1970).

14. H. S. 182, J784, "Residual Stress Measurement by X-Ray Diffraction," Soc. Autom. Engr., (1971), SAE, New York.

15. R. E. Ogilvie, "Stress Measurement With the X-Ray Spectrometer," M.S. Thesis, Mass. Inst. Tech., (1952).

16. W. Parrish and H. Van Olphen, "X-Ray Diffractometry for Complex Powder Patterns," Prog. Anal. Chem., 1, (1968), Plenum Press.

17. C. Jatczak and H. Zantopulos, "Systematic Errors in Residual Stress Measurements Due to Specimen Geometry," Adv. in X-Ray Analysis, 14, (1970), Plenum Press, p. 360.

18. H. Muro, J. Sadaoka, and M. Tokoda, "Examination of Correction Factors in X-Ray Residual Stress Measurements of Hardened Steel," Journal Society Mat. Science, 15, No. 159, 1966.

19. S. Aoyama, K. Satta and M. Tada, "The Effect of Setup Errors on the Accuracy of Stress Measured With the Parallel Beam Diffractometer," J. Soc. Mat. Sci., Japan, 17, No. 183, 1968, p. 1071.

X-RAY DIFFRACTION RESIDUAL STRESS ANALYSIS USING HIGH PRECISION CENTROID SHIFT MEASUREMENT TECHNIQUES - APPLICATION TO URANIUM-0.75 WEIGHT PERCENT TITANIUM ALLOY

W. E. Baucum and A. M. Ammons

Union Carbide Corporation-Nuclear Division, Oak Ridge

Y-12 Plant, P.O. Box Y, Oak Ridge, Tennessee 37830

ABSTRACT

The standard three-point parabola technique has been used extensively in residual stress analysis for peak location with precisions (standard deviation) of peak shift measurements in the range of 0.01° to 0.03° (2θ). While this level of precision produces useful data for peaks at high diffraction angles, it becomes untenable when a material with high elastic modulus is encountered whose only suitable peak falls below approximately 140° 2θ. Uranium-0.75 weight percent titanium alloy, which is the material of interest in this investigation, falls into this category.

A technique has been developed by which diffraction peak centroid shifts are determined with a precision of 0.002° of diffraction angle ($\Delta 2\theta$). The method consists of step-scanning over the peak of interest to acquire data in digital form. A small computer is then used to locate the diffraction line by calculating its centroid from digital data.

The use of a special probe consisting of four distance transducers (LVDT) to achieve precise alignment of the specimen surface on a Siemens back-reflection stress diffractometer will also be described.

A calibration experiment was performed on uranium-0.75 titanium tensile specimens in which the X-ray diffraction response was measured as a function of various levels of applied stress in order to determine the experimental elastic moduli. It was found that repetitive analyses on a single area of a cylinder had a standard deviation on the measured diffraction centroid shift of 0.002 degree which is approximately an order of magnitude smaller than that achieved by the standard parabola fitting technique. Thus, the residual stress can be defined to a precision of $\sim$ 2,000 psi, as opposed to 30,000 psi, at best, were the measurements obtained by the conventional procedure.

INTRODUCTION

X-ray diffraction has been used extensively in past years to nondestructively measure residual stresses in a wide variety of materials and has, in general, performed this task well. Some materials, however, are more amenable to this method because better precision can be obtained on these than others. The main properties which affect the measurement precision are the elastic constants of the material, the diffraction angle of a suitable resolvable diffraction peak, and the intensity of this peak. A material which does not lend itself to residual stress analysis by X-ray diffraction would typically have a high modulus of elasticity along with low Poisson's ratio and its only suitable diffraction peak falling below $140° \ 2\theta$. The purpose of this paper is to describe a method by which such materials, a uranium-0.75 weight percent titanium alloy in this case, may be analyzed with sufficient precision to be useful, even on large samples. The usefulness of the analysis will be demonstrated by the presentation of data from a precision experiment and a calibration experiment.

EXPERIMENTAL METHOD

There are essentially three methods used extensively for stress analysis by X-ray diffraction--two-exposure, $\sin^2\psi$, and extrapolation. (1, 2)

The method used in this study was the standard two-exposure technique with modifications to increase precision. Basically, the angular positions of a back-reflection Bragg diffraction peak are measured with the sample in two orientations with respect to the X-ray beam. First, the sample is positioned such that the diffraction vector is normal to the surface; secondly, the X-ray beam is rotated by an angle, ψ, with respect to its former position. The angular difference, $\Delta 2\theta$, between the positions of the Bragg peak for these two orientations is related to stress by the equation,

$$\sigma = \frac{E \cot \theta}{(\nu+1)\sin^2\psi}(\Delta 2\theta) = k(\Delta 2\theta), \qquad (1)$$

where

- σ   is the stress,
- E   is the Young's modulus,
- θ   is the diffraction angle,
- Δ2θ is the change in 2θ,
- ν   is the Poisson ratio,
- k   is the stress factor = $\frac{E \cot \theta}{(\nu+1)\sin^2\psi}$, and
- ψ   is the rotational angle between first and second measurement.

The most common method used for peak location is the three-point parabola technique.(3-6) This technique mathematically fits a parabola to three data points near the top of a peak and then locates its center. Only a few authors give their measurement precision, but the standard deviations seem to fall in the range of 0.01° to 0.03° Δ2θ, depending on peak breadth and intensity. (3,4,7) The overall precision of the stress analysis is dependent on the measurement of Δ2θ and thus on the precision with which the angular position of a peak can be determined. Using the three-point parabola technique, it has been possible to determine the shift in the broad peaks of uranium alloys with a precision (standard deviation) of 0.03° Δ2θ or, in terms of residual stress, a precision of ∼ 30,000 psi for the U-0.75 weight percent Ti alloy. Since this precision is relatively poor, it was essential to develop a more precise method of peak location.

Experimentation with various methods of peak location indicated that peak centroids could be located very precisely. An annealed gold powder, glued onto a thin sheet of Mylar, was analyzed repetitively, and it was found that the centroid shift of the (422) peak could be reproduced to a standard deviation of 0.002° Δ2θ. This was obtained by calculating the total centroid of peak and background. The precision using the peak centroid (background subtracted out) was 0.02° Δ2θ. This difference is probably due to the fact that it is difficult to precisely define the background, since the background intensity is small compared to the total intensity in the peak. Thus, the total centroid was used in this study to precisely locate the diffraction peaks.

The uranium peak of indices (331) and (135) (the two cannot be resolved) which falls at approximately 131.6° 2θ using CuKα radiation was chosen for analysis since it appeared to have the best peak-to-background ratio in the back-reflection region, even in deformed material. A step-scan was made from 129 degrees to 134.5 degrees in increments of 0.02 degree with the intensity at each increment recorded on paper tape for beam inclination angles of ψ = 0 and ψ = -32 degrees.

Equipment

The equipment required for this method of analysis may be grouped for discussion into five parts. First, the X-ray generation equipment consisted of a Siemens Crystalloflex IV X-ray generator with Cu target X-ray tube. Secondly, the angular positioning equipment consisted of a Siemens back-reflection diffractometer for examination of large samples and an indexer-stepping motor system to permit electronic control of the angular position of the diffractometer detector. Thirdly, the X-ray detection and counting system consisted of a scintillation detector, linear amplifier, pulse height analyzer, timer, and scaler. Fourthly, the data output system consisted of printout interface from timer and scaler to Teletype which both typed and punched data on paper tape. Fifthly, the data reduction system consisted of a high-speed paper tape reader interfaced to a PDP 8/I computer which read the data and calculated centroids of the peaks.

A flow diagram given in Figure 1 shows how the system was assembled. With X-rays impinging on the sample and the detector at the starting 2θ position, a timed count is begun on the counter-timer. At the end of the preset time or count interval, the counter-timer sends the data to the Teletype via the printout interface and, at the same time, signals the controller which, in turn, tells the indexer to advance the stepping motor by the preset angular increment. The indexer and printout interface then signal the controller that the stepping motor and Teletype have both finished their jobs, and the controller initializes and begins another time count. The next count cannot begin until

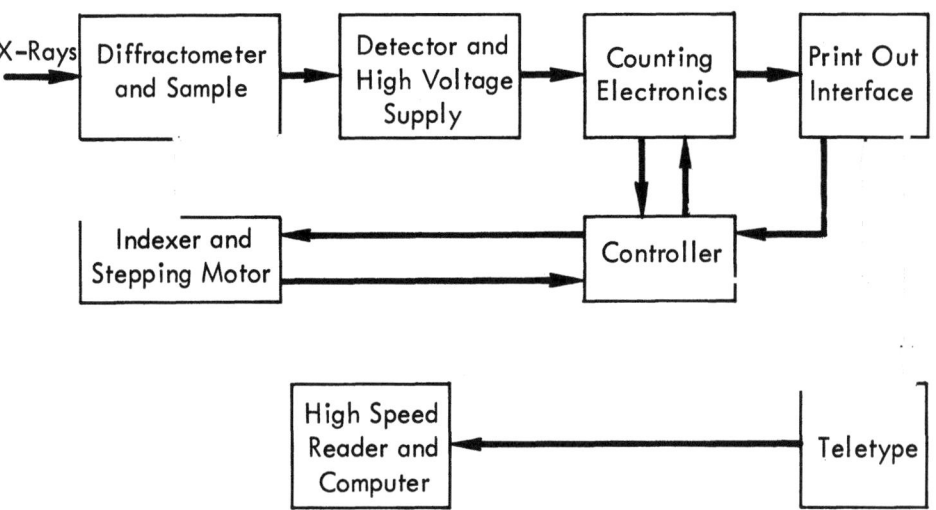

Figure 1: EQUIPMENT FLOW DIAGRAM

both the stepping motor and Teletype are ready for the next point. This cycle continues until the final 2θ angle has been reached. Both the preset time or count and the angular increment may be arbitrarily fixed by the experimenter. The punched paper tape from the Teletype is fed to the computer for centroid calculation.

To obtain good analysis precision, one must be able to align the sample (in this case, a large cylinder) reproducably with respect to the incident X-ray beam and the diffractometer since deviations in the alignment cause error in the stress measurement. To do this for a large sample was very difficult and required the design and use of a special alignment probe which is attached to the diffractometer. The probe consisted of four LVDT's (linear distance transducer) positioned in a circular fashion and separated from each other by 90 degrees of rotation (see Figure 2). The reading from each LVDT is sampled individually using a switching box. To align the sample, one positions it such that any two LVDT's which are separated by 180 degrees of rotation show the same distance reading. There is a pointer at the center of the LVDT circle to indicate the center of the analysis area. The probe is removed during the actual analysis.

Centroid Calculations

The method of centroid calculation is based on Simpson's Rule for a parabolic approximation to the area under a curve. Referring to Figure 3,

$$\bar{X} = \frac{\sum_{i=1}^{k} M_i}{\sum_{i=1}^{k} A_i}. \tag{2}$$

Now

$$A_i = \int_{X_i-h}^{X_i+h} I(x)dx \approx \frac{h}{3}[I(X_i-h) + 4I(X_i) + I(X_i+h)], \tag{3}$$

and

$$M_i = \int_{X_i-h}^{X_i+h} XI(x)\,dx \approx \frac{h}{3}[(X_i-h)I(X_i-h) + 4X_iI(X_i) + (X_i+h)I(X_i+h)]; \tag{4}$$

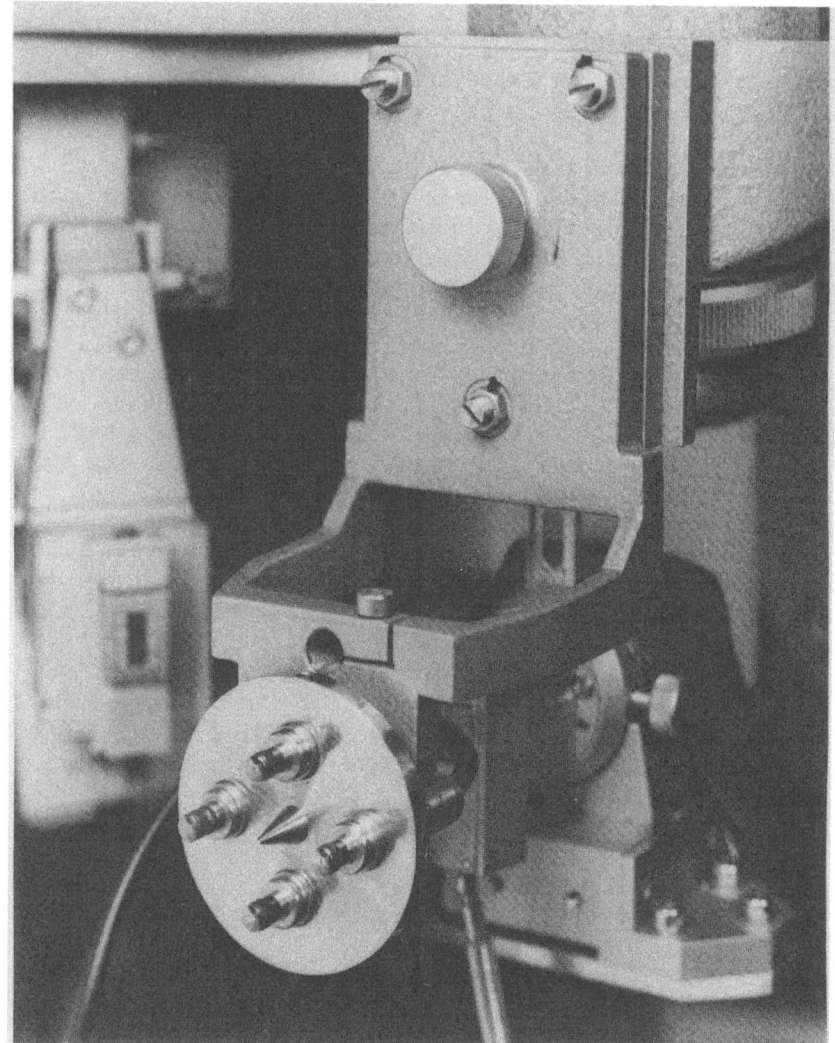

Figure 2: LVDT ALIGNMENT PROBE

thus,

$$\bar{X} = \frac{\sum_{i=1}^{k} [(X_i-h)I(X_i-h) + 4X_iI(X_i) + (X_i+h)I(X_i+h)]}{\sum_{i=1}^{k} [I(X_i-h) + 4I(X_i) + I(X_i+h)]}, \qquad (5)$$

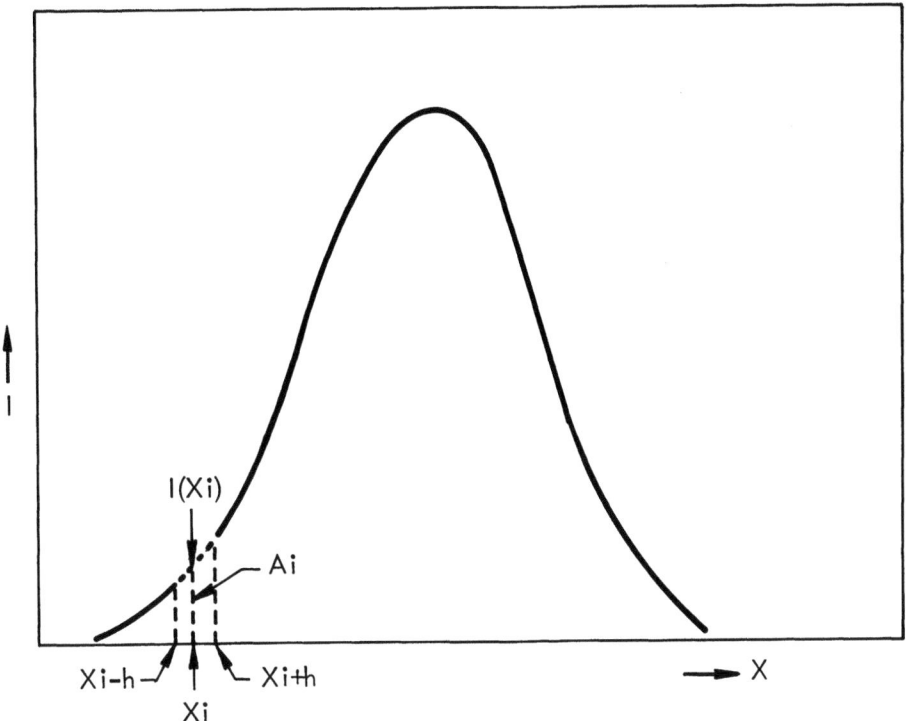

Figure 3: ILLUSTRATION OF CENTROID CALCULATION

where

$$k = \frac{(\text{no. of data points}) - 3}{2} + 1, \qquad (6)$$

since three data points are required for the first incremental area and two each for the rest. The centroid calculation is much too laborious to perform by hand, and thus, a Focal language program was written for a minicomputer (8K core memory) which first corrected each data point for Lorentz polarization and absorption (8) and then calculated $\bar{X}$ according to Equation (5).

RESULTS AND DISCUSSION

Once developed, this method was then tested on the material of interest, U-0.75 weight percent Ti alloy, to show that a viable analysis of such "hostile" materials could be performed. The test consisted of these items: the determination of analysis precision and determination of the diffraction behavior of the material under known applied stresses.

Precision

In order to determine the significance of a difference between any two measured values of residual stress, it is necessary to know the precision of analysis on a single measurement. This may be done by making a series of measurements on one sample and calculating the experimental standard deviation on a single measurement.

A cylinder of U-0.75 w/o Ti (8-inch diameter by 8-inch length), which had been heated for one hour at 800°C and water-quenched, was obtained for the precision experiment. The cylinder was lightly etched over a one-inch diameter spot to remove the deformed surface layer and analyzed seven times for residual stress. The results are presented in Table 1. The cylinder was moved and realigned in the diffractometer between each analysis. The standard deviation of the resulting peak shifts was calculated to be 0.002° $\Delta 2\theta$. This is the overall standard deviation, including error due to both peak location and sample alignment along with any other instrumental errors that may have been present. This is more than an order of magnitude better precision than was obtained using the parabola fitting technique (2,000 psi instead of 30,000 psi for U-0.75 w/o Ti alloy).

Table 1

RESULTS OF PRECISION EXPERIMENT

Analysis Number	Observed Shift (Degrees)
1	0.026
2	0.029
3	0.026
4	0.031
5	0.027
6	0.031
7	0.029

Calibration

It can be seen from Equation (1) that both E and ν of the U-0.75 w/o Ti alloy must be known in order to calculate from a measured $\Delta 2\theta$. Although mechanically measured values of these quantities are available, they have been found to vary somewhat from those determined from X-ray diffraction stress analysis.(3) It was, therefore, necessary to measure their values experimentally by X-ray diffraction in order to determine the absolute residual stress values. Equation (1) shows that the value, k, is simply the slope of a plot of $\Delta 2\theta$ versus applied stress and thus can be measured experimentally by applying known stresses to a specimen of the material of interest.

The tensile specimen for the calibration experiment was fabricated by machining a 7-inch long by 1-inch wide strip from a 0.050 inch sheet. The specimen was subsequently heat-treated at 800°C for one hour and water-quenched. A small tensile machine with calibrated load meter was then positioned so as to align the tensile specimen with respect to the stress diffractometer. The specimen was realigned after each change in applied stress in order to correct any movement due to the change in applied stress level. The gold on Mylar reference sample (Table 2) was applied to the specimen surface, using Vaseline, and analyzed to determine the peak shift due to instrumental factors. This shift (0.003° $\Delta 2\theta$) was subtracted from the data obtained in the calibration experiment. Known stresses ranging from 8.67 to 73.25 ksi were applied to the specimen and centroid shift measured at each value.

Table 2

PRECISION EXPERIMENT ON GOLD MYLAR SAMPLE

Analysis Number	Centroid Shift (Degrees $\Delta 2\theta$)
1	- 0.001
2	- 0.003
3	- 0.003
4	- 0.007
5	- 0.005
6	- 0.003
7	- 0.001
8	- 0.001
	$\bar{X}$ = - 0.003

A linear curve was least-squares fitted to the data and its slope calculated to obtain $k = 9.81 \times 10^5$ psi/$\Delta 2\theta$ (Figure 4). The standard deviation on the slope was 3 percent of the value which indicated a good fit of the data to the linear curve. The Y-intercept of the curve was assumed to indicate a residual stress of -32,000 psi originally present in the tensile specimen (3), as might be expected from the quenching treatment that the alloy was given before analysis.

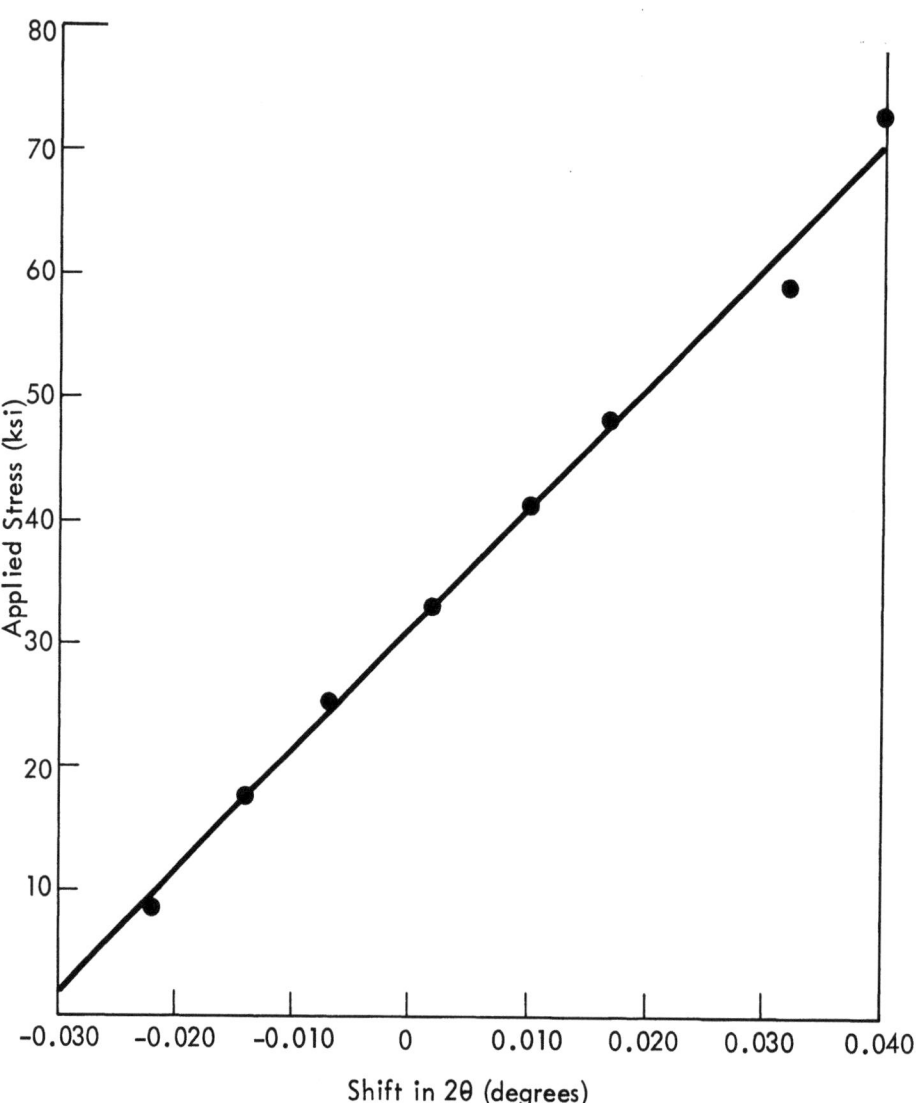

Figure 4: RESULTS OF CALIBRATION EXPERIMENT

The good fit of the data to a linear curve indicates that the stress-strain relationship followed Hooke's law and that good precision was obtained in locating the broad diffraction peaks. The value obtained for k is actually an average value arising from the two unresolved reflections (331 and 135) which may be individually shifting by different amounts since the structure of the material is orthorhombic. The linearity of the observed curve, however, indicates that this fact should not disturb the measurements and the average value of k may be used.

CONCLUSION

High precision in diffraction peak location is required in order to make a useful residual stress analysis of materials having high elastic moduli and only a low angle diffraction peak suitable for analysis. The widely used parabola fitting technique has not given sufficient precision, and thus, a method was developed to measure the peak centroid as a means for precisely locating diffraction peaks. A precision (standard deviation) of $0.002°$ for measuring centroid shift was obtained on both Au and a large U-0.75 w/o Ti alloy cylinder using the LVDT probe for sample alignment. This is approximately an order of magnitude better than could be obtained using the three-point parabola fitting technique. In addition, a calibration experiment was performed on the alloy to determine the stress factor, k, with good precision. Thus, a material which normally would have been unamenable to stress analysis by X-ray diffraction has been successfully analyzed.

ACKNOWLEDGEMENTS

The authors acknowledge the capable assistance of A. L. Coffey for basic development of the centroid calculation computer program and T. L. Williams for design of the controller module in the step-scan system.

REFERENCES

1. "X-Ray Measurements of Surface Residual Stresses in Cold Worked α-Brass", W. Wallace and T. Terada, <u>Advances in X-Ray Analysis</u>, 14, C. S. Barrett, J. B. Newkirk, C. O. Ruud, Plenum Press, New York, pp. 389-407; 1971.

2. "X-Ray Stress Analysis", Eckard Macherauch, <u>Proceedings of Second SESA International Congress on Experimental Mechanics</u>; 1965.

3. *Elements of X-Ray Diffraction*, B. D. Cullity, Addison-Wesley Publishing Company, Reading, MA, pp. 447-449.

4. "X-Ray Diffraction Study of Residual Macrostresses in Shot-Peened and Fatigued 4130 Steel", A. L. Esquivel and K. R. Evans, *Experimental Mechanics*, pp. 496-503; November, 1967.

5. "Experimental Methods of X-Ray Stress Analysis", D. M. French and B. A. McDonald, *Experimental Mechanics*, pp. 456-461; October, 1969.

6. "Computer Programs of Analysis of Diffractometer Data in Residual Stress Measurement", W. Wallace and T. Terada, *Metallography*, $\underline{4}$, pp. 339-342; 1971.

7. "Errors in Residual Stress Measurements Due to Random Counting Statistics", Carol J. Kelley and M. A. Short, *Advances in X-Ray Analysis*, $\underline{14}$, C. S. Barrett, J. B. Newkirk, C. O. Ruud, Plenum Press, New York, pp. 377-388; 1971.

8. *Measurement of Stress by X-Ray*, Information Report No. TR-182, Society of Automotive Engineers, New York 17, New York.

9. "Evaluation Truncation Methods for Accurate Centroid Lattice Parameter Determination", J. Taylor, M. Mack, and W. Parrish, *Acta Crystalligrapica*, pp. 1229-1245; 1964.

AUTHOR'S COMMENTS ON QUESTIONS RAISED

The major question raised at the end of the oral presentation concerned the method of choosing appropriate limits of diffraction angle between which to integrate the diffraction profile. It has been asserted that special truncation methods, along with corrections for instrumental and specimen observations, must be used to obtain the absolute centroid location of a diffraction profile.[9] Granted: however, this method of residual stress analysis does not require a knowledge of the absolute centroid location; in fact, for this particular case where the diffraction profile consisted of two unresolved reflections, the true centroid position is rather meaningless. Instead, what is needed is the relative difference in the diffraction profile position for the two incident beam inclination angles. Any point may be used to measure this shift so long as it is very reproducible (σ of 0.001° in the present study). Calibration data obtained by the same approach from material under known stress loads allows definition of the stress in an unknown. There is no necessity or assertion that the centroid position found by this method is the absolute position of the true centroid.

The integration limits were determined by finding points on either side of the diffraction profile ($\psi = 0°$) which appeared to be at background level and proceeding 0.5° 2θ further away from peak on either side to obtain the limits. Since the maximum shift of the profile was $\sim 0.1°$ $\Delta 2\theta$, this procedure assured that the limits chosen would remain at background level with a profile shift due to residual stress. One might argue that the parabola fitting technique would be equally reproducible if a large number of data points were used, as with the centroid method. To test this hypothesis, parabolas were least-squares fitted to the upper 20 percent of the diffraction peaks obtained from the data used in the previously described calibration and precision experiments, and the respective standard deviations were approximately a factor of three greater than those calculated for the centroid technique. The difference probably lies in the fact that a particular shape for the profile must be assumed in the parabola fitting technique; whereas, no such assumption is necessary for the centroid method.

The results of the calibration experiment which show a linear dependence of profile shift with applied stress, as predicted by Hooke's law, demonstrates that this technique of centroid location is adequate for residual stress analysis.

AN X-RAY AMORPHOUS SCATTERING INVESTIGATION OF THE CORROSION OF A POTASSIUM SILICATE GLASS $K_2O-3SiO_2$

R. W. Gould and M. S. Hill

University of Florida

Gainesville, Florida 32611

ABSTRACT

Structural modifications occurring in corroded potassium silicate glasses were studied using x-ray amorphous scattering techniques. Pair function distribution curves are compared and the differential method is used to follow the structural changes. The structure appears to approach that of vitreous silica after long leaching times.

INTRODUCTION

Vitreous silica and other glassy structures have been extensively studied over the past 40 years by x-ray diffraction techniques. Fourier integral analysis was introduced in 1927 by Zernike and Prins [1] and later (1930) by Debye and Menke [2]. This technique allowed the transformation of experimental x-ray intensity data into the well known radial distribution function (rdf). Warren, Krutter and Morningstar [3] made the first extensive study of the structure of vitreous silica in 1936 using the (rdf) technique. Later, in 1969, Mozzi and Warren [4] reinvestigated the vitreous SiO_2 structure using an improved experimental technique and a new distribution function concept known as the pair function distribution (PFD).

Alloy glasses were also investigated at an early date using the (rdf) technique, (Valenkov and Porai Koshits) [5] in 1936 and Hartlief [6] in 1938. These workers introduced the concept of subtracting (rdf) curves to investigate the effects of alloying additions.

This paper describes the structural changes occurring in a potassium silicate glass of the composition $K_2O-3SiO_2$ which has undergone corrosive leaching of the potassium.

THEORY OF AMORPHOUS SCATTERING

The pair function distribution technique is discussed rigorously by Warren [7,8] and the results are given below in equation (1).

$$\sum_{u.c.} \sum_i N_{ij}/r_{ij} P_{ij}(r) = 2\pi^2 r \rho_e \sum_{u.c.} Z_j = \int_0^{km} k\, i(k) \exp(-\alpha^2 k^2) \sin r\, k\, dk \quad (1)$$

where u.c. denotes one unit of composition and $k = 2\pi s$. Mozzi and Warren [4] applied this technique successfully to a reevaluation of the structure of vitreous silica. The experimental intensity values are used to calculate the Fourier integral on the righthand side of equation (1). The left side of equation (1) is then calculated assuming various structural models for the glass until a good fit between calculated and experimental PFD's is obtained. The various terms found in equation (1) are defined in reference [4]. In the present study only the righthand side of equation (1) was utilized since a differential method was used.

EXPERIMENTAL PROCEDURES

This paper is taken from a thesis by M. S. Hill [9] in which three potassium silicate glass compositions were investigated. Only the composition $K_2O-3SiO_2$ will be discussed in detail here.

Alloy Preparation and Corrosion

The alloy $K_2O-3SiO_2$ was prepared by melting reagent grade potassium carbonate and vitreous silica in a platinum crucible. After a 24-hour homogenization at 1000°C, the samples were crushed and ground to pass a 200 mesh screen and stored in a dessicator. Corrosive leaching of potassium was conducted at room temperature using distilled water whose pH was adjusted to 7.0 using 0.5 Molar hydrochloric acid. Five leaching experiments were performed with the leaching times being cumulative. Table 1 below describes these experiments.

Table 1

Water Corrosion of $K_2O-3SiO_2$ at pH = 7.0

Leaching Bath	Time (Hrs.)	Cumulative Time (Hrs.)
1	1	1
2	12	13
3	12	25
4	12	37
5	24	61

The powder was dried and pressed into a pellet after each leaching treatment and structurally examined by the x-ray technique to be described below. The validity of using pressed powder specimens was established early in the program as a result of a structural comparison of vitreous SiO_2 (bulk) with vitreous SiO_2 (pressed powder). The composition of the leached powders was determined by an amorphous peak shifting technique [10] and x-ray spectroscopy.

X-ray Diffraction Procedures

$CuK\alpha$ and $MoK\alpha$ x-rays were used to obtain the structural data. The copper data was taken from a General Electric spectrogoniometer having a curved LiF primary beam monochromator. A continuous slow scanning x-ray technique was used and the data was corrected for air scatter, background, polarization, and compton scattering. This corrected data was subsequently hand normalized about its independent atomic scattering factor curve. Instrumental broadening was neglected because the Sauder method, when applied, [4] indicated only a neglible correction factor.

The molybdenum data was obtained from a Norelco high angle goniometer equipped with a graphite diffracted beam monochromotor and a scintillation detector. The measured intensity was corrected for air scatter, background and polarization and was then hand normalized about its independent atomic scattering factor curve. Compton scattering was neglected because it was efficiently removed with the slit system used [11]. Multiple scattering was also neglected owing to its small value [7,8].

The value of the product of the sharpening and convergence factor $\exp(-\alpha^2 k^2)/g^2[k]$ in equation (1) was chosen to be unity at $k = 0$ and 0.1 at $k = k_{max}$. The normalized intensity (Ieu/N) as a function of s was used to calculate the integral on the right side of equation (1), using a multi-element Fourier transform program [9]. The computer output plot usually contained extraneous ripples at low values of r due to experimental intensity errors, normalization errors and termination satellites. When the ripples appeared to be unduly large, the data was renormalized and retransformed. A Kaplow correction [12] procedure was not employed.

EXPERIMENTAL RESULTS

Normalized Diffraction Data for $K_2O-3SiO_2$ and three other glasses, including pure SiO_2, is shown in Figure 1. Figure 2 shows the effect of the five leaching treatments on the normalized x-ray data for the $K_2O-3SiO_2$ composition. Note in both Figures 1 and 2 the ordinate has been offset for each curve in order to clarify the presentation.

Calculated PFD Curves for $K_2O-3SiO_2$

The experimental PFD curves obtained from the normalized data given in Figure 2 are shown in Figure 3. As before, each curve is slightly offset vertically. Peaks below one Angstrom are caused by series termination effects and are neglected in the discussion.

Table 2 is a summary of the interatomic distances obtained for the potassium silicate glasses examined in this study. In addition, this table includes a comparison of the SiO_2 results of this study with those of Mozzi and Warren [4] which are accepted as being correct.

Differential Curves for the System $K_2O-3SiO_2$

Figure 4 shows the differential curves produced by subtracting the PFD curve for pure SiO_2 from the PFD curves of $K_2O-3SiO_2$ after the various leaching treatments shown in Table 1. The peaks in the PFD differential curves were interpreted in a manner similar to that proposed by [5,6] and the results are given in Table 3.

SUMMARY AND CONCLUSIONS

The following conclusions can be drawn from the results of the present investigation:

1. Diffracted x-ray intensity data for bulk and powder samples of vitreous silica appeared similar. The only difference was a slight loss in intensity in the main maximum for the pressed powder sample. The PFD curves for the two samples appeared similar and unchanged. When studying the structure of glass with the

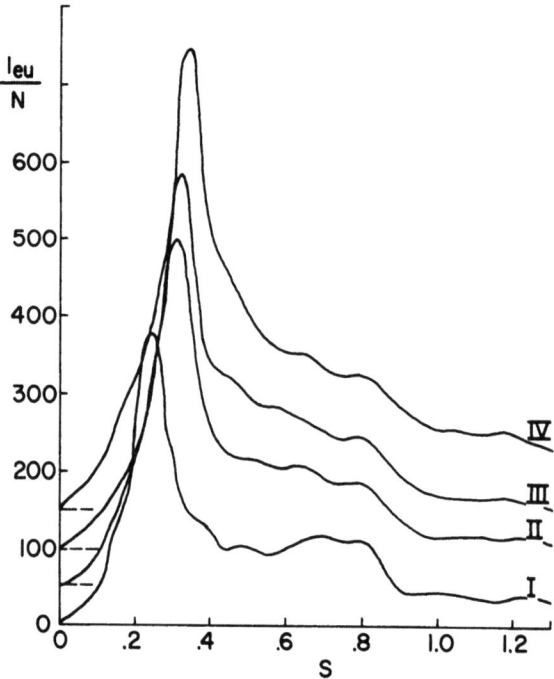

Figure 1. A comparison of the normalized X-ray data for K_2O-$2SiO_2$ (IV), K_2O-$3SiO_2$ (III), K_2O-$4SiO_2$ (II), and SiO_2 (I).

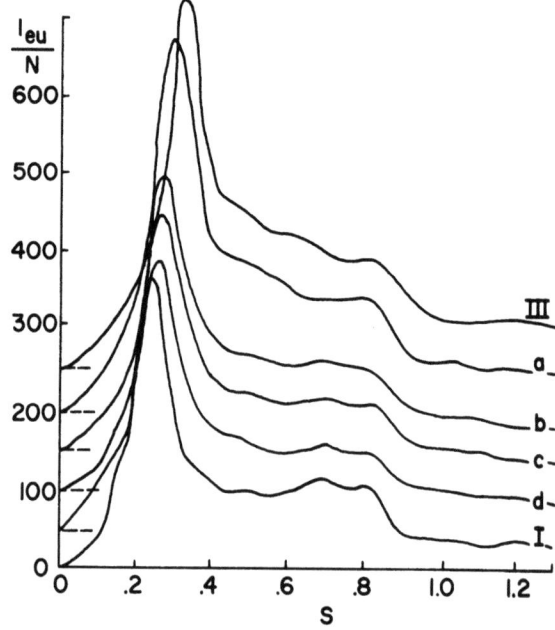

Figure 2. A comparison of the normalized X-ray intensity data for K_2O-$3SiO_2$ and the same compound leached as described in Table 1. The bottom curve labeled I is for SiO_2. The curve labeled (a) is the first leaching.

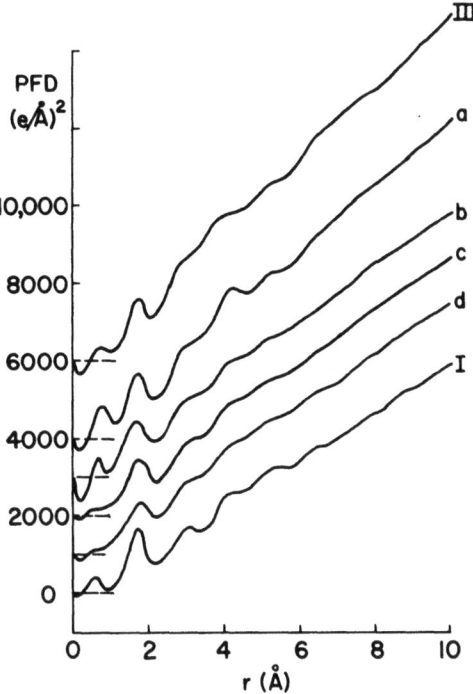

Figure 3. Experimental PFD curves for the compound K_2O-$3SiO_2$ and the same compound leached as described in Table 1. The bottom curve labeled I is for SiO_2. The curve labeled (a) is the first leaching.

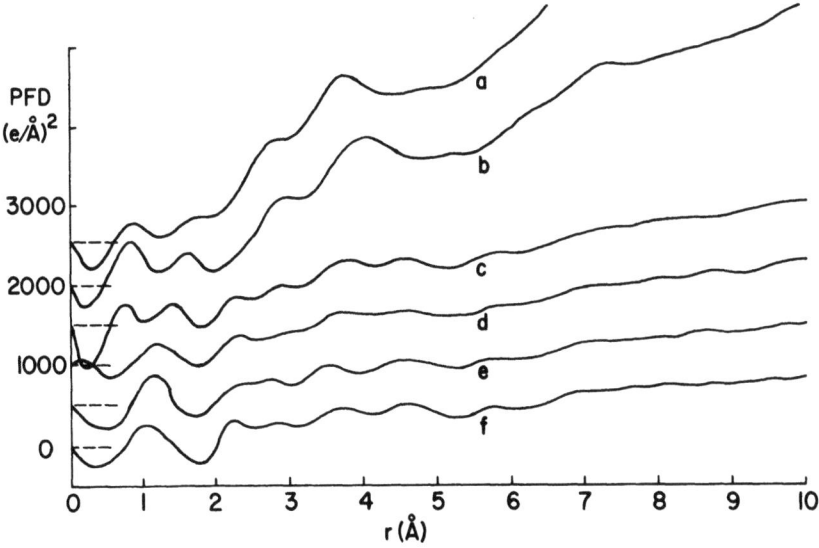

Figure 4. The application of the differential method to the PFD curves for the compound K_2O-$3SiO_2$ before and after leaching. Curve (a) represents the non-leached sample, and curves (b) through (f) represent the leaching as described in Table 1 in the text.

Table 2

Interatomic Distances in Potassium-Silicate, Leached
Potassium-Silicate, and Silica Glass
(Å)

	$Si-1^{st}O$	$Si-1^{st}Si$	$Si-2^{nd}O$	$Si-2^{nd}Si$ $O-2^{nd}O$	$Si-3^{rd}O$
SiO_2 (a)	1.62	3.12	4.15	5.10	6.40
SiO_2 (b)	1.64	3.10	4.20	5.15	6.45
$K_2O-4SiO_2$	1.67	(3.07)	4.10	(5.15)	----
Leach # 1	1.65	3.05	4.15	5.15	----
2	1.67	3.10	4.15	5.20	----
3	1.65	3.10	4.15	5.15	6.50
4	1.65	3.10	4.15	5.15	6.40
5	1.65	3.10	4.15	5.15	6.40
$K_2O-3SiO_2$	1.70	(2.95)	4.10	----	----
Leach # 1	1.70	3.10	4.15	5.35	----
2	1.66	3.10	4.20	5.20	----
3	1.65	3.10	4.20	5.20	----
4	1.65	3.10	4.15	5.15	6.45
5	1.65	3.10	4.15	5.15	6.45
$K_2O-2SiO_2$	1.75	(3.00)	4.10	----	----
Leach # 1	1.70	3.10	4.10	5.15	----
2	1.65	3.10	4.15	5.20	----
3	1.65	3.10	4.15	5.15	6.45
4	1.65	3.10	4.15	5.15	6.45

(a) Mozzi and Warren (1969)
(b) Present Investigation

Table 3

Interatomic Distances in Potassium-Silicate and
Leached Potassium-Silicate Glass Obtained by
the Differential Method
(Å)

Peak	# 1	# 2	# 3	# 4	# 5	# 6
K_2O-$4SiO_2$	2.90	3.75	4.85	----	(7.00)	8.00
Leach # 1	2.90	3.75	4.55	----	7.20	----
2	2.85	3.65	4.50	5.80	----	----
3	2.90	3.65	4.45	----	----	(8.00)
4	2.95	3.65	4.50	(5.80)	7.10	8.00
5	2.90	3.55	4.55	5.80	(7.00)	8.00
K_2O-$3SiO_2$	2.80	3.65	----	----	(7.00)	----
Leach # 1	2.90	3.75	(4.55)	----	(7.10)	----
2	2.90	3.70	4.55	----	(7.00)	8.00
3	2.85	3.70	4.55	----	(7.00)	8.00
4	2.85	3.60	4.50	5.85	7.10	8.00
5	2.80	3.60	4.55	5.75	(7.00)	8.00
K_2O-$2SiO_2$	(2.90)	3.65	----	----	(7.00)	----
Leach # 1	2.85	3.60	4.75	5.80	----	8.00
2	2.85	3.60	4.50	5.75	----	7.95
3	2.85	3.60	4.50	5.75	(7.10)	8.00
4	2.85	3.60	4.50	5.75	6.95	7.95

aid of x-rays, the pressed powder samples appear to be satisfactorily representative of the structure found in bulk glass.

2. A peak shift in the main maximum from the x-ray data to a higher s value occurred when vitreous silica was alloyed with potassium oxide. In addition to the peak shift, the other maxima became less distinct with increasing potassium content. Upon removing the potassium from the glass system by a number of leaching experiments, the peak position of the main maximum shifted to lower s values. The main maximum for all systems ceased to shift at an s value of 0.263 (vitreous silicas' main maximum occurs at

0.243). This is indicative of residual potassium remaining in the glass system. X-ray spectrographic analysis of the system confirmed this result. However, spectrograph analysis indicated the amount of potassium oxide in the system to be about half that given by the main maximum shifting technique.

3. In order to insure that the powders of the various systems were completely leached, the powders were placed in a neutral solution of distilled water. After 48 hours no potassium was detected in the solution, thus the x-ray diffraction patterns are caused by homogeneously leached potassium silicate glass and not heterogeneous particles. Note that the powder size varied from ten microns to 100 microns, with the majority of the particles being less than 50 microns (as observed in the scanning electron microscope).

4. The interatomic distances for vitreous silica obtained in the present investigation are in fair agreement with the distances obtained by Warren and Mozzi (1970). In no instance was there a variance of more than 0.05Å.

5. The interatomic distances for silica in $K_2O-4SiO_2$, $K_2O-3SiO_2$ and $K_2O-2SiO_2$ glasses varied slightly. The greatest change of 0.1Å occurred in the Si-1stO distance for $K_2O-2SiO_2$. The other distances remained almost constant. The potassium had the effect of broadening and decreasing the intensity of the peaks in the distribution curve. The distribution curve approached that of the vitreous silica when the potassium was being removed from the glass systems.

6. Application of the differential method to the potassium silicate glass systems yielded distinct peaks representative of potassium related interatomic distances. The peaks increased in sharpness with increasing potassium content. The first three peaks occurred at 2.85, 3.75, and 4.80Å. Hartlief (1938) reported distances of 2.33, 3.50, and 4.70Å, corresponding to the interatomic distances for K-O, K-Si, and K-K, respectively. When the potassium was removed from the glass systems, the distances obtained by the differential method changed slightly. The distances of the first three peaks decreased with decreasing potassium content. In the figures for the differential method applied to the leached systems, it was observed that the formation of a peak at 2.30Å occurred with decreasing potassium content. The same distances (2.30, 2.80, 3.70, 4.55) appeared in the differential curve for the final leached systems. Only the magnitude of the peaks changed significantly. This indicates that some residual potassium remains in the final leached glass.

7. The final leached glass structure resembles that of vitreous silica with some residual potassium in the system. This residual potassium has very little effect on the silica tetrahedra, but

slightly broadens its nearest neighbor peaks in the distribution curves.

ACKNOWLEDGEMENTS

The authors would like to acknowledge the financial assistance of the Office of Naval Research, ONR Contract N00014-68-A-0173-0006.

REFERENCES

1. F. Zernike and J. A. Prins, "Die Beugung von Röntgenstrahlen in Flüssigkeiten als Effekt der Molenkülanordnung," in Zeitschrift für Physik, 41, 184 (1927).

2. P. Debye and H. Menke, "Bestimmung der inneren Struktur von Flüssigkeiten mit Röntgenstrahlen," in Physikalische Zeitshrift, 31, 797 (1930).

3. B. E. Warren, H. Krutter and O. Morningstar, "Fourier Analysis of X-ray Patterns of Vitreous SiO_2 and B_2O_3," in Journal of American Ceramic Society, 19, 202 (1936).

4. R. L. Mozzi and B. E. Warren, "The Structure of Vitreous Silica," in Journal of Applied Crystallography, 2, 164 (1969).

5. N. Valenkov and E. Porai-Koshits, "X-ray Investigation of the Glassy State," in Zeitschrift für Kristallographie, 95, 195 (1936).

6. G. Hartleif, "Beiträge zur Struktur des Kieselglases und der Kalisilikatgläser," in Zeitschrift für Anorganische Chemie, 238, 353 (1938).

7 B. E. Warren, "Exact Method for an Amorphous Sample With More Than One Kind of Atom," in X-ray Diffraction, p 135-42, Addison-Wesley Publishing Co., Reading, Mass. (1969).

8. B. E. Warren, "Amorphous Scattering," Lectures at the University of Florida (Spring 1969).

9. M. S. Hill, "An X-ray Investigation of Interatomic Distances in Potassium Silicate Glasses," M.S. Thesis, University of Florida (1972).

10. D. B. Nash, "New Technique for Quantitative SiO_2 Determinations of Silicate Materials by X-ray Diffraction Analysis of Glass," in Advances in X-ray Analysis, 7, 209 (1964).

11. C. V. Gokularathnam, "X-ray Diffraction Analysis of Structural Changes in Vitreous Silica," Ph.D. Dissertation, University of Florida (1971).

12. R. Kaplow, S. L. Strong and B. L. Averbach, "Radial Density Functions for Liquid Mercury and Lead," in Physical Review, <u>138</u>, A1336 (May 31, 1965).

A REVIEW OF X-RAY DIFFRACTION METHODS

FOR DIFFUSION STUDIES

        J. A. Carpenter, Jr.* and D. R. Tenney**

   * Physics Research
     Chrysler Corporation
     P.O. Box 1118 (CIMS 418-38-04)
     Detroit, Michigan 48231

  ** Metallurgical Engineering
     Virginia Polytechnic Institute and
        State University
     Blacksburg, Virginia 24061

ABSTRACT

X-ray diffraction methods for studying solid-state diffusion are reviewed. Because of the lack of penetration, such methods are suited for diffusion zones spanning only a few microns. Most involve analyses of one, or more, (hkl) "band" of intensities spread in 2θ as a result of the corresponding lattice parameter spread associated with compositional inhomogeneities. Further, many are non-destructive, making it possible to follow the progression of diffusion with time in the same specimen.

INTRODUCTION

X-rays have found use in the field of solid-state diffusion since the beginning of the modern era (circa 1930, e.g. 1-6) of such studies. The electron beam microprobe is probably the most recognized wherein spatial variations are determined by fluorescence. But, techniques based on diffraction, though they offer many advantages, are not as well known (perhaps because the concepts and techniques of the two fields, x-ray diffraction and diffusion, tend to be so specialized that it is rare to find workers well versed in both). The purpose of this paper is,

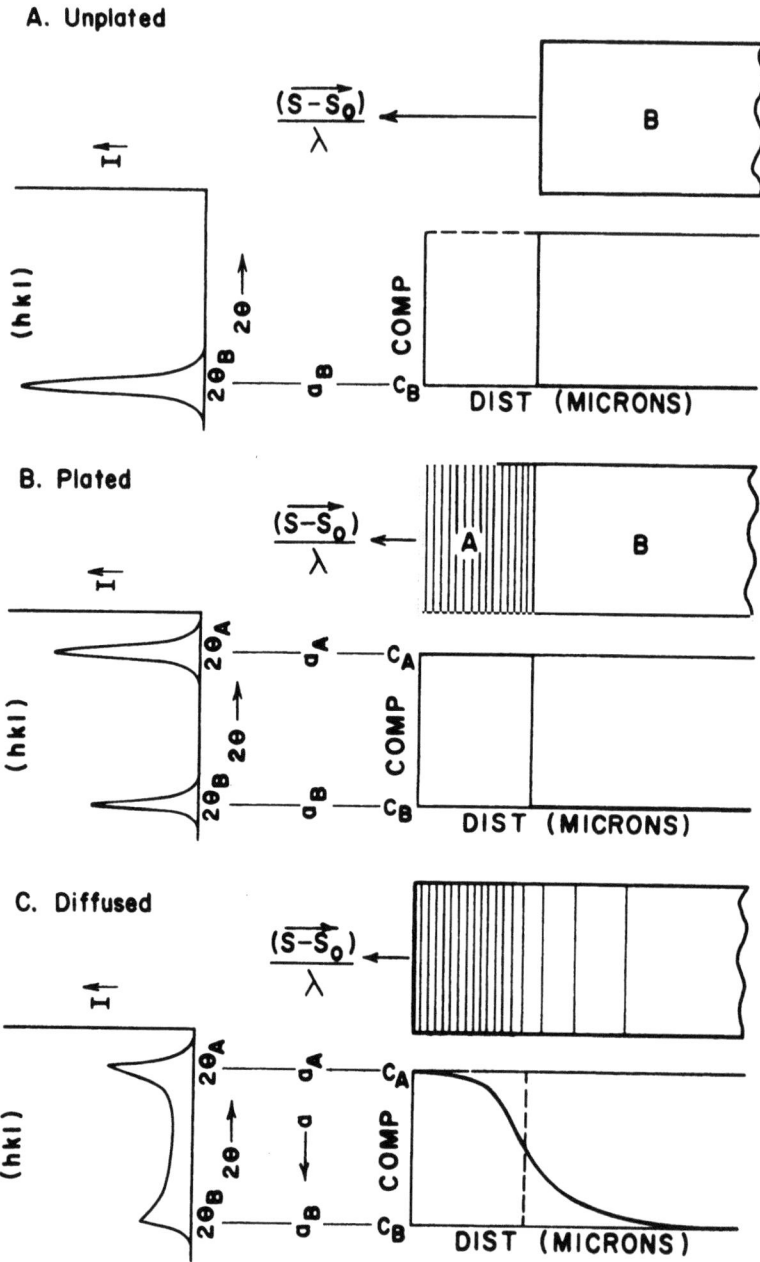

FIGURE 1. Three stages in an x-ray diffraction investigation of solid state diffusion. See text.

therefore, to review such diffraction methods for studying solid-state diffusion in the hope of fostering a better mutual understanding of the two disciplines.

The bases of the phenomena to be discussed may be illustrated in Figure 1. Three states of an idealized diffusion specimen are depicted in the upper rights of Figures 1(a), (b) and (c); the corresponding composition (COMP) versus distance (DIST) curves and the intensity (I) versus diffraction angle (2θ) curves are shown at the lower rights and lefts, respectively. For the sake of demonstration, it is assumed that the specimen face is scanned with a beam of monochromatic x-rays in such a way that the diffraction vector $(\bar{S}-\bar{S}_o)/\lambda$ is always perpendicular to the face. $\bar{S}$ and $\bar{S}_o$ denote the direction vectors of the diffracted and incident beams, respectively, and λ is the wavelength of the radiation.

In Figure 1(a) the bare substrate, composed entirely (for simplicity) of B atoms, produces a peak at diffraction angle, $2\theta_B$, as dictated by λ, the Miller indices, (hkl), of the reflection and the lattice parameter, a_B. In Figure 1(b) the B substrate has been plated with an overlay composed of A atoms; a new peak appears at $2\theta_A$ commensurate with the lattice parameter, a_A. The intensity of the B peak is reduced as both the incident and diffracted beams must pass through the A overlay in going to and from the B substrate. In Figure 1(c), the diffused specimen is shown. The two separate peaks of Figure 1(b) are replaced by a continuous "band" of intensity spread between $2\theta_A$ and $2\theta_B$ as a result of the corresponding spread in lattice parameter which now exists because of diffusion.

The admittedly unusual arrangement of the diffraction profiles with respect to the corresponding composition profiles in Figure 1, is designed to stress a most important basic concept, namely, that the diffraction angle, 2θ, is the measure of alloy composition, c, through the lattice parameter, a. The intensity at the 2θ, relative to the intensities at all other angles in the band, is then the measure of the location of that composition in the diffusion zone. Note, too, the units of "microns" on the abscissae of the composition profiles. Because of absorption, only compositions located no deeper than about $2/\mu$ below the surface of the specimen (where μ is the linear absorption coefficient for the radiation being employed) can be detected. Therefore, as a general rule of thumb, diffusion zones on the order of from one to twenty microns are the province of x-ray diffraction methods.

EARLY APPLICATIONS

Between about 1930 and 1950 x-ray diffraction methods were used by various workers (1-12), some now recognized among the pioneers of the field of solid-state diffusion, for example, Jost (2-3), Matano (4-6) and Kirkendall (8). Many of these early applications have been reviewed elsewhere (e.g. 13, 14), and hence are only mentioned here. (However, the interested reader may find some of these papers of value particularly with regard to the chronology of the early progress in the field of diffusion.) For the most part, these early methods simply involved the more obvious use of diffraction as a means of identifying compositions alone. Here attention will concentrate on more modern methods whereby both diffusion zone compositions and the corresponding locations are determined.

RUDMAN

In 1960 P. S. Rudman (15) introduced a technique based on the realization that, absorption effects duly accounted for, the intensity, $I(\theta_c)$, diffracted by any composition, c, at angle, θ_c, is proportional to the number of unit cells, $N(c)$, having that composition. The following expression may be written:

$$N(c) = K \cdot Q(\theta_c) \cdot I(\theta_c), \tag{1}$$

where K is a constant and $Q(\theta_c)$ is a compilation of calculable terms dependent upon θ_c. An observed intensity profile, $I(\theta)$, as shown at the top of Figure 2 was converted to a plot (center) of $N(c)$ versus c. Defining an "x-ray equivalent penetration distance" as

$$x(c) = \int_0^c N(c) \, dc \, / \, \int_0^1 N(c) \, dc, \tag{2}$$

a plot (bottom) of c versus $x(c)$ was determined from the $N(c)$ versus c curve.

The parameter $x(c)$ is not a true distance and so the bottom curve of Figure 2 is not a true composition profile. However, it was used to determine the "degree of interdiffusion," defined as the ratio of the net number of atoms having crossed the Matano interface at some time to the total number that will have crossed when complete homogeneity of a sample is achieved. The number, M_t, having crossed at any time, t, may be calculated from the bottom curve of Figure 2 as

$$M_t = \int_{c(x')}^1 x \, dc = \int_{c(x')}^0 x \, dc \tag{3}$$

where $c(x')$ designates the composition at the Matano interface.

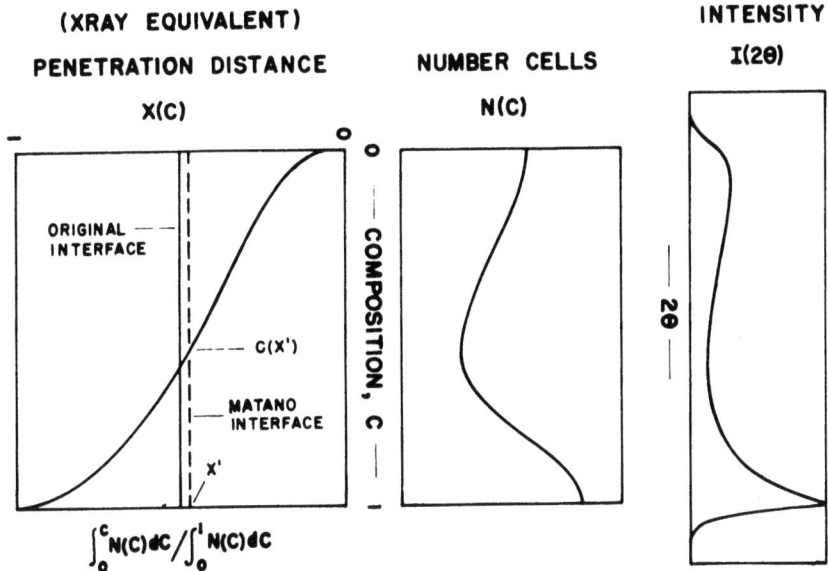

FIGURE 2. Rudman method. Diffraction profile (top) is used to calculate number of cells versus composition (center) which in turn determines the variation of composition with "x-ray equivalent penetration distance" (bottom).

All of the work by Rudman (15) and Fisher and Rudman (16) was concerned with diffusional homogenization of Cu-Ni powder compacts. Measurements were carried out by means of a diffractometer and the resulting profiles were separated into $K\alpha_1$ and $K\alpha_2$ components prior to analysis. Samples of the same particle sizes were annealed for different times in the temperature range from 750 to 1050°C, analyzed, and the "degree of interdiffusion" determined. From these data it was possible to extract a value of the activation energy for the dominant diffusion mechanism. It was concluded that from 750 to 950°C, the predominant mechanism was grain boundary diffusion whereas above it was volume diffusion.

A more recent application of Rudman's method to diffusion in the Cu-Ni compacts was made by Heckel (17) who also used an electron microprobe. The results of Rudman's x-ray and the microprobe methods were in good agreement. Contrary to Rudman's findings, though, Heckel concluded that the rate-controlling step in the homogenization of Cu-Ni powders in the same range of temperatures was volume and not grain-boundary diffusion.

Still another application was made by Tronsdal and Sorum (18, 19) who used a significantly different experimental approach. Instead of a powder specimen, a thin foil was prepared by electrodepositing an unspecified thickness of one component onto a ten micron thick foil of the other component. A special two-crystal diffractometer was used whereby it was possible to superimpose the $K\alpha_1$ and $K\alpha_2$ components rather than separating them as Rudman did. Because of the thin-foil type specimen, it was also possible to carry out selective-absorption transmission radiographic studies thereby obtaining pictorial information concerning the distribution of the diffusion components at various points in the grain structure of the specimen.

Besides the fact that the plot of c versus "x-ray equivalent penetration distance" is not a true diffusion profile, the factor limiting the general application of the Rudman method is absorption. The method is most valid where all diffracting regions have nearly the same linear absorption coefficients or if the regions are small compared to the depth of penetration of the x-ray beam.

PINES

Several x-ray diffraction methods for studying diffused specimens were developed by various Russian authors, the more general being those developed under B. Ya. Pines (20-26) between 1956 and 1962. The methods of Levitskaya and Vodop'yanova (27) and Fogel'son (28), though they appeared subsequent to Pines, are more limited in scope.

The type specimen used by Pines was a flat circular disc of metal, 7-10 mm. in diameter and 2-4 mm. thick, onto which a 1 to 15 micron layer of another metal was either electroplated or simply impressed. The specimen was mounted on the outer rim of and with its face perpendicular to a diameter of, a 72 mm. cylindrical camera specially constructed so that the fine focus of an x-ray tube was located also at the rim of the camera but diametrically across from the specimen. During exposure the specimen was rotated about that diameter. The film resulting from each exposure was converted into intensity versus 2θ measurements by means of a microphotometer. In all cases only one reflection was analyzed. No attempt was made to separate the band into its spectral components but a partial

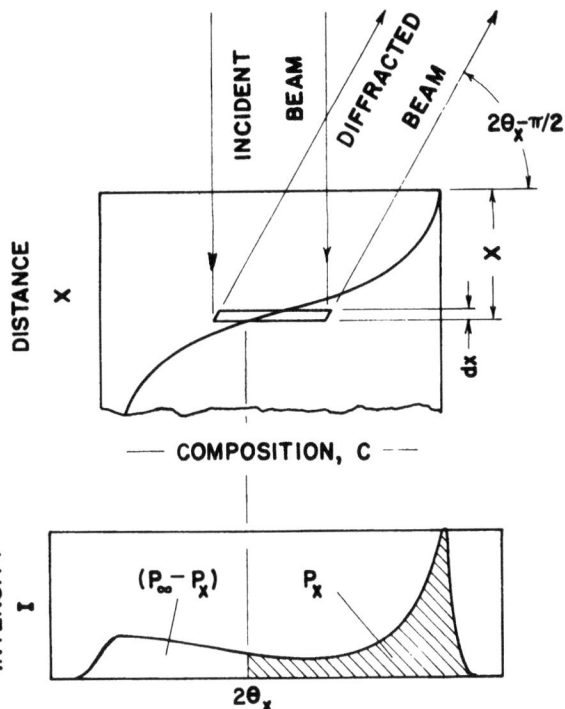

FIGURE 3. Pines method. Experimental geometry (above) and areas of intensity profile analyzed (below). See text.

correction for instrumental broadening was applied at the ends of each band (20).

Pines' analyses began with the following expression:

$$dP = \eta \cdot I_o \cdot q(2\theta) \cdot \exp[-k(2\theta)\bar{\mu}x] \, dx \qquad (4)$$

where dP is the integrated intensity diffracted at 2θ by the small differential volume of thickness, dx, located a distance, x, below the specimen surface. The average absorption coefficient of all the material above x is denoted as $\bar{\mu}$; $k(2\theta) = 1 + (1/\cos 2\theta)$; $q(2\theta)$, a compilation of other angular dependent terms and I_o and η, the incident intensity and scaling factor, respectively.

Invariably, I_o and η were removed by making certain assumptions as noted below.

In Pines' first analysis assumptions were made that q, $\bar{\mu}$ and k were constant throughout the specimen. Integration from $x = 0$ to $x = \infty$ yielded an expression for P_∞, the total integrated intensity diffracted by the diffused specimen, namely

$$P_\infty = \eta \cdot I_o \cdot q \cdot [\bar{\mu} k]^{-1} \qquad (5)$$

Similar integration from $x = 0$ to $x = x$ under the same assumptions yielded P_x, the integrated intensity from all the material above x, namely,

$$P_x = \eta \cdot I_o \cdot q \cdot [\bar{\mu} k]^{-1} \cdot [1 - \exp(-k\bar{\mu} x)]. \qquad (6)$$

P_x was taken proportional to the area under the profile lying between the end of the band corresponding to the surface composition and $2\theta_x$, the position corresponding to the composition at x, while P_∞ was taken proportional to the total area under the entire x-ray profile as indicated in Figure 3.

The problem, of course, was to determine the location, x, corresponding to the composition, c. Dividing equation 6 by equation 5, yielded a simple expression for x

$$x(c) = -\ln[1 - P_x/P_\infty] \, [\bar{\mu} k]^{-1} \qquad (7)$$

independent of the parameters η, I_o and q. P was measured for each selected composition and, using appropriate estimates for $(\bar{\mu} k)$, the location determined. Strictly, the $(\bar{\mu} k)$ used should have been the same for all compositions but in practice it was different for each, apparently being taken as the average between $(\bar{\mu} k)$ for the surface composition and that for the composition being considered.

The assumptions that q, $\bar{\mu}$ and k were constant are reasonable for many alloy systems but break down in others, resulting in a lack of general applicability of this first method. The limitations imposed by the assumptions were recognized by Pines and his co-workers with the result being the appearance in 1958 of a more refined theory (22) to account for couples of elements of widely different absorption and diffraction properties.

A more detailed form of equation 4 was used, namely,

$$dP = \eta \cdot I_o \cdot q'(2\theta) \cdot f^2(2\theta) \cdot \exp\left[-\int_0^x k(x)\mu(x)dx\right] dx \qquad (8)$$

with

$$q'(2\theta) = \Omega(2\theta) \cdot N^2(2\theta) \cdot (pS^2)_{hkl}. \qquad (9)$$

$\Omega(2\theta)$ represents a compilation of "a polarization factor and functions of θ" (reference 22, page 620); $N(2\theta) = 1/V_o$, the number of units cells per unit volume; p is the recurrence or multiplicity factor for the (hkl) diffraction planes being considered; and S^2 is the constant portion of the structure factor, $S^2 f^2(2\theta)$ where $f^2(2\theta)$ is the square of the atomic scattering factor. It will be noted that the term $q'(2\theta)$ here is identically $q(2\theta)$ of equation 4. Also, in equation 8, $\mu(x)$ is the linear absorption coefficient of the material at, not above, the location, x.

Equation 8 was first solved for the case of massive, or infinitely thick, specimens of pure A or pure B,

$$P_{A(B)} = \eta \cdot I_o \cdot q'_{A(B)} \cdot [\mu k]^{-1}_{A(B)} \cdot f^2_{A(B)}. \qquad (10)$$

$P_{A(B)}$ denotes the experimentally measured integrated intensities from either pure A or B in the (hkl) reflection. It was assumed that the atomic scattering factor for an A-B alloy of composition, c, can be written in terms of the scattering factors of the pure elements, namely,

$$f^2(2\theta) = f^2_A + c(f^2_B - f^2_A). \qquad (11)$$

Solving equations 10 for both f^2_A and f^2_B and using them in equation 11 yielded a new form for equation 8

$$dP(c) = \emptyset(c) \exp[-\int_0^x k(x)\mu(x)dx] \, dx \qquad (12)$$

where

$$\emptyset(c) = \Omega(c) \cdot N^2(c) \cdot (pS^2)_{hkl} \cdot \left\{\left(\frac{P\mu k}{\Omega N^2 pS^2}\right)_A \right.$$

$$\left. \left[\left(\frac{P\mu k}{\Omega N^2 pS^2}\right)_B - \left(\frac{P\mu k}{\Omega N^2 pS^2}\right)_A\right]\right\}. \qquad (13)$$

Note the loss again of the unknown parameters η and I_o but note, too, that this loss is accomplished at a price, namely, the necessity of making measurements on standards of pure A and pure B in addition to those on the sample itself.

Dividing both sides of equation 12 by $\emptyset(c)$ and integration from the surface location, $x = o$, where $c = c_o$, to a given location, x, where the concentration is c, yielded

$$\int_{c_o}^{c} \emptyset^{-1}(c) \, dP(c) = \int_{o}^{x} \exp[-\int_{o}^{x} k(x)\mu(x)dx] \, dx \qquad (14)$$

The left side of this equation can be determined experimentally. Inspection of $\emptyset(c)$, as defined by equation 13, shows that all parameters can either be determined experimentally or calculated knowing only the composition, c, and, of course, the proper values of p and S. Thus, the experimental I (intensity) versus $2\theta(c)$ diffraction profile can be converted to a curve $I(c)/\emptyset(c)$. The area under such a curve from the upper limit of the profile corresponding to the surface composition to the location $2\theta_x$ is the integral of the left-hand side of equation 14 for the given composition, c.

The right-hand side of equation 14 is not known, though. Remembering that the purpose is to identify the value x where the composition is c, it is, therefore, required that an integration limit be found. This cannot be done in general without knowing the form of the variation of μ with x. It is assumed that $\mu = f(c)$ is known, hence, to solve for x explicitly would require that $c = f(x)$ be known which is, of course, exactly what is to be determined.

A method of successive approximations was used to find x for each c. As a first approximation, $\bar{\mu}k$ was assumed constant, yielding an explicit formula for x

$$x(c) = -\ln \left[1 - \bar{\mu}k \int_{c_o}^{c} \emptyset^{-1}(c) \, dP(c) \right] [\bar{\mu}k]^{-1} \qquad (15)$$

from which a first approximate curve of c versus x could be constructed. Using this, the right side of equation 14 was evaluated directly as a function of x and new pairs of c and x found such that the right side of equation 14 equaled the left side, thus yielding a second approximation to the true c versus x curve. This second approximate curve was used to evaluate the right side of equation 14 again yielding a third approximation and so on through successive approximations until the true c versus x curve was identified, presumably by observation of no change in successive curves.

It was claimed that by evaluating the integral in the log term of equation 15 numerically, the first approximation could be made close to the true curve. By imagining the specimen broken up into tiny discrete blocks of uniform composition and evaluating

$$\bar{\mu}k \int_{c_o}^{c} \emptyset^{-1}(c) \, dP(c)$$

as

$$\sum_i \bar{\mu}k(c_i) \int_{c_{i-1}}^{c_i} \phi^{-1}(c)\, dP(c)$$

(where $\bar{\mu}k$ is taken as a constant only over the small block c_{i-1} to c_i, rather than over the entire range from c_o to c), the value of $x(c)$ calculated from equation 15 was found to be close to the true final value.

LEVITSKAYA AND VODOP'YANOVA

Levitskaya and Vodop'yanova (27) in 1962 and Fogel'son (28) in 1968 detailed methods for measuring a diffusion coefficient by analyzing only the extreme of the intensity profile corresponding to the compositions nearest the specimen surface. There was no attempt made in either case to delineate the entire composition curve. It was assumed *a priori* that diffusion in the specimen could be adequately described by a single diffusion coefficient independent of concentration. Obviously, these methods were more limited than Pines' but, on the other hand, these did attack perhaps the most troublesome region of an intensity profile from a diffusion specimen - the ends. In addition, these methods, especially Fogel'son's, represent first attempts at simulation, namely, trying to fit an observed experimental x-ray profile to a calculated diffusion curve, essentially the reverse of Pines' or Houska's (see below) methods.

An x-ray diffraction profile from the end regions of a diffusion specimen may be distorted as a result of instrumental broadening, as shown in Figure 4. Instrumental broadening causes the predominating intensity from the surface concentration, c_o, to overlap into neighboring regions. The result of this broadening is, first, a lowering of the intensity, $I(c_o)$, at $2\theta(c_o)$; second, raising of intensities in regions closer in toward the center of the band; and, third, perhaps a shift in the location of the point of maximum intensity from $2\theta(c_o)$ to $2\theta_{max}$, where the intensity is I_{max}. Of course, all regions are affected by instrumental broadening but it is found that regions near the center are little affected. In the center the intensity of one region with respect to another is about the same so that whatever intensity one region loses by broadening is compensated for by its neighbors.

The analysis by Levitskaya and Vodop'yanova (27) was limited to the very earliest stages of diffusion in a plating of original thickness, L_o, when solute material had not yet penetrated very far toward the outer surface. A certain composition, c', located in the diffusion zone a distance, L, from the outer surface or a

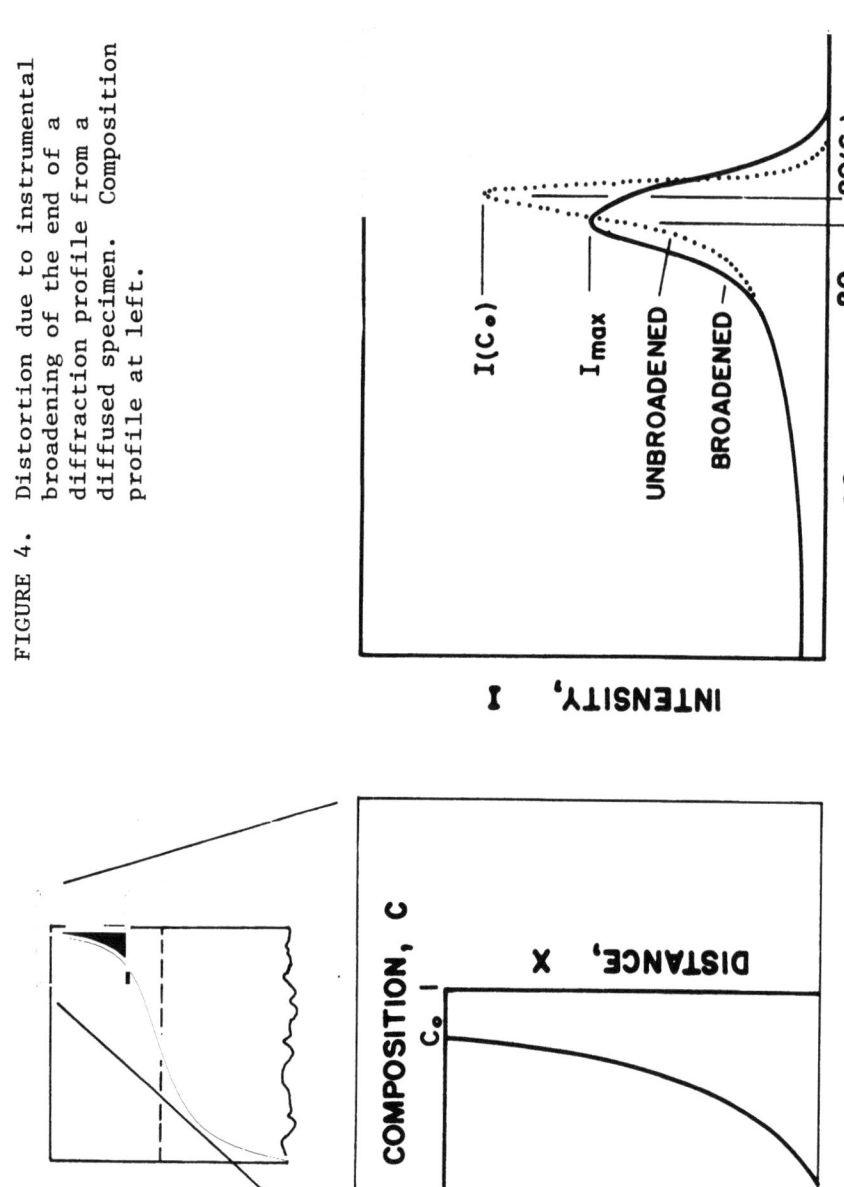

FIGURE 4. Distortion due to instrumental broadening of the end of a diffraction profile from a diffused specimen. Composition profile at left.

distance, x', from the original interface, was defined as that having an angular position on the x-ray profile displaced by one full-width at half-maximum from the position commensurate with the initial composition of the plating. It was assumed that compositions angularly displaced farther than a half-width produce little or no distortion of the big peak due to whatever pure material remained. Comparing the intensity, P_L, contained between the end of the band and the position associated with c', with the original intensity, P_{Lo}, from the undiffused plating and knowing L_o, x' could be found from

$$x' = L_o + [\mu k]^{-1} \ln[1-(P_L/P_{Lo})\{1 - \exp(-\mu k L_o)\}]. \tag{16}$$

This could then be used in the well-known solution of Fick's Second Law for diffusion in a doubly-infinite system, namely

$$c'/c_o = 1/2 \{1 - \text{erf}[x'/(2\sqrt{Dt})]\} \tag{17}$$

from which D could be extracted, all other quantities being known.

FOGEL'SON

Fogel'son (28) treated the case of diffusion in a doubly-finite system. The sample was seen as originally two slabs, the thickness of the solute-rich slab at the surface being L_o, the thickness of the other being L_1. The solution for diffusion in such a doubly-finite system may be written as

$$c = (c_o/2) \sum_{n=-\infty}^{n=+\infty} \text{erf}\,[\{2n - (x-L_o)/(L_o+L_1)\}\lambda\,] \tag{18}$$

$$- \text{erf}\,[\{2n - (x+L_1)/(L_o+L_1)\}\lambda\,]$$

where $\lambda = (L_o+L_1)/(2\sqrt{Dt})$. This expression was used to calculate a c versus x diffusion profile for each of several different values of Dt which was then imagined divided into a number of small equal composition intervals. Each small interval produced an integrated intensity which spread in 2θ according to a Cauchy distribution with a half-width, σ. This half-width was taken as that for the distribution of the intensity from the solute-rich slab prior to diffusion, the maximum of that distribution being I^o. At any given 2θ position, then, the instantaneous intensity, $I(2\theta)$, was the sum of contributions from all slabs in the diffusion zone, namely,

$$I(2\theta)/I° = [1 - \exp(-\mu k L_o)]^{-1} \qquad (19)$$

$$\sum_n \frac{\exp(-\mu k x_{n-1}) - \exp(-\mu k\, x_n)}{1 + 2\sigma^{-1}(2\theta - 2\theta_n)^2},$$

2θ being the position associated with the n-th interval in the zone. A curve of $I(2\theta)/I°$ was constructed versus 2θ, the maximum at the solute-rich end being denoted as $I_{max}/I°$. Values at $I_{max}/I°$ were tabulated as functions of $\sqrt{Dt}$. Having found $I_{max}/I°$ experimentally and knowing t, it was a simple matter to find the proper D.

HOUSKA

The most recent (and perhaps most extensive) work has been carried out by C. R. Houska and co-workers (29-35). The samples were usually flat, copper discs, three-fourths inch in diameter and from one- to three-eighths inch thick onto which one to fifteen microns of another metal was electroplated. The composites were annealed in vacuum at a temperature within the volume diffusion range so that compositional variations occurred only perpendicular to the original interface and the specimen surface. X-ray diffraction measurements were carried out by means of diffractometers on at least two reflections. For the purpose of analysis, the diffused specimen was imagined broken into a series of small volume elements each spanning a small composition range. The location of any given element was determined by comparing the intensities the element diffracted in the two (or more) reflections measured. The theoretical basis of this approach was discussed in detail by Houska (29).

In Figure 5, one such element is shown (exaggerated) spanning the composition range, c_1 to c_2, located between depths, x_1 and x_2, respectively, and having an average composition, $\bar{c}$, located at depth, $\bar{x}$. Also shown are the intensities, P^i and P^j, it diffracts in two reflections, i and j. The "effective volume", ΔV^g, of the element is

$$\Delta V^g = A_o \cdot g \cdot \sin^{-1}\theta \, |\Delta a (dx/da)| , \qquad (20)$$

where $A_o \sin^{-1}\theta$ is the specimen area exposed to a beam of cross-sectional area, A_o; Δa, the lattice-parameter spread associated with the composition interval; and (dx/da), the inverse change of the lattice parameter with distance across the interval. It will be recognized that $|\Delta a(dx/da)|$ is, to a good approximation, identical to the thickness $|x_1 - x_2|$ of the element. The factor g is the volume fraction of the material in the element properly oriented to diffract in the given reflection (hence ΔV^g must be called an "effective volume").

The intensity diffracted by such an element depends not only on its volume, as expressed by equation 20, but also by its location beneath the specimen surface. A central most premise of Houska's technique is that the location of the element can be determined by accounting for absorption in the material above. The proper absorption terms are

$$\exp[-k\int_0^{\bar{x}} \mu(x)dx = \exp(-k\bar{\mu}\bar{x}), \tag{21}$$

where

$$\bar{\mu} = (\bar{x})^{-1} \int_0^{\bar{x}} \mu(x) \, dx. \tag{22}$$

In the above, $\bar{\mu}$ is meant to denote an average value for all the material above $\bar{x}$ whereas $\mu(x)$ signifies the manner in which the linear absorption coefficient varies with distance.

Combining equations 20 and 21 according to standard x-ray theory, a general expression for the intensity diffracted by the element results:

$$P(\theta) = K_1 K_2(\theta) \cdot g \cdot |dx/da| \cdot \exp[-k\bar{\mu}\bar{x}], \tag{23}$$

where K_1 is a compilation of known constants and $K_2(\theta)$, a compilation of calculable θ-dependent terms. If $P(\theta)$ can accurately be measured in several reflections, $\bar{\mu}\bar{x}$ may be found as the slope of a plot of $\ln[P(\theta)/K_2(\theta)]$ versus k, the intercept of which is $\ln[K_1 g |dx/da|]$.

Alternately, if $P(\theta)$ is measured in only two reflections, $\bar{\mu}\bar{x}$ may be found according to the expression

$$\bar{\mu}\bar{x} = \ln[(P^i K_2^j k^i)/P^j K_2^i k^j)] \, (k^i - k^j)^{-1}. \tag{24}$$

These procedures may be applied to as many elements as desired except at the very ends of the diffusion zone where equation 23 no longer applies as the term $|dx/da|$ goes to infinity.

Knowing $\bar{\mu}\bar{x}$ for a series of chosen intervals of different average compositions, it remains to extract $\bar{x}$. Like Pines, a method of successive approximations is used. It is assumed at first that $\bar{\mu}$ can be calculated as

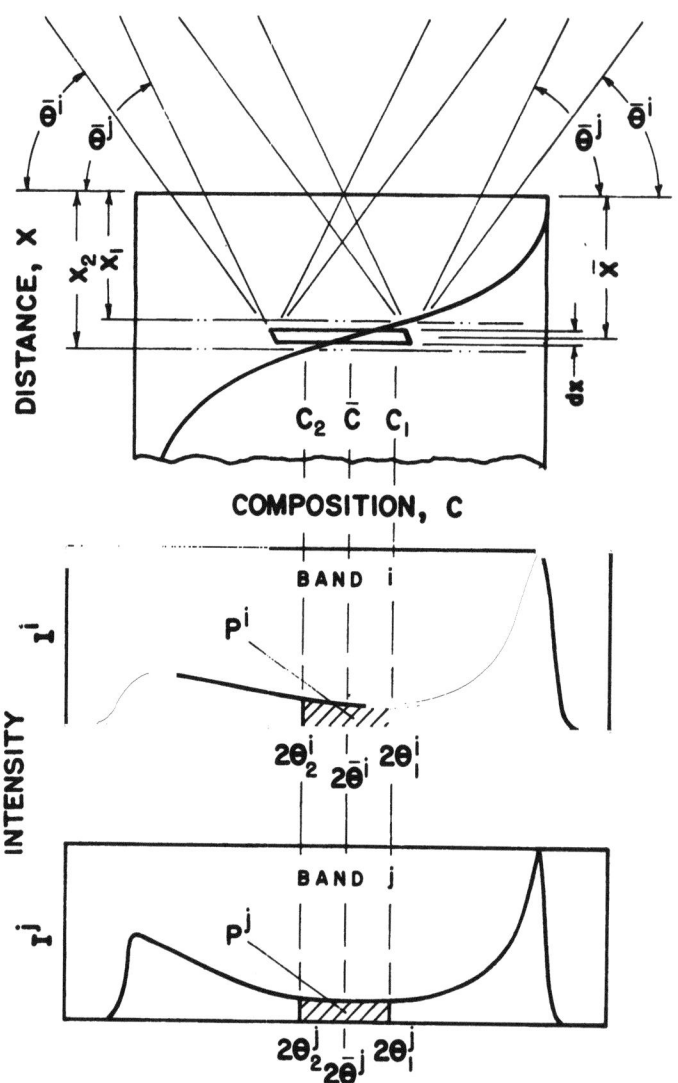

FIGURE 5. Houska method. The diffusion profile in the specimen (above) is determined by analyzing two or more diffraction profiles (center, below).

$$\bar{\mu} \simeq (\mu_o + \mu_x)/2, \qquad (25)$$

i.e. a simple average of μ_o, the linear absorption coefficient of the surface composition at $x=o$, and μ_x, that for the composition being considered at x. Using this, a first approximate composition profile is calculated from which $\mu(x)$ in equation 22 may be evaluated in turn yielding a better $\bar{\mu}$, and hence a better $\bar{x}$, for each composition. This process is repeated until successively generated profiles change little.

The initial work employing this approach was done by Bales (30) and Braski (31) who studied diffusion in samples of nickel plated on polycrystalline copper substrates or copper plated onto polycrystalline nickel substrates. The latter work was performed by Tenney, Carpenter and Talty (32-34) who plated nickel or palladium onto faces of single crystals of pure copper. In this latter work Tenney and Carpenter (32, 33) found that the nickel deposits could be plated with the same (111) epitaxial orientation as the copper substrate. Thus, all regions of the sample could be treated from a diffraction standpoint as a single crystal permitting investigations of the intensity distribution in the ω-direction resulting from non-uniform stress, small particle size or simple mosaic tilting. By making an ω-scan, or "rocking curve", after tuning in at a given 2θ position associated with a given composition, it was possible to assess the structure of the specimen at that point in the diffusion zone. By making rocking curves for other compositions, it was possible to determine the variation of diffusion-induced substructure simultaneously with the composition profile.

This technique is probably the most generally applicable method to appear to date. The primary advantage is that measurements may be made based solely on a single mounting of a specimen; standard samples are not required as in Pines' methods. In addition, this theory treats the θ and c dependencies of parameters such as atomic scattering factors and linear absorption coefficients in a more realistic way. The prime limitation of this, as well as Pines', approach is the basic assumption that diffusion occurs only in a direction perpendicular to the surface and original interface so that all planes parallel to the surface and interface are iso-concentration planes. Unnam, Carpenter and Houska (35) have recently attacked the problem of x-ray diffraction from specimens in which the diffusion is non-planar, though through methods based on simulation of, and not direct analyses of, experimental x-ray profiles.

SUMMARY

There have been numerous applications of x-ray diffraction methods to the study of solid-state diffusion. Of those methods which have appeared, three are of prime interest. The Rudman method has application to studies of diffusional homogenization of powder metal compacts. The Pines and Houska methods are more generally applicable to study of planar diffusion. Of these two, the latter is probably the most general.

Some advantages of x-ray methods for studying diffusion are:

1. They are non-destructive. A given sample may be repeatedly annealed at a given temperature and analyzed after each anneal in order to watch the progress of diffusion with time.

2. Accurate composition versus distance profiles can be mapped for diffusion zones of extremely small dimensions, for example, 1 micron or less, without making any prior assumptions about the constancy of the diffusion coefficient.

3. The amounts of materials required to produce a diffusion specimen are small. Thus, the diffusion of precious or rare metals may be studied at reasonable cost.

4. The times of diffusion are short. As the total diffusion zone is limited, accurate measurements may be made on samples diffused for minutes or hours rather than days or months.

5. The diffusion-induced substructure can be assessed in the case of single crystal specimens. Being non-destructive, these methods permit an _in situ_ measurement of the variation of the substructure with distance in the diffusion zone with no worry about distortion of the substructure resulting from any sectioning technique.

The major deterrent to the use of these methods seems to be a general lack of knowledge about their existence. It is hoped that this review can help to remedy this situation. Such methods may represent a significant new contribution to the study of diffusion and, as such, an entirely new realm for the application of x-ray analysis.

ACKNOWLEDGEMENTS

One of us, JAC, Jr., wishes to thank Chrysler Corporation for its encouragement and generous financial support of the publication of this paper. DRT acknowledges the support of the National Science Foundation during the period this paper was being written.

REFERENCES

1. C. F. Elam, "The Diffusion of Zinc in Copper Crystals," J. Inst. Met. 43, 217-235 (1930).

2. W. Jost, "Die Diffusionsgeschwindigkeit von Kupfer in Gold," Zeit. Phys. Chem. 16B, 123-128 (1932).

3. W. Jost, "Die Diffusionsgeschwindigkeit einiger Metalle in Gold und Silber," Zeit. Phys. Chem. 21B, 158-160 (1933).

4. C. Matano, "X-ray Studies on the Diffusion of Copper into Nickel," Memoirs of the College of Science of the Kyoto Imperial University 15, 351-353 (1932).

5. C. Matano, "Further X-ray Studies in the Diffusion of the Nickel-Copper System," Memoirs of the College of Science of Kyoto Imperial University 16, 249-259 (1933).

6. C. Matano, "X-ray Studies on the Diffusion of Metals in Copper," Japan J. of Phys. 9, 41-47 (1934).

7. L. C. Hicks, "An X-ray Study of the Diffusion of Chromium into Iron," Trans. AIME 113, 163-178 (1934).

8. E. Kirkendall, L. Thomassen and C. Upthegrove, "Rates of Diffusion of Copper and Zinc in Alpha Brass," Trans. AIME 133, 186-203 (1939).

9. V. G. Mooradian and J. T. Norton, "Influence of Lattice Distortion on Diffusion in Metals," Trans. AIME 117, 89-97 (1935).

10. A. H. Sully, "A Simple Method for the Study of Metallic Diffusion in Certain Binary Alloy Systems," J. Sci. Instr. 22, 244-245 (1945).

11. J. DuMond and J. P. Youtz, "An X-ray Method of Determining Rates of Diffusion in the Solid State," J. Appl. Phys. 11 357-365 (1940).

12. P. Duwez and C. B. Jordan, "Application of the Theory of Diffusion to the Formation of Alloys in Powder Metallurgy," Trans. ASM 41, 194-212 (1949).

13. W. Seith and T. Heumann, Diffusion of Metals: Exchange Reactions, AEC-tr-4506, USAEC translation of: Diffusion in Metallen: Platzwechselreaktion, Springer-Verlag (1955).

14. S. D. Gertsriken and I. Ya. Dekhtyar, Solid State Diffusion in Metals, AEC-tr-6313, USAEC translation of publication under same name by State Publishing House for Physical-Mathematical Literature (1962).

15. P. S. Rudman, "An X-ray Diffraction Method for the Determination of Composition Distribution in Inhomogeneous Binary Solid Solutions," Acta Cryst. 13, 905-909 (1960).

16. B. Fisher and P. S. Rudman, "X-ray Diffraction Study of Interdiffusion in Cu-Ni Powder Compacts," J. Appl. Phys. 32, 1604-1611 (1961).

17. R. W. Heckel, "An Analysis of Homogenization in Powder Compacts Using the Concentric-Sphere Diffusion Model," Trans. ASM 57, 443-463 (1964).

18. G. O. Tronsdal and H. Sorum, "An X-ray Diffraction Method for the Study of Interdiffusion in Metals," Physica Norvegica 1, 141-144 (1962).

19. G. O. Tronsdal and H. Sorum, "Interdiffusion in Cu-Ni, Co-Ni and Co-Cu," Phys. Stat. Solidi 4, 493-498 (1964).

20. B. Ya. Pines and E. F. Chaikovskii, "X-ray Diffraction Determination of the Heterodiffusion Coefficients in Alloys Forming Solid Substitution Solutions," Translation from Russian by Morris D. Friedman, Inc., New York, 7 pages Original Russian source: Dokl Akad. Nauk SSSR 111 1234-1237 (1956).

21. E. F. Chaikovskii, "Certain Data on the Coefficients of Heterodiffusion in Cu-Ni, Fe-Ni and Fe-Cr Alloys with Deformed and Undeformed Crystal Lattices," Proc. Acad. Sci. USSR-Phys. Chem. Section 112, 123-127 (1957).

22. B. Ya. Pines and I. V. Smushkov, "X-ray Determination of the Coefficients of Heterodiffusion in Alloys Whose Components Differ Considerably in X-ray Absorption," J. Tech. Phys. (USSR) 28, 619-625 (1958).

23. B. Ya. Pines and I. V. Smushkov, "X-ray Determination of the Coefficients of Heterodiffusion in Cr-Mo and Ni-W Systems," J. Tech. Phys. (USSR) 28, 626-631 (1958).

24. B. Ya. Pines and I. V. Smushkov, "X-ray Investigation of Heterodiffusion in Cu-Ni Alloys," Sov. Phys.-Sol. State 1, 858-863 (1959).

25. B. Ya. Pines and E. F. Chaikovskii, "An X-ray Investigation of the Kinetics of Reactive Diffusion in the Al-Sb System," Sov. Phys.-Sol. State 1, 864-869 (1959).

26. B. Ya. Pines, I. G. Ivanov and I. V. Smushkov, "Partial Diffusion Coefficients and Self-Diffusion Coefficients in Copper-Nickel Alloys," Sov. Phys.-Sol. State 4, 1882-1890 (1962).

27. M. A. Levitskaya and N. A. Vodop'yanova, "X-ray Determination of the Diffusion Coefficients by the Method of Double Thin Metallic Layers," Sov. Phys.-Sol. State 4, 458-460 (1962).

28. R. L. Fogel'son, "Method of Determining the Coefficient of Diffusion by X-ray Studies," Fiz. Metal. e Metallog. 25, 492-496 (1968).

29. C. R. Houska, "X-ray Diffraction from a Binary Diffusion Zone," J. Appl. Phys. 41, 69-75 (1970).

30. T. T. Bales, "An X-ray Diffraction Technique for Determining the Concentration Gradient Existing Beneath Thin Films," M.Sc. thesis Virginia Polytechnic Institute, May, 1965, 50 pages.

31. D. N. Braski, "An X-ray Diffraction Method for Studying Small Diffusion Zones," M.Sc. thesis Virginia Polytechnic Institute, May, 1965, 54 pages.

32. D. R. Tenney, J. A. Carpenter and C. R. Houska, "X-ray Diffraction Technique for the Investigation of Small Diffusion Zones," J. Appl. Phys. 41, 4485-4492 (1970).

33. J. A. Carpenter, D. R. Tenney and C. R. Houska, "Method for Determining Composition Profiles and Diffusion-Generated Substructure in Small Diffusion Zones," J. Appl. Phys. 42, 4305-4312 (1971).

34. P. K. Talty and D. R. Tenney, "X-ray Diffraction Investigation of Bi-metallic Diffusion Zones in the Cu-Pd System," to be published in Met. Trans.

35. J. Unnam, J. A. Carpenter and C. R. Houska, "X-ray Diffraction Approach to Grain Boundary and Volume Diffusion," J. Appl. Phys. 44, 1956-1967 (1973).

POLE FIGURE RANDOM INTENSITY CALCULATION USING

POWDER INTEGRATED RATIOS

Carlos Sergio Viana and Gustau Ferran

COPPE-Universidade Federal do Rio de

Janeiro - Brasil - c.p. 1191 ZC 00

ABSTRACT

This paper describes a method for automatic quantitative pole figure plotting up to 70°, using only one sample and Schulz reflection technique. Random intensities are calculated for the usual planes of iron, using the ratios of calculated powder integrated intensities of these planes to the intensity of a high multiplicity factor plane, the random intensity of which had been obtained by integration up to 70° over the pole figure, using Bragg and Packer's method; the latter integration shows a decreasing error when the multiplicity factor increases. With this method it is possible to normalize the pole figures without using a physical standard and to reduce greatly the time to obtain a quantitative pole figure.

INTRODUCTION

The production of a textureless sample to normalize direct pole figures is not easy. Several workers have used metal powders embedded in plastic, sintered and, whenever possible, a sample cycled around a phase transformation temperature (1). None of these samples, however, can yield satisfactory (hkl) random intensities because they are structurally different from the textured material. Even a perfectly random oriented material will introduce important errors if it is used to normalize cold worked and fully recrystallized materials (2).

Bragg and Packer (3) calculated the (hkl) mean intensities by superposing a polar net on complete (hkl) pole figures and integrating over the whole net, using the expression

$$\bar{I}_{hkl} = \frac{\sum\limits_{0}^{\Pi/2} \sum\limits_{0}^{2\Pi} I(\alpha,\beta) \sin \alpha \, \Delta\alpha\Delta\beta}{2\Pi} \qquad (1)$$

This technique, when improved by the use of a computer and with intensity measurements at equal angular intervals, gives mean values of high precision and allows the self normalization of pole figures. However, this process requires complete pole figures, obtained in general using reflection and transmission techniques. The use of transmission technique has some disadvantages: (i) the intensities must be corrected for absorption; (ii) it is necessary to match reflection and transmission intensities; (iii) thin samples are needed, which often interfere with material grain size, giving imprecise intensity measurements; and (iv) even using a high power x-ray generator (1.5 KVA) these intensities are much smaller than the reflected ones, introducing error.

Considering that in a large number of cases, pole figures up to 70° of latitude would be satisfactory, the objective of this work is to describe a procedure to obtain quantitative pole figures. This method calculates random mean intensities to normalize (hkl) pole figures, using only the reflection technique up to 70°, an equation derived from Bragg and Packer's work, and mean intensity ratios of the same diffraction patterns.

THEORETICAL DEVELOPMENT

In a texture goniometer the reciprocal lattice halfsphere is generally scanned following two paths, viz., spiral or concentric circles up to 90°. Using a technique to measure intensities at constant angular intervals, the mean intensity may be calculated by,

$$\bar{I}_{hkl} = \frac{\sum\limits_{i=1}^{N} I_i(\alpha,\beta) \, \Delta L_i(\alpha,\beta)}{L} \qquad (2)$$

Equation (2) is derived from Eq. (1), where $I_i(\alpha,\beta)$ is the intensity corrected for background and absorption, $\Delta L_i(\alpha,\beta)$ is the arc length in which $I_i(\alpha,\beta)$ was measured, N is the total number of measurements and L is the length of the scanned path.

If the values obtained using Eq. (2) are actually mean values, a ratio of one such value to another must be equal to the integrated intensity ratio for the same planes, obtained from the powder of the material or calculated by the classical formula,

$$\frac{I^t_{hkl}}{I_o K} = p \, |F|^3 \exp(-2M) \, (LP). \qquad (3)$$

Thus one must have

$$\frac{\bar{I}_{(hkl)i}}{\bar{I}_{(hkl)j}} = \frac{I^t_{(hkl)i}}{I^t_{(hkl)j}} = \frac{I^p_{(hkl)i}}{I^p_{(hkl)j}} \qquad (4)$$

where I^t is the theoretical integrated intensity, I^p is the powder integrated intensity and $\bar{I}$ is given by Eq. (2).

On the other hand one can expect that the mean intensity calculated up to 70° approaches the value calculated up to 90°, using Eq. (2), as the multiplicity factor increases. If the crystalline structure has high multiplicity factors (for instance in the cubic system) there may be a plane (HKL), for which

$$\frac{\sum_{i=1}^{N} I_i(\alpha,\beta)\,\Delta L_i(\alpha,\beta)}{L^{90}} = \frac{\sum_{i=1}^{M} I_i(\alpha,\beta)\,\Delta L_i(\alpha,\beta)}{L^{70}} \qquad (5)$$

where $N > M$

So, with $\bar{I}_{HKL}$ and the ratios (4) one can normalize pole figures up to 70°, without using any physical standard and avoiding the transmission technique.

EQUIPMENT AND SAMPLES

Philips x-ray diffraction equipment was used, including a high voltage generator (PW 1011), a texture goniometer, scintillation detector, samplifier-analyzer and a table printer (Victor model). Filtered Mo Kα radiation, with a voltage of 44 KV and 34 mA tube current, was used in all the experiments. The receiving slits were 4.9 x 4.0 mm for reflection and 4.9 x 2.15 mm for transmission, the goniometer radius being 183 mm.

In order to have a high multiplicity factor, a BCC metal (0.03% carbon steel) was used in the form of cold rolled (70% thickness reduction) sheet, to minimize the grain effects.

EXPERIMENTAL

Complete quantitative pole figures were obtained using the reflection and transmission techniques. In the reflection as well as in the transmission methods the reciprocal lattice halfsphere was scanned in concentric circles 5° of latitude apart from 5° up to 90°.

A FORTRAN (4) program was developed to (i) correct the intensity values for background and absorption, (ii) match reflection and transmission intensities and (iii) calculate mean values up to 70°, and up to

90° using Eq. (2). The same program normalizes the intensities by the 90° mean value and prints the pole figure. In an IBM/360 computer it takes 1 min. 50 sec. per pole figure (see Fig. 1).

Integrated (hkl) peak intensities were measured using filings from the same steel, in the same texture goniometer and with the same slits. The peaks were scanned several times and the background measured at both sides of the peaks. Theoretical intensities were calculated by Eq. (3) considering the steel as pure iron.

RESULTS AND DISCUSSION

Columns 1 and 2 of Table I give the plane and the multiplicity factor. Columns 3 and 4 give the intensity mean values calculated by Eq. (2) for complete ($\alpha = 90°$) and partial ($\alpha = 70°$) pole figures respectively; Column 5 gives the percent difference between these two. In Column 6 and 7 the mean values of the theoretical and powder integrated intensities are given, respectively. This Table shows that the relative difference decreases as the multiplicity factor increases and that for plane (123) it is only a few units percent. Also it should be noticed that the partial (200) and (222) mean values are greater than the complete ones; this is so because these diffractions have their principal high density regions in back reflection. Of course these differences may vary with the material and its texture, but this variation should be smaller for a plane of higher multiplicity factor.

On Columns 2, 3, and 4 of Table II are the intensity ratios of diffractions (200), (222), and (110) to the (123), calculated using Columns 6, 7, and 3 of Table I, respectively. It should be noticed that the agreement between theoretical and pole figure calculated values is better than between theoretical and the powder ones. This may be due to the bad quality of the powder sample used.

Column 2 of Table III shows the mean (hkl) intensities recalculated multiplying the theoretical ratios of Table II by the mean (123) intensity measured up to 70°. Column 3 shows the percent differences between these values and their corresponding pole figure mean intensities up to 90°. As can be seen, these differences are smaller than those of Table I, for (hkl) intensities up to 70°; this shows the consistency of the method.

Now using the pole figure calculated ratios (from Table II, Column 4) and performing the same multiplications, a new set of mean (hkl) intensity values is found (Column 4) and the differences become still smaller, as can be seen comparing Columns 3 and 5 on Table III. This better agreement could be expected because the complete pole figure mean intensities take in account factors that are not assumed for Eq. (3), viz., (i) the theoretical ratios are obtained by assuming a parallel beam, whereas in the texture goniometer the divergence of the incident beam is 2.1°, (ii) the measuring technique does not completely scan the reciprocal lattice half-sphere, (iii) the Bragg-Brentano diffraction condition, generally assumed

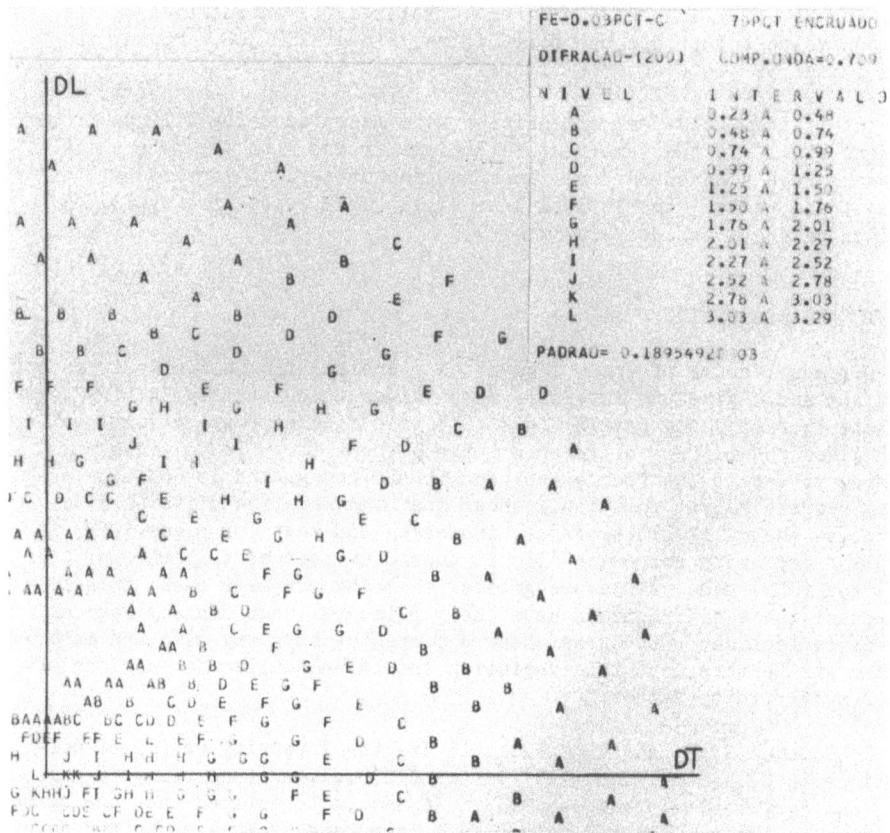

Figure 1. Pole Figure as Printed
(Real Radius = 15 cm)

Table I

Line	Multiplicity factor	Pole figure mean intensity up to 90°	Pole figure mean intensity up to 70°	% difference	Theoret. intensity	Powder intensity
200	6	2737	3406	-24.5	137826	9853
222	8	481	560	-16.4	21532	1373
110	12	17038	15530	+ 8.8	851769	79541
123	48	1778	1747	+ 1.7	90185	8214

Table II

Intensity ratios	Calculated Theoretically	Measured from powder sample	Pole figure calc. up to 90°
$\frac{200}{123}$	1.53	1.20	1.54
$\frac{222}{123}$	0.24	0.17	0.27
$\frac{110}{123}$	9.44	9.68	9.58

Table III

Line	Calculated mean intensity, theoret. ratios	% difference	Calculated mean intensity, complete pole fig. ratios	% difference	Pole figure mean intens. up to 90°
200	2669	+2.5	2689	+1.8	2737
222	422	+13.9	472	+1.9	481
110	16499	+3.2	16737	+1.8	17038

for calculating the theoretical ratios is no longer valid when tilting the sample in the texture goniometer and (iv) the large width of the receiving slit of the goniometer gives a poor resolution.

CONCLUSION

By the simple method described it is possible to produce quantitative pole figures by reflection up to 70° (that could be extended up to 80° easily), without an actual random sample. This method is based upon the measurement of mean intensity up to 70° of a high multiplicity factor diffracting plane and the available value of intensity ratios. The accuracy of the mean intensities used for normalization can be improved when the intensity ratios are obtained from the complete pole figure mean intensities, instead of the theoretical powder intensities. A disadvantage of this method is that it requires one to obtain one set of complete pole figures.

This procedure reduces greatly the time to produce the quantitative pole figure and should avoid the errors of secondary and primary extinction appearing when the structural perfection of the physical random sample is different from the textured one.

REFERENCES

1. K. Aoki, S. Hayami and M. Matsuo, "Improvement of the Accuracy in Representation of Conventional Pole Figures," in J. B. Newkirk and G. R. Mallett, Editors, <u>Advances in X-Ray Analysis</u>, Vol. 10, p. 342 (1967).

2. W. B. Hutchinson, T. W. Watson and I. L. Dillamore, "Improved Drawability Through Control of Textures," J. I. S. I., Nov. 1968, p. 1479 (1968).

3. R. H. Bragg and C. M. Packer, "Quantitative Determination of Preferred Orientation," J. Appl. Phys., $\underline{35}$, p. 1322 (1964).

4. C. S. Viana, Thesis "Metodo automático para traçado de figuras de polo", COPPE, Centro Tecnologia, Universidad Federal do Rio de Janeiro, May, 1972.

X-RAY EMISSION FROM LASER-PRODUCED PLASMAS

C. M. Dozier, P. G. Burkhalter, B. M. Klein,
D. J. Nagel and R. R. Whitlock
Naval Research Laboratory
Washington, D. C. 20375

ABSTRACT

Intense x-rays are emitted by plasmas formed when subnanosecond laser pulses are focused onto materials. Plasmas produced by pulses containing up to 100 J can re-emit over ten percent of the energy as x-rays above about 1.0 keV. These plasmas may be useful flash x-ray sources.

INTRODUCTION

Extremely hot, dense plasmas are produced when energetic, subnanosecond laser pulses are focused onto target materials. Such plasmas are of interest in the search for a nuclear fusion power source [1]. A large fraction of the laser pulse energy which is absorbed in a laser-plasma is re-emitted as radiation at wavelengths different than the laser light. This re-emitted radiation provides one of the best indicators, or diagnostics, of conditions in the plasma. When power densities in the focused laser beams exceed 10^{14} watts/cm^2, the x-ray emission is intense and x-ray measurements are the prime diagnostic tool. Plasma characteristics such as temperature and density can be derived from x-ray data.

The results obtained from x-ray diagnostic measurements of laser-produced plasmas at the Naval Research Laboratory (NRL) and at other laboratories [2,3,4] suggest that these plasmas may be useful as flash x-ray sources for other experiments. The plasmas are small, intense x-ray sources with spectra near one keV which are characteristic of the target materials. The x-ray pulse duration is similar to the laser pulse length, namely about 1.0 nanosecond.

In this paper, the emphasis is on the laser-plasmas as a source of x-rays rather than as a medium for the production of fusion power. First, the physical processes involved in the production and emission of laser plasmas are outlined. Then the laser, target, and x-ray instrumentation used in the work is described. Various measured characteristics of x-ray emission from laser-plasmas are summarized prior to a discussion of potential applications.

THE PRODUCTION OF X-RAYS IN LASER-PRODUCED PLASMAS

It is amazing that x-ray photons with energies in excess of a thousand electron volts can be produced by focusing light photons with energies of about one electron volt onto target materials. It is even more striking that the x-ray production is fairly efficient. What are the physical processes which lead to this conversion of energy in the form of light photons to x-rays?

Figure 1 shows schematically the laser pulse interaction with a target at the beginning, middle, and end of the laser pulse. As shown in Figure 1a, the laser pulse is focused to a diameter of about 50 μm at the target. The initial portion of the pulse, or a small prepulse, vaporizes the target, creating the plasma.

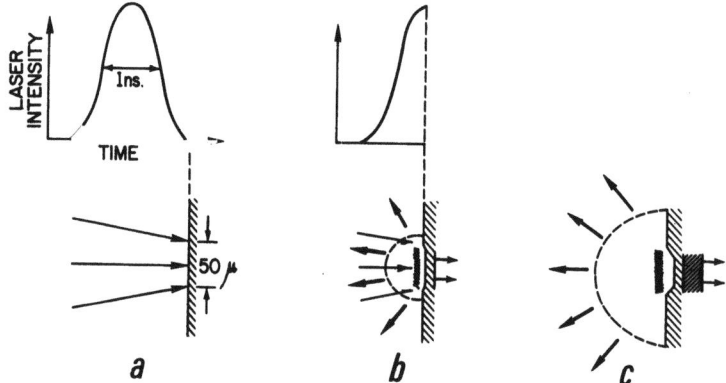

Figure 1. Laser-plasma production, heating and emission. (a) Initial portion of laser pulse, focused to 50 μm at target, creates plasma. (b) Central portion of pulse heats plasma to x-ray emitting temperatures. (c) After pulse has passed, plasma rapidly expands and x-ray emission is terminated.

Plasmas can be considered as mixtures of gases: one gas is the free electrons; the other gases are ions of the target material. The free electrons in the expanding plasma absorb the energy of the incident photons by inverse bremsstrahlung. This deposition is most efficient in a region, shown as a dark band on Figure 1b, where the electron density is optimum for a given laser light wavelength. For Nd:glass lasers, (λ = 1.06 μm) this optimum electron density is $10^{21}/cm^3$. At power densities greater than 10^{14} watts/cm^2, these photon-electron interactions occur repeatedly during the laser pulse, raising the average kinetic energy of the electrons to about a kilovolt. The corresponding temperature of the electrons is about 10^7 °K.

As the electron energy and plasma temperature increase (Figure 1b), the collisions between electrons and ions in the plasma strip the ions further and excite the remaining outer shell electrons. The number of electrons stripped from the ions depends on the energy transferred to the electron gas and the energy required to remove electrons from the ions. With present laser powers, low-Z elements can be stripped to the K shell and heavier elements into the L, M, or N shells. Unlike conventional electron-excited sources in which x-rays are produced by inner shell ionization, x-rays produced in these plasmas are predominantly from transitions of the outer shell electrons in the stripped ions. For practical purposes to obtain x-rays from transitions to the K-shell, ions must be stripped until the K-shell electrons are the only ones remaining. During nanosecond laser pulses, the electrons and ions cannot move large distances and the plasma remains dense. Stripped ions can be excited many times by electron collisions and large x-ray yields per ion result.

Once the laser pulse has passed, as shown in Figure 1c, the plasma continues to expand. Recombination of ions and electrons as the plasma expands and cools yields further x-ray emission. Also indicated in the figure is a shock wave which is formed in response to the plasma blowoff and propagates into the target material.

The efficiency with which x-rays are produced is dependent on the laser pulse energy, pulse shape, the wavelength and quality of the beam; the focusing by the lens; and the composition, construction, and orientation of the target. One of the overall aims of the present work is to learn how to vary these parameters to control and optimize the x-ray emission for particular applications.

INSTRUMENTATION

Laser and Target Facility

NRL's high-powered Nd:glass laser is a high quality, one-of-a-kind pulsed laser [5] as is each of the large lasers used in fusion research. The laser, shown in Figure 2, consists of an oscillator which produces a 1.06 μm gaussian light pulse of 0.25 or 0.9 nanosecond. The pulse is amplified by a linear chain of Nd:glass laser amplifiers until energies up to about 100 J/pulse are obtained. If desired, after amplification pulses can be frequency-doubled with a KDP crystal to 5320 Å. In addition to the single pulse operation, a small variable prepulse can be provided to initiate the plasma production before the main pulse arrives. Isolation devices prevent light reflected from the target from being reamplified to the point of damaging the laser components.

Figure 2. NRL high-power Nd:glass laser. Pulse emitted in direction of arrow. Laser amplifiers, A, are shown; Nd:YAlG oscillator is to the right, outside field of view.

The target facility, shown in Figure 3, is located about 100 feet from the laser. Targets are mounted in a 26 port vacuum chamber. The large number of ports permits simultaneous x-ray, plasma, and neutron measurements. Focusing of the laser beam is done with f 15 or f 1.9 lenses.

Figure 3. Target facility. Laser beam enters from right and is focused by lens, L, onto target in chamber.

X-Ray Instrumentation

The x-ray diagnostics measure the source size, the spectra, time duration, and intensity of the x-ray emission. Instrumentation, shown schematically in Figure 4, includes pinhole cameras, crystal and grating spectrographs, and active detectors such as pyroelectric detectors and calorimeters.

X-ray pinhole photographs, taken with a 25 µm platinum pinhole, provide information on the size of the plasma and the optimum placement of the target relative to the lens. Collimators made of fine tubes can also be used to give an image of the source.

Unlike steady-state x-ray measurements, the crystal in the spectrograph cannot be rotated during the short pulse; however, a selected energy band can still be measured with the crystal since divergent radiation from the source strikes the crystal over a range of Bragg angles. Typically, KAP crystals (2d = 26.6 Å) are used to cover the energy range from 800 eV to 2500 eV. Kodak No-Screen films used to record the spectra were protected by light-tight, 25 µm Be foils.

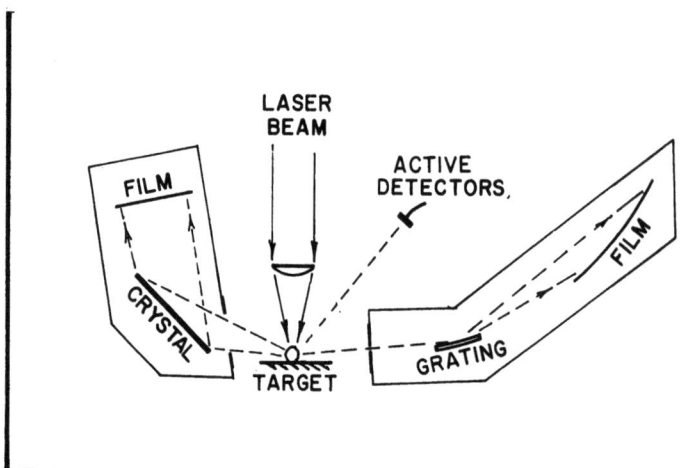

Figure 4. Schematic of x-ray diagnostic instrumentation.

The grazing-incidence grating spectrograph [6], on loan from the National Aeronautics and Space Administration, Goddard Space Flight Center, was used to record vacuum ultraviolet and soft x-ray spectra. The 1200 line/mm grating covered the range of 35 eV to over 1300 eV which overlaps that accessible with the KAP crystals. Kodak 101 plates recorded the spectra.

Ultra-fast and time integrating detectors measure the time history and intensity of the x-rays. A commercial lithium-tantalate pyroelectric detector, normally used for infrared measurements is used for time history measurements. A calibrated thermopile calorimeter provides intensity measurements.

CHARACTERISTICS OF THE X-RAY EMISSION FROM PLASMAS

In this section the measured characteristics of the x-ray emission from laser-produced plasmas are briefly described.

Source Size

One of the simplest methods of determining the source size is to take a pinhole picture. The technique is insensitive to wavelengths above the window cutoff of the film wrapping; in this case, about 800 eV. Figure 5 shows x-ray pinhole pictures taken at right angles to the laser beam and parallel to the target surface.

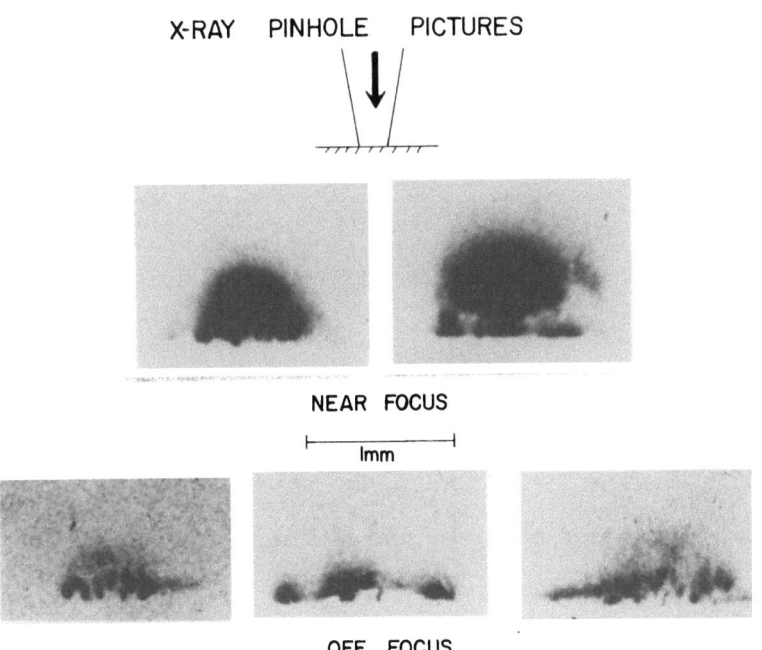

Figure 5. X-ray pinhole pictures with various focusing conditions.

The top two pictures show the general appearance of the plasmas when the target is at, or near, the focal plane of the lens. The lower three pictures show plasmas in which the target was not placed near focus. Spotty, irregular plasmas are formed and the x-ray yield is greatly reduced when the target is off focus. Pictures taken with other filters indicate that the most intense radiation, when the target is at focus, comes from a region 50 to 100 μm in diameter, about equal to the diameter of the focal spot [7].

Spectra

X-ray and vacuum ultraviolet spectra with energies less than 5 keV have been measured for many elements across the periodic table. Most of the work has been done with the 0.9 nsec pulses of 1.06 μm light. Less extensive measurements have been made with laser pulses at 5320 Å, 0.25 nsec in length and/or preceded by a prepulse. Figure 6 shows spectra of highly ionized Al, Zn, and Gd in the range of 1.0 to 2.5 keV obtained with a few joules incident on pure metal targets. The Al spectrum is from helium-

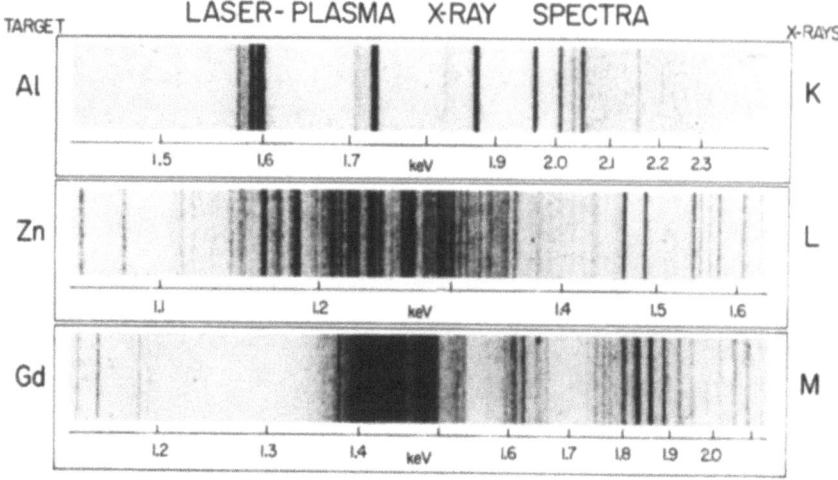

Figure 6. Examples of K, L, and M line spectra from laser-produced plasmas.

like, and hydrogen-like ions. That is, the ions are stripped to the K shell and have only two or one electrons remaining. The Zn spectrum shows stripping into the L shell, i.e., 20 or more electrons have been stripped from the ions. Similarly, the Gd spectrum shows ionization into the M shell, i.e., 36 or more electrons are missing.

Similar spectra have been obtained in many other elements. With a single pulse K spectra up to S (Z = 16), L spectra up to Br (Z = 35), and M spectra to Er (Z = 68) have been observed. The addition of a prepulse was found to significantly improve the energy coupling. K spectra of Ti (Z = 22) were excited with this technique. The addition of the prepulse is an effective way to produce a hotter more energetic plasma and x-ray emission.

Temporal Behavior

The emission characteristics of the x-rays are expected to follow the laser pulse behavior closely [8]. Since the laser pulses are of subnanosecond in duration, the x-ray emission will be approximately the same duration. Unfortunately, x-ray detectors with less than a nanosecond time resolution are not readily available, and the exact temporal behavior cannot be determined at this time. A pyroelectric detector which has 0.9 nanosecond rise time has been used to confirm the short duration of the x-ray pulse. Figure 7

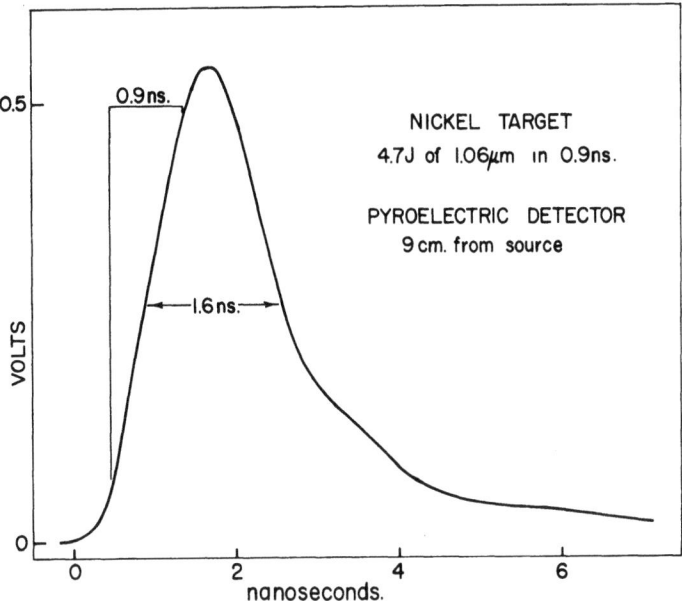

Figure 7. Time duration of x-ray pulse as measured with pyroelectric detector with 0.9 nsec risetime.

shows the signal from this detector. Deconvolution of the detector risetime from the observed signal sets an upper bound on the duration of the radiation at about one nanosecond.

Intensity

The intensity of the x-rays emitted by the plasmas depends on the efficiency with which the laser energy is coupled to the plasma. Effects caused by varying the focusing conditions are evident in the pinhole pictures in Figure 5. Other variables, such as differences in the target surface and the beam intensity distribution in the focal spot region, can also affect the plasma and are much less amenable to measurement. Attempts to standardize the laser pulse characteristics, focusing conditions, and target preparation have not eliminated all of these variations, and a fair amount of scatter still remains in the intensity measurements. Averaged results for Al and Cu plasmas, created with a single pulse, are shown in Figure 8. The percent conversion factor is the fraction of the energy in the laser pulse re-emitted as x-rays above about 800 eV, assuming the plasma radiates isotropically. These data are indicative of the large fraction of the laser beam energy re-emitted as x-rays even though shot-to-shot variations occur. By comparison, less than 0.1 percent of the energy in an electron beam is converted to x-rays in conventional x-ray tubes.

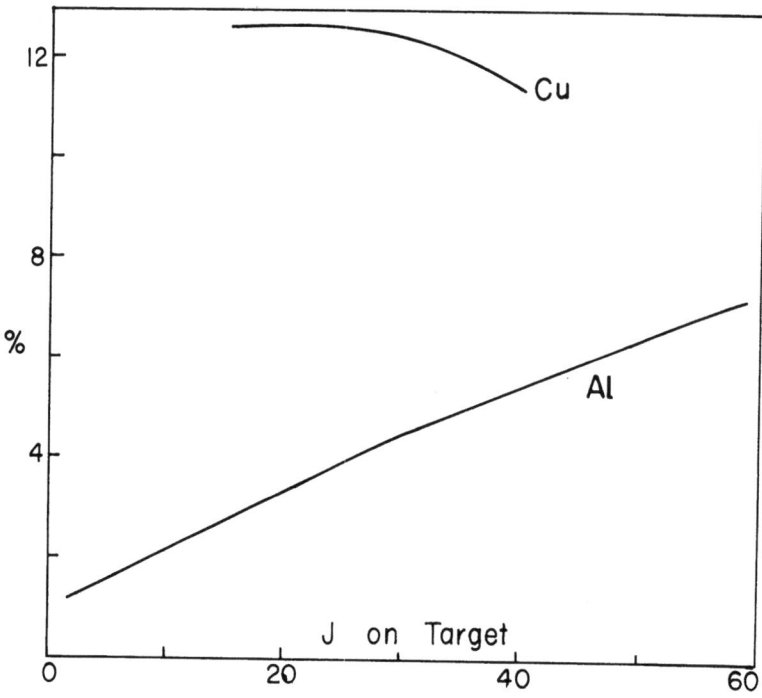

Figure 8. Conversion efficiency of laser pulse energy into x-rays above ~ 800 eV assuming isotropic emission.

APPLICATIONS OF LASER-PRODUCED PLASMA X-RAY SOURCES

For the immediate future, x-rays from laser-produced plasmas will continue to be a prime diagnostic tool for fusion research involving these plasmas. A better understanding of the formation, heating, and emission of these plasmas is still needed. Whether or not these laser-plasmas find wider acceptance as a flash x-ray source will depend on several factors: (1) The lasers used to produce x-ray emitting plasmas must be competitive with other flash x-ray sources; (2) Harder spectra are desirable for most projected uses and (3) applications which can take advantage of such a source must be developed.

The design and construction of less expensive lasers with the characteristics necessary to excite keV spectra seems to be technologically possible. In fact, very high-powered lasers may not be necessary. Many of the elements can be excited with relatively modest power levels. Magnesium spectra, showing stripping into the K-shell, for example, have been observed with pulses from a small commercial laser having an 18 nanosecond pulse [9]. However, most lasers do not have pulses of ~ 1 nanosecond which are necessary to generate x-ray emitting plasmas efficiently. It is anticipated that one nanosecond lasers with outputs of a few joules will be available soon and that they will become cheaper as development proceeds.

The keys to creating hotter plasmas which result in more energetic spectra are (a) coupling more of the laser pulse energy into the plasma and (b) insuring that the energy is re-emitted rather than being wasted in plasma expansion. The use of prepulses has already permitted some increased energy deposition. Other pulse shaping techniques may also be useful. Different configurations of the targets may also be useful in assisting in the plasma heating. Energy losses due to expansion, for example, may be eliminated partially by focusing the laser beam into a small hole in the target. Sideways motion of the plasma would be limited, and denser, potentially hotter plasmas may be possible. The use of very small targets with only enough mass to produce a plasma of the optimum density may also help produce more energetic spectra.

Laser plasmas have several present and anticipated uses. These sources are ideal for atomic and astrophysical studies. They are providing a better understanding of atomic processes; ionization and emission parameters are useful in understanding the x-radiation from extraterrestrial sources. In fact, spectra obtained with the grating spectrograph have already been used by astrophysicists to identify several unidentified lines in solar x-ray spectra [10]. Laser-produced plasmas provide the astrophysicist with an excellent means of simulating in the laboratory many conditions that exist in solar and stellar sources.

Other experiments requiring flash x-ray sources can also use radiation from laser plasmas. Most often these experiments have had to be performed with high energy bremsstrahlung sources. Laser-plasmas can provide an intense, low energy x-ray source for these experiments. These plasmas are the only intense, nanosecond source of x-rays around one keV. A few examples of experiments that may find laser-plasma sources useful can be mentioned. Radiography of projectiles and during transient phenomena has been an important use of flash sources [11]. With low energies available in laser-plasma sources, less dense materials, such as other plasmas, can be examined. Flash x-ray sources have been used to excite transient phenomena, like the short-lived excitons in alkali halide color centers [12]. Also studies of shock compression in materials using diffraction techniques [13] may find the intense line radiation from laser plasmas valuable. Since the x-rays are characteristic of the target materials, the possibility of using these plasmas for chemical analysis has been considered.

Although lasers have been used to excite optical spectra for this purpose [14, 15], the extension of this technique to the x-ray region does not appear to have any clear advantages. Complications due to the many lines in the spectra and the unknown quantity of material evaporated from the target and contributing to the x-ray emission override most of the advantages that may be gained.

The ultimate use of laser-produced plasmas will depend on their usefulness for specific tasks, and the advantages they offer in accomplishing these tasks. Their best applications appear to be a source for atomic spectroscopy and the study of transient phenomena.

ACKNOWLEDGEMENTS

This work is possible because of the laser development efforts of J. M. McMahon and O. C. Barr. U. Feldman and G. A. Doschek operated the grating spectrograph and interpreted the spectra from this instrument. We thank them, and also J. Stamper for assistance with the target facility, T. deRiex and L. Scott for operating the laser, and E. Turbyfill and J. Cheadle for taking some of the data. The U. S. Atomic Energy Commission and the Defense Nuclear Agency have supported the work.

REFERENCES

1. J. Nuckolls, L. Wood, A. Thiesen, and G. Zimmerman, "Laser Compression of Matter to Super-High Densities: Thermonuclear (CTR) Applications," Nature 239, 139-142 (1972).

2. J. Mallozzi, H. M. Epstein, R. G. Jung, D. C. Applebaum, B. P. Fairand, and W. J. Gallagher, "X-Ray Emission from Laser Generated Plasmas," Vol. I and II, Final Report to ARPA under Contract DAAH01-71-C-055Q, Battelle (1972).

3. S. W. Mead, R. E. Kidder, J. E. Swain, F. Ranier, and J. Petruzzi, "Preliminary Measurements of X-Ray and Neutron Emission from Laser-Produced Plasmas," Appl. Optics 11, 345-352 (1972).

4. J. F. Kephart, R. P. Godwin, and G. H. McCall, "High-Resolution X-Ray Spectroscopy of Bremsstrahlung from Laser Produced Plasmas," Bull. A.P.S. 17, 971 (1972).

5. J. M. McMahon and O. C. Barr, "Glass Laser System Used Routinely for Target Irradiation," Proc. 17th. Ann. Conv. SPIE. Sem. on Laser Tech. II, San Diego, Aug. 1973 (to be published).

6. W. E. Behring, L. Cohen, and U. Feldman, "The Solar Spectrum: Wavelengths and Identifications from 60 to 385 Angstroms," Astrophysical Jour. 175, 493-523 (1972).

7. J. F. Holzrichter, C. M. Dozier, and J. M. McMahon, "X-Ray Point Source Projection Photography with a Laser-Produced Source," Appl. Phys. Letts. (to be published Dec. 1973).

8. J. A. Stamper, O. C. Barr, J. Davis, G. A. Doschek, C. M. Dozier, U. Feldman, B. M. Klein, W. M. Manheimer, E. A. McLean, J. M. McMahon, N. K. Windsor, and F. C. Young, "Laser-Matter Interaction Studies at NRL," to be published in the Proceedings of the Third Workshop on "Laser Interactions and Related Plasma Phenomena," (1973).

9. T. N. Lee and D. J. Nagel, "K X-Ray Emission from Laser-Produced Mg Plasma," Bull. A.P.S. 18, 684 (1973).

10. U. Feldman, G. A. Doschek, D. J. Nagel, W. E. Behring, and L. Cohen, "Transitions of Fe XVIII and Fe XIX Observed in Laser-Produced Plasmas," The Astrophysical Jour. 183, L43-L45 (1973).

11. F. J. Grundhauser, W. P. Dyke, and S. D. Bennett, "A Fifty-Millimicrosecond Flash X-Ray System for Hypervelocity Research," Proceeding of 5th Inter. Congress on High Speed Photog., J. S. Courtney-Pratt, Editor, p. 149-153, SMPTE (1962).

12. R. G. Fuller, R. T. Williams, and M. N. Kabler, "Transient Optical Absorption by Self-Trapped Excitons in Alkali Halide Crystals, Phys. Rev. Lett. 25, 446-449 (1970).

13. Q. Johnson and A. C. Mitchell, "First X-Ray Diffraction Evidence for a Phase Transition during Shock-Wave Compression," Phys. Rev. Lett. 29, 1369-1371 (1972).

14. S. D. Rasberry, B. F. Scribner, and M. Margoshes, "Laser Probe Excitation in Spectrochemical Analysis. I: Characteristics of the Source," Appl. Opt. 6, 81-86 (1967).

15. S. D. Rasberry, B. F. Scribner, and M. Margoshes, "Laser Probe Excitation in Spectrochemical Analysis. II: Investigation of Quantitative Aspects," Appl. Opt. 6, 87-93 (1967).

CALCULATION AND MEASUREMENT OF INTEGRAL
REFLECTION COEFFICIENT VERSUS WAVELENGTH
OF "REAL" CRYSTALS ON AN ABSOLUTE BASIS

D. B. Brown, M. Fatemi and L. S. Birks

Naval Research Laboratory

Washington, D. C. 20375

ABSTRACT

A method for calculation of the integral reflection coefficient of crystals of intermediate perfection is introduced. This method can greatly reduce experimental effort for the selection and calibration of crystals. It also serves as a conceptual framework for studies of mosaic block structure and of crystal modification. Good agreement between calculated and experimental values of the integral reflection coefficient is shown for, (a) LiF crystals of two degrees of perfection, (b) elastically bent quartz, and (c) 001, 005, 006, and 007 diffraction from KAP. Zachariasen's division of crystals into two types is extended. It is concluded that the integral reflection coefficients for 200 LiF cannot be raised to the ideally imperfect limiting values.

INTRODUCTION

Most research directed toward the optimization of spectrometer crystals has been conducted in a rather Edisonian fashion. This tendency has resulted from the lack of an adequate theoretical framework to tie together diverse experimental evidence and to suggest promising areas of inquiry. We believe that we can now give more precise answers to such questions as:

- Can spectrometer crystals be divided into several recognizably distinct types?
- What are the key crystal parameters controlling resolution and diffracting power?
- In what ways can the diffracting power of a crystal be increased, and what are the side effects of such an increase?

In this paper we shall report on a treatment of crystal diffraction power which we have found helpful in answering these and other related questions.

In 1967 Zachariasen (1) introduced a treatment of x-ray diffraction which he hoped would handle crystals with properties lying within the full range between the perfect and ideally imperfect limiting cases. This theory was approximate, partly phenomenological, and somewhat intuitive. Zachariasen's treatment was directed primarily toward the relatively small spherical crystals commonly used by crystallographers. His results have been widely used by crystallographers for the treatment of extinction corrections in structure refinements.

We have now re-worked his approach for two geometries commonly utilized in spectrometers, namely:

- the semi-infinite plane slab in Bragg (i.e. reflection) geometry,
- the convex-curved crystal wherein the region which is diffracting may be considered to be a slab in Laue (i.e. transmission) geometry.

The resulting formulation has been found useful in two ways. First, it permits an approximate calculation of the integral diffracting power, thus eliminating or greatly reducing the amount of experimental work needed to select and calibrate crystals. Secondly, it has provided a conceptual and calculational tool useful in our work on the modification and improvement of crystal diffraction properties.

THE FORM OF THE CALCULATION MODEL

The basic mathematical form of our calculational model is given by the following equations.

$$\mathcal{P} = \mathcal{P}_K \, y(x) \qquad (1a)$$

$$x \propto Q \, \alpha \, \overline{T} \qquad (1b)$$

Using these equations we shall discuss briefly the key parameters of the problem and indicate how they are put together in a relatively simple fashion. Note that the integral diffracted power, $\mathcal{P}$, is equal to the product of two things. First, $\mathcal{P}_K$, the integral diffracted power in the kinematical (zero extinction) approximation. And secondly, an extinction factor y, which is a function of the extinction parameter x. The mathematical form of the extinction factor $y(x)$ is dependent on the diffraction geometry. Thus, we have used different forms of $y(x)$ for plane slabs in Bragg geometry and for convex-curved crystals.

The extinction parameter x is proportional to the product of three things: (a) The first is the parameter Q which commonly appears in diffraction theory. This may be considered to be the diffraction efficiency of a crystal in the absence of absorption and extinction. (b) The second, α, is a factor which contains information about the degree of imperfection of the crystal. It is dependent on the size of the mosaic blocks which make up the crystal and also on the degree of angular misorientation of these mosaic blocks. To a first approximation α is inversely proportional to the diffraction peak breadth. (c) The third factor, $\bar{T}$, is an effective pathlength for the diffracted radiation. It is related to the crystal size. It contains an allowance for the fact that not all of the crystal "sees" the radiation because of the effects of x-ray absorption.

The discussion of Eqns. 1 just concluded constitutes a complete statement of the basic structure of our problem; though, of course, it would take considerably more space to spell out the detailed implications of what has been said. We should like at this point to re-emphasize the fact that the solutions we obtain are approximate solutions in the following sense. Our calculational method gives exact solutions for certain limiting cases. For intermediate cases it gives reasonable results. The degree of reasonableness of these answers has been put to the test of experiment.

CALCULATIONS VS. EXPERIMENT

We shall now show the results of several experimental tests of our calculations of the integral diffracted power.

As our first test we have considered a quite perfect crystal of LiF, and the results are given in Fig. 1. In this figure the integral reflection coefficient, R, is plotted vs. wavelength. The solid lines show the ideally mosaic and perfect crystal limiting cases for purposes of orientation. The open circles are experimental results. The dashed lines represent calculations for mean mosaic block sizes of 0.2 and 0.3 cm. These are the lower and upper limits of the sizes of the naturally occurring subgrains as revealed using dislocation etch pitting. Note that these comparisons of R values are absolute – no arbitrary fitting parameters have been used.

As our second test we have considered a relatively imperfect specimen of LiF, and the results are given in Fig. 2. Here, again, the ideally mosaic and ideally perfect limiting cases are shown for comparison. This crystal was produced by Birks and Seal (2) using a process they called "flexing." The crystal was bent and then un-bent in order to increase the dislocation con-

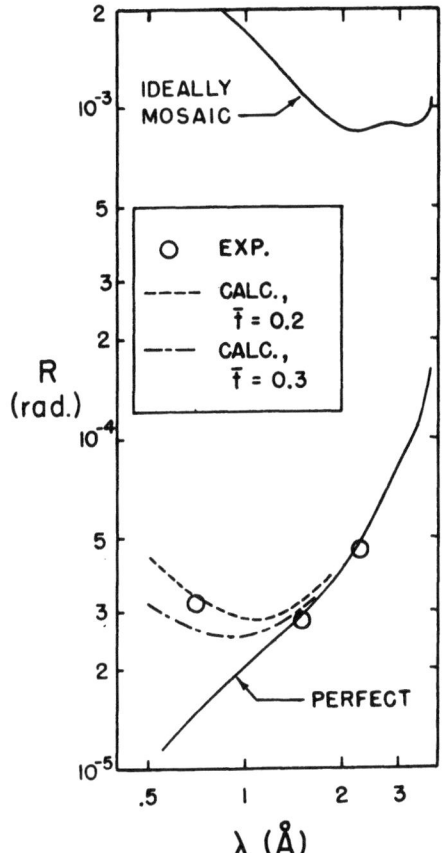

Fig. 1. The integral reflection coefficient vs. wavelength for relatively perfect LiF.

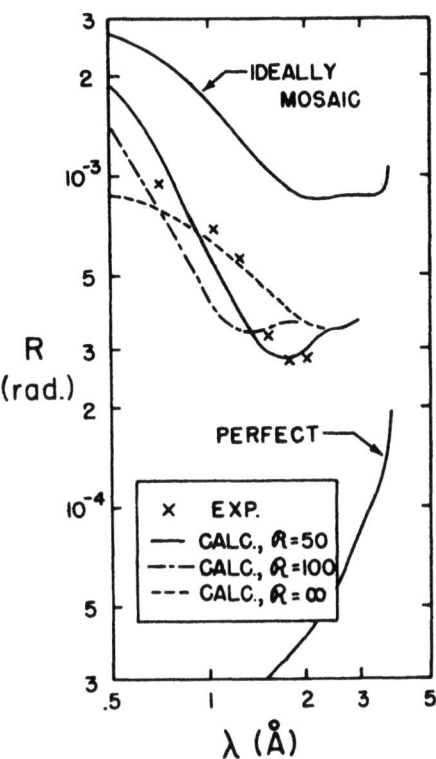

Fig. 2. The integral reflection coefficient vs. wavelength for relatively imperfect LiF.

centration, and thus increase the R values. The x's represent experimental values due to Vierling et al. (3). It is suspected that the crystal was not flattened completely after the flexing process, i.e. that it retains some residual warping, or waviness, or both. The dashed line shows calculated R values assuming a flat crystal; the dash-dot-dash line shows a calculation assuming an average bending radius of 100 cm; the solid line shows a calculation assuming an average bending radius of 50 cm. All of these calculations were based on a mosaic block size of 2.7 microns. Further, they were based on an angular misorientation of the mosaic blocks having a Gaussian distribution with a full-breadth-at-half-maximum of a little over one minute. These values for the mosaic block size and misorientation were obtained using dislocation etch pitting and a simple model for the formation of mosaic blocks from dislocation arrays.

Note that we have now considered a relatively perfect crystal and a relatively imperfect crystal. Further, in the second example we have shown data for a crystal which had undergone plastic deformation. In Fig. 3 we show results for a crystal of elastically bent α-quartz. The integral reflection coefficient has been plotted versus the reciprocal of the radius of bending. The circles represent data taken by White (4). This is a rather interesting experiment. As the crystal is bent to a smaller radius, x-ray extinction is reduced and thus the integral reflection coefficient is increased. Thus the R value is continuously and reversibly variable as a function of the bending radius. All data points are for Mo Kα radiation and use the 20.2 planes of quartz. The solid line gives the calculated results. It will be observed that the experimental data lie above the calculated values by about 20%. In fact, the two experimental values corresponding to the smallest bending radii lie above the theoretical maximum, given by the dashed line. This suggests to us some systematic error in the experimental data.

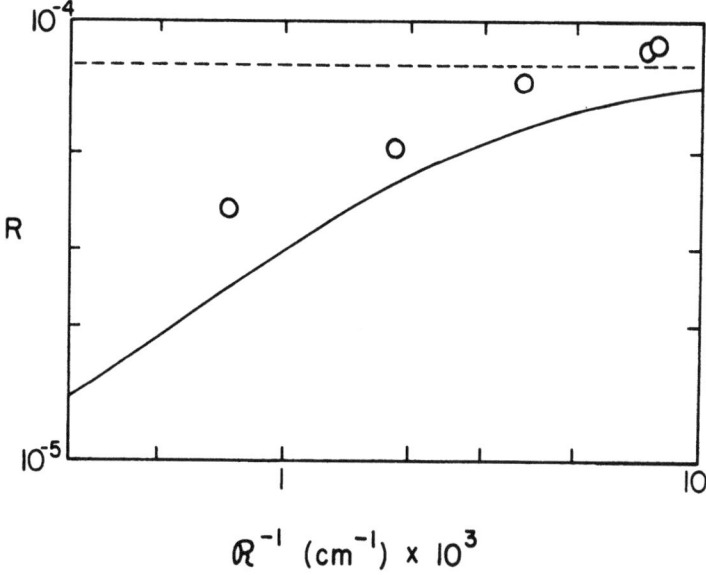

Fig. 3. The integral reflection coefficient versus the reciprocal of the bending radius for elastically bent quartz.

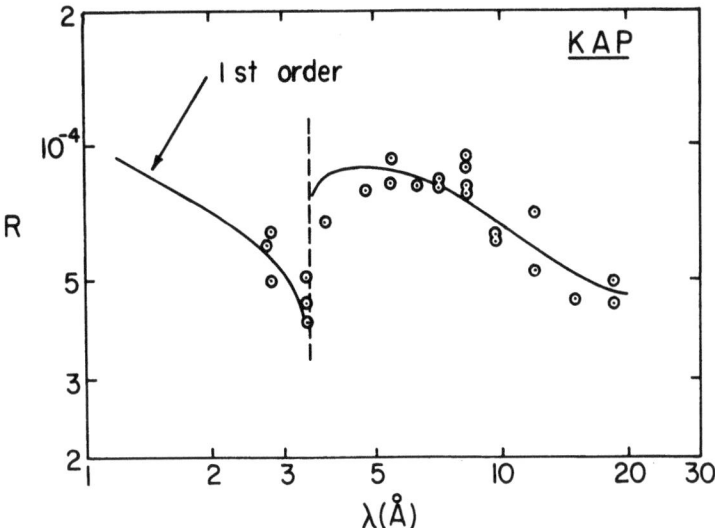

Fig. 4. The integral reflection coefficient vs. wavelength for 001 KAP.

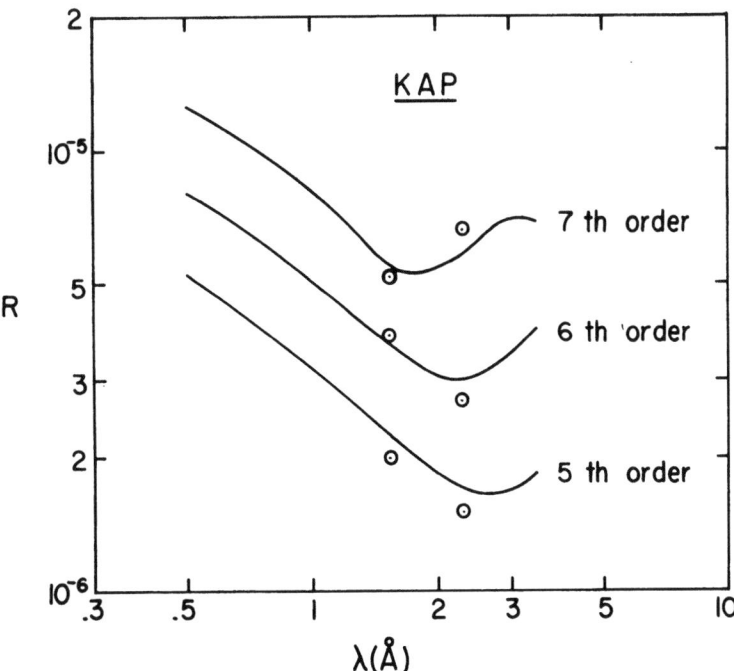

Fig. 5. The integral reflection coefficient vs. wavelength for 005, 006, and 007 KAP.

As a final example we will show you a rather different test of our method. This will be a comparison of calculations with experimental R values for several orders of diffraction from KAP. Fig. 4 shows calculated single crystal spectrometer R values for the 001 reflection of KAP. The open circles are data due to Burkhalter and co-workers of our laboratory (5). The turning down of the R values on both sides of the potassium K edge is due to anomalous scattering. The calculations were based on a mosaic block size of 10 microns. Fig. 5 shows a similar comparison for the double-crystal spectrometer R values for the 5th, 6th, and 7th order reflections of KAP. Again a mosaic block size of 10 microns was assumed.

CRYSTAL TYPES

In crystals of LiF, secondary extinction is controlled primarily by the mosaic block misorientation, and is nearly independent of the mosaic block size. Zachariasen has called these Type I crystals. It is most probable, however, that KAP is not of this type but rather of Zachariasen's Type II. Crystals of Type II are those for which secondary extinction is primarily controlled by the mosaic block size, and is nearly independent of the mosaic block misorientation.

It is very likely that Zachariasen's division of crystals into two types, based on their extinction properties, can be considerably extended. An interesting by-product of our analysis is that mosaic blocks formed by dislocation arrays lead (with a few improbable exceptions) to crystals of Type I; the reason is that dislocation arrays contribute significantly to mosaic block misorientation. Possible candidates for the generators of Type II mosaic block structure are twin boundaries, stacking faults, and growth horizons. We have mentioned that for Type I LiF we have obtained a good comparison between theory and experiment using mosaic block parameters deduced using dislocation etch pitting. To the best of our knowledge an analogous test has not been made for Type II crystals such as KAP. The 10 micron block size which we used to get good agreement for four different orders of KAP R values has not been independently confirmed. This is an experimental problem which needs further attention.

It is interesting that Birks, a number of years ago, noted two types of spectrometer crystals (6). For his first type (including the alkali halides and the metals) surface abrasion followed by etching results in an increased integral diffracting power. For his second type (including such crystals as fluorite, calcite, topaz, and quartz) abrading and etching does not alter the integral diffracting power. It seems that this crystal classification is closely related to the Type I and Type II crystals of Zachariasen.

CRYSTAL MODIFICATION

It has been common practice with LiF crystals to increase their R values by using mechanical treatments such as surface abrasion. It is interesting to inquire whether or not LiF can be made ideally imperfect by such means. If we require that the R values be reduced from the ideally imperfect value by no more than 10% it follows that a dislocation density of about 5×10^{10} cm^{-2} is needed. This dislocation density is a factor of 5000 greater than that which Birks and Seal obtained in flexed LiF, and a factor of 50 to 500 greater than that which they observed at the surface of abraded LiF. Moreover, the diffraction peak breadth-at-half-maximum implied by a dislocation density of 5×10^{10} cm^{-2} is 0.8 degrees. Thus, it appears that it should be very difficult to push the 200 reflection from LiF up to the ideally mosaic limit by adding additional dislocations. Moreover, if those high dislocation densities were possible, the resulting resolution and selectivity would be unsatisfactory for most crystal spectrometer applications.

It is not known whether the R values of Type II crystals can be optimized by some physical treatment of the crystal. Our knowledge of the mosaic block structure of Type II crystals is sufficiently limited to make this a matter only of interesting speculation at this time.

SUMMARY

In summary, we have developed a technique, based on the mosaic block model, which allows calculation of the integral diffracted power for plates in Bragg geometry lying anywhere in the range from ideally imperfect to perfect, and in the range from flat to highly curved. This technique, developed following a method of Zachariasen, has been designed to give exact results in the perfect and ideally mosaic limits. Between these limits it gives results which agree with experiment both in terms of absolute intensity and in the variation of R with wavelength. Using this technique we now have sound reasons for knowing what can be done and what should be done in the search for optimum crystals.

REFERENCES

1. W. H. Zachariasen, "A General Theory of X-Ray Diffraction in Crystals," Acta Cryst. 23, 558-564 (1967).
2. L. S. Birks and R. T. Seal, "X-Ray Properties of Deformed LiF," J. Appl. Phys. 28, 541-543 (1957).

3. J. Vierling, J. V. Gilfrich, and L. S. Birks, "Improving the Diffracting Properties of LiF," Appl. Spectry. **23**, 342-345 (1969).

4. J. E. White, "X-Ray Diffraction by Elastically Deformed Crystals," J. Appl. Phys. **21**, 855-859 (1950).

5. P. G. Burkhalter, R. R. Whitlock, J. V. Gilfrich, and L. S. Birks, Naval Research Laboratory, unpublished data.

6. L. S. Birks, "Electron Probe Microanalysis," p. 75 ff., John Wiley (1963); or p. 43 ff., John Wiley (1971).

X-RAY PRODUCTION CROSS SECTIONS FOR Ti, Co, Ge, Rb AND Sn BY 16 - 44 MeV OXYGEN ION BOMBARDMENT*+

R. P. Chaturvedi**
T. W. Bonner Nuclear Laboratories, Houston, Texas

J. L. Duggan
Oak Ridge Associated Universities, Oak Ridge, Tennessee

T. J. Gray
North Texas State University, Denton, Texas

C. C. Sachtleben
Hastings College, Hastings, Nebraska

J. Lin
Tennessee Technological University, Cookeville, Tennessee

ABSTRACT

Absolute K-shell ionization cross sections were measured for Ti, Co, Ge, Rb, and Sn for incident oxygen ions from 16-44 MeV. The x-rays were measured with a high resolution Si(Li) detector (166 eV at 5.9 keV). All of the data represents cross section measurements for thin targets. The measured cross sections for these elements are compared to the theoretical predictions of the Binary Encounter Approximation (BEA). K_α/K_β ratios and energy shifts were also extracted from the data. The experimental data are compared to measured cross sections for other elements to give an overview of the systematics for oxygen ion induced x-ray production cross sections in this energy range. Some comment will also be given in regard to the use of oxygen ions to measure the parameters associated with ion implanted semiconductors.

* Experiments were performed on the Tandem VandeGraaff Accelerator at the Oak Ridge National Laboratory, Oak Ridge, Tennessee

+ Travel funds were provided by the U.S.A.E.C.

**On leave from the State University of New York at Cortland, New York

INTRODUCTION

The characteristic x-rays of elements have been produced by charged particle bombardments for more than six decades. Chadwick(1) used protons and alpha particles as the projectiles to study ion induced x-ray processes. Coates (2) was the first to use heavy ions ($M>M_\alpha$) to induce characteristic x-rays in Al, S, Br, Pb, and Mo. This idea was rediscovered by Armbruster (3) [Also Armbruster, et.al.(4)]. Specht (5) used heavy fission fragments to excite x-rays in several elements. He was able to study almost all the features observed in the heavy ion induced x-ray spectra of elements.

Recently several review papers [F. W. Saris (6), P. Richard (7), and J. D. Garcia, et.al.(8)] summarize the available experimental and theoretical results on heavy ion-atom collisions. These measurements involve direct observation of either characteristic x-rays or Auger electrons that result from the filling of an inner shell vacancy. From these data, it may be concluded that the inner-shell excitation mechanism in heavy-ion-atom collisions is different from that for proton impact. In the case of heavy-ion-atom encounters, it is supposed that during the collision a short-lived quasi-molecule is formed and inner-shell electrons are promoted because energy-level crossing occurs as the projectile approaches the target atom.

Briefly speaking, the important distinctions of heavy-ion-atom collisions are: (A) The cross sections depend critically on the relative binding energies of the collision partner; more so at the lower energies (<100 keV); (B) The cross sections are much larger than for incident protons or alpha particles of comparable velocities; (C) A measurable shift in energy and line broadening is observed in Si(Li) x-ray spectra; (D) The appearance of non-characteristic x-ray transitions; (E) The dependence of fluorescent yield on the energy of the ions; for this reason the cross sections for <u>x-ray productions</u> are distinguished from the <u>cross-section for vacancy production</u> in the atomic shells of the target; (F) In high resolution experiments, satellites [D. Burch, et.al.(9)] and hyper-satellite spectral lines [P. Richards, et.al.(10)] appear. These observations as well as the energy shift in the characteristic x-rays are interpreted as resulting from multiple inner shell ionization during the heavy-ion-atom collisions; (G) A general disagreement between the heavy-ion-atom interaction data and the predictions of Binary Encounter Approximation (BEA) [Garcia (11)] as well as Plane Wave Born Approximation (PWBA) [Merzbacher and Lewis (12)]. Both these theories have remarkable applicability in the case of protons and alpha particles. The agreement between theory and experiment becomes even more impressive when binding and coulomb deflection effects are included [Basbas, et.al.(13)].

An interesting fact emerges when one examines the data for 8-30 MeV oxygen ions incident on Mn, Ni, Cu, Zn, Ge, and Sn [Sachtleben, et.al.(14)]. The cross section falls approximately on an emperical curve, representing a plot of the reduced parameters $U_k^2 \sigma_k / Z_1^2$ and $E/\lambda U_k$ (where U_k is the K-shell binding energy, σ_k is the K-shell vacancy production cross section, Z_1 is the atomic number of the oxygen ions, E is the energy of the projectile, and λ is the ratio of the oxygen mass to the electron mass.) Furthermore, as the energy of the oxygen ion increases, the difference between the experimental and the theoretical predictions of BEA becomes smaller. This most probably happens because at the higher energies the oxygen ions are nearly stripped of their outer electrons and, therefore, behave like a point charge. With this as a background we have measured the x-ray production cross-section of Ti, Co, Ge, Rb, and Sn. An attempt has also been made to investigate the possibility of the oxygen-ion-atom encounter in measuring the parameters associated with the ion implanted semiconductor by studying characteristic x-rays of Zn in a Si matrix.

EXPERIMENTAL

The beam of oxygen-ions was extracted from the Oak Ridge National Laboratory Tandem-VandeGraaff. It was focused on the target by using two tantalum apertures of 2 mm diameter and separated by a distance of 47.2 cm. The second aperture was 30.5 cm from the target.

Due to the inherent problem of preparing a uniform target and the determination of the charge of oxygen ions, the 19° Rutherford scattered oxygen beam and the induced x-rays were measured simultaneously. To calculate the energy loss of the ions within the target, the thicknesses were measured by 2-MeV protons and ^{241}Am alpha particles. Finally, the energy loss was calculated using stopping power data from the Nuclear Data tables.

The intrinsic efficiency of the detector as well as the absorption of the ion induced x-rays in the range 4.5 to 32 keV were determined experimentally by positioning calibrated radioactive sources, inside the beam tube at the same location and size as the beam spot on the target. The sources included were ^{48}V, ^{54}Mn, ^{57}Co, ^{65}Zn, ^{109}Cd, ^{137}Cs, ^{203}Hg, and ^{241}Am. A 10 mil thick mylar was introduced between the target and the detector to reduce bremsstrahlung background.

The spectra of x-ray pulses and oxygen-ion detector signals were stored in 1000 channels of a 16,000 channel analyzer. The latter was interfaced with the CDC 3200 computer for analyzing and plotting. The dead time in either spectra was kept below 5 per cent. Figure 1 shows a typical spectrum of Ni K_α and K_β x-rays produced by 8- and 20-MeV oxygen ions. In the figure the relative shifts of the K_α and K_β lines can be seen, in going from incident oxygen energies from 8-20 MeV. These energy shifts will be discussed in a later section of this paper.

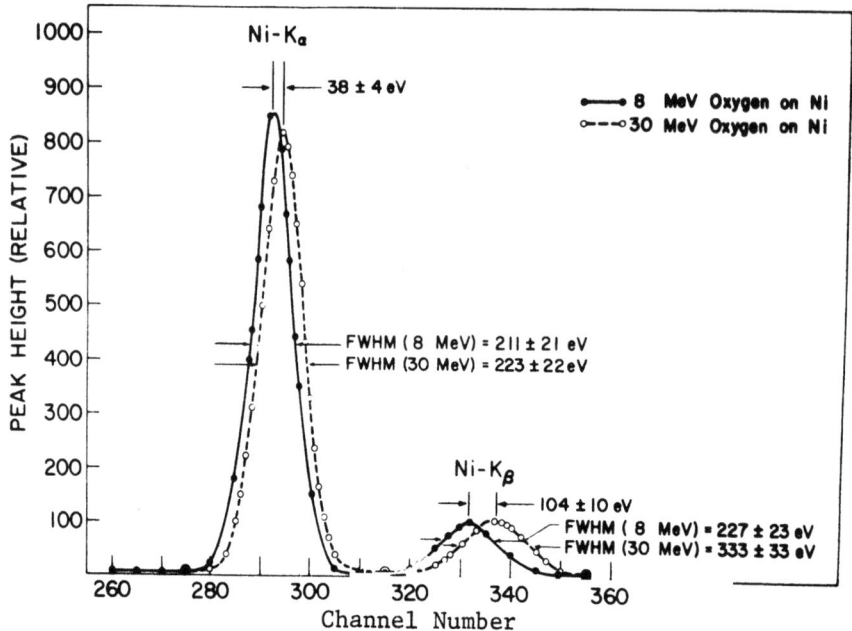

Figure 1 K-X-Ray Energy Shift for 8 and 30 MeV Oxygen Impact on Nickel

RESULTS AND DISCUSSION

It can be assumed that for the elements and energies studied in this paper that the oxygen elastically scattered cross section is approximately Rutherford at a Laboratory angle of 19°; [J. J. Simpson, et.al.(15)]. Therefore, the x-ray production cross section σ_x in terms of Rutherford scattering cross section σ_R (barns) is given as:

$$\sigma_x = \sigma_R \frac{N_x}{N_0} \frac{\Omega}{\varepsilon} \qquad (1)$$

where N_x is the number of x-rays observed, N_0 is the number of elastically scattered oxygen ions, Ω is the oxygen ion-detector solid angle in steradians, and ε is the total x-ray detector efficiency. The measured value of Ω was 9.11×10^{-4} sr. Equation (1) eliminates error in the cross section data which occurs due to the uncertainties in measuring the thickness of a target and the number of charged particles impinging on it. The main experimental error is due to the efficiency of the x-ray detector which is estimated to be about 10 per cent. The results of the measured ionization cross sections for Ti, Co, Ge, Rb, and Sn are given in Table I.

TABLE I

(Cross Section for K-Shell Ionization Productions in Ti, Co, Rb, Ge, and Sn*)

Cross sections in barns; BEA (Binary Encounter Approximation Theory

Element		16 MeV	20 MeV	24 MeV	28 MeV	32 MeV	36 MeV	40 MeV	44 MeV
Ti	Expt.	5,220	13,730	26,200	49,500	78,420	114,200	166,400	193,200
	BEA	14,610	25,330	37,540	50,240	63,520	77,740	92,030	104,100
Co	Expt.	384.7	1,267	2,090	3,910	6,434	10,290	14,300	20,170
	BEA	1,977	3,599	5,706	8,241	11,130	14,310	17,730	21,290
Rb	Expt.	18.4	34.5	---	108	203	332	495	682
	BEA	66.7	133	---	365	529	727	936	1,232
Ge	Expt.	52.0	115	222	393	630	1,060	1,850	2,420
	BEA	286	642	1,110	1,630	1,920	3,020	3,940	4,790

		8.6 MeV	13.6 MeV	18.7 MeV	23.8 MeV	28.8 MeV
Sn	Expt.	.035	.291	1.09	2.79	4.55
	BEA	.300	1.14	3.14	7.98	14.6

*Standard deviation ± 15%

There is a slight problem that is present when the experimental results are compared to the theory. The theoretical cross sections predict the total ionization of the K-shell. In the laboratory the x-ray production cross-sections are measured. Hence, it is necessary to divide the x-ray production cross-section by the fluorescence yield (ω_k) to obtain the ionization cross section (σ_I). The problem comes in that it has been shown that the value of ω_k is dependent on the velocity of the incident heavy ion [Bhalla and Hein (16)]. For the comparison shown in Table I, the values of ω_k which were used are the latest tabulation by Bambynek, et.al.(17). These are: 0.221, 0.366, 0.554, 0.679, and 0.850 for Ti, Co, Ge, Rb, and Sn respectively.

Figure 2 shows a plot of the K-shell ionization cross sections for the above elements with the predictions of the Binary Encounter Approximation.

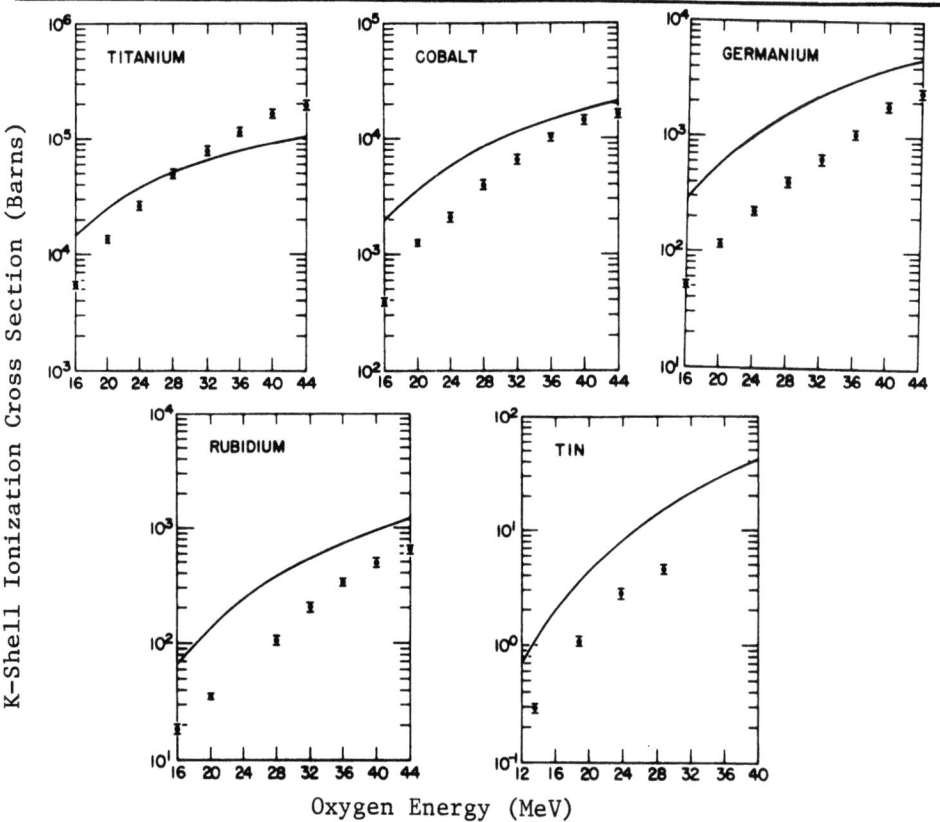

Figure 2 K-Shell Ionization Cross-Sections Induced by Oxygen Ion Impact

(The solid curves are predictions of the Binary Encounter Approximation)

For low values of $E/\lambda U_k$, the experimental results are considerably lower than the predicted values. This difference becomes smaller for $E/\lambda U_k \geq 0.1$ which is in agreement with other published results [T.M. Kavanagh, et.al.(18); Sachtleben, et.al.(14)].

Figure 3 shows the variation of K_α/K_β ratio as a function of $E/\lambda U_k$ which includes the present and previous data [Sachtleben, et. al.(14)] of this group. It indicates a continuous decrease of K_α/K_β ratio from approximately 7.5 to 5.0 as $E/\lambda U_k$ goes from 0.03 to 0.12; beyond which it is not changing so sharply.

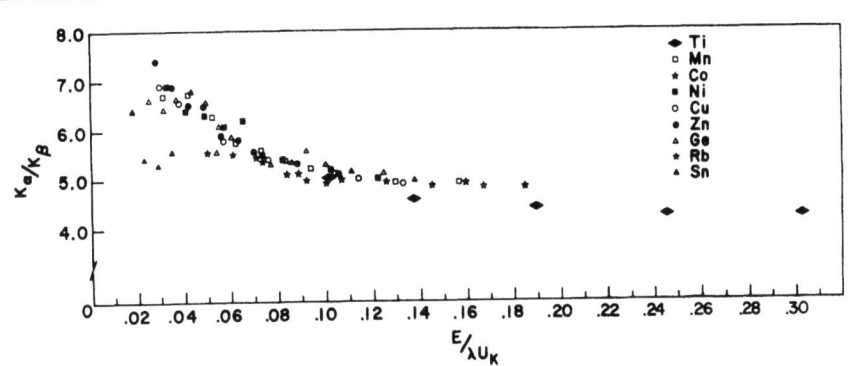

Figure 3 K_α/K_β Ratios for Oxygen Bombardment on Ti, Mn, Co, Ni, Cu, Zn, Ge, Rb, and Sn

The change in K_α/K_β is also reflected via the energy shifts ΔE_α and ΔE_β in the characteristic energies as shown in Figure 4. Burch and Richard (19) found that for 15 MeV oxygen ion impact on Ca and V, shifts for E_α and E_β are 54 ± 6 and 52 ± 6 for E_α and 158 ± 6 and 156 ± 6 for E_β respectively. This compares favorably with our E_β shift of 151 ± 6 for Ti caused by 16 MeV oxygen impact; ΔE_α being 32 ± 6 eV is, however, in disagreement with these data. Saltmarsh, et.al.(20) have measured ΔE_α and ΔE_β for Ti in the same projectile energy range as in the present case and they obtained comparably higher results. Our results agree with Saltmarsh, et.al., in one qualitative aspect; that is the maximum shift in both cases does not occur at $E/\lambda U_L = 1$ for Ti, Co, Ge, Rb, and Sn. The maximum shift occurs at $E/\lambda U_L \simeq 1.6, 1.3, 1$ for Ti, Co, Ge respectively. This is in accordance with the fact that the impact parameter of the projectile being less than the Bohr radius of L shell, effectively increases U_L, more for lower Z than for the higher Z elements. [G. Basbas, et.al.(21,22)].

Figure 5 shows the variation of σ_x as a function of Z, ($22 \leq Z \leq 51$) for the fixed bombarding energy of the oxygen ion. The agreement between various groups is quite good, particularly Richard's data on Copper [as quoted by Garcia, et.al.(8)].

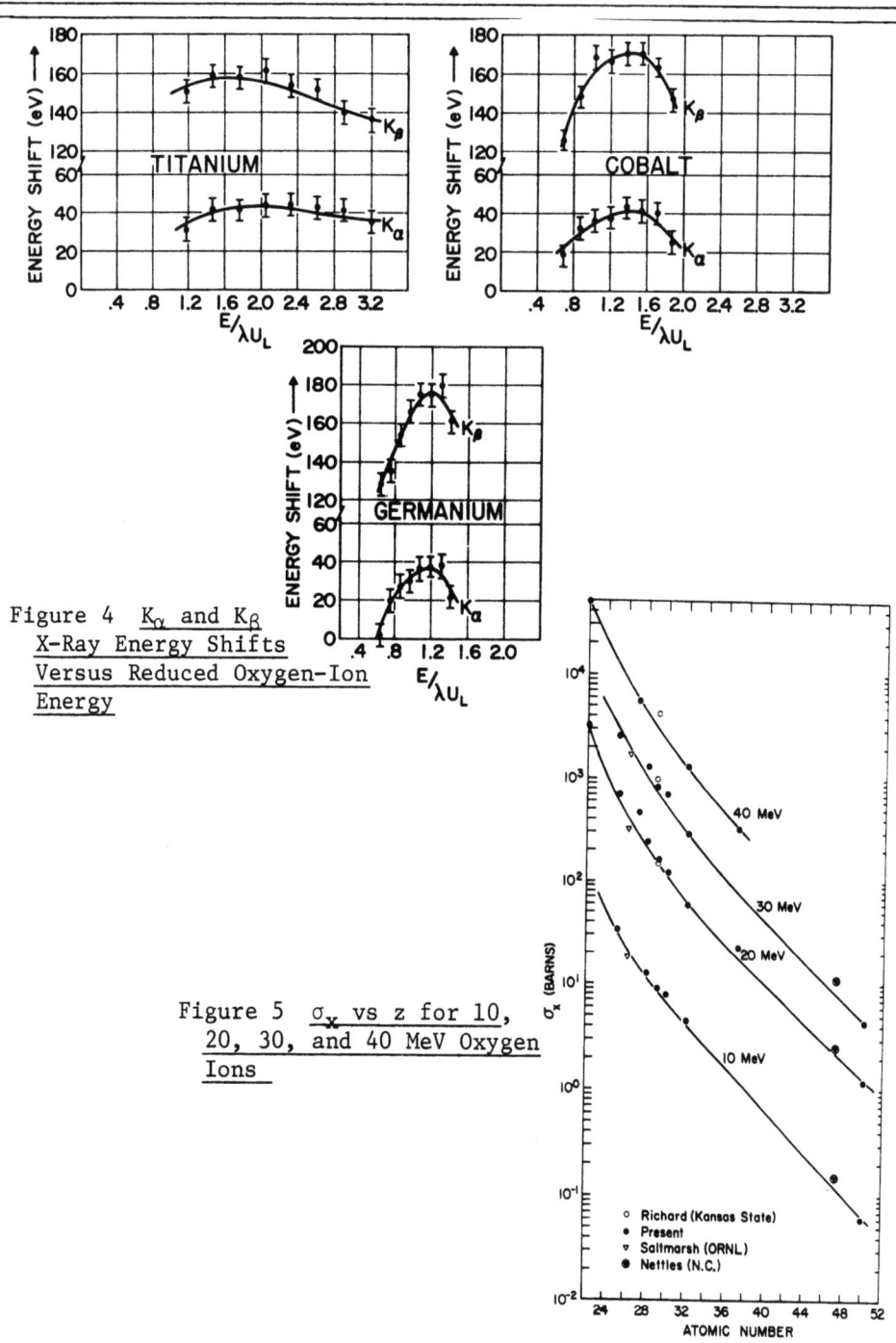

Figure 4 K_α and K_β X-Ray Energy Shifts Versus Reduced Oxygen-Ion Energy

Figure 5 σ_x vs z for 10, 20, 30, and 40 MeV Oxygen Ions

Figure 6 shows a comparison of the experimental points for nine elements plotted with the BEA theory. For low values of $E/\lambda U$ the experimental results are more than an order of magnitude lower than the BEA predictions. The difference, however, becomes smaller as $E/\lambda U$ increases, which is in agreement with the measurements of Kavanagh, et.al.(18).

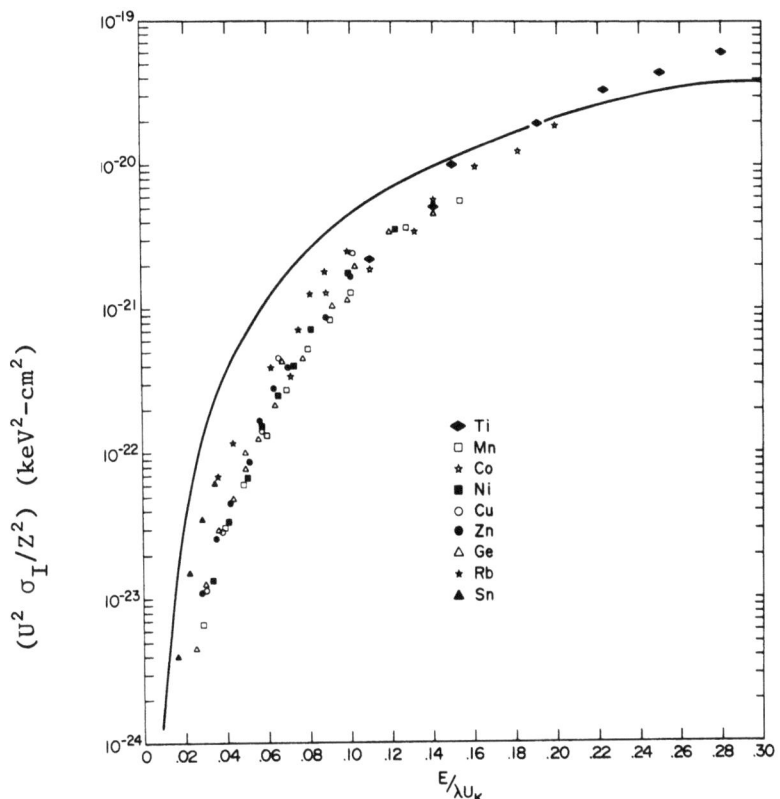

Figure 6 K-Shell Ionization by Oxygen Impact
(The solid curve is the prediction of the Binary Encounter Approximation)

Evaluation of Ion Implanted Semiconductors by X-Ray Measurements

For these measurements a Zinc implanted semiconductor was furnished by Texas Instruments. Zn ions in concentrations of 3.1×10^{15} atoms/cm^2 in a 1 mm thick silicon matrix was used. The target was evaluated with both 1.4 MeV protons and 24 MeV oxygen ions. For the proton spectra the peak to background measurement for the K lines was 15:1; while for oxygen bombardment it was 2 ± 0.5:1. The decrease in the signal-to-noise ratio with the increasing

Z for the bombarding particle is a direct result of the increased background from the charged particle induced bremsstrahlung originating in the thick host material, i.e. silicon. This can be understood in terms of the stopping power for an ion of energy E, mass M, and atomic number Z. For protons and oxygen ions, the ratio of stopping powers Sp and So respectively is given:

$$\frac{So}{Sp} \approx \frac{M_o}{M_p} \frac{Z_o^2}{Z_p^2} \frac{E_p}{E_o} \approx 60 \quad (2)$$

for the 1.40 MeV protons and 24 MeV oxygen ions. The measured ratio of the x-ray production cross section at these energies is approximately 20:1 for σ_x (oxygen)/σ_x (proton). As the stopping power for these energies is governed by the atomic processes, then it is expected that from this rather crude model the signal-to-noise ratio should decrease by a factor of 3 for 24 MeV oxygen ions over that observed for 1.4 MeV protons. However, spectral broadening of the K_α components in x-ray spectrum which are due to the multiple ionization causes a further deterioration of the signal-to-noise ratio by a factor of 2. Hence, from these calculations, it is expected that the observed signal-to-noise ratio would be 2.5:1 for 24 MeV oxygen ions when compared to a value of 15:1 for 1.4 MeV protons.

It is, therefore, concluded that for the purpose of analysis of trace elements in infinitely thick targets such as silicon, it is advantageous to use the lower energy protons instead of higher energy oxygen bombardment.

REFERENCES

1. J. Chadwick, "The γ Rays Excited by the β Rays of Radium," Phil. Mag. 24, 594 (1912)

2. W. M. Coates, "The Production of X-Rays by Swiftly Moving Mercury Ions," Phy. Rev. 46, 542 (1954)

3. P. Armbruster, "Ionisierung innerer Elektronenschalen bei der Abbremsung von Spaltprodukten," Z. Physik 166, 341 (1962)

4. P. Armbruster, E. Röckl, H. J. Specht, and A. Vollmer, "Die Untersuchung fast-adiabatisdner Stöbe mit Spaltprodukten," A. Naturforsch, 19A, 1301 (1964)

5. H. J. Specht, "Ionisation innerer Elektronenschalen bei fast-adiabatischen Stöben schwerer Ionen," Z. Physik 185, 301 (1965)

6. F. W. Saris, "Characteristic X-Ray Production in Heavy-Ion-Atom Collisions," The Physics of Electronics and Atomic Collisions,

Ed. by T. R. Grover and F. J. De Heer, North Holland Publishing Co., (1972), pg. 181.

7. P. Richards, "X-Ray Production by Heavy Ions," In Proceedings of the International Conference on Inner Shell Ionization Phenomena and Future Applications, Atlanta, Georgia, 1972, Edited by R. W. Fink, S. T. Mahson, J. M. Palms, and P. V. Rao; U.S.A.E.C., Oak Ridge, Tennessee (1973)

8. J. D. Garcia, R. J. Fortner, and T. M. Kavanagh, "Inner Shell Vacancy Production in Ion-Atom Collisions," Rev. Mod. Phy., Vol. 45, 111 (1973)

9. D. Burch, P. Richards, and R. L. Blake, "Resolved Structure in Fe K_α X-Rays Produced by 30-MeV Oxygen Ions," Phy. Rev. Letters 26, 1355 (1971)

10. P. Richards, W. Hodge, and C. Fred Moore, "Direct Observation of K_α Hypersatellites in Heavy-Ion Collisions," Phy. Rev. Letters 29, 393 (1972)

11. J. D. Garcia, "Inner-Shell Ionizations by Proton Impact," Phy. Rev. A, 1, 280 (1970); "X-Ray Production Cross Sections," Phy. Rev. A, 1, 1402 (1970); "X-Ray Production by Alpha Particle Impact," Phy. Rev. A, 4, 955 (1971)

12. E. Merzbacher and H. W. Lewis, "X-Ray Production by Heavy Charged Particles," Handbuch Der Physik (Springerverlag, Berlin) 34, 166 (1958)

13. G. Basbas, W. Brandt, and Roman Laubert, "Universal Cross Sections for K-Shell Ionization by Heavy Charged Particles I. Low Particle Velocities," Phy. Rev. A 7, 983 (1973)

14. C. C. Sachtleben, J. L. Duggan, and R. P. Chaturvedi, "X-Ray Production Cross Sections for Mn, Ni, Cu, Zn, Ge, and Sn by 8-30 MeV Oxygen Ion Bombardment," Bull. Am. Phy. Soc., 18, 103 (1973)

15. J. J. Simpson, J. A. Cookson, D. Ecclesmall, and M. O. L. Yates, "Coulomb Excitation of Ti^{50}, Cr^{52}, and Fe^{54} First Excited States," Nucl. Phy. 62, 385 (1965)

16. C. P. Bhalla and M. Hein, "Theoretical K-Shell Fluorescence Yield of Multiply Ionized Neon," Phy. Rev. Letters 30, 39 (1973)

17. W. Bambynek, B. Crasemann, R. W. Fink, H. U. Freund, H. Mark, C. D. Swift, R. E. Price, and P. V. Rao, "X-Ray Fluorescence Yields, Auger, and Coster-Kronig Transition Probabilities," Rev. Mod. Phy. 44, 716 (1972)

18. T. M. Kavanagh, R. J. Fortner and R. C. Der, "Production of Characteristic X-Rays by Heavy Ion Bombardment," In Proceedings of the International Conference on Inner-Shell Ionization Phenomena and Future Applications, Atlanta, Georgia, 1972, Ed. by R. W. Fink, S. T. Mahson, J. M. Palms, and P. V. Rao U.S.A.E.C., Oak Ridge, Tennessee (1973)

19. D. Burch and P. Richards, "X-Ray Spectra from Oxygen-Ion Bombardments on Ca and V at 15 MeV," Phy. Rev. Letters 25, 983 (1970)

20. M. J. Saltmarsh, A. Van der Woude and C. A. Ludermann, "Energy Shifts and Relative Intensities of K X-Rays Produced by Swift Heavy Ions," Phy. Rev. Letters 29, 329 (1972)

21. G. Basbas, W. Brandt, and Roman Laubert, "Z_1 Dependence of K-Shell X-Ray Production by Heavy Charged Particles," Phy. Letters 34A, 277 (1971)

22. G. Basbas, W. Brandt, R. Laubert, A. Ratkowski, and A. Schwarschild, "Projectile Charge Dependence of K-Shell Ionization by Swift Light Nuclei," Phy. Rev. Letters 27, 171 (1971)

SOME BIOMEDICAL APPLICATIONS OF CHARGED-PARTICLE-INDUCED X-RAY

FLUORESCENCE ANALYSIS

J. L. Campbell, A. W. Herman, L. A.
McNelles, B. H. Orr and R. A. Willoughby

University of Guelph, Guelph
Ontario, N1G 2W1, Canada

ABSTRACT

X-ray fluorescence induced by charged particles has been employed in trace element analysis of both animal and human blood, tissue and bone samples. Preparation techniques included microtome slicing and wet digestion in nitric acid, internal chemical standards being used in the latter case.

Most of the specimens arose from a study of interactions between the toxic elements lead and zinc in growing foals; this was motivated by reports of sickness and death in foals raised near lead-zinc smelters. The cause of toxicity in animals from environmental pollution is often attributed to single factors, whereas in reality interactions among many factors, including a variety of toxic and nutrient trace elements, should be considered.

A variety of spectra are presented and elemental concentrations derived. Agreement between the X-ray data and atomic absorption spectrophotometry is encouraging. The results demonstrate the potential of particle-excited X-ray fluorescence as a broad-range analytical technique for the study of trace element interactions.

INTRODUCTION

Use of charged particle beams in X-ray fluorescence analysis is receiving widespread attention on account of the potentially high signal-to-background ratio, rapidity of data accumulation and wide-range sensitivity. Several workers (1), (2), (3) have recently reached the common conclusion that 2-4 MeV protons are preferable to other particles on a variety of grounds.

We have employed proton-excited XRFA to study blood, liver, kidney and bone specimens from foals bred at the Ontario Veterinary College. The work is part of a study (4) of lead and zinc toxicity, originally prompted by reports of lameness and death among foals reared near lead/zinc smelters. Its primary objectives were to determine separately the clinical signs of lead and of zinc poisoning and the signs due to simultaneous uptake of toxic amounts of both, and to deduce the nature of any interaction between the two elements. Nine foals were used, basal diets being supplemented by toxic amounts of ZnO in three cases, $Pb(CO_3)_2$ in two, and both compounds in three; the final animal was a control.

After necropsy, analysis was performed by Cominco Ltd., Trail, B.C. using atomic absorption spectrophotometry (AAS); iron, zinc, copper and lead levels were determined in various specimens. A parallel analysis using proton-excited XRFA was carried out in order to compare results with AAS and to examine a wider range of trace elements. The main purpose was to ascertain the overall accuracy of the XRFA method as an analytical tool in a real, rather than a contrived situation, and to assess its potential for future trace element interaction studies similar to the present one.

EXPERIMENTAL ARRANGEMENT

Figure 1 illustrates the vacuum target apparatus (constructed totally from aluminum) employed in these measurements. An incident proton beam entering from the left is defined and aligned by means of three graphite collimators A, B, C. Induced X-rays emerging from a sample mounted on the target ladder are detected by means of a Kevex guard-ring Si(Li) detector (220 eV at 5.9 keV) 7cm distant from the target. A high purity aluminum collimator is interposed between the detector and target such that the detector only views the sample itself. A unique feature of the chamber is the addition of a 0.020" tubular aluminum cold trap maintained at liquid nitrogen temperature via a copper cold finger. The trap enshrouds almost the entire target area thereby condensing hydrocarbons in the immediate vicinity. The vacuum obtained was $\sim 10^{-5}$ torr. A more detailed account of the target chamber design, beam handling and inter-related experimental parameters is given elsewhere (3).

SAMPLE PREPARATION

For purposes of comparison, two methods of preparation of tissue samples were adopted. The first involved a simple approach whereby any chemical treatment was minimized; it consisted of

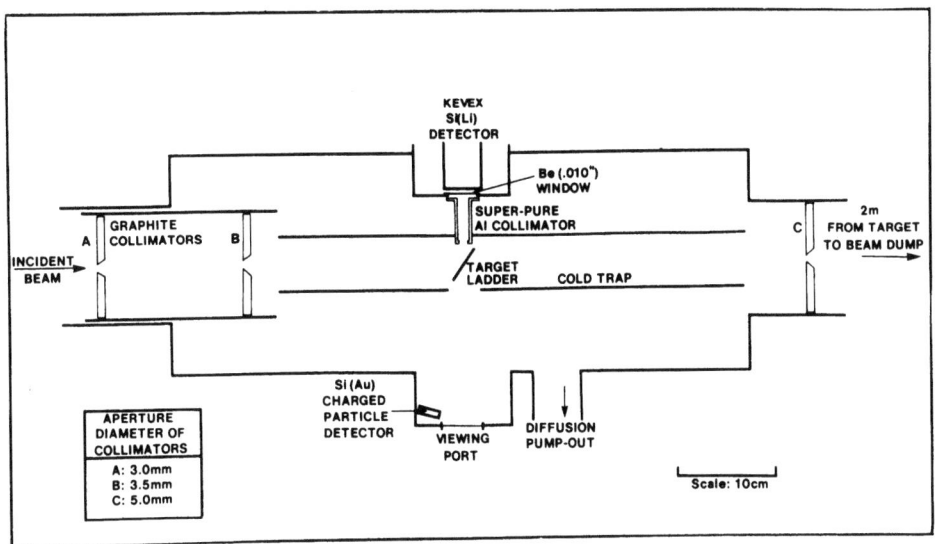

Figure 1. Vacuum target chamber

slicing 10 microns of frozen tissue with a microtome and mounting the slice on a carbon foil of thickness 20 µg/cm^2. The choice of carbon foils as target backings and their trace impurities have been discussed extensively elsewhere (5). This procedure involved minimal risk of contamination.

The second approach involved wet digestion of tissue (∼0.25gms) with high purity nitric acid (Ultrex brand) and also included addition of an internal standard solution containing strontium (∼20 ppm), silver (∼190 ppm) and cadmium (∼240 ppm). A drop of the mixture was subsequently placed on the carbon foil and allowed to dry. The risk of trace metal contamination is perhaps greater with this procedure; however the overall efficacy of both methods could be easily evaluated upon final analysis. Bone samples were prepared solely by wet digestion and blood targets were fabricated into thin, dry self-supporting films.

DISCUSSION OF EXPERIMENTAL RESULTS

Protons in the energy range 2-3 MeV were obtained from the 3MeV Van de Graaff accelerator at McMaster University, Hamilton, Ontario. With the beam currents employed (≲ 1µa) there was no apparent loss of material or damage to the targets. Volatilization of elements from the targets was monitored at various beam currents up to 1 µa as a function of time by observing cumulative X-ray intensities at 5-minute intervals; no measurable losses were observed. Sample X-ray spectra of liver, kidney and bone are presented in Figures 2-5

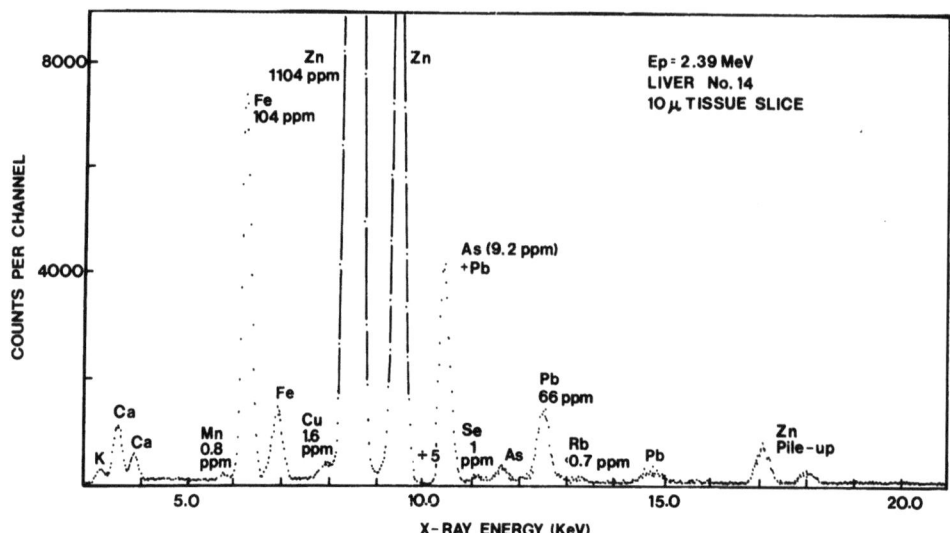

Figure 2. X-ray spectrum of liver of foal 14

for a proton energy of 2.39 MeV. A large variety of elements was observed in the spectra, viz chromium, manganese, iron, nickel, copper, zinc, arsenic, selenium, bromine, lead, rubidium and molybdenum. Figure 2 is the X-ray spectrum of a microtome slice of liver tissue from foal number 14. Figures 3, 4 represent the wet digested tissue samples of the kidney cortex and kidney medulla of

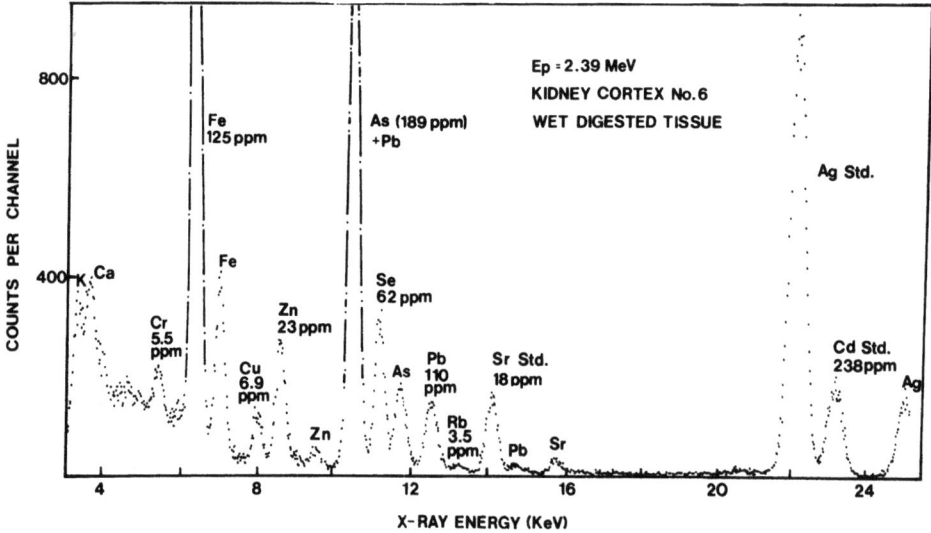

Figure 3. X-ray spectrum of kidney cortex of foal 6

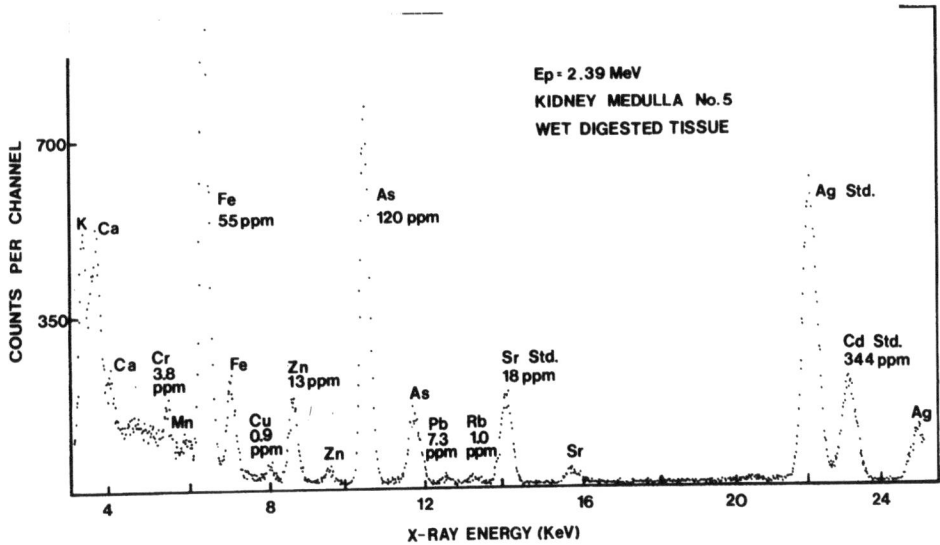

Figure 4. X-ray spectrum of kidney medulla of foal 5

foals 6 and 5 respectively in the lead-fed group. Figure 5 represents a wet digested bone sample from a foal of the lead-fed group.

Measurement of the concentrations of these trace metals was performed for both modes of sample preparation. Since an internal standard could not be easily incorporated in the microtome tissue

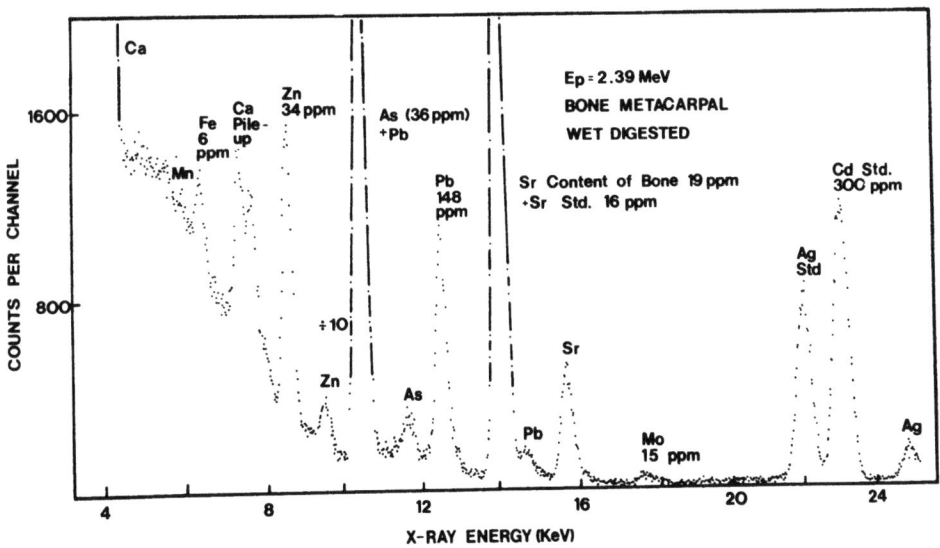

Figure 5. X-ray spectrum of bone from lead-fed foal

slices, the concentration of the most dominant element, generally iron, was determined from isotope-excited X-ray fluorescence using a ^{109}Cd source (20 mCi). The concentrations of the remaining trace elements could be calculated from the proton-excited X-ray intensities relative to their corresponding known iron values. Analysis of the wet digested tissue samples was a simpler procedure due to the internal standards of strontium, silver and cadmium whose concentrations were precisely predetermined. That Sr and Cd were useful standards was established by noting that the Sr/Cd X-ray intensity ratio was constant for a wide variety of specimens; the Ag intensity fluctuated and therefore was not utilised. In bone of course, only the Cd could be used.

These calibration techniques are discussed extensively elsewhere (6) full attention being given to excitation cross-sections, fluorescence yields and detection efficiencies. Reduction of the large volume of data was accomplished using a PDP-9 computer, which derived concentrations from the raw spectra using the above parameters.

The results of this analysis were compared with the AAS data for iron, copper, zinc and lead (4). Table 1 represents a compilation of the data for liver samples only and for trace metals of iron, copper, zinc, and lead for the corresponding group of foals. Overall agreement with AAS is quite good with, perhaps, slightly better agreement between wet digested tissue analysis and atomic absorption. This may be due to the fact that wet digestion involves larger samples which are thus more representative than 10 micron tissue slices. However, differences should not be great since elemental distribution in the liver is quite uniform. Table 2 is a compilation of the data for the kidney cortex and kidney medulla and for zinc and lead only. Again overall agreement with AAS is quite good for the cortex, with generally better agreement for the wet digested samples. In the case of the kidney medulla there is some disagreement; however the medulla is not a well defined region of the kidney and uniformity from one tissue sample to the next is not expected.

The original study (4) of toxicity in foals reproduced the clinical symptoms of separate lead and zinc poisoning in accordance with the lead and zinc concentrations. However, it was discovered that the nature of the interaction between toxic amounts of lead and zinc together is one where the toxic amounts of zinc prevent the development of the clinical signs of lead poisoning despite the increased retention of lead in the liver and kidney. Many such interactions exist between toxic and nutrient elements in metabolic systems and the data for other elements made available by this analysis are now being correlated.

Some extreme concentrations of arsenic ranging up to 800 ppm

TABLE 1: TRACE METAL CONTENT OF LIVER[a]

	Pb Fed Animals			Pb and Zn Fed Animals				Zn Fed Animals				Control
	5[b]	6	x̄	9	13	14	x̄	1	2	4	x̄	8
Iron												
*	120	152	141	133	194	142	156	194	117	147	153	114
**	153	114	134	126	157	104	129	153	165	170	163	102
+	145	120	132	140	140	90	123	195	150	120	155	90
Zinc												
*	46	59	53	818	100	950	623	1334	1130	1709	1391	85
++	55	55	55	676	157	1104	646	1426	1190	1857	1489	87
+	45	55	50	785	102	1200	696	1370	1300	1900	1527	78
Copper												
*	5.2	13.1	9.1	4.0	7.4	5.4	5.6	7.9	6.7	9.9	8.1	4.2
++	7.2	14.2	10.7	1.8	9.2	1.6	4.2	2.6	-	8.5	-	8.4
+	6.5	14.3	10.0	3.5	7.4	2.7	4.5	5.6	4.8	5.4	5.3	5.3
Lead												
*	16	26	21	66	34	51	50	1.8	1.6	6.5	2.2	1.3
++	26	26	26	80	44	66	63	<2.0	12	9	-	<0.7
+	20	33	27	70	42	48	53	0.9	4.5	7.0	4.1	1.0

[a] Values in ppm, net weight basis
[b] Number refers to identification of each foal
* Proton-excited X-ray fluorescence analysis of wet digested tissue
** Isotope-excited X-ray fluorescence analysis of microtome tissue slice
\+ Atomic absorption spectrophotometry
++ Proton-excited X-ray fluorescence analysis of microtome tissue slice

TABLE 2(a) : TRACE METAL CONTENT OF KIDNEY CORTEX[a]

Pb Fed Animals			Pb and Zn Fed Animals				Zn Fed Animals				Control
5[b]	6	x̄	9	13	14	x̄	1	2	4	x̄	8
Zinc											
19*	23	21	526	22	437	328	296	737	676	569	33
25**	33	29	528	46	272	282	353	634	528	505	36
20+	20	20	115	26	490	210	295	710	580	511	27
Lead											
74	110	92	-	220	139	-	1.0	3.6	6.1	3.6	<1.2
88	168	128	462	340	1400	734	2.5	4.0	1.8	2.8	<5.5
55	86	71	185	300	150	211	1.9	5.0	6.5	4.5	1.5

TABLE 2(b) : TRACE METAL CONTENT OF KIDNEY MEDULLA[a]

Pb Fed Animals			Pb and Zn Fed Animals				Zn Fed Animals				Control
5	6	x̄	9	13	14	x̄	1	2	4	x̄	8
Zinc											
13	13	13	256	10	143	136	296	452	360	369	13
22	12	17	232	23	74	110	191	62	142	132	20
15	15	15	65	10	260	112	490	350	350	397	12
Lead											
7.3	1.1	4.2	2.8	1.5	6.7	3.7	2.6	<1.0	1.0	-	<0.4
4.5	1.3	2.9	60	27	0.7	29	<1.3	<0.2	<5.5	-	<17
20	25	23	75	47	45	56	0.9	2.5	4.0	2.5	0.5

[a] Values in ppm, wet weight basis
[b] Number refers to identification of each foal
* Proton-excited X-ray fluorescence analysis of microtome tissue slice
** Proton-excited X-ray fluorescence analysis of wet digested tissue
\+ Atomic absorption spectrophotometry

were discovered in both liver and kidney of all the foals as illustrated in Figures 2-4. Values of 10-15 ppm are considered lethal for these young foals. A systematic search of the procedures employed revealed that immediately following necropsy, a solution of glutaraldehyde plus buffer was injected which then permeated the tissue and arrested further decay. It was determined by isotope induced X-ray analysis that the solution contained large amounts of arsenic and hence proved to be the source. Upon analysis of the selenium data it was determined that most tissue samples contained normal traces of this essential element of the order of ∼0.5 ppm. However foals numbered 5,6,9,13,14 registered levels ranging from 10-40 ppm in the kidney cortex only. The common denominator for this group is the lead feed which contained lead carbonate. Many lead compounds including mined lead contain traces of lead selenite and hence the selenium observations are not unexpected.

HUMAN LIVER AND LUNG ANALYSIS

Proton-excited X-ray fluorescence has been extended in our laboratory to the study of human liver and lung inflicted with cancer. Figure 6 represents the X-ray spectrum of a sample of cancerous human lung in which traces of manganese, iron, nickel, copper, zinc, lead, selenium, bromine and rubidium were measured using internal standards of strontium and cadmium. The essential trace elements such as manganese, iron, copper and zinc involved in the normal synthesis of tissue and cells are being evaluated for a variety of cancerous and normal samples for comparison.

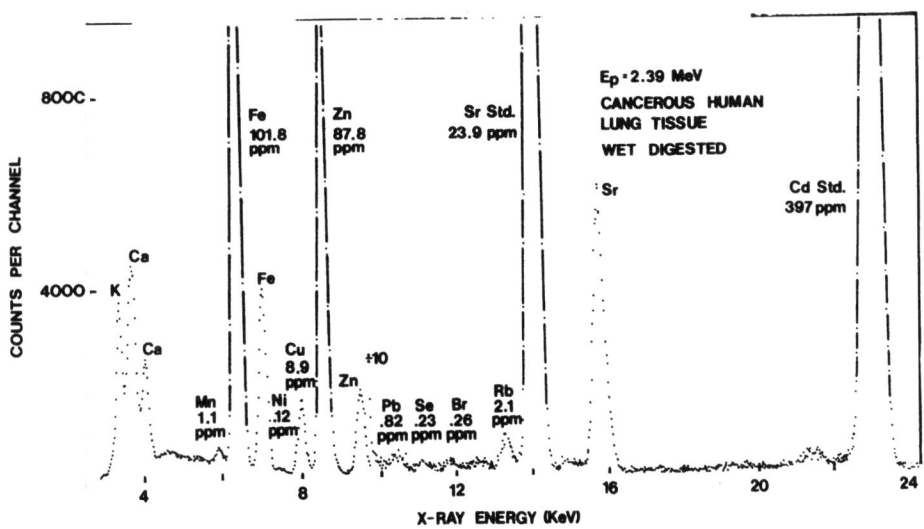

Figure 6. X-ray spectrum of cancerous human lung

CONCLUSIONS

Broad range trace element analysis using proton-excited X-ray fluorescence appears to have considerable potential as an analytical tool for rapid and accurate measurement. Accuracy of measurement has been investigated here by comparison with an established analytical technique. The large number and availability of small accelerators to outside users is also facilitating X-ray fluorescence for many small laboratories. Most biomedical samples can be prepared with minimal contamination by wet digestion, and addition of internal standards appears to be a satisfactory calibration technique. More rigorous tests, involving analysis of NBS standard reference materials are in progress, and preliminary results (6) are encouraging.

ACKNOWLEDGEMENTS

This work was supported by the Canadian National Health Grants Programme (Project 605-7-744). We are indebted to the staff of the McMaster Accelerator Laboratory for their assistance.

REFERENCES

1. J.J. Kliwer, J.J. Kraushaar, R.A. Ristinen, H. Rudolph and W.R. Smythe, "Trace Element Analysis by Observation of Characteristic X-rays", Bull. Am. Soc. $\underline{17}$, 545 (1972).

2. J.A. Cooper, "Comparison of Particle and Photon-excited X-ray Fluorescence applied to Trace Element Measurements of Environmental Samples", Nucl. Instr. Meth. $\underline{106}$, 525-538 (1973).

3. A.W. Herman, L.A. McNelles and J.L. Campbell, "Choice of Physical Parameters in Charged-particle-induced X-ray Fluorescence Analysis", Int. J. App. Rad. in press.

4. R.A. Willoughby, E. MacDonald, B.J. McSherry and G. Brown, "Lead and Zinc Poisoning and the Interaction between Lead and Zinc Poisoning in the Foal", Can. J. Comp. Med. $\underline{36}$, 348-359 (1972)

5. A.W. Herman, L.A. McNelles and J.L. Campbell, "Target Backings for Charged-particle-induced X-ray Fluorescence Analysis", Nucl. Instr. Met. $\underline{109}$, 429-437 (1973).

6. J.L. Campbell, A.W. Herman, B.H. Orr and L.A. McNelles, "Specimen Preparation and System Calibration in Proton-excited X-ray Fluorescence Analysis", to be submitted for publication (1973).

QUALITATIVE ANALYSIS OF THE KOSSEL BACK REFLECTION

PATTERN FROM SELECTED SEMICONDUCTORS

Robert L. Fitzpatrick

Motorola Semiconductor Products Division

Phoenix, Arizona 85008

ABSTRACT

The value of the divergent beam method as a powerful tool for investigating crystal perfection over rather wide limits has been known since the analytical work of K. Lonsdale. This method has been treated experimentally and theoretically by numerous authors and was popularized by Sigmund Weissmann. Although not nearly as useful as the reflection and transmission topographic techniques of Berg-Barrett and Lang for crystal characterization, it is <u>potentially</u> very useful for three dimensional quantitative strain analysis. The "technique phase" as Harvey Yakowitz remarked, is essentially over; however, refinements and optical improvements are certain to evolve if quantitative measurements are to become useful. Very short exposures and the non-destructive nature of the reflection divergent beam - Kossel method is a useful survey tool that has gained limited acceptance.

Presently, the reflection mode is useful in the qualitative survey of numerous semiconductor materials and processes, some of which have been considered from time to time by various authors. A range of crystal perfection, as well as some surface modifications, are surveyed. Kossel reflection pattern artifacts and irregularities are discussed.

INTRODUCTION

Prior to and during general applications of the Berg-Barrett and Lang topographic techniques for localized crystal perfection, strain and misorientation, a number of integrated intensity measurements were made using the Bragg (1) (2) (3) (4) and Laue (5) arrangements from specific crystal volumes. A "merit factor" was assigned (5) in relation to the integrated intensity obtained from each specific volume in the crystal (the higher intensity, the less perfect the volume, etc.). Investigators earlier had observed that the "perfect" volume integrated intensity was proportional to the structure factor F and that imperfect volumes diffracted up to $[F]^2$; hence, intensity variations (imperfection, strain and misorientation) can easily be observed. Topographic techniques put physical meaning to enhanced diffracted intensity observed during the crystal alignment prior to "mapping." These methods have been improved by using multiple crystal topographic methods.

Highly diverging x-ray radiation can be utilized, particularly to characterize crystals of "intermediate" crystal perfection as suggested by K. Lonsdale (6), who pointed out that when a crystal is highly perfect, primary extinction limits the reflection volume to a very thin layer near the surface. Conversely, when a crystal is grossly imperfect, the diffracted beams are attenuated by secondary extinction and do not form sufficiently sharp cones and disappear in the background irradiation of the film.

Small elastic stresses present in the crystal cause small changes in the lattice 'd' spacing of certain planes (hkl) and produce corresponding shifts in the line positions on the film; thus, it is possible to study their magnitude and distribution in single crystals. James (2) first coined the term "Kossel lines" after Kossel, who first produced them.

Generation of a pseudo-Kossel back reflection trace is sketched in Figure 1, wafer surface is parallel with the film. For example, CuKα from a point source -S- strikes the plane (h,k,l) at a Bragg angle θ_B and is diffracted as a cone about its normal. Let ϕ be the angle between the diffracting plane and the crystal surface nearest -S-. The semiapex angle of the cone is equal to $(90-\theta_{hkl})$. A film placed as in Figure 1 will intercept diffracted and scattered radiation and the respective (hkl) traces (7) generated are:

$\phi < \theta_B$ Ellipses occur.   (1)

$\phi = 0$ A circle is obtained (diffracting (2) planes are perpendicular to the axis of the incident beam). The largest (2θ) trace will appear nearest the center of the film.

$\phi = \theta_B$ A parabola occurs (highly unlikely) (3)

$\phi > \theta_B$ Does not exist in back reflection (4)

The back reflection pattern may be expected to contain largely elliptical conics and perhaps some circular traces, with only a remote chance of finding a parabolic trace.

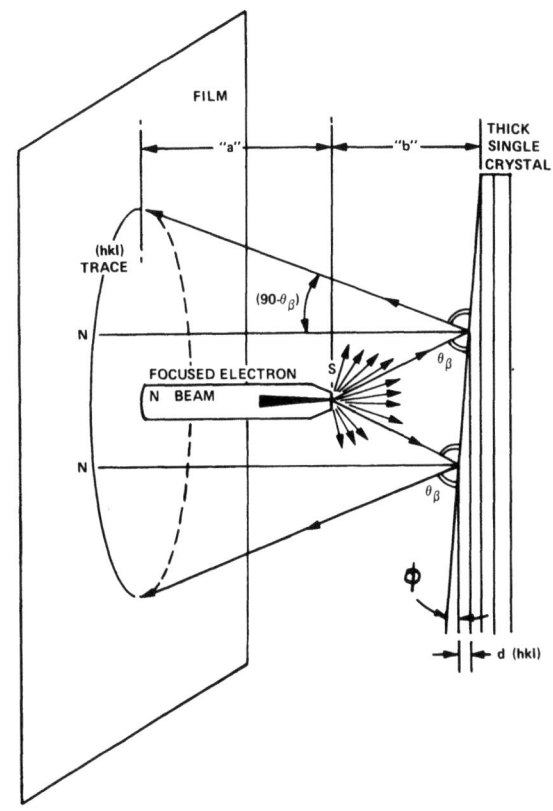

Figure 1. Generation of a Pseudo Kossel Trace in Back Reflection

Lonsdale suggested use of the stereo projection to represent the expected pattern and further, choice of radiation if the plot with a suitable wavelength is demonstrated. The choice of radiation is critical if special "lens patterns" or intersections are desired for lattice parameter measurements (7). Kossel back reflection pattern of (001), and (111) oriented silicon crystals diffracted with CuKα_1, α_2 radiation are shown in Figure 2.

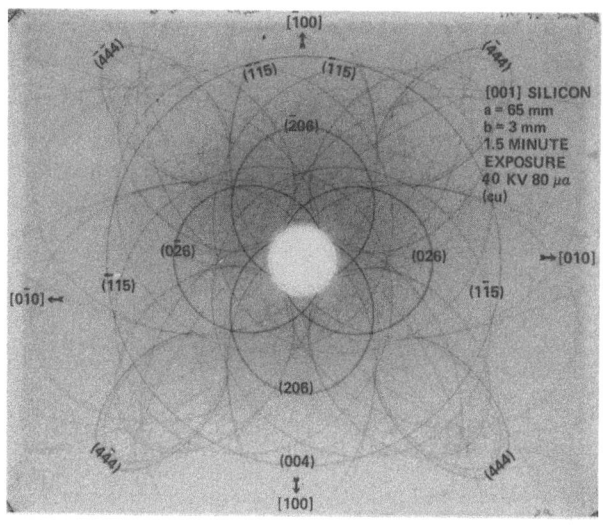

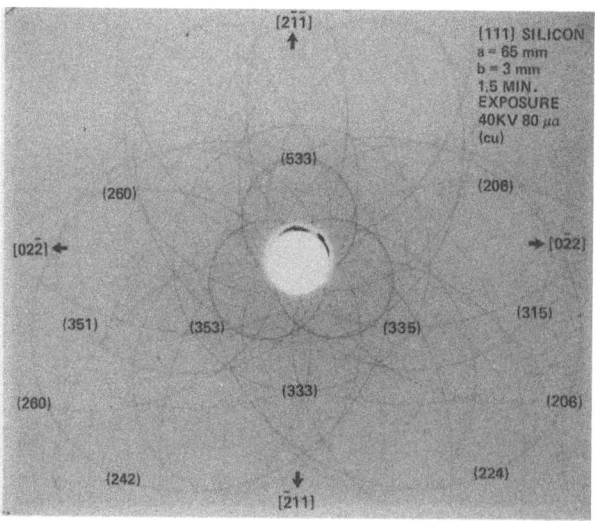

Figure 2. Kossel Pattern of (001) and (111) Oriented Silicon with CuK$\alpha_1\alpha_2$ Diffracting

EXPERIMENTAL

The divergent beam method and the production of pseudo Kossel pattern has been known for some time. Standard diffraction equipment was used by Geiseler (8), et al around 1947 to produce such a pattern from a fluorescing target place near the sample. Exposures ranged from 8 to 16 hours with extraneous Laue "spots" also present.

It was not until the advent of the micro focus source of high intensity that increasing use has been made of the divergent beam-Kossel method, a large source that gives the effect of broadening. Two types of micro sources have produced "Kossel" patterns. 1) The electron micro probe and 2) the micro focus x-ray generator. Both are point sources of approximately 1.0 and 10μ diameter (electrons and x-ray respectively); however, the method of sample irradiation and production of Kossel and pseudo Kossel patterns as expected are fundamentally different. With the micro probe "Kossel" patterns were first obtained by Castaing (9), who suggested the potential of the method.

The micro focus generator became practical with the electron lens design of Imura (10). The divergent beam, Kossel applications, especially three dimensional strain analysis, was popularized by Weissmann (11).

Figure 3. Micro Focus Generator With Back Reflection Optics Modified

The Rigaku B-3 micro focus generator, used during the course of these experiments, is shown in Figure 3 with the back reflection optics modified. The Rigaku-Kossel copper transmission target provides a 10μ spot and diffraction approaching 160° (2θ) at the cassette.

The beam is aligned (after standard alignment) with the lens current OFF. By maximizing intensity of a spot on a fluorescence screen pressed against the target (micro positioning of filament), lens current is adjusted for maximum spot intensity. Penumbra was checked and found to be minimal.

Superficial (hkl) trace discontinuities are produced by a physical discontinuity of the specimen in the diffracting plane caused by a small sample (reduce b), surface topography, and local lattice misorientation.

The extended target absorbs skew radiation (see [444] traces of Figure 2a) and is minimized for the largest useable "b" value. The Ray Diagram, Figure 4, shows a functional relationship of θ_B, (35 and 70°) "a" and "b" common to back reflection experiments of normal incidence. The (004) trace of Figure 2a was obtained from a diffracted circle ~9mm in diameter.

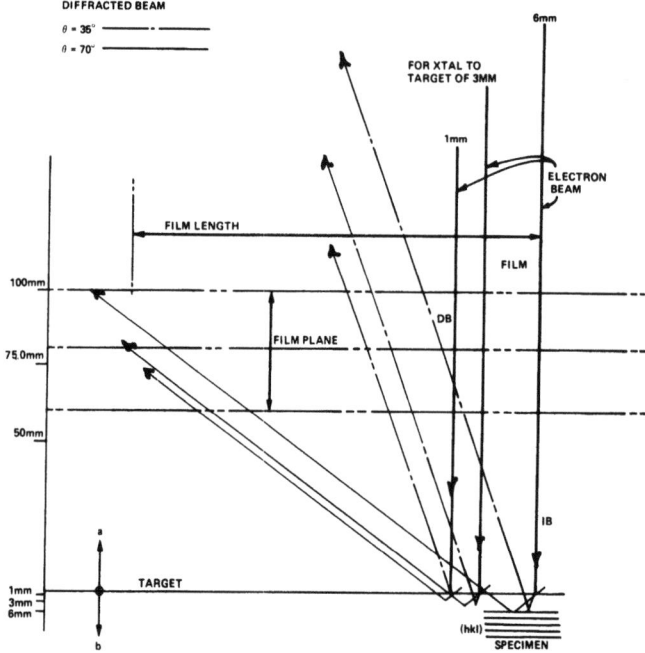

Figure 4. Ray Diagram Relating θ_B, b, and a

The copper target was generally operated at 40KV and 80μa (specimen fluorescence may reduce trace intensity). The "a" values ranged from 50 to 90mm and "b" values from 0.5-6mm, the lens was often set at 234mm. Increasing "a" enlarges the pattern, while increasing "b" enlarges the pattern and irradiated area. No effort was made to eliminate CuKβ etc. Numerous exposures, using several filaments, averaged 1.5 minutes each and were generally recorded on Kodak SB54 single coated blue sensitive medical x-ray film (28.3 x 23.3cm exposed) of low isotropic film shrinkage.

The Rigaku generator optical layout is satisfactory for small research crystal surveys, after the camera mount is aligned and bolted to the table top; however, back reflection work with larger crystals requires modification similar to that shown in Figure 3. For transmission work, the optical layout is not convenient or satisfactory and requires other modification.

APPLICATIONS

Numerous materials and several processes have been surveyed with the back reflection Kossel method. Several pseudo Kossel patterns obtained are included to illustrate somewhat the applicability of the method.

At first some variation in exposure was observed (formulated by Yakowitz [12]). These differences were investigated with variations in bulk crystal perfection and surface modification of silicon (including sawn, half and two μ diamond lapped, step etched and cleaved surfaces). Two extremes were compared during a single exposure; limited exposure latitude, however, was obtained; sawn surface work damage gave visual evidence of line broadening of all traces. Crystal perfection line width, as expected, is immeasurable. The Kossel Pattern (hereafter shown split) in Figure 5a shows the 0.6° mosaic spread of a graphite, and in 5b, a near perfect (001) silicon crystal.

Gross discontinuities (not artifacts) unrelated to strain are another means of detecting crystal perfection with this macro technique. More useful than the conventional Laue technique to survey research grown crystals, the large area Laue technique (13) is shown in Figure 6a, adjacent to the Kossel pattern 6b of Czochralski grown GaP; multiple grain growth is easily detected.

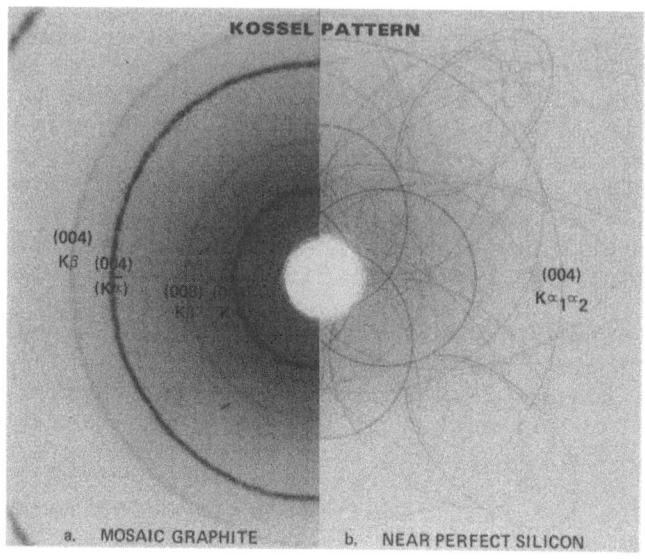

a. Mosaic Graphite   b. Near Perfect Silicon
Figure 5. Kossel Pattern

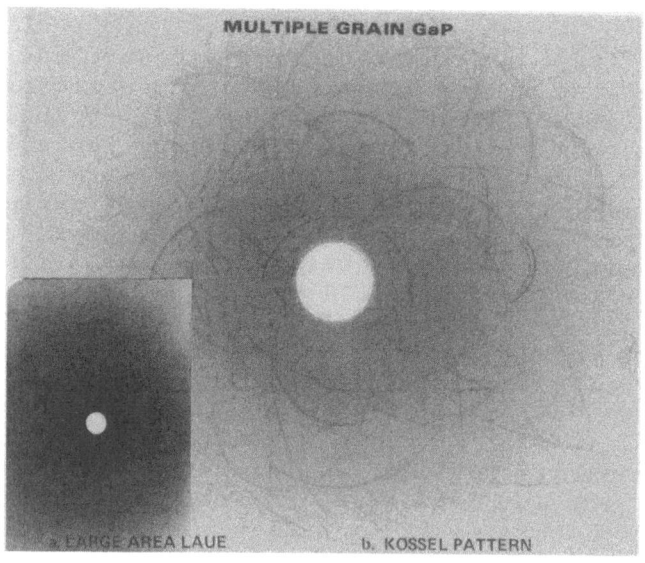

a. Large Area Laue   b. Kossel Pattern
Figure 6. Multiple Grain GaP

Working with suspected highly strained ideal mono crystal experimental mediums, line profile discontinuity directly related to strain or local lattice misorientation (11) was not observed. Volumes expected to be highly strained were compared to expected low strain volumes of equal contour, including the seed-growth interface junction and crystal bottoms jerked from the melt of GaAs and $LiNbO_3$ as well as large facets of vapor phase grown [111] silicon.

In addition to bulk characterization, thin crystalline films on substrates offer interesting mediums of investigation, especially for interfacial strain characterization.

The Kossel Pattern, Figure 7a, of a 9μ germanium doped silicon layer on (001) silicon, strikingly shows the orientation of the epitaxial layer with respect to the substrate even with the (004) trace. This Δd difference of ∼0.0054Å, larger than the $α_1$-$α_2$ separation is clearly resolved, and furthermore, irregularities in the epitaxial outer traces suggesting reduced crystal perfection and strain are in agreement with topography. Figure 7b shows a Kossel pattern of 20μ (001) silicon on (1̄102) sapphire. Orientation of the epitaxy with

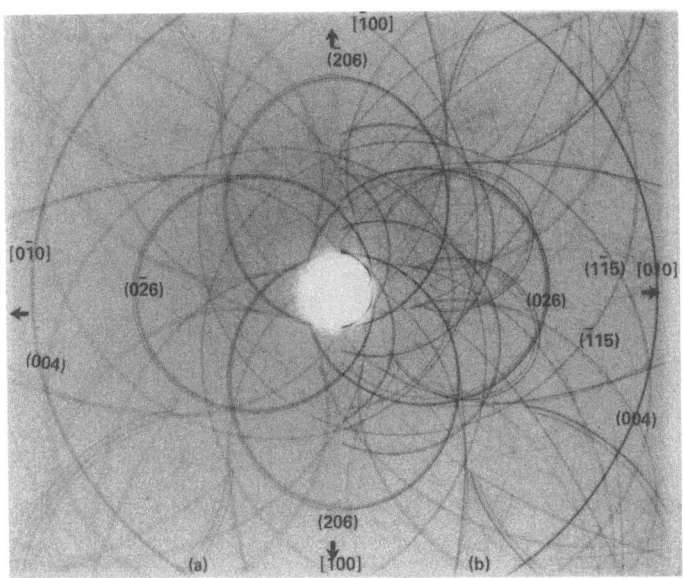

a. Germanium Doped Silicon Epitaxial on (001) Silicon b. Silicon Epitaxy on (1̄102) Sapphire

Figure 7. Kossel Pattern

respect to substrate is less obvious, however, it is determined easier than with the Laue method. Other patterns were obtained from thin ∿0.5μ silicon layers and with (111) spinel substrates. Observations of α_1 and α_2 separation in several planes suggested that the thicker layers and growth on (111) spinel were of slightly higher relative perfection in agreement with topography and the literature.

A Kossel pattern of an extremely warped (r=3.7cm, t=0.022cm) silicon wafer ("curl" toward beam) in the [$\bar{1}$00] direction is compared to flat (001) silicon in Figure 8. In addition, the pattern of the back side (also different) shows the (004) trace major axis now larger than the radius of the unstrained flat wafer in a [$\bar{1}$00] direction. Though an unreasonable curvature, it does show modified pattern continuity for r and t.

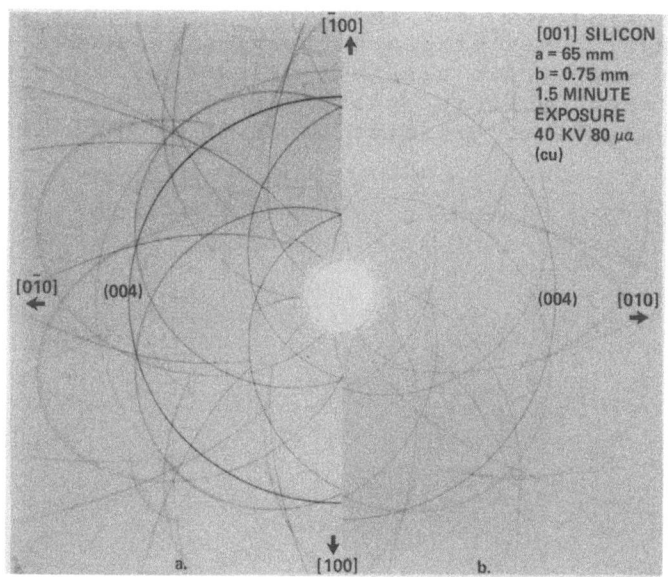

a. Warped Wafer   b. Flat Wafer
Figure 8. Kossel Pattern of (001) Silicon

Pattern overlay surveys of flat and processed silicon with single crystal and preferred orientations grown and deposited, of slight curvature, did not prove useful, primarily because of the difficulty of maintaining a single target to specimen distance.

Future surveys include the transmission mode, improved alignment and a precise selected area diffraction method (localizing spatially the divergent cone). Lattice parameter measurements of selected materials are of intermediate interest; hopefully, three dimensional quantitative strain analysis of bulk and thin film deposited material interfaces will prove useful.

ACKNOWLEDGEMENTS

The author wishes to thank Tom N. Tucker for reviewing the paper and furnishing the various work damage silicon surfaces. Czochralski pulled $LiNbO_3$ crystals were kindly provided by Don R. Clement and the art and photography work by Glen K. Richter.

REFERENCES

1. C. G. Darwin, "The Theory of X-Ray Reflexion, I & II," Phil. Mag. 27, 315-333, 675-690, (1914).

2. R. W. James, "The Optical Principles of X-Ray Diffraction in Crystals," London (1950).

3. J. R. Patel, R. S. Wagner and S. Moss, "X-Ray Investigation of the Perfection of Silicon," Acta Met. 10, 759-764 (1962).

4. J. R. Carruthers, R. B. Hoffman and J.D. Ashner, "X-Ray Investigation of the Perfection of Silicon," J. Appl. Phys. 34 3388-3393 (1963).

5. L. V. Azaroff and W. L Bragg, "Elements of X-Ray Crystallography," McGraw Hill, 250-251 (1968).

6. K. Lonsdale, "Divergent-Beam X-Ray Photography of Crystals," Phil. Trans. Roy. Soc., London, Ser. A 240, 219-250 (1947).

7. H. Yakowitz, "The Divergent Beam Technique," ADVANCES IN ELECTRONICS AND ELECTRON PHYSICS, Supplement 6, Electron Probe Microanalysis, Ed. A. J. Tousimis and L. Marton (1969) Academic Press, Inc. 361-431.

8. A. H. Geiseler, J. K. Hill, and J. B. Newkirk, "Divergent Beam X-Ray Photography with Standard Diffraction Equipment," J. Appl. Phys. Vol. 19, 1041-1049 (1948).

9. R. Castaing
   Thesis, University of Paris (1951).

10. T. Imura, "A Study of Deformation of Single Crystals by the Divergent Beam Method," Part I, II and III, J. Jap. Inst. of Metals, Vol. 16, 10- (1952).

11. T. Ellis, L. F. Nanni, A. Shrier, S. Weissmann, G. E. Padawer, and N. Hosokawa, "Strain and Precision Lattice Parameter Measurements by the X-Ray Divergent Beam Method, I and II," J. Appl. Phy., Vol. 35, 11, 3364 (1964).

12. H. Yakowitz and D. L. Vieth, "Exposure Time Relationings for Kossel Micro Diffraction Photographs," J. of Research NBS Vol. 696, 213-216 (1965).

13. J. Angilello, "Large Area X-Ray Back Reflection Technique for Crystal Surface Examination," Norelco Reporter, Vol. XV No. 1, pp. 15, 27 (1968).

AN EXPERIMENTAL EVALUATION OF THE ATOMIC NUMBER EFFECT

L. Parobek and J.D. Brown
Faculty of Engineering Science and Centre for
Interdisciplinary Studies in Chemical Physics
The University of Western Ontario
London, Canada

ABSTRACT

A method for measuring the atomic number effect is developed using a sandwich sample technique. The depth distributions of x-ray production, $\phi(\rho z)$ curves, have been measured for a zinc tracer in aluminum, copper, silver and gold matrices at 30, 25, 20 and 15 keV. The $\phi(\rho z)$ curves were measured using a Cambridge Microscan 5 in which the electron beam is normal to the sample surface and the x-ray take-off angle is $75°$.

Samples of the low concentrations of copper (~1 Weight %) in aluminum, nickel, silver and gold were prepared. For each alloy system (for example, Cu - Al), three different concentrations of copper were prepared. The intensity ratios from the sample to the pure element (standard) for each system have been plotted against concentration. At such low concentrations of copper the relation between this ratio and concentration is linear. The slopes of the curves have been compared to the equivalent factors obtained as ratios of the area under $F(\rho z)$ curves for aluminum, silver and gold to the area under $F(\rho z)$ curve for copper, respectively. The $F(\rho z)$ curves are obtained from $\phi(\rho z)$ curves; $F(\rho z) = \phi(\rho z) \exp(-\mu \rho z \csc \psi)$ where μ is mass absorption coefficient.

Comparisons are made between these experimental data and the current methods of calculating the atomic number effect.

INTRODUCTION

The sandwich sample technique has been used by Castaing and

Descamps (1) to determine the absorption correction and Brown (2) used the technique for determining the depth distribution of secondary x-ray production. The same technique can be used to determine the atomic number effect in systems where secondary x-ray production is negligible.

If layers of identical thickness of a single tracer element are deposited within a series of sandwich samples having matrix elements of different atomic numbers then differences in the measured depth distributions will be due to the differences in backscattering and stopping power of the matrix elements. These differences are just those ascribed to the atomic number effect. This method is illustrated in Figure 1. The atomic number correction factor is directly related to the ratio of the areas under the $\phi(\rho z)$ curves. The purpose of this paper is to show that this technique gives good results in determination of the atomic number effect and furthermore provides valuable experimental data needed when theoretical approaches (Diffusion Model, Monte Carlo Calculation or Transport Equation) are used to calculate the various corrections.

EXPERIMENTAL MEASUREMENTS

Four samples were prepared for the measurements of the depth distribution curves. The matrix elements were aluminum, copper, silver and gold and the tracer element was zinc. Because of the flourescence of the zinc $K\alpha$ line by the L lines of gold, the atomic number effect in this system cannot be determined until the flourescence contribution is removed. Therefore this system will not be considered further. The matrix elements were chosen, in part, because of their x-ray properties; in part, because of the ease of evaporation. All evaporations in preparing the sandwich samples were carried out from a heated tungsten filament in an oil diffusion pumped system.

The tracer and layer thicknesses were determined by weighing and x-ray spectrometry. Agreement between the two methods was within one percent. The depth distribution curves were measured at 30, 25, 20 and 15 keV electron energies in a Cambridge Microscan -5 in which the electron beam is normal to the sample surface and the x-ray take-off angle is 75°.

For comparison with the atomic number data from the $\phi(\rho z)$ curves, homogeneous Cu - (Al, Ni, Ag) alloys were prepared from metals of at least 99.99% purity. For each alloy system three different concentrations were prepared in the concentration range of 0.5 - 1.5 weight percent.

To prepare Cu-Al alloys a powder mixture of aluminum and copper

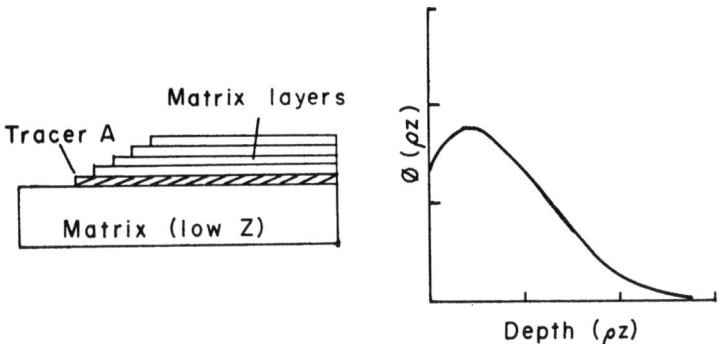

(a) Low Z matrix, Tracer A.

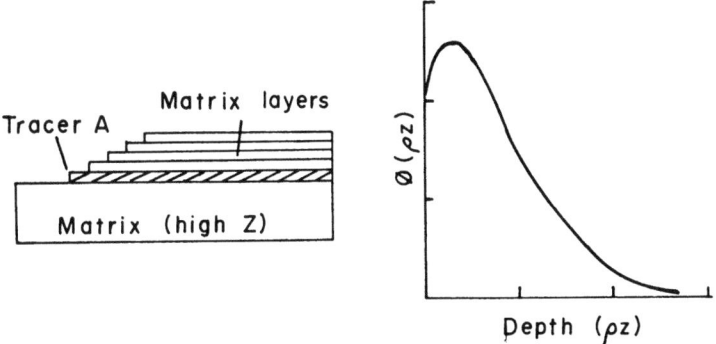

(b) High Z matrix, Tracer A.

Figure 1. Use of the Sandwich Sample Technique to Determine the Atomic Number Effect.

TABLE 1. K_{Cu}^1 Values, CuKα, Silver Matrix

E_o (keV)	F(ρz)	Alloys
15	1.14 ± 0.01	1.13 ± 0.01
20	1.06 ± 0.01	1.07 ± 0.01
25	1.03 ± 0.01	1.02 ± 0.01
30	0.97 ± 0.01	0.96 ± 0.01

was melted in a carbon crucible under hydrogen atmosphere. The samples were homogenized in evacuated quartz tubes at 580°C. The copper-nickel alloys were prepared by leviation melting of nickel and copper in an induction furnace under inert atmosphere and then homogenized in vacuum at 900°C. The induction furnace also was used to prepare copper-silver alloys. The copper and silver were melted in a carbon crucible under inert atmosphere, then homogenized in vacuum at 750°C.

The CuKα intensity from the alloys was measured at 15, 20, 25 and 30 keV electron energies in the Cambridge Microscan 5 using a LiF analysing crystal. From each alloy five measurements of x-ray intensity were taken at five different points. These intensities have been compared to the intensity from pure copper as standard. Backgrounds were measured off peak 2°2θ and from the same elements. The intensity from the pure copper standard for each electron energy was adjusted to about 10,000 counts/sec, to obtain approximately the same standard deviation in intensity measurements for all electron energies. All intensity measurements were corrected for drift and dead time. The standard deviation of the intensity measurements for all alloys was less than 1%.

The true concentration of copper in the alloys was determined by atomic absorption spectrometry. Repeated determinations in all cases gave a difference of less than 1%.

The experimental results from the alloys were compared to the results from the depth distributions through the following equation.

$$K_{Cu}^1 = \frac{K_{Cu}}{C_{Cu}} = \frac{\int_0^\infty F_i(\rho z)\, d\rho z}{\int_0^\infty F_{Cu}(\rho z)\, d\rho z} *$$

where K_{Cu} is the ratio of the intensity from the alloy to pure element standard and C_{Cu} is the concentration of the copper in the alloy. The F(ρz) curves are related to the ϕ(ρz) curves: F(ρz) = ϕ(ρz) exp(-μρz cscψ). The F(ρz) curves represent the origin of x-rays which escape from the sample. A comparison of the two methods for measuring K_{Cu}^1 is given in Tables 1 and 2.

---

*i = Al, Ag

TABLE 2. K_{Cu}^1 Values, CuKα, Aluminum Matrix

E_o (keV)	F(ρz)	Alloys
15	0.780 ± 0.008	0.780 ± 0.008
20	0.808 ± 0.008	0.810 ± 0.008
25	0.819 ± 0.008	0.835 ± 0.008
30	0.830 ± 0.008	0.842 ± 0.008

CALCULATION OF CORRECTION FACTORS

The objective of the atomic number correction procedure is to estimate the factor relating the intensity ratio to concentration, C_A, in the expression

$$K_A = C_A \frac{\int_0^\infty \phi_{AB}(\rho z) \exp(-\chi_{AB}\rho z) \, d(\rho z)}{\int_0^\infty \phi_A(\rho z) \exp(-\chi_A \rho z) \, d(\rho z)} \quad (1)$$

where $\chi = \mu \csc \psi$, μ = mass absorption coefficient and ψ = x-ray take-off angle.

This expression is exact in the absence of fluorescence effects and includes two main corrections, atomic number and absorption correction.

Many methods have been proposed to evaluate the correction factor linking measured intensities to concentration in Equation (1). Poole (3) has collected experimental x-ray data for 229 standard samples of known composition. Histograms for each method were prepared showing the relative errors between calculated and known composition. The best results were obtained by using the Duncumb and Reed correction. In this method the absorption and the atomic number corrections are evaluated separately.

$$K_A = C_A \frac{R_{AB} \int_E^{Ex} \Psi_{AB}(E)/S_{AB}(E) \, dE}{R_A \int_E^{Ex} \Psi_A(E)/S_A(E) \, dE} \times \frac{(\text{Abs. Factor}) AB}{(\text{Abs. Factor}) A} \quad (2)$$

where R is a factor which takes account of the loss of ionization due to backscatter, $\Psi(E)$ is the ionization cross-section for the radiation concerned and $S(E)$ is the stopping power of the target for electrons of energy E, R is taken from the curves produced by Bishop (5) and S is evaluated from the expression

$$S = \text{const.} \, (\overline{Z/A}) \, (I/\overline{E}) \ln (1.166 \, \overline{E}/J) \quad (3)$$

where J is the mean ionization potential for the target material and is evaluated from an empirical relation

$$J/Z = 15.05 \left[1-e^{-0.072Z}\right] + (42/Z^{Z/10}) - (Z/400) \quad (4)$$

Equation (4) was evolved to give values of J which resulted in the best possible fit between corrected analytical results and known concentration for selected alloys--i.e. microanalytical results have been used to determine the apparent variation of the physical parameter J. The absorption correction is carried out using Philibert's $f(\chi)$ formula as modified by Duncumb to take account of overvoltage.

The comparison of the atomic number factors

$$\left(\frac{R_{AB}\int_E^{Ex} \Psi_{AB}(E)/S_{AB}(E) \, dE}{R_A \int_E^{Ex} \Psi_A(E)/S_A(E) \, dE}\right)$$
calculated according to this method

and experimental factors $\left(\dfrac{\int_0^\infty \phi_B(\rho z) \, d(\rho z)}{\int_0^\infty \phi_A(\rho z) \, d(\rho z)}\right)$ obtained from $\phi(\rho z)$

curves are shown in Tables 3 and 4. Macres and Wolf (6) have been claiming that atomic number correction can be successfully calculated by using diffusion model approach. Factors calculated using this correction procedure are also shown in Tables 3 and 4.

TABLE 3. Atomic Number Correction Factors
Silver Matrix, CuKα

E_o (keV)	Experimental	D. & R.	Diff. Model
15	1.16 ± 0.01	1.148	1.065
20	1.10 ± 0.01	1.121	1.039
25	1.10 ± 0.01	1.104	1.032
30	1.05 ± 0.01	1.094	1.027

TABLE 4. Atomic Number Correction Factors
Aluminum Matrix, CuKα

E_o (keV)	Experimental	D. & R.	Diff. Model
15	0.780 ± 0.008	0.790	-
20	0.808 ± 0.008	0.819	0.899
25	0.819 ± 0.008	0.834	0.925
30	0.829 ± 0.008	0.847	0.946

DISCUSSION OF THE RESULTS

The intensity generated in the sample for a small

concentration of the tracer element in the matrix is represented by the total area under the depth distribution curves. Thus, for example, the three curves for the zinc tracer represent the distribution of x-ray production of zinc in aluminum, copper and silver. By preparing alloys containing small concentrations of the tracer element in the matrix elements, the x-ray intensity escaping from the samples will be related to areas under the $F(\rho z)$ curves. To be able to compare these results, the ratio of the areas under $F(\rho z)$ curves for aluminum and silver to the area under $F(\rho z)$ curve for copper, respectively, has to be compared with K_{Cu}^1 factors obtained from the alloys. The comparison of K_{Cu}^1 values obtained from $F(\rho z)$ curves and from alloys is shown in Tables 1 and 2. The agreement between values is very nearly within the range of the experimental errors.

The comparison of the atomic number factors obtained from $\phi(\rho z)$ curves to the same factors obtained by using Duncumb's and Reed's correction procedure and Diff. Model approach is shown in Table 3 for silver matrix and in Table 4 for aluminum matrix. The best agreement between experimental results and theoretical results in both cases, silver and aluminum matrix, was when Duncumb's and Reed's correction procedure was used. This confirms that $\phi(\rho z)$ curves were measured with reasonable accuracy because Duncumb and Reed used an experimental data to determine the apparent variation of the parameter J in expression (3). Atomic number factors calculated according to the Diff. Model are not in as good agreement with the experimental values.

CONCLUSIONS

By comparing the two independent measurements of the atomic number effect, we have proved that the $\phi(\rho z)$ curves have been determined with good accuracy. Thus the $\phi(\rho z)$ curves can be successfully used to determine both atomic number and absorption correction. These data ($\phi(\rho z)$ curves) are especially important when theoretical approaches are used to calculate the various corrections. Duncumb's and Reed's atomic number correction is perhaps the most reliable correction so far developed with agreement to within 5% with the experimental measurements. Hopefully, by developing an empirical analytical function which could describe the depth distribution of the x-ray production as a function of E_o, E_K, Z and α both the atomic number and absorption correction could be established with greater accuracy.

ACKNOWLEDGMENTS

We would like to thank the National Research Council of

Canada whose financial support has made this work possible.

REFERENCES

1. R. Castaing and J. Descamps, J. Phys. Radium <u>16</u>, 304 (1955).

2. J.D. Brown, Doctorate Thesis, University of Maryland (1966).

3. D.M. Poole, "Quantitative Electron Probe Microanalysis," K.F.J. Heinrich, Editor, National Bureau of Standards Special Publication 298, p. 93 (1968).

4. P. Duncumb and S.J.B. Reed, "Quantitative Electron Probe Microanalysis," National Bureau of Standards Special Publication 298, p. 133 (1968).

5. H.E. Bishop, "Optique des Rayons X et Microanalyse," Hermann, Paris (1967).

6. R.C. Wolf and V.G. Macres, "Electron Probe Microanalysis," A.J. Tousimis and L. Marton, Editors, Academic Press, p. 73 (1969).

X-RAY EMISSION FROM THIN FILM MATERIALS

P. A. Stine, S. J. Hruska, and G. L. Liedl

Purdue University

West Lafayette, Indiana 47907

ABSTRACT

A theoretical model and experimental measurements of the X-ray emission from thin films generated by an electron beam were analyzed and compared. Ionization correction factors for the thin films and for substrate effects were developed to calculate intensity ratios of film and film on substrate to pure bulk intensities.

The fraction of electrons transmitted through gold, silver, and copper with energy greater than the critical excitation energy was measured as a function of thickness and primary electron energy. The characteristic X-ray intensities from gold on bismuth, palladium, nickel and aluminum substrates; copper on palladium, nickel and aluminum substrates; and silver on aluminum substrates were measured and intensity ratios to pure bulk intensities were compared to the model prediction. The prediction of the X-ray intensity ratios were approximated by a simple electron balance model for the ionization correction factors and the consideration of electron energy distributions did not alter the predictions significantly.

An empirical conversion parameter, following the bulk procedures of Ziebold and Ogilvie, was determined experimentally for alloy thin films of gold and silver on oxidized silicon substrates for films between 500 and 2000 Å. The prediction of the parameter from the intensity model is in fair agreement with the measured values. However, the empirical conversion parameter could prove to be useful for thin film analysis.

INTRODUCTION

The generation of X-rays by electron excitation has been studied extensively for X-ray emission from bulk materials (materials whose thickness is sufficient for the electrons to lose all their incident energy). The results of these studies have been used for the application of electron excited X-ray emission for quantitative chemical analysis as developed originally by Castaing (1) and reviewed recently by several authors including Philibert (2). The current knowledge of the X-ray emission processes by electron excitation allows the calculation of reasonable quantitative data for bulk samples. Although the extension of this analysis technique for the chemical analysis for thin films should provide a convenient analytical technique for the difficult problem of the quantitative chemical analysis of these films, it has only recently been applied to thin films. Compositions of thin films are frequently specified on the basis of other parameters such as deposition rates, thermodynamic properties, gas phase composition of the reactants, etc. While these indirect measurements may be adequate for process control they frequently provide misleading results.

The application of electron excited X-rays to thin film analysis for special cases has been made by a number of authors including Theisen (3), Marshall and Hall (4), and Djuric and Cerovic (5) who have limited themselves to very thin films. Each of these authors assumed that the films were so thin that absorption and fluorescence could be neglected. These approaches are useful for the analysis of films whose thickness is approximately 1000 Å or less. Another approach has been taken by Sweeney, Seebold and Birks (6), Crockett and Davis (7), and Hutchins (8). These authors use the distribution in depth of characteristic X-rays that has been experimentally determined. This approach is limited by the present knowledge of the experimental distribution whose accuracy is most uncertain for thin layers. Further, Colby (9) has proposed an analysis using low electron energies to reduce the beam penetration. This approach allows the electrons to lose their energy within the thin film and bulk analysis techniques may be applied. However, the low incident electron energy limits the number of elements whose useful characteristic X-rays can be excited and limits the analysis to samples with low atomic number elements.

This study was undertaken to provide information on the prediction of the X-ray excitation from thin films by some of the basic ionization phenomena. In particular, the influence on the ionizations by the electrons backscattered from and transmitted through the film, and by the substrate are considered. Also, an empirical conversion parameter for thin films, following the bulk procedures of Ziebold and Ogilvie (10) is considered.

Ionization Model

The total number of ionizations may be approximated by considering the sum of the contribution from direct ionizations and indirect ionizations. This approach although not adequate for absolute intensities has been useful for relative intensities such as being considered here.

Following the approach of Green and Cosslet (11) with some modifications, the total direct ionizations produced by each electron is:

$$N_D^A = \eta_a^A N_o/A_A \int_o^{\rho_A t} Q_A \, d(\rho_A X) \qquad (1)$$

where $Q_A = K/EE_c \ln E/E_c$ with $K = 21.8 \times 10^{-20}$ and η_a^A is the fraction of electrons absorbed in the sample with energy above the critical excitation energy, N_o is Avogadro's Number, A_A is the mean atomic weight of the sample, ρ_A is the density of the sample, E is the electron energy at a position within the sample, and E_c is the critical ionization energy. Equation (1) may be integrated using the Thomson-Whiddington energy loss equation

$$E_o^2 - E^2 = c\rho X$$

where E_o is the initial electron energy and c is the Thomson-Whiddington constant to give

$$N_D^A = 4N_o K\eta_a^A/(cA_A E_c^A)[E_o(\ln E_o - 1) - (E_o^2 - c\rho_A t)^{\frac{1}{2}} (\ln(E_o^2 - c\rho_A t)^{\frac{1}{2}} - 1)$$

$$- \ln E_c (E_o - (E_o^2 - c\rho_A t))] \qquad (2)$$

The fraction of electrons absorbed, η_a^A, with energy above E_c for a thin film on a substrate can be approximated by assuming single reflection of the electrons at the two interfaces to give

$$\eta_a^A(\rho_A t) = 1 - \eta_B^A - \eta_T^A + \eta_T^A r_s (1 - \eta_B^A)$$

where η_B^A and η_T^A are the backscatter fraction and transmitted fractions repectively from the film as a function of mass thickness and r_s is the bulk backscatter fraction of the substrate. For a bulk sample, η_T^A is zero and the fraction of absorbed electrons is simply one minus the backscatter fraction for a bulk sample.

The indirect ionizations from the continuous spectrum may be estimated from Dyson (12) and Springer and Aachen (13) as

$$N_I^A = 0.85 \times 10^{-6} Z_A(r_x - 1)/r_x \int_{E_C}^{E_o}(E_o/E - 1)dE \quad (3)$$

where Z is the mean atomic number of the sample, and r_x is the ratio of the absorption coefficient on the high energy side of the absorption edge to that on the low side.

The intensity from the thin film is proportional to the sum of the two types of ionizations with a correction for absorption as

$$I_{AB}^A = (C_A N_D^A + N_I^A) \, ABS(\chi_{AB}^A, t) \quad (4)$$

where C_A is the weight fraction of element A and

$$ABS = 1/\chi_{AB}^A[1 - \exp(-\chi_{AB}^A, t)]$$

with

$$\chi_{AB}^A = (\mu/\rho)_{AB}^A \, \rho_{AB} \, \csc \alpha$$

and α is the take-off angle of the spectrometer. The relative intensity ratio of the thin film to bulk intensity, K_A, may be calculated using a bulk thickness determined from the Thomson-Whiddington energy loss with $E = E_c$ as

$$K_A = I_{AB}^A/I_{Pure\,A}^A \quad (5)$$

QUANTITATIVE MODEL

A thin film model following the approach of Ziebold and Ogilvie for bulk alloys can be approximated from the above ionization model. It has been found experimentally that the calibration curves of binary systems of bulk samples satisfy the relation

$$C_A/K_A = a_{AB} + (1 - a_{AB})C_A \quad (6)$$

where a_{AB} is the conversion parameter for the intensity from element A in an A-B binary system. From equation (6), the value of the limit of the ratio C_A/K_A as C_A approaches zero is the conversion parameter a_{AB}. A thin film intensity ratio, K_A, may be defined as the ratio of the intensity from element A in the binary thin film to the intensity of a thin film of pure A of the same thickness. The thin film parameter, a'_{AB}, may be obtained from the limit of

C_A/K_A assuming that indirect ionizations are negligible and that the backscatter electron fractions, transmitted electron fractions, and the density of the binary system are all linear functions of the pure element quantities weighted by the weight fractions of the elements to yield:

$$a'_{AB} = \frac{N_D^A(E_c^A, E_o, \rho_A t, \eta_a^A) \, ABS(\chi_A^A, t)}{N_D^A(E_c^A, E_o, \rho_B t, \eta_a^B) \, ABS(\chi_B^A, t)} \qquad (7)$$

A conversion parameter using a bulk pure element as a standard may also be obtained following a similar approach.

Experimental

The first part of the experimental study was the measurements of the X-ray emission from pure thin films over a range of thicknesses and on various substrates. These experimental studies were performed at 20, 25 and 29 KV in an ARL-AMX Electron Microprobe to determine the fraction of electrons transmitted with energy greater than the critical excitation energy and the characteristic X-ray intensities. The trace layer and substrate combinations analyzed during this study were gold on bismuth, palladium, nickel and aluminum; copper on palladium, nickel, and aluminum; and silver on aluminum.

The films were prepared by vapor deposition simultaneously on metallographically polished metal substrates, collodion, and a masked glass slide. The collodion supported films were mounted on copper grids and the collodion was dissolved in amyl acetate to provide the unsupported thin films. Film thickness was measured with a multiple beam interferometer on both the glass slide and the metal substrates.

All electron current and X-ray intensity measurements were taken with a sample current 0.03 μ-amps on a bulk sample of the thin film material. The characteristic X-ray lines measured were Au $L\alpha$ and Cu $K\alpha$ with a LiF crystal and Ag $L\alpha$ with an ADP crystal. Each intensity was taken as an average of five fixed time measurements and was corrected for background and absorption.

The total fraction of electrons transmitted through the thin films with energy greater than the critical excitation energy, η_T, was measured for a constant beam current using a Faraday cage with a retarding grid similar to one described by Cosslett and Thomas (14). The measured values of η_T with thin film thickness are given in Table I. The estimated errors in the measurements were ± 100 Å in thickness and ± 2% in electron transmission.

Table I

Transmitted Electron Fractions Through Au, Ag, and Cu With Energies Above Critical Excitation Energy

Au (Å)	20KV	25KV	29KV	Ag (Å)	20KV	25KV	29KV	Cu (Å)	20KV	25KV	29KV
407	0.57	0.60	0.68	289	0.79	0.82	0.88	651	0.61	0.63	0.71
531	0.43	0.49	0.56	513	0.62	0.72	0.76	1027	0.49	0.53	0.65
699	0.37	0.40	0.50	1040	0.42	0.52	0.53	3266	0.22	0.25	0.34
758	0.31	0.37	0.46	3097	0.16	0.22	0.27	5935	0.11	0.16	0.24
1264	0.25	0.32	0.38	5380	0.08	0.16	0.17				
3215	0.06	0.13	0.17	6920	0.03	0.11	0.10				
5398	0.0	0.03	0.08								
7181	0.0	0.0	0.01								

The X-ray intensities from pure gold, silver, and copper unsupported thin films and all trace layer and substrate combinations mentioned previously were measured and compared to the intensity of the pure bulk material. These intensity ratios were then compared to the predicted ratio calculated from equation (5) with the absorption correction applied to the measured intensity ratios instead of the calculated values. For the calculated ratios, the transmitted fractions measured in this study were used and the backscatter fractions were calculated from Bishops (15) data. The absorption data was taken from Birks (16) and bulk backscatter data from Bishop. The total relative errors in the measured intensities ratios due to counting statistics, thickness errors, and absorption coefficient were estimated to be less than 8% for any measurement.

The second part of the experimental study was the preparation of gold-silver thin film alloys and the X-ray measurements for the empirical composition conversion parameter. Films of 500, 1000, 1500, and 2000 Å in thickness, each with compositions of 0, 20, 40, 60, and 100 atomic percent gold were deposited onto oxidized silicon wafers. The films were deposited in a dual source vacuum evaporator with crystal control monitors on each source. The composition of each film was determined by the mass of the elements deposited as measured by the crystal monitors which had been calibrated previously. Film thickness was measured by a stylus technique using a Dektak.

All X-ray emission measurements were obtained using the same procedure as described in the first part of the experiment with an operating voltage of 25 KV.

Results and Discussions

The fractional electron transmission data for electrons with energy above the critical energy given in Table I are consistent with the results of Cosslett and Thomas (14) who measured the fractional electron transmission of electrons with energy above 50 eV. The transmission fractions of this study are consistently below those of Cosslett and Thomas as expected because of the different energy ranges considered. Transmission fractions with energy above 50 eV also were measured in this study and were within experimental errors of Cosslett and Thomas's data.

Typical results of the measured versus calculated intensity ratios for gold films on bismuth and aluminum substrates are shown in Figures 1 and 2. These figures show surprisingly good agreement between the simple theoretical curve and the measured values as a function of primary beam energy and substrate effects. The largest variation between the measured and calculated values is seen in Figure 2 for an aluminum substrate with a primary beam voltage of 29 KV. This difference is most likely due to the simplifying assumptions in the model and uncertainties in the transmitted and backscatter electron data. For example, the uncertainty in the calculated bulk thickness is greater for a deeper penetration of the electrons, i.e., for a higher primary voltage, and a greater uncertainty in the fraction of absorbed electrons is observed for the more transparent films. The comparison of the intensity ratios for gold without a substrate and on palladium and nickel substrates showed agreement between the measured and calculated intensity similar to that illustrated in Figures 1 and 2. A consideration of electron energy distributions did not alter the prediction significantly.

Figure 3 shows the comparison between the calculated and measured data for silver on an aluminum substrate. The correlation is similar to that found for the gold films with the largest differences observed for the higher operating voltage.

An example of the copper intensity ratios is shown in Figure 4 for copper films on palladium. A much larger difference is observed between the measured and calculated values for all operating voltages. Similar differences were also observed for the copper films on nickel and aluminum substrates. The reasons for the greater difference in the measured to calculated ratios for the copper film as compared to the gold and silver films is not readily apparent. However, it should be noted that the K-lines were measured for copper and L-lines for the other two. Any difference in the ionization cross section for different characteristic lines

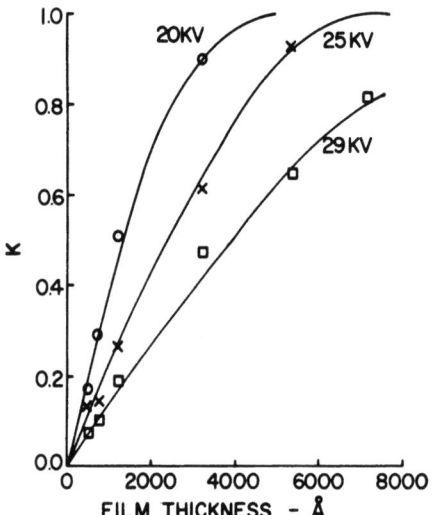

Figure 1: Intensity Ratios of Gold on Bismuth to Bulk Gold as a Function of Thickness and Operating Voltage

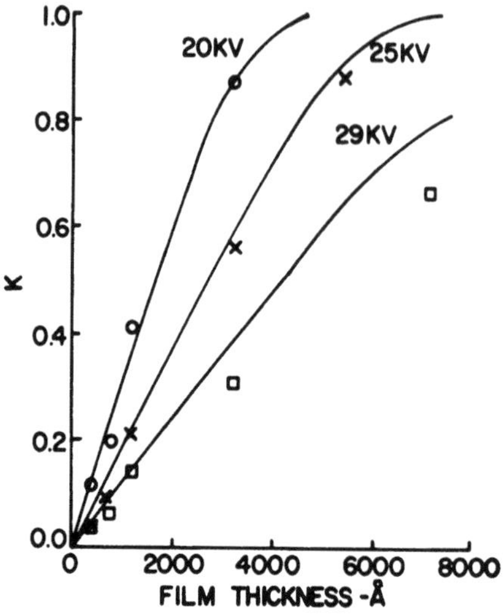

Figure 2: Intensity Ratios of Gold on Aluminum to Bulk Gold as a Function of Thickness and Operating Voltage

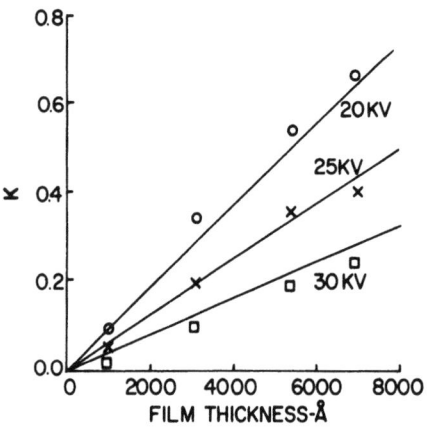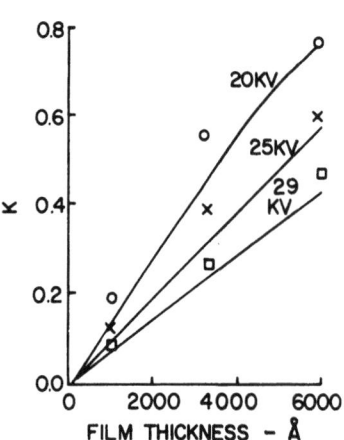

Figure 3: Intensity Ratios of Silver on Aluminum to Bulk Silver as a Function of Thickness and Operating Voltage

Figure 4: Intensity Ratios of Copper on Palladium to Bulk Copper as a Function of Thickness and Operating Voltage

Table II

Thin Film Composition Conversion Parameters for Gold-Silver Alloys with $E_o = 25$ KV

t(Å)	a_{AgAu}		a_{AuAg}	
	Measured	Calculated	Measured	Calculated
1000	0.90	0.60	2.18	2.10
1500	0.85	0.95	2.07	1.84
2000	0.82	1.27	2.01	1.70
		1.43		1.24

was not included in the model. Also, more problems were encountered in the deposition of the copper films than for the gold and silver films. Problems in film adherence of copper on some substrates were observed and during the measurements of film thickness a high surface roughness was observed on the copper film.

The thin film composition parameters following the approach of Ziebold and Ogilvie (10) for bulk alloys are given in Table II for silver-gold alloys. Both the measured values and calculated values from equation (7) of the parameter are given for 1000, 1500, and 2000 Å films for an operating voltage of 25 KV. A larger thickness dependence for the calculated values is shown in Table II than observed from the actual measurements. In fact, the parameter for silver radiation, $a_{Ag,Au}$, shows an increasing trend with thickness for the calculated values as compared to a slight decreasing trend for the measured values. However, the bulk value for the silver radiation is greater than the thin film value and one would expect an increasing trend of the parameter with thickness. Although the agreement between the measured and calculated values varies up to 30%, the mean values compare favorably over the thickness range. If one uses the mean values over the thickness range considered, the error in the parameter would yield a composition uncertainty of approximately 10%. This uncertainty is well within the range for use in many thin film alloy studies.

CONCLUSIONS

The authors conclude that the prediction of the X-ray intensity produced by electron bombardment in thin films to that produced in bulk standards may be approximated adequately for many studies using a simple electron balance. The use of more accurate ionization cross section functions and stopping power relations that are available in the literature should be considered even though numerical integrations would have to be employed. From the results of this study, the quantitative determination of thin film composition from known ionization parameters can not be expected to yield results as accurate as bulk determinations. However, the thin film single parameter conversion approach appears to be valid for the system studied and should be investigated further. In particular, the consideration of the bulk element as a standard with the calculation of the intensity of a thin film of the same thickness as the alloy film should be investigated. Also, other alloy systems need to be investigated to give confidence to the approach.

ACKNOWLEDGEMENTS

This work was partially supported by the Advanced Research Projects Agency (IDL Program DAHC-0213) and the National Science Foundation (MRL Program GH33574).

REFERENCES

1. Castaing, R., "Application of Electron Beams to a Method for Local Chemical and Crystallographical Analysis," O.N.E.R.A. Publ. No. 55, (1951).
2. Philibert, J., "Electron Probe Microanalysis. Quantitative Analysis," *Fifth International Congress of X-Ray Optics and Microanalysis*, (Springer-Verlag, New York, 1969), p. 114.
3. Theisen, R., *Quantitative Electron Microprobe Analysis*, (Springer-Verlag, Inc., New York, 1965), pp. 20-22.
4. Marshall, D. J., and Hall, T. A., "Electron-probe X-ray Microanalysis of Thin Films," *X-Ray Optics and Microanalysis, Fourth International Conference*, (Hermann Press, Paris, 1966), pp. 374-81.
5. Djuric, B., and Cerovic, D., *Fifth International Congress of X-Ray Optics and Microanalysis*, (Springer-Verlag, New York, 1969), pp. 180-86.
6. Sweeney, W. E., Seebold, R. E., and Birks, L. L., "Electron Probe Measurements of Evaporated Metal Films," *J. Appl. Phys.* **31**, 1061 (1960).
7. Crockett, G. H., and Davis, C. D., "Coating Thickness Measurements by Electron Probe Microanalysis," *Brit. J. Appl. Phys.* **14**, 813 (1963).
8. Hutchins, G. A., *The Electron Probe* (McKinley Heinrich, Wittrey, Eds.), John Wiley and Sons, New York, (1966), pp. 390-404.
9. Colby, J. W., "Quantitative Microprobe Analysis of Thin Insulating Films," *Advances in X-Ray Analysis,* **11**, (University of Denver, Plenum Press, New York, 1968), pp. 287-305.

10. Ziebold, T. O., and Ogilvie, R. E., "An Empirical Method for Electron Microanalysis," Anal. Chem., 36, 322-327 (1964).
11. Green, M., and Cosslett, V. E., "The Efficiency of Production of Characteristic X-Radiation in Thick Targets of a Pure Element," Proc. Phys. Soc. (London), 78, 1206 (1961).
12. Dyson, N. A., "The Continuous X-Ray Spectrum from Electron Opaque Targets," Proc. Phys. Soc., 73, (1959), 924-936.
13. Springer, G., "The Correction for Continuous Fluorescence in Electronprobe Microanalysis," 1967, Fortschritte der Mineralogie, Vol. 45, pp. 103-124.
14. Cosslett, V. E., and Thomas, R. N., "Multiple Scattering of 5-30 KeV Electrons in Evaporated Metal Films."
    I. Total Transmission and Angular Distribution, Brit. J. Appl. Phys., 15, 883 (1964).
    II. Range Energy Relations, Brit. J. Appl. Phys., 15, 1283 (1964).
    III. Backscattering and Absorption, Brit. J. Appl. Phys., 16, 779 (1965).
15. Bishop, H. E., "Calculations of Electron Penetration and X-Ray Production in a Solid Target," Brit. J. Appl. Phys., 1, Series 2, 673 (1968).
16. Birks, L. S., Electron Probe Microanalysis, (Interscience, New York, 1963), pp. 201-237.

AUGER ELECTRON EMISSION MICROGRAPHY AND MICROANALYSIS OF SOLID SURFACES

K. Hayakawa, H. Okano, S. Kawase and S. Yamamoto

Central Research Laboratory, Hitachi Ltd,

Kokubunji, Tokyo 185, Japan

ABSTRACT

An electron probe Auger emission microanalyzer has been constructed. The instrument is composed of an electromagnetic focussing primary probe column and a cylindrical mirror electron energy analyzer. By using this instrument, Auger electron spectroscopy studies have been carried out in the modes of both emission microanalysis and emission micrograph. The feasibility of this method is investigated through its application to the study of iron surface.

INTRODUCTION

When element analysis of solids is planned non-destructively, as well as in microareas, it has been customary to use characteristic X-rays and/or Auger electrons emitted from atoms in solids. As electrons, being signal carriers, have definite electronic charge and mass characteristics, their energy value is substantially effected in the motion in solids by inelastic scattering effects. Thus, electron emission signals are predominant from very thin layers, of about 10 Å range. This is one of the characteristic differences in Auger emission analysis as compared with X-ray analysis.

In a high speed energy scanning Auger electron spectroscopy (AES) method,[1-2] we apply new functions of a microprobe primary electron beam and the scanning of the irradiating position on a specimen surface.[3] This analytical instrument is an electron probe microanalyzer using Auger electron emission instead of characteristic X-ray emission.[4-5]

The present work is concerned with the construction of an electron probe Auger emission microanalyzer as an analytical instrument of solid surfaces and for use in investigation of its feasibilities through its application to the studies of iron surfaces.

MICROANALYZER

Figure 1 shows the schematic construction of the electron probe Auger emission microanalyzer used in the present study. This instrument was designed to clarify the fundamental performance of the Auger emission microanalysis.

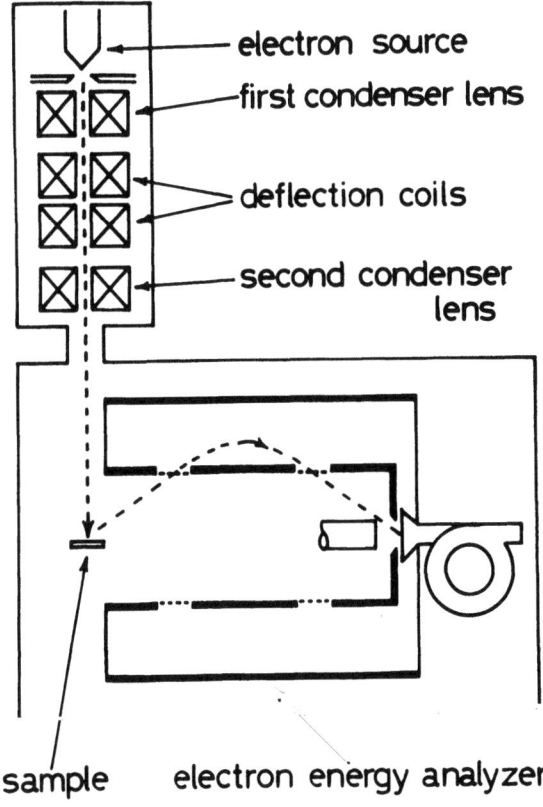

Fig. 1. Schematic construction of an electron probe Auger emission microanalyzer.

The primary beam column consists of an electron gun, two stage electromagnetic condenser lenses and two pairs of X-Y deflection coils. The electron energy analyzer is the cylindrical mirror type (CMA)[2]. The primary electron energy can be chosen in the energy range from 1 KeV to 10 KeV. Using a tungsten hair pin cathode, the spot size of the primary beam on the specimen surface is about 2 μmϕ with the beam current of the order of 10^{-7}A. All parts except for lenses and coils are enclosed in a vacuum, which is evacuated by ion-sputtering pumps. The vacuum is attained in the range of 10^{-9} Torr.

METHOD OF MEASUREMENTS

Figure 2 shows a schematic diagram of Auger emission microanalysis method. Auger emission spectra are measured in the mode of the phase sensitive detection by superimposing sinusoidal perturbing potential to the outer electrode of the CMA[1]. In this case the spectra become the first derivative form of the electron energy distribution with respect to the electron energy.

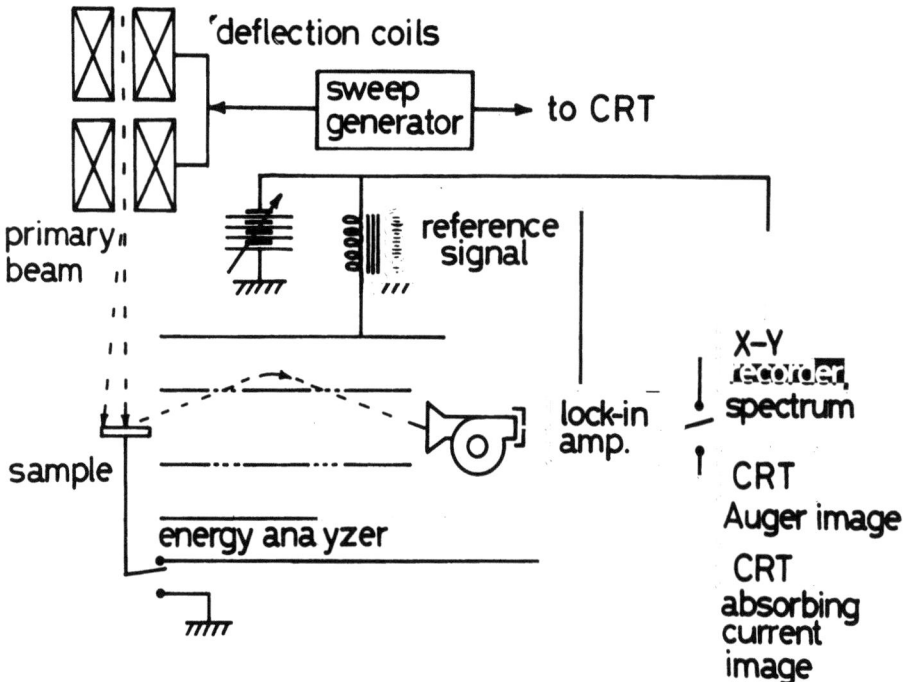

Fig. 2. Schematic diagram of Auger emission microanalysis method.

Auger emission micrograph(AEM) are recorded in the following way. The parameters of the CMA are fixed to pass electrons having desired energy corresponding to a spectral line emitted from the specimen. At the same time the output signal of a lock-in amplifier is fed to the brightness modulator of a CRT. Synchronizing the scanning of the probe position on the specimen surface with that of a spot of the CRT, brightness distribution patterns of the emission signal are obtained. This pattern corresponds quantitatively to the spatial distrbution of the element on the specimen surface under investigation. In the present experiment, the number of raster lines is about 400 in a frame and the recording time is about four minutes per frame.

To measure microanalysis spectra, analyzing points are selected in the absorbing current image(ACI). Fixing the probe to a selected point, Auger spectra are obtained from the area 1-2 µmϕ of the selected point.

In investigating flat surface specimens, the absorbing current images give much greater contrast than the secondary electron images. The number of electronic charges absorbed in the specimen by the primary beam irradiation is varied place to place, due to the variation of chemical composition in the surface layer. This suggests that the contrast in the ACI plays a subsidiary role in the distribution and the sorts of element species.

EMISSION MICROGRAPH

The five photographs shown in Fig. 3(a) are Auger emission micrographs and an absorbing current image obtained from the same area of iron specimen surface. Brightness contrasts in these AEMs indicate the intense emission of Auger electrons from the corresponding regions of the specimen.

Contrast patterns in the iron and oxygen images are approximately identical. The bright patterns in the carbon image correspond approximately to the dark areas in the iron and oxygen images. In the sulphur image the contrast of some particular regions is very high.

These features indicate that some oxygen compounds are formed on the surface layer of the specimen and that carbon atoms are segregated outside of the oxygen compounds.

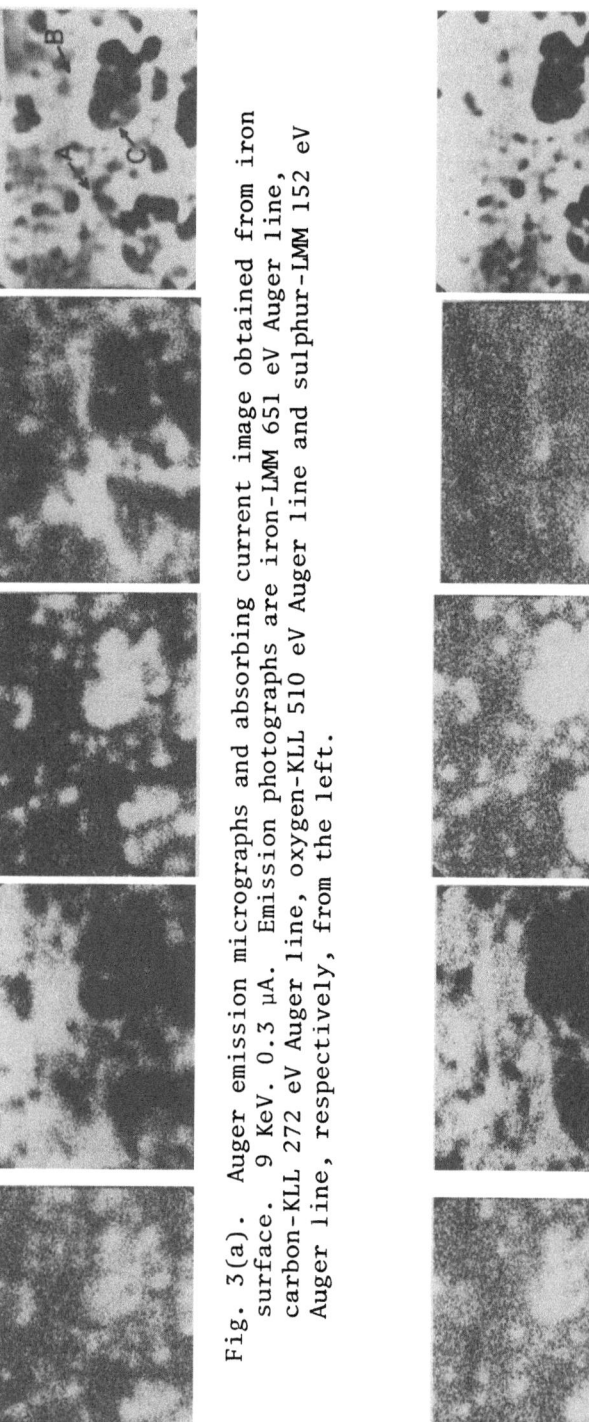

Fig. 3(a). Auger emission micrographs and absorbing current image obtained from iron surface. 9 KeV. 0.3 μA. Emission photographs are iron-LMM 651 eV Auger line, carbon-KLL 272 eV Auger line, oxygen-KLL 510 eV Auger line and sulphur-LMM 152 eV Auger line, respectively, from the left.

Fig. 3(b). Auger emission micrographs and absorbing current image obtained from the iron surface after resistively heated at 300°C for two minutes in the vacuum.

Effect of Heat-Treatment

Figure 3(b) shows Auger emission micrographs and absorbing current image after the specimen (Fig. 3(a)) was resistively heated at 300°C for two minutes. In Fig. 3(a) and (b), the observed area is approximately identical. After heat-treatment, the shape of the sulphur image exhibits remarkable changes and the carbon image also changes to some extent. The contrast pattern of the iron and oxygen images was not destroyed by this heat-treatment. From these pattern observations, sulphur and carbon atoms participating in the changes in contrast are presumed to be adsorbed loosely by the specimen surface. With increasing heating temperatures, drastic changes of contrast patterns were observed even in the case of iron and oxygen.

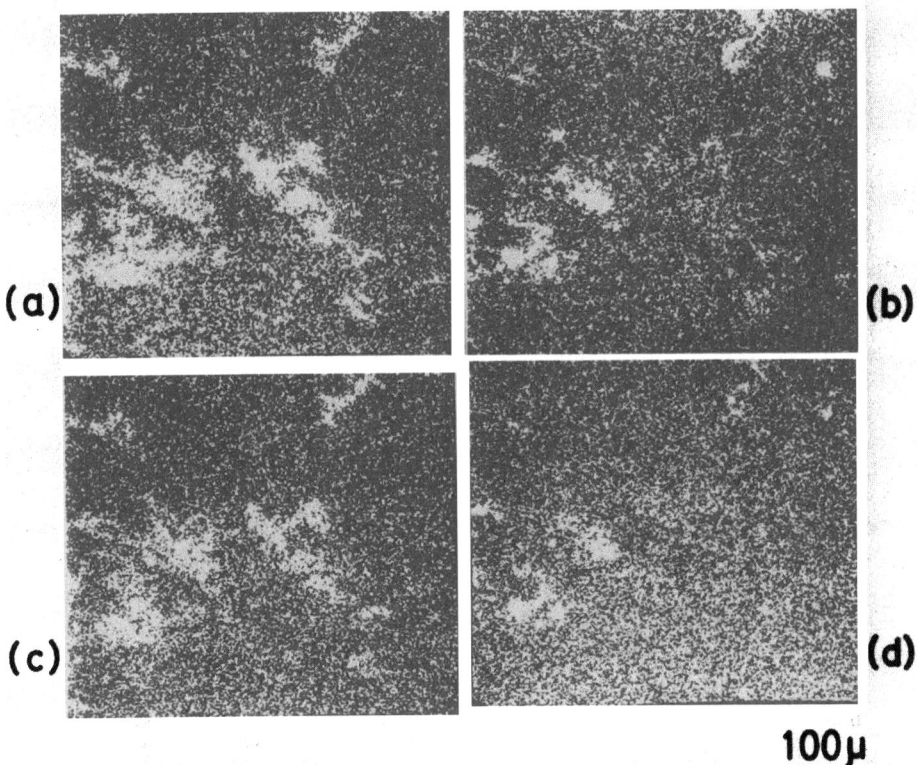

Fig. 4. Auger emission micrographs obtained from iron surface showing effect of argon ion etching by 1 KeV, 5 μA/cm^2 beam for 30 minutes. (a) oxygen-KLL 510 eV AEM before etching, (b) sulphur-LMM 152 eV AEM before etching, (c) oxygen-KLL AEM after etching and (d) sulphur-LMM AEM after etching.

Effect of Ion Etching

By adding an ion sputtering gun to the Auger emission microanalyzer, the effect of ion etching may be revealed in situations by the Auger emission micrographs as well as the absorbing current images. By these modes of observation, the shape of the impurity clusters exsisting in the surface layer can be observed in the depth direction. The ion gun incorporated was PHI model 04-131, and a 1 KeV argon ion beam was used for etching.

Figure 4 shows the effect of the argon ion etching on the sulphur segregation formed in the oxide region of iron specimen surface. In the emission micrographs shown in Fig. 4, the bright areas are reduced after the ion etching. This effect is more noticeable in the case of sulphur. The photographs shown in Fig. 4 are examples of the results of heavy ion etching. A low magnification is also used in order to emphasize the change of the images due to the ion etching. By recording the emission micrographs in the intermediate stages of etching, it will be possible to reconstruct three dimensional distributions of sulphur and oxygen atoms in the specimen.

MICROANALYSIS

Figure 5 shows Auger emission microanalysis spectra obtained from an iron surface, whose analyzing points are indicated by the marks A-C in the absorbing current image in Fig. 3. In these spectra, spectral lines emitted from iron, sulphur, carbon and oxygen atoms are conspicuous. In each spectrum it is clear that the intensities of iron and oxygen spectral lines show fair correlation with each other and those between oxygen and carbon lines are inversely correlated. These intensity relations are consistent with the image contrast shown in Fig. 3.

In the present Auger emission microanalysis method, the analyzing area is evaluated to be about 2 µmϕ.

INTERPRETATION OF IMAGE CONTRAST

In Fig. 3, the intensity patterns of the iron and oxygen AEMs are observed, for the most part, to be identical. As observed from the AEMs, a thin layer of carbon atoms is adsorbed in the area outside of the iron oxide. The iron Auger electron emission from the underlying surface is suppressed by this adsorbed layer. Therefore, the intensity patterns of the iron and oxygen AEMs possess identical features.

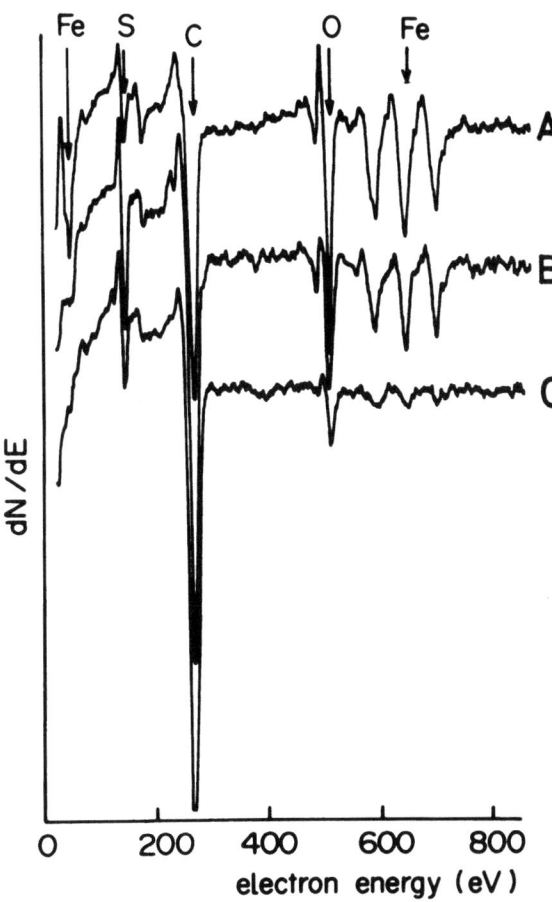

Fig. 5. Auger emission microanalysis spectra obtained from iron surface. Analyzing points A-C are indicated in the absorbing current image in Fig. 3(a).

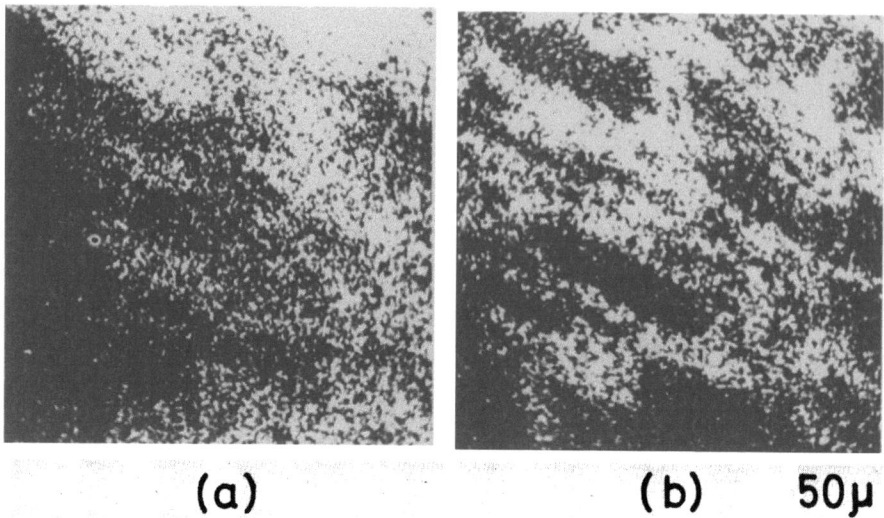

Fig. 6. X-ray emission images obtained from iron surface. 10 KeV, 0.65μA. (a) Iron Kα X-rays, (b) oxygen Kα X-rays.

Figure 6 shows X-ray emission images concerning iron and oxygen obtained from the same iron specimen surface. The intensity patterns of these images show a complementary relation. Number of iron atoms per unit volume in the region of iron oxide is less than that of the iron region. Therefore, the iron characteristic X-ray emission intensity is effectively reduced in the region of the iron oxide. In the X-ray case, the thin layer of adsorbed carbon contributes little to forming the image contrast.

As described above, the difference in the image contrast between the electron case and the X-ray case is related to the depth of the signal emission.

DISCUSSION

In the preceding sections various characteristics concerning the Auger emission microanalyzer have been demonstrated. The functions of the instrument confirmed in the present work are;

(i) the analyzing area is reduced of the order of a few times 10^{-4} in comparison with the current Auger electron spectroscopy method,

(ii) Auger emission micrographs obtained by several kinds of Auger spectral lines enable us to determine the distribution of the chemical compositions on the surface layer of solids,

(iii) the halftone contrast in the AEM provides a semi-quantitative evaluation of the distribution of surface atoms and

(iv) combination with the ion etching technique opens the possibility of three dimensional element analysis of solids with a resolution of 2 μmϕ in area and 10 Å in depth.

In performing the AEM patterns shown in Fig. 3 in a few minutes per frame, about 10^{-7} A primary current is required. Figure 7 shows a calculation of the spot size of the primary beam vs the primary electron energy with a parameter of the primary current in the case of a tungsten hair pin cathode[3]. In the calculation, the brightness and half angle of the primary probe were assumed to be 10^5 A/cm^{2}sr. at 100 KeV and 10^{-2} rad..

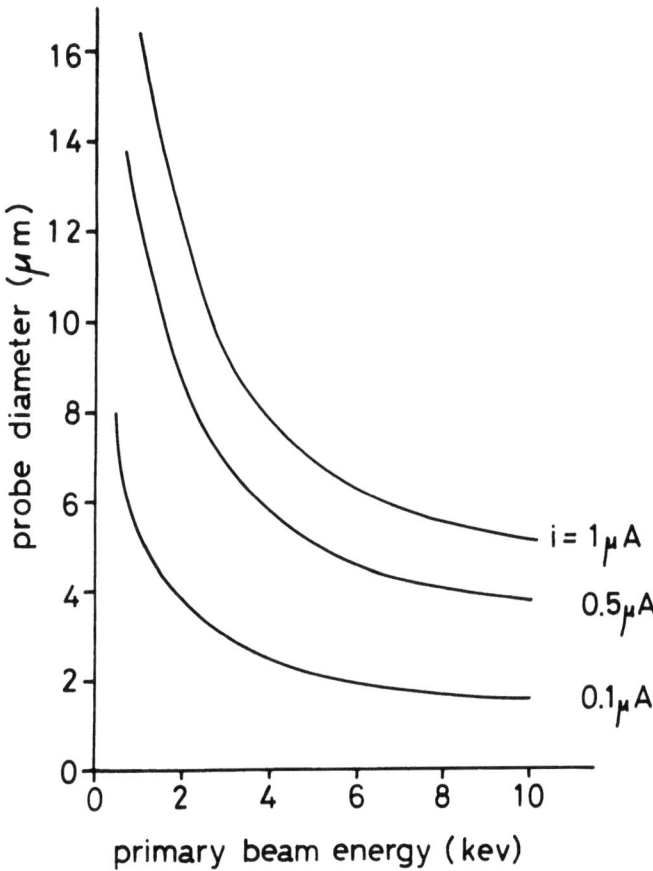

Fig. 7. Calculation of electron probe size vs probe energy with a parameter of the probe current in the case of tungsten hair pin cathode.

At a primary energy of 9 KeV and a current of 1×10^{-7} A, the spot size is evaluated to be 1.5 µmϕ. The resolution of the present Auger emission microanalysis is not less than this value. Improvement of the resolution may be attained by decreasing the primary current.

REFERENCES

1. L. A. Harris, "Analysis of Materials by Electron-Excited Auger Electrons," J. Appl. Phys. 39, 1419 (1968).

2. P. W. Palmberg, G. K. Bohn and J. C. Tracy, "High Sensitivity Auger Electron Spectrometer," Appl. Phys. Lett. 15, 254-255 (1969).

3. K. Hayakawa, H. Okano, S. Kawase and S. Yamamoto, "Auger Electron Emission Micrographic Studies of Cleavage Surface of Graphite Single Crystal," J. Appl. Phys. 44, 2575-2580 (1973).

4. N. C. MacDonald and J. R. Waldrop, "Auger Electron Spectroscopy in the Scanning Electron Microscope: Auger Electron Images," Appl. Phys. Lett. 19, 315-318 (1971).

5. L. A. Harris, "Auger Electron Analysis in a Scanning Microanalyzer," GE Technical Information Series No. 71-C-273, Sept. 1971.

A COMBINED PHOTOELECTRON/X-RAY FLUORESCENCE SPECTROMETER

H. K. Herglotz and D. R. Lynch

Engineering Physics Laboratory

E. I. du Pont de Nemours & Co. (Inc.), Wilmington, Delaware

ABSTRACT

Facilities for energy-dispersive x-ray fluorescence analysis have been added to an advanced version of the high-sensitivity ESCA (electron spectroscopy for chemical analysis) instrument described at the 1972 Denver Conference. Since the excitation mechanisms for electron emission and x-ray fluorescence are the same, the instrument's powerful source of primary x-rays is an asset to both types of spectroscopy. The geometrical arrangement of source, electron analyzer, and x-ray detector permits easy change from one mode of operation to the other without change of sample. While ESCA is valuable for the analysis of light elements and of surfaces apart from the bulk, x-ray fluorescence is useful for the analysis of bulk or substrate. The high excitation power makes the instrument useful also for trace analysis in solid or liquid samples. Modifications that could further enhance the usefulness of the instrument are described.

INTRODUCTION

When x-rays ionize core levels of atoms in a sample, events occur that are of interest to the analytical chemist and physicist. Best known of these events and one exploited by x-ray spectroscopists is the emission of a characteristic x-ray quantum, in the rearrangement of the ionized atom. ESCA (electron spectroscopy for chemical analysis) utilizes the photoelectron for the same analytical purpose, and advanced ESCA instrumentation was able to fill some blind spots in x-ray spectroscopy (XRS), primarily the analysis of elements below atomic number 11. ESCA also has considerable merits of its own: The information it yields about bonding is superior to that obtainable by XRS, and its ability to deal with a surface layer a few angstroms thick makes it unique for surface analysis. An instrument combining the methods of surface and bulk analysis would undoubtedly enhance the information that could be extracted from a sample. This paper describes such an instrument.

INSTRUMENT

Our ESCA instrument was reported at last year's Denver Conference on Applications of X-Ray Analysis (1). It owed its high sensitivity to a powerful source of aluminum radiation and a high-luminosity electron energy analyzer (2). Since that time we have been using an advanced version of this instrument that incorporates a superior electron energy analyzer (3), one having higher sensitivity and resolution than the original. The excellent performance of this analyzer made possible the compact Du Pont ESCA-650 commercial instrument described elsewhere (4). It is our larger laboratory instrument, however, on which incorporation of XRS facilities, described in this article, was demonstrated. It could also be incorporated, after some modification, in the ESCA-650.

Either an energy-dispersive or a wavelength-dispersive spectrometer would be suitable for combining with the ESCA instrument. We decided on energy-dispersive instrumentation because of simplicity and because it best suited our prime application. The spectrometer selected from a commercial supplier has a beryllium window on a gate valve, which gives the option of operating the solid-state detector without an absorbing window between sample and detector. This option will be discussed in more detail later.

Figure 1 shows a cross section of the ESCA instrument to which the x-ray spectrometer is attached. Arrangement of sample, source and electron analyzer under a common stainless steel bell jar made it possible to move the electron analyzer horizontally and to open a direct optical path from the sample surface to the solid-state detector on top of the bell jar. Collimators over the sample and under the detector prevent any spurious x-ray quanta from impinging on the detector. The switch from ESCA to the XRS and back is easily done from outside the vacuum chamber via a hand-operated rotating feedthrough.

The unavoidably large distance between sample and Si(Li) detector is contrary to the customary close coupling and large acceptance angles

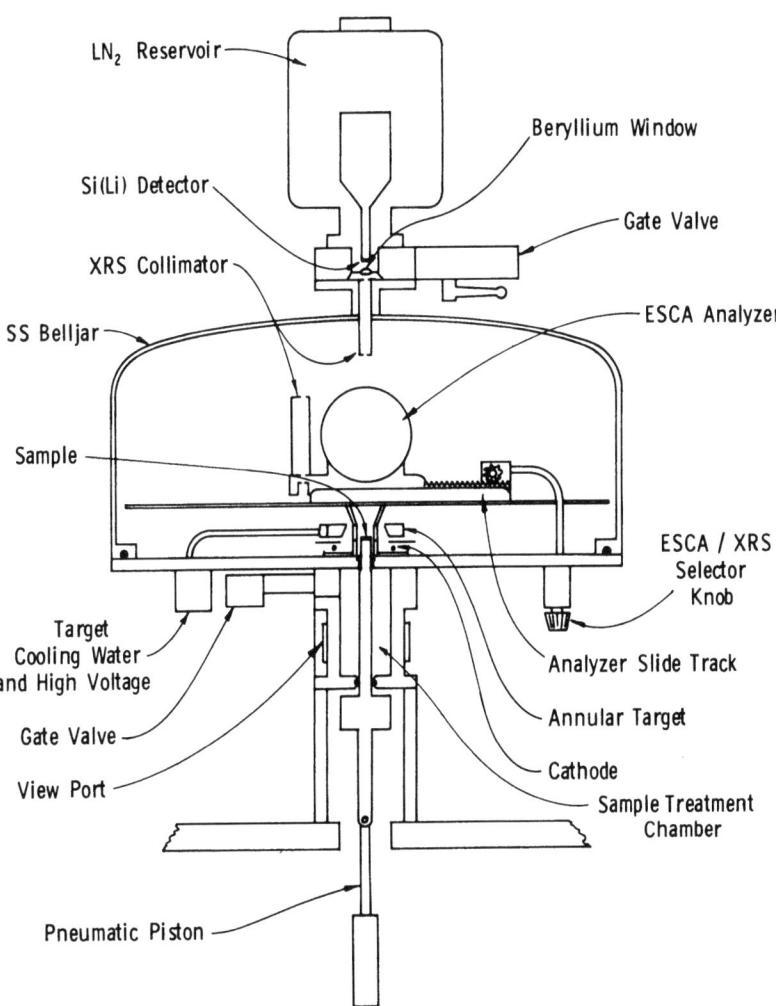

Fig. 1. Combined ESCA/XRS instrument.

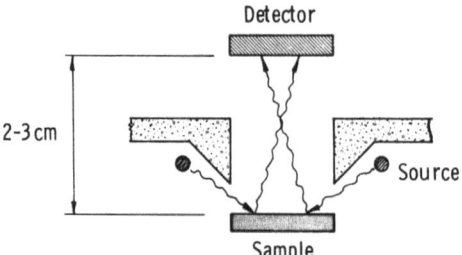

Fig. 2. Schematic x-ray fluorescence arrangement with isotopic excitation source.

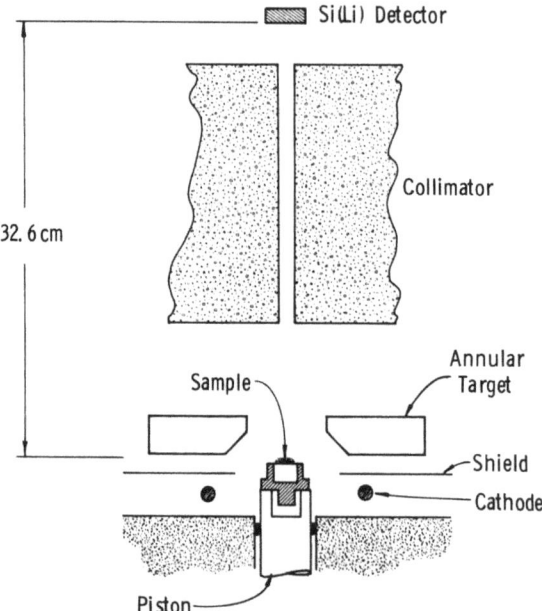

Fig. 3. X-ray fluorescence excitation in ESCA instrument.

allowed by the nature of solid-state detectors (Fig. 2). These advantages make it possible to excite samples by isotopic excitation sources. In our instrument we have an extremely strong x-ray excitation source, with an annular target surrounding the sample (Fig. 3), that can be operated at up to 20 kV and 300 mA. The strength of the source more than compensates for the large sample-to-detector distance and for the unfavorable excitation characteristics of an aluminum target. Since bremsspectrum is responsible for exciting x-ray spectra of most elements, a high-atomic-number element would be desirable because the integral bremsspectrum intensity is proportional to the atomic number of the target. On the other hand, ESCA enforces use of a low-atomic-number element with a narrow monochromatic line. Based on the drawings of Figs. 1 and 2, the acceptance angle of the detector in our arrangement is only $\sim 5 \times 10^{-3}$ that of a closely coupled arrangement with an isotopic excitation source. But a relatively strong (100 mC) isotopic source emits only 3.7×10^9 photons/sec, while the corresponding figure for a 300-mA x-ray source is $\sim 10^{15}$ (5). Therefore, a geometrical loss by a factor of 10^3 still leaves us with very powerful excitation, as our experience has confirmed.

EXAMPLES OF APPLICATION

The merits of combining surface and bulk analysis are demonstrated in Fig. 4. Coated sheet metal was easily identified. ESCA used for analyzing the coating showed fluorine, carbon, oxygen, and titanium; x-ray fluores-

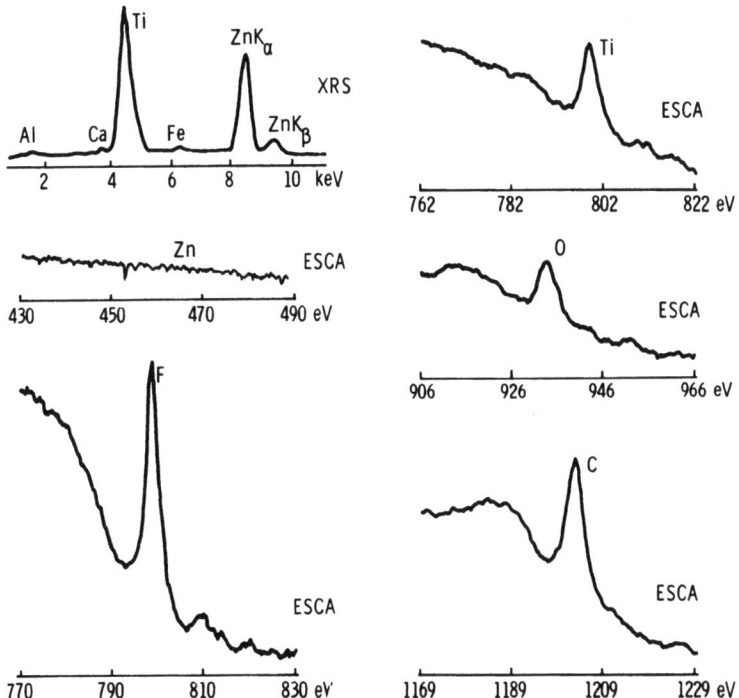

Fig. 4. ESCA and x-ray spectra of a coated metal. Count rate of highest peak: XRS = 6.5 kilocounts (kc)/100 sec; ESCA = 5 kc/sec.

cence produced strong signals from titanium and zinc and a weak signal from iron. The sample was thus identified as galvanized steel with a fluorocarbon coating and titanium dioxide pigment. No quantitative analysis was sought in this case.

Figure 5 demonstrates combined x-ray/ESCA analysis of contaminants in water. In this case, 1 µl of a solution containing 1000 ppm of cobalt and sodium is deposited on a 6-µm-thick Mylar® polyester film substrate. The sample holder is designed so that only fluorescent radiation from the sample can be seen by the detector (Fig. 6). It sits on the tip of a pneumatically operated piston that injects it from the sample conditioning chamber (10^{-4} to 10^{-5} torr) via a gate valve into the instrument chamber (10^{-7} torr). This system was previously described in detail (1). X-ray spectroscopy permits quantitative evaluation of cobalt, as well as chlorine introduced as anion with the sodium chloride. ESCA is effective in detecting and quantitatively determining sodium.

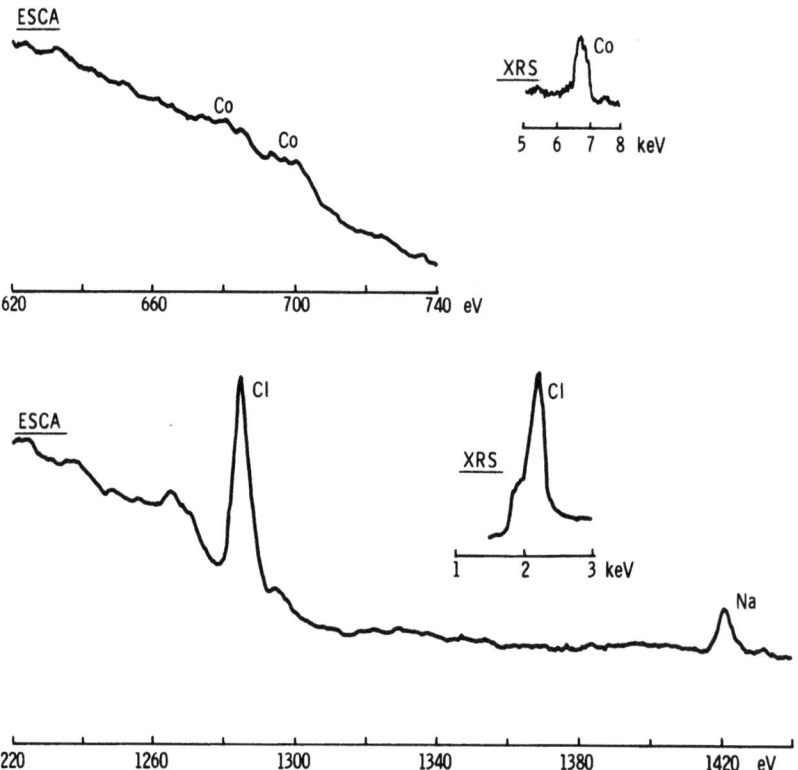

Fig. 5. Spectra of aqueous solutions recorded with combined ESCA/XRS instrument. Count rate of highest peak: XRS = 700 counts/200 sec; ESCA = 2.8 kc/sec.

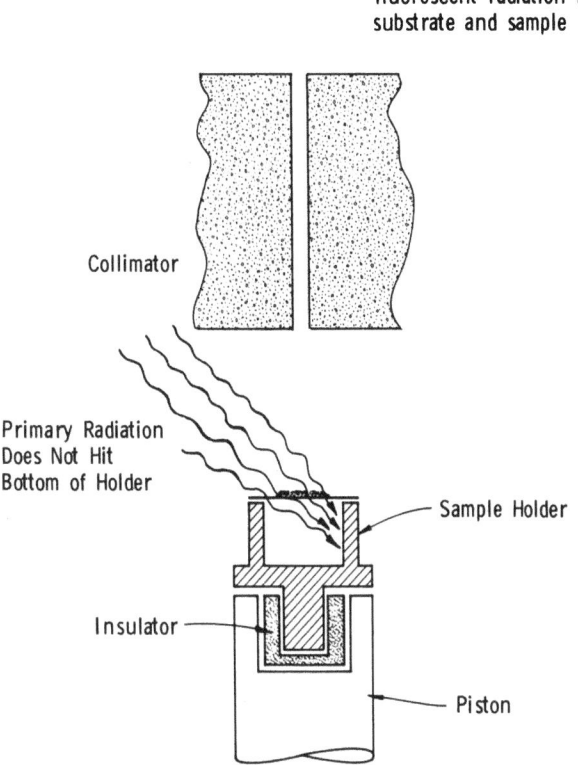

Fig. 6. Sample holder for combined ESCA/XRS instrument.

Figure 7 shows ESCA and XRS spectra of a TiO_2/carbon black mixture. The ESCA spectra suggest the absence of TiO_2, but XRS of the same sample allows us to quantitatively analyze (by calibration) the TiO_2 content in the mixture.

In all these cases, the combination of x-ray and electron spectroscopy is more powerful than either method alone. A word needs to be said, however, about quantitative analysis by ESCA. The shallow escape depth (less than 100 Å) of the photoelectrons causes the signal strength to depend on the way the sample is spread on the substrate. This effect is comparable to, but far more severe than, the particle-size effect in x-ray spectroscopy. Matrix and instrument effects are comparable in both methods. It would exceed the scope of this publication to describe them in detail; suffice it to say that a dedicated effort can produce quantitative results in most cases, and nearly always is the combination of the two methods more informative than either one by itself.

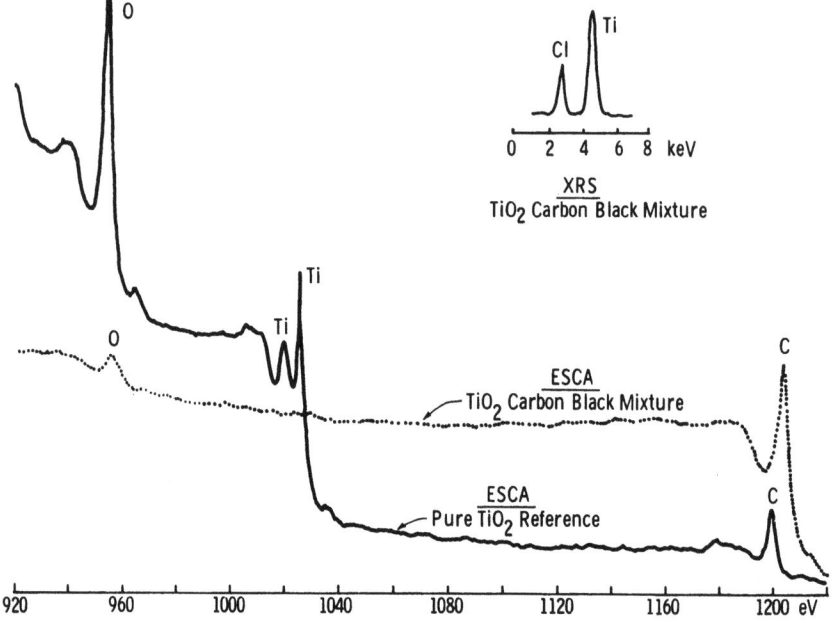

Fig. 7. Electron and x-ray spectra of TiO$_2$/carbon black mixture. Count rate of highest peak: XRS = 9 kc/100 sec; ESCA = 19 kc/ sec.

FUTURE WORK

Work to date has shown the concept of a combined ESCA/XRS instrument to be feasible and useful. We shall continue our experimentation with concepts and equipment additions that could further enhance the instrument's usefulness. Some of the modifications under study or in the planning stage are briefly mentioned in the following.

- Tertiary Excitation. The arrangement in Fig. 8 has been used for efficient excitation of a particular element with considerable background reduction. Bremsspectrum from the aluminum target (primary radiation) excites characteristic radiation of the perforated tube (here copper) serving as a secondary emitter, which cannot be seen by the detector. The (secondary) copper radiation in turn excites the cobalt to be analyzed in the

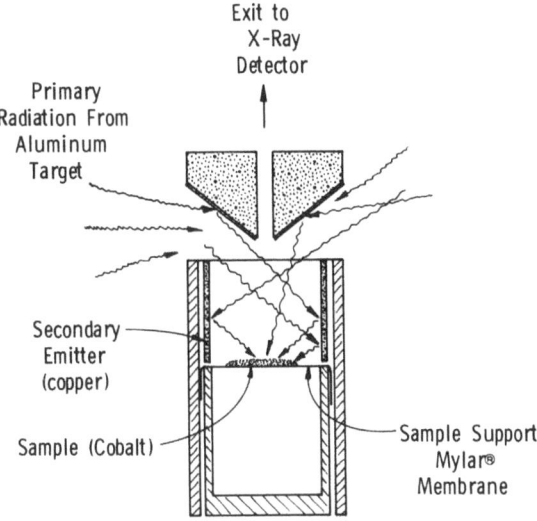

Fig. 8. Sample holder for tertiary excitation.

sample (tertiary excitation). It is easy to choose a secondary emitter with a characteristic radiation slightly more energetic than the absorption discontinuity of the element to be analyzed. The secondary emitter can be changed without breaking the vacuum. Initial results with this arrangement show promise.

- Extension to Low-Atomic-Number Elements. Carbon x-ray quanta of 284 eV, for example, are energetic enough to generate carrier pairs in the Si(Li) detector. They are not detected, however, because of absorption in the beryllium window over the Si(Li) detector and low-energy background and noise, which swamp the detector. Part of this low-energy background comes from secondary events generated by high-energy photons.

We have eliminated absorption in the beryllium window by using the previously mentioned commercial gate-valve arrangement over the detector (6). When the gate valve is opened (Fig. 1), the bell jar vacuum of 10^{-7} torr provided by a turbo-pump extends into the detector. High-energy photons are kept away from the detector by a total reflection collimator (Fig. 9) that acts as a low pass filter and only passes radiation below

~ 1.5 keV in reasonable intensity (7,8). Preliminary experiments have shown that electric noise obscures the low-element signal. Efforts will therefore be concentrated on eliminating the noise.

- Liquid Samples. A sample container was constructed to fit on the top of the sample injection piston. Liquid is separated from the high vacuum of the sample chamber by a Mylar® window. Scouting experiments with this arrangement were successful in analyzing a nickel solution; improvement is possible by adapting the geometry of the transparent sections of the sample holder to the flux of exciting radiation from the annular x-ray target.

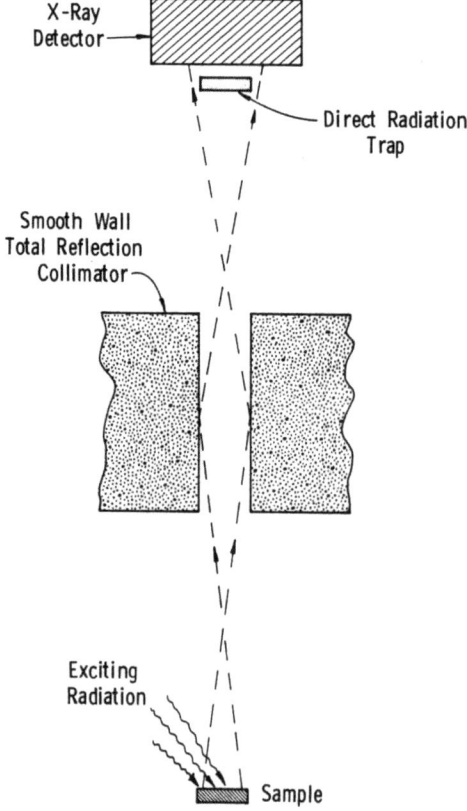

Fig. 9. Total-reflection low-energy pass filter.

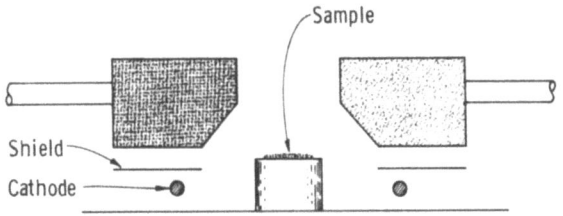
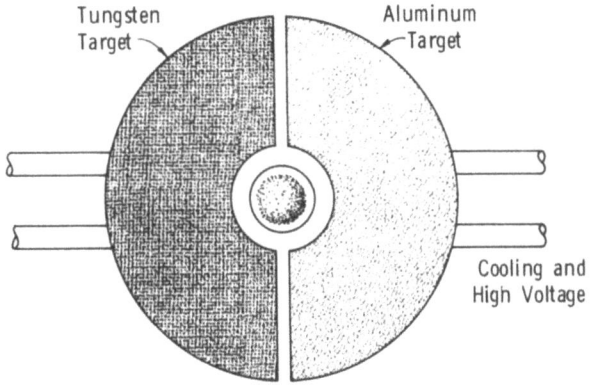

Fig. 10. Dual-element target for ESCA or XRS excitation.

- Heavy-Element Target. ESCA depends on the narrow excitation lines from an aluminum or magnesium target with little background producing bremsspectrum; XRS of all elements is best served by a heavy element (W, Rh) with a strong bremsspectrum. The conflicting requirements of the two methods can be satisfied by a composite target consisting of two electrically independent halves (Fig. 10). The positive 20-kV high voltage can then be switched to either aluminum or tungsten for excitation of the sample for ESCA or XRS. We plan to incorporate this arrangement.

- "In-Situ" Dual Analysis. The sample treatment chamber described in ref. (1) allows thermal or chemical treatment of samples, immediately followed by injection into the spectrometer, where the effects of the treatment on surface and bulk can be measured by ESCA and XRS without exposure of the sample to the atmosphere. It is our intent to make use of this unique possibility in the future.

CONCLUSIONS

There is considerable advantage in a combined ESCA/XRS instrument over separate spectrometers because of

- possibility of "in-situ" analysis of a sample by both methods;
- reduced cost of dual instrument vs. two devices.

There is no interference between the two modes of operation. Therefore, the inferiority of "universal" instruments to specialized devices is not encountered here.

REFERENCES

1. R. D. Davies, H. K. Herglotz, J. D. Lee, and H. L. Suchan, "A High-Sensitivity ESCA Instrument," in L. S. Birks, et al., Editors, Advances in X-Ray Analysis, Vol. 16, p. 90-101, Plenum Press (1973).

2. J. D. Lee, "A New Electrostatic Energy Analyzer," Rev. Sci. Instr. 43, 1291 (1972).

3. J. D. Lee, "A Nondispersive Electron Energy Analyzer for ESCA," Rev. Sci. Instr. 44, 893 (1973).

4. "New Instruments" section in Rev. Sci. Instr. 44, 91 (1973).

5. D. B. Brown and J. V. Gilfrich, "Measurement and Calculation of Absolute X-Ray Intensities," J. Appl. Phys. 42, 4044 (1971).

6. Kevex Corporation, Burlingame, California.

7. H. K. Herglotz, "Paraffin Mirrors for Ultrasoft X-Rays," Nature 214, 263 (1967).

8. H. K. Herglotz (to Du Pont), "X-Ray Spectrograph Apparatus Using Low Angle X-Ray Reflecting Units and Means to Vary the X-Ray Incidence Angle," U.S. Patent 3,418,466, Dec. 24, 1968.

A SPHERICALLY BENT CRYSTAL X-RAY SPECTROMETER WITH VARIABLE CURVATURE

Donald L. Parker

St. Mary's University

San Antonio, Texas 78284

ABSTRACT

The design and performance of a spherically bent crystal x-ray spectrometer with variable curvature are given. A thin crystal with the diffracting planes parallel to the face is mounted on a vacuum chuck consisting of an O-ring in a brass mounting. A controlled partial vacuum is applied behind the crystal to cause spherical deformation of the lattice. Thus, rays from a point source on the focusing circle are diffracted to a line image also on the focusing circle. The differential pressure is automatically varied such that the source-to-crystal and crystal-to-image distances are equal and constant for all Bragg angles and hence the simple θ-2θ motion of a one flat crystal spectrometer is used.

The data are accumulated by a scanning proportional counter tube placed behind a vertical slit (perpendicular to the scattering plane) located at the image line. The fixed chord length is 22 cm and the instrument is designed to scan from zero up to 120° 2θ. Crystals are easily interchanged and the automatic vacuum regulator has sufficient flexibility to allow tailoring the spherical bending to crystals of materials of various thicknesses. The resolution is easily adjusted by either the size of the x-ray source or the width of the detector slit. The performance of the spectrometer has been evaluated by characteristic x-rays produced by various samples placed in a demountable x-ray tube. The main advantages of this three-dimensional focusing instrument are the very high signal-to-noise ratio and the very low levels of x-ray flux required.

INTRODUCTION

The advantages of bent crystal spectrometers are well known; however, scanning instruments using bent crystals with fixed curvature are necessarily awkward since they require the source-to-crystal distance and/or the detector slit-to-crystal distance to be varied if a large spectral region is to be scanned. The spherically bent crystal x-ray spectrometer described here offers two unusual features:

(a) Three-dimensional focusing of x-rays from a point source (or from a distributed source between the crystal and a virtual point source).

(b) Continuous variation of the lattice curvature such that the point source-to-crystal and detector slit-to-crystal distances are constant and equal.

The relation of the focal circle to the spectrometer is shown in the "double exposure" diagram of Figure 1.

DESIGN

Although the spectrometer is adaptable to any thin disc shaped crystal with suitable elastic properties, the only type studied to date are silicon wafers as widely used in the semiconductor industry. The crystals are mounted on an O-ring in a crystal holder. The spherical deformation is achieved with a controlled partial vacuum on the backside of the wafer. Several crystals, 34 mm and 50 mm in diameter (using different size holders), were repeatedly deformed to radii of curvature as small as 30 cm. The deformation is completely reversible except when a wafer occasionally shatters under the extreme strains. Upon shattering, the crystals disintegrate into a powder except for a few fragments coming from the edge overhanging the O-ring where the strain is not so great. A wafer has never been observed to shatter at a differential pressure less than that previously applied to that same specimen. The ultimate radius of bending of a particular wafer is very sensitive to fine abrasions on its backside and, although no systematic study was made, seemed to be relatively independent of crystal diameter or thickness. Of course, the differential pressure required for a given curvature has a strong dependence on the wafer thickness.

Prior to actual construction of the spectrometer, two different tests were made using a sealed copper x-ray tube and a pinhole as a point source. In the first test, using a 20 micron pinhole, a series of x-ray photographs was made of the direct beam incident

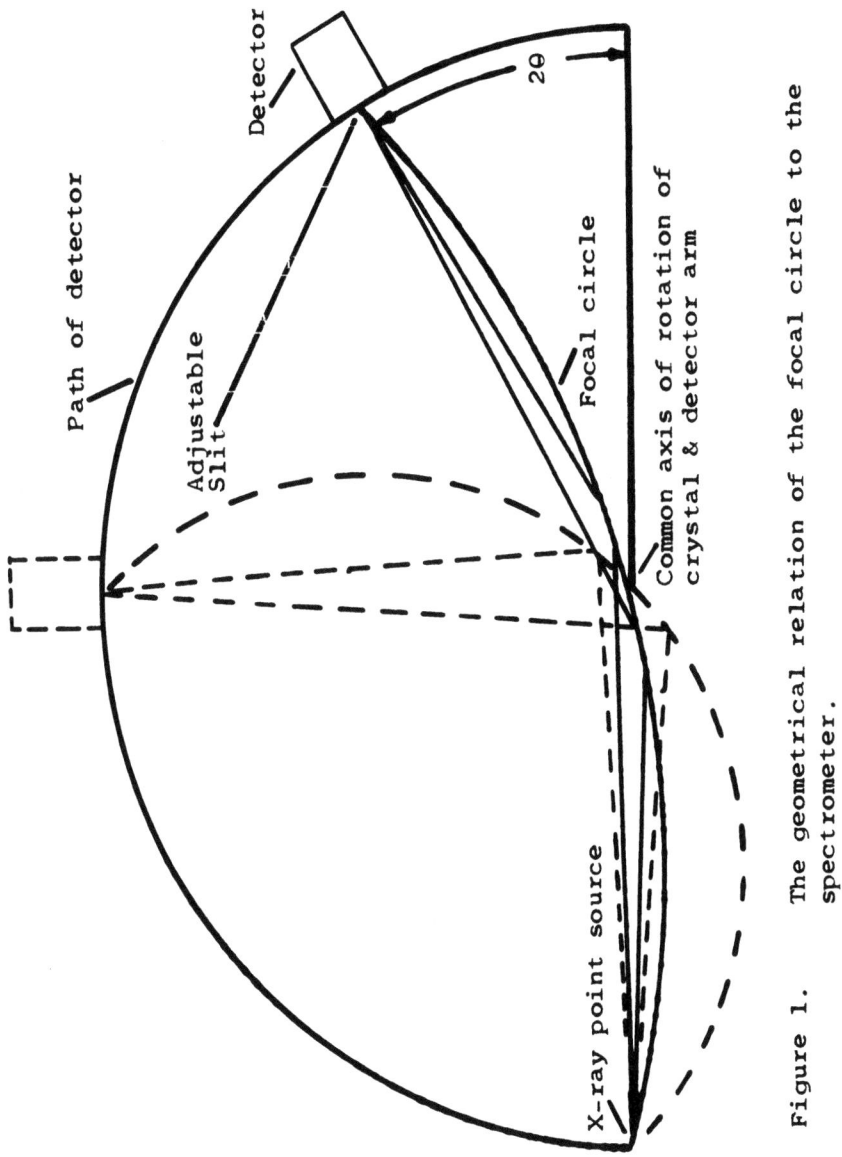

Figure 1. The geometrical relation of the focal circle to the spectrometer.

upon the crystal; the diffracted beam very near the crystal; and the diffracted beam at the image line. These photographs were made for both the first order diffraction peak for $CuK\alpha_1$ (~28 degrees 2θ) and for third order (~95 degrees 2θ). The photographs of the beam immediately before and after diffraction confirmed that all portions of the x-ray cone striking the crystal were participating in the diffraction. The photographs of the image lines were very straight and very sharp, on the order of 20 microns wide for the first order case. Also, the line image was found to become shorter at higher diffracting angles as expected.

The second test consisted of determining the relation between the differential pressure for optimum focusing of the x-rays and the Bragg angle. These data are shown in Figure 2. A 100 micron pinhole was used as the point source and a 127 micron slit was placed in front of the detector. The pinhole and slit were each 22 cm from the axis of crystal rotation. For each datum point, the detector slit was placed at twice the Bragg angle and held stationary as the crystal was rotated slowly (via a long lever arm and a micrometer driven by a synchronous motor) through the diffraction peak for the continuum radiation. Several scans were made for each position of the detector slit as the pressure was varied. As the pressure was varied, the diffraction peaks increased to a maximum (which is assumed to give optimum focusing) and then decreased. The error bars in Figure 2 indicate the range of pressures over which the diffraction peaks were within 20% of the maximum height for each angle. The angular widths (FWHM) of the diffraction peaks were 75 arcseconds of crystal rotation for all Bragg angles. It should be mentioned that for the smaller Bragg angles (< 20 degrees) the crystal did not intercept all rays from the source, thus the crystal edges (where the deformation is not spherical) actually participated in the diffraction. Had the angular width of rays from the source been limited to those striking near the center of the crystal, the error bars for the smaller angles would, no doubt, be smaller.

CONSTRUCTION

In order to extend the range of the spectrometer in the softer x-ray range, it is built in a sealed aluminum chamber with a plexiglass lid for helium gas flushing. The detector arm, which carries the proportional counter tube and slit, is driven by a large worm gear. The crystal holder is constrained to move at one-half the angular speed of the detector arm by equal length arms and a sliding pivot on the crystal holder. The worm gear is driven either manually by a crank or a synchronous motor on the outside of the chamber. An odometer allows the detector arm

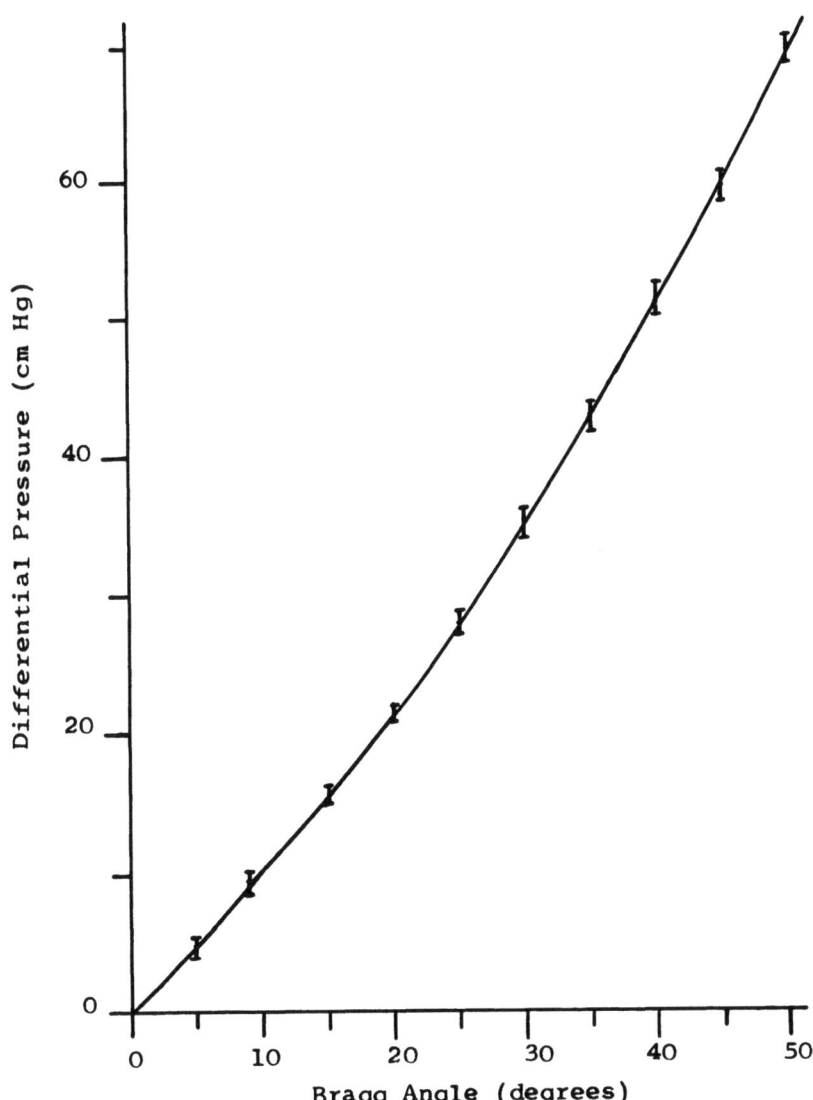

Fig. 2 Differential pressure vs. Bragg Angle for optimum focusing for a silicon crystal 34 mm in diameter and 0.33 mm thick with the 111 planes parallel to the face. Error bars contain pressures giving peak counts greater than 80% of maximum. Chord length is 22 cm with a 100 μm pinhole source and 127 μm detector slit.

position to be read to an accuracy of 0.01 degree. All rotary motions are supported by ball bearings and the sliding pivot was carefully hand lapped for a minimum of friction and angular error. Backlash is on the order of 0.02 degrees 2θ and the tracking error is about the same as determined by the angular positions of known peaks.

The spectrometer chamber is mounted on a commercial machinist rotary table which has x-y translations as well as rotation. A metal bellows with a flange and mylar window connects the chamber to the demountable x-ray tube. The bellows allows small adjustments of the chamber with respect to the x-ray source. The rotary table allows fine adjustment of the x-ray incident angle without changing the relationship between the lattice planes and the detector slit. The x-y translations are used for fine adjustment of the source-to-crystal distance and the x-ray takeoff angle.

Careful machining procedures assure that the rotary table, crystal holder, and detector arm all rotate about the same vertical axis. Tilt adjustment about a horizontal axis is made with a small micrometer head. Some compromise in the location of these axes is necessary since ideally, they should go through the center of the crystal. This is, of course, impossible to satisfy for all angles since the center of the crystal moves relative to the holder. It is this compromise which causes the tracking error cited above because the error is unmeasurable over short scans of a few degrees.

The demountable x-ray tube is built into a two-inch copper T-section which is connected to a vacuum pump station. The anode, with a tungsten inlay, and cathode were taken from a small dental x-ray tube with provision added to replace the filament. The anode is beveled at 23 degrees with respect to the tube axis. Windows are located on opposite sides of the tube: one perpendicular to the tube axis for simultaneous analysis with an energy dispersive detector and one inclined at 26 degrees which connects to the crystal spectrometer through the bellows. Thus the x-ray mean takeoff angle is adjustable from zero to about six degrees. The electron focal spot is a horizontal rectangle about 1x3 mm.

The controlled partial vacuum required for optimum focusing is provided by the electronic regulator described in the reference. A synchro, which rotates at the same angular speed as the crystal holder through a parallel arm linkage, provides the regulator with the angular position information. The regulator causes the absolute pressure behind the crystal to vary essentially linearly with the crystal rotation over a range of about sixty degrees. Both the slope and intercept of the pressure versus angle relation are independently adjustable. The nonlinear relation between the differential pressure for optimum focusing and the Bragg angle requires that the regulator slope and intercept controls be readjusted when changing from very low to very high Bragg angles. Also daily fluctuations in atmospheric pressure require frequent readjustments of the regulator controls. This is easily accomplished, however, simply by maximizing the count rate at each end of the angular region to be scanned just prior to the scan.

PERFORMANCE

The initial application of the spectrometer has been to detect various chemical elements in trace amounts through their characteristic x-rays produced by electron bombardment. Samples are placed directly on the x-ray tube anode by either of two methods: powdered samples are dusted onto a thin film of silicon grease on the anode and liquid samples containing dissolved materials are evaporated on the anode with a heat lamp.

With the powder-on-silicon grease technique the vacuum deteriorates as the x-ray tube current is increased due to vaporization of the grease. However, after a few minutes of operation (typically at 25 Kv and up to 100 microamperes) the vacuum stabilizes and drops to a reasonable value. The x-ray spectra observed from a sample mounted in this fashion do not change even after several hours of operation. Thus, it appears that the electron beam "welds" the sample to the tungsten anode. With nonconducting powders there are instabilities in the tube current during the initial operation period. In the initial alignment of the spectrometer characteristic K_α x-rays from pure powder samples (or compounds) of all the elements from titanium through zinc were observed. These known peaks along with the numerous L lines from the tungsten inlay were used for adjustment of the spectrometer. In view of the application of the spectrometer, no attempt was made to determine the ultimate resolving power of the instrument. However, with a 40 micron detector slit the K_α lines are easily resolved for all of the above elements. During most of the runs for trace element analysis, the detector slit was set at 250 microns.

In order to determine the sensitivity of the powder method for detection of traces of various elements, known samples were prepared using silica gel as a matrix. A very careful series of runs with zinc and silica gel indicated that the sensitivity of this method is only slightly below 0.1% zinc. Some remarks are in order regarding the difficult problem of defining the limit of detection. All the data referred to in this paper were taken with a continuous scan at 0.5 degrees 2θ per minute with a single channel pulse height analyzer, ratemeter, and pen recorder. "Limit of detection" used here is not meant to be precise and means simply that the particular peak of interest is less than about 20% above the background. A stepping motor drive with an automatic multichannel scalar (or even a manual scan with perhaps 100 sec counts taken at each position) with an appropriate statistical analysis could considerably lower the limits claimed here.

Runs were also made on NBS powdered orchard leaves containing 0.03% iron, 0.009% manganese, and 2% calcium. The iron was easily detectable but the manganese was just about at the limit of detection. The calcium peak gave over 500 counts/second with a background of 6 counts/second with the tube operating at only 5 microamperes. The 25 kv limit imposed by the power supply and tube design used here is apparently considerably below the optimum for detection of Kα x-rays from elements heavier than perhaps iron.

Although the powder sample technique does not seem to have any particular merit as a new analytical tool, the results obtained with the residue from the evaporated liquids is considerably more interesting. Multiply diluted samples containing water soluble compounds of cobalt, iron, copper, and chromium were placed directly on the anode with a microliter syringe and evaporated with a heat lamp. Thus the quantities of the metals in the residue were accurately known. Concentrations in the liquid specimens varied from 0.1 to 20 ppm. Figure 2 shows the results obtained when 50 μl of water containing 120 nanograms of chromium was evaporated directly on the tungsten inlay. More recent runs with smaller water drops (and hence residue spots more closely conforming to the electron focal spot) place the limit of detectability of chromium by this method at no more than 10 nanograms. Similar results were obtained for iron; however, the copper and cobalt sensitivities were noticeably poorer. Control samples of pure water mounted by exactly the same procedure produced no trace of any impurities. There is, however, a tendency for an iron impurity to appear with this tube after several hours of operation. Apparently, there is sputtering of the iron cathode material to the anode. The iron impurity buildup, after a few hours of operation after cleaning, corresponds to about 20 nanograms of iron. The evaporated samples produced no deterioration of the vacuum or current instabilities in the tube. It should be mentioned that the 0.1 ppm concentrations used in these experiments by no means represents the limit of sensitivity of this technique. The ultimate sensitivity is limited only by the amount of impurities introduced with the particular preconcentration technique used.

CONCLUSIONS

The efficiency of this three dimensional focusing instrument

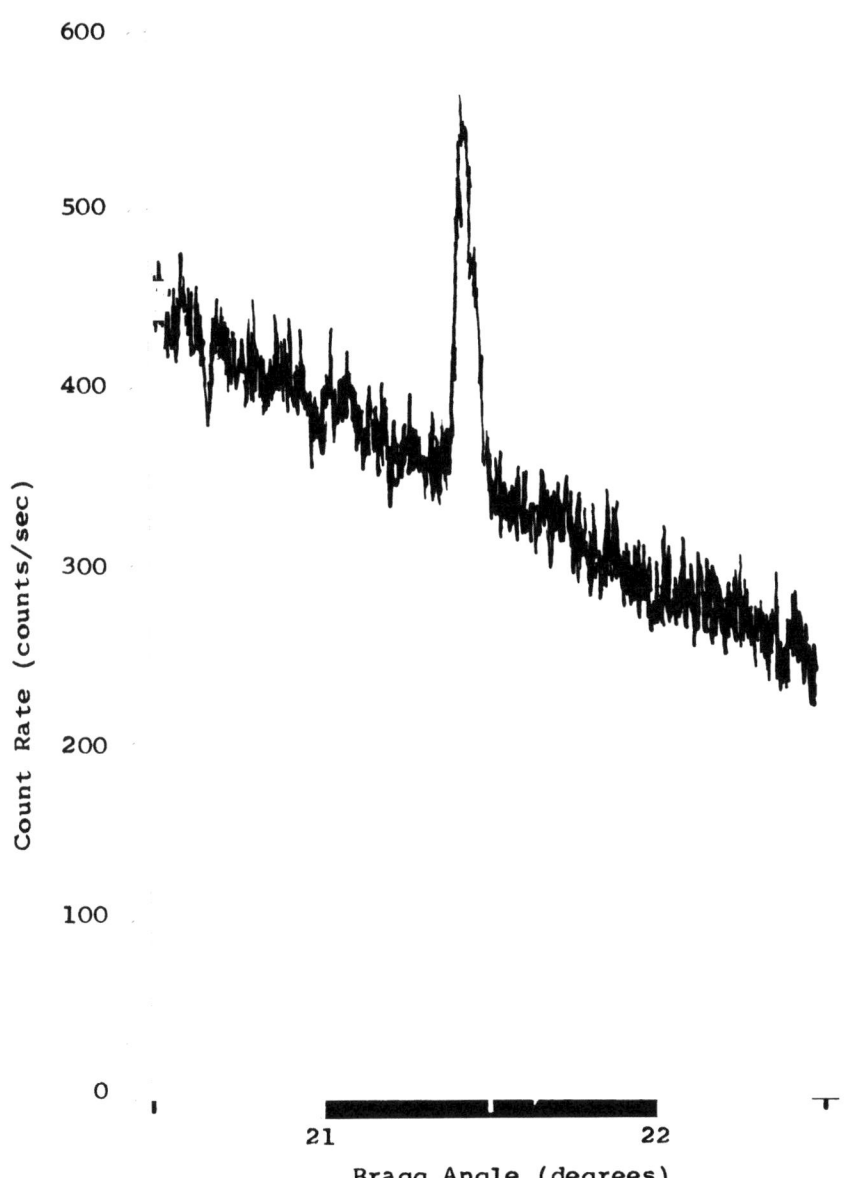

Fig. 3 Cr Kα peak obtained from a diluted solution of Cr O$_3$ and distilled water evaporated directly on the x-ray tube anode. The Sample contained 120 nanograms of chromium of which approximately one third was within the electron focal spot.

for a small or point source seems to be well established. For example, the run displayed in Figure 3 was made at 25 kv and 50 microamperes of tube current. This combination of high count rate, low x-ray source intensity, and moderately high resolution would be impossible with a flat crystal instrument or even a cylindrical instrument. The silicon crystals used offer a number of advantages not the least of which is their uniformly high quality and very low price. Also, the complete absence of second order diffraction from these crystals allows more effective pulse height discrimination against simultaneous higher order diffraction.

The reader may have noted that the spherical deformation could also be achieved with a controlled partial vacuum in the chamber and a vacuum behind the crystal. This variation with very thin crystals (requiring low pressures) with larger d spacings would allow the technique to be used well into the soft x-ray range.

Among other possible applications of the spherically bent instrument is microanalysis since it is ideally suited to point source radiation. And, of course, there are numerous applications which could utilize advantageously the very intense monchromatic radiation achievable in the line image when a powerful sealed tube is used as a source.

ACKNOWLEDGEMENTS

The author wishes to thank John Frisbie and Bruce Reinhard for their considerable help in constructing the spectrometer and x-ray tube. The author also gratefully acknowledges the financial support from the Robert A. Welch Foundation and the Texas Natural Resources Foundation.

REFERENCES

D. L. Parker, "An Electronic Partial Vacuum Regulator," Rev. Sci. Inst. $\underline{43}$, 1103 (1972).

MEASUREMENT OF THE X-RAY SENSITIVITY OF SILICON DIODES IN THE ENERGY REGION 1.8 to 5.0 keV

Jacque J. Hohlfelder

Sandia Laboratories

Albuquerque, New Mexico 87115

ABSTRACT

Planar, silicon diodes have been calibrated in the energy region 1.8 to 5.0 keV using low-energy, continuous-mode x rays. The spectral energy distribution of each of six x-ray spectra used in the calibration was measured using a non-dispersive Si(Li) x-ray spectrometer. X-ray spectral fluxes were measured using a xenon-filled, parallel-plate ionization chamber. A chopper-wheel and phase-sensitive detector combination was used to enhance the signal-to-noise of the silicon diode signals in order to measure the x-ray induced diode currents, $\sim 10^{-10}$ A., in the presence of diode dark currents of order, 10^{-5} A. Measured data were consistent with diode silicon dead layers of from 0.3 to 0.4 μm. Average agreement to within five percent was found between measured diode sensitivities and calculated diode sensitivities. The measured diode x-ray sensitivities are compared with independently-measured diode x-ray sensitivities of similar diodes.

INTRODUCTION

Planar, silicon diodes have been employed as radiation diagnostic tools for low energy, pulsed x-radiation because of their short charge collection time, high charge collection efficiency, and high photon absorption efficiency at energies below 10.0 keV. In the energy region around 2.0 keV the diffused silicon diode's x-ray characteristics depend critically upon the diode's dead layer. The diode x-ray calibration was performed in the energy region 1.8 to 5.0 keV in order to characterize the x-ray absorption in the diode dead layer.

METHOD

The silicon, PIN type, diodes calibrated were planar, diffused, and totally depleted (1). The manufacturer's diode specifications include a 100 mm^2 active area, a 250 μm depletion depth, and an effective dead layer of less than 0.5 μm silicon. Diode leakage currents at room temperature ranged from 1.0 to 20.0 μA.

Low energy, x-radiation was produced by fluorescing targets with intense, low-energy, bremsstrahlung and characteristic copper radiation. Each x-ray spectrum was determined from measurements made using a Si(Li), non-dispersive, x-ray spectrometer and a pulse height analyzer system. The magnitude of the flux incident upon each diode was calculated from flux measurements made using a xenon-filled ionization chamber. The current which each characteristic x-ray spectrum induced in each silicon diode was measured by chopping the incident x-ray beam and using phase-sensitive detection. The dead layer was calculated for each diode from the best mathematical fit of the experimentally measured diode currents to a theoretical model.

LOW ENERGY X-RAY MACHINE

The x-ray source used in this calibration was a low-energy, high-intensity, high spectral purity x-ray machine (2); it is similar in design to an x-ray system at the Los Alamos Scientific Laboratory. It incorporates the cathode-anode geometry due to Henke (3) and employs a water-cooled copper anode. Schematics showing the experimental configurations are shown in Figures 1, 2, and 3. For each fluorescent target the x-ray machine was operated with the anode power dissipation ranging from 1.7 to 2.2 kW in order to maximize the x-ray output. The power output stability remained better than one percent. Six fluorescent targets were employed to provide predominantly characteristic K or L fluorescent radiant x-ray spectra whose mean energies ranged from 2.0 to 4.5 keV. The important x-ray machine operating parameters and fluor characteristics are shown in Table 1.

MEASUREMENT OF THE X-RAY SPECTRA

The spectral energy distribution for the radiation from each fluor was determined through the use of a Kevex, Si(Li) non-dispersive x-ray spectrometer. A Kevex electronics subsystem was used to process the pulses from the spectrometer preamplifier. The amplified pulses were energy-analyzed by means of a Hewlett-Packard multichannel analyzer system. Data output was provided to a plotter, oscilloscope display, and a teletype. The system resolution

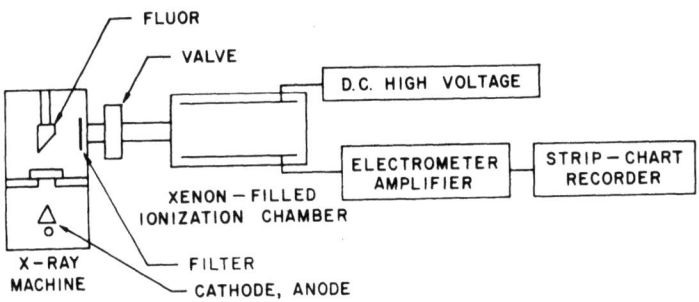

FIGURE 1. Experimental Configuration - Measurement of X-Ray Flux

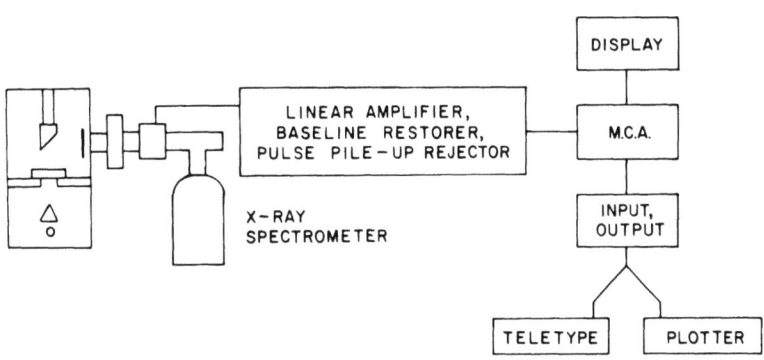

FIGURE 2. Experimental Configuration - Measurement of X-Ray Spectrum

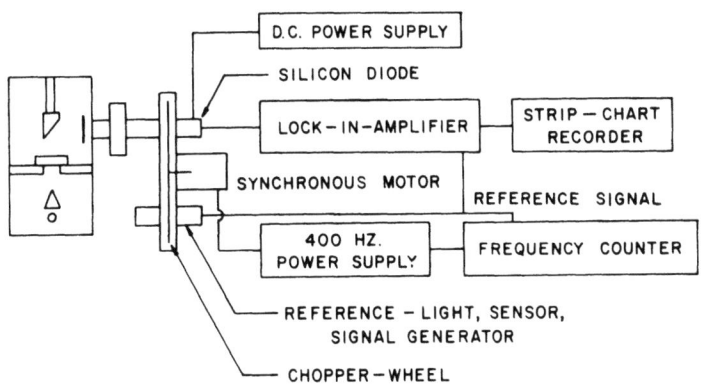

FIGURE 3. Experimental Configuration - Measurement of Silicon Diode Current

TABLE 1

X-RAY MACHINE AND SPECTRAL PARAMETERS

X-Ray Machine Voltage–keV	X-Ray Machine Anode Current–mA	Fluor, Filter Material	Average Spectral Energy–keV	Approximate Spectral Purity	Spectral Intensity at Fluor–Watt/Steradian
5.0	350	Silicon, none	2.04	0.53	2.96×10^{-5}
6.0	350	Zirconium, none	2.50	0.56	3.09×10^{-5}
7.0	300	Molybdenum, none	3.26	0.52	4.87×10^{-5}
8.0	250	Silver, none	3.92	0.54	7.04×10^{-5}
8.0	250	Tin, none	4.18	0.59	6.98×10^{-5}
10.0	200	Titanium, Titanium	4.80	0.77	9.09×10^{-5}

was measured to be 235 eV for 5.895 keV x rays. A system energy calibration was performed using known K and L line energies from the six fluorescers. Measured spectral data were corrected for the effects of x-ray absorption losses within the spectrometer. Some x-ray spectral characteristics are listed in Table 1.

MEASUREMENT OF X-RAY FLUX

For each of the fluorescers the intensity of the x-ray spectrum was calculated from the ionization chamber currents. These currents were measured using a xenon-filled, parallel-plate, ionization chamber placed 28.2 cm from the fluorescer center. The ionization chamber's xenon pressure was 600 torr. At this pressure the ionization chamber was nearly totally absorbing (97 percent) for x rays in the energy region of interest.

The x-ray transmission of the chamber's 0.001 inch thick beryllium window was measured at three energies. The window's areal density determined from these data was used to correct for window absorption at all energies. The ionization chamber's saturation curves were measured to insure that the ionization chamber's collection efficiency exceeded 0.99 for each spectral flux measurement. The calculations of spectral flux which included corrections for losses due to xenon L-fluorescence were performed using a value of 21.5 joule/coulomb for $\overline{W}$, the average energy per ion-electron pair in xenon (5). The calculations assumed isotropic point source radiation from the fluor, but the legitimacy of this assumption is not critical since both the ionization chamber and the silicon diode view the same solid angle from the fluor. Values of the spectral flux per unit solid angle are tabulated in Table 1.

MEASUREMENT OF X-RAY INDUCED DIODE CURRENTS

Phase-sensitive detection was used to measure the x-ray induced silicon diode currents (on the order of 10^{-10} A) in the presence of the diode dark currents (on the order of 10^{-5} A). The x rays incident upon the diode were "chopped" at a fixed frequency, and the resulting diode signal was measured at that frequency using a lock-in-amplifier.

The "chopper" employed in the diode calibration was a motor-driven chopper-wheel. The wheel, fabricated from an aluminum disc, had circular holes placed in it at uniform intervals along a circular center-line, coincident with the disc's center. The chopper-wheel was rotated at 8,000 R.P.M. by means of a direct-drive, synchronous motor. The motor was powered by a 750 V.A., single-phase, 400 Hz supply. The measured short-term frequency stability of the combined power supply, motor, and chopper-wheel was better than five parts in 10^5. A reference signal for the lock-in-amplifier was obtained from a light and an optically-coupled photodiode sensor which were mounted on the chopper-wheel support frame diametrically opposite to the silicon diode being calibrated. A light-tight housing enclosed the silicon diode, the reference sensor system, and the chopper wheel. The diode viewed the x rays through a beryllium window which was mounted in the housing and which separated the chopper-wheel assembly from the x-ray machine vacuum. The housing was filled with helium during diode calibration; the x-ray transmission losses through the helium from the beryllium window to the diode surface were less than one percent. Additionally, the synchronous-motor mount was cooled with chilled nitrogen in order to prevent heat transfer from the motor to the diode.

The signal from the silicon diode was detected through the use of a Princeton Applied Research preamplifier and lock-in-amplifier. The lock-in-amplifier was calibrated at the chopper frequency using its internally-generated reference square-wave. The lock-in-amplifier's direct current output was recorded on a strip-chart recorder. A mechanical shutter on the x-ray machine enabled the lock-in-amplifier measurements of the diode signals to be made with and without incident x rays, thus, compensating for some synchronous background effects.

Measurements of the x-ray induced diode currents for each diode were performed on different days, and these measurements were interspersed with ionization chamber measurements of the spectral intensities. No long term time-dependent effects were found.

THEORY

In order to determine the silicon diode dead layer from the measured diode currents, it is necessary to calculate the effective detector area exposed to the incident x rays as a function of the angular displacement of the chopper-wheel. This function enables calculation of the direct current in the diode corresponding to the a-c component measured at the chopper-wheel fundamental-frequency by means of the lock-in-amplifier.

The angular dependence of the area of the apertured diode illuminated by the x rays (assumed to be plane-parallel) from the fluor and collimated by the chopper-wheel is described by:

$$f(x) = \sum_{i=0}^{\infty} a_i \cdot \cos(iwx) \qquad (1)$$

where w is the angular fundamental-frequency of the chopped waveform. The root-mean-square of the fundamental-frequency component of the "chopped" waveform, RMS, is related to $f(x)$ by:

$$\text{RMS} = \frac{w}{\sqrt{2}\,\pi} \int_0^{\frac{2\pi}{w}} \cos(wx) \cdot f(x)\, dx. \qquad (2)$$

Denote by DC, the maximum exposed area of the apertured-diode chopper-wheel assembly. The ratio: DC/RMS is:

$$\frac{DC}{RMS} = \frac{\text{maximum exposed area}}{\frac{w}{\sqrt{2}\,\pi} \cdot \int_0^{\frac{2\pi}{w}} \cos(wx) \cdot f(x)\, dx} \qquad (3)$$

For the chopper-wheel used in this calibration, the radius of the chopper hole, circular center-line was 2.292 inches; sixteen circular holes of diameter 0.500 inches were placed at uniform intervals along this center-line. The diameter of the diode aperture was 0.418 inches. Using these parameters Equation (3) yields DC/RMS = 2.99.

The correctness of the assumptions underlying the chopper-wheel analysis was examined by measuring the ratio, DC/RMS, experimentally using light. The result depended upon the geometry of the light source. The value, DC/RMS = 4.77, which was measured using a plane parallel, but non-uniform, light beam was used in the calculations.

The fraction of the current which is induced in an apertured, planar, silicon diode due to a uniform, incident beam of chopped x rays, and which is measured by the lock-in-amplifier at the chopper waveform fundamental frequency is:

$$i_t(L) = \frac{\text{SUM}(L) \cdot B \cdot \left(\frac{RMS}{DC}\right)}{W_s} \qquad (4)$$

In Equation (4), B is the effective solid angle subtended by the apertured detector at the source, W_s is the average energy per electron-hole pair in silicon (W_s = 3.64 joule/coulomb), (RMS/DC) is the inverse of the quantity given by Equation (3), and SUM(L) is the integral shown in Equation (5).

$$\text{SUM}(L) = \int_0^\infty S(E) \cdot \text{FIL}(E) \cdot \text{PIN}(E,L)\, dE \qquad (5)$$

where S(E) is the x-ray spectral output of the source, FIL(E) is the effective x-ray filtration in the x-ray path from the source to the diode, and PIN(E,L) is the theoretical silicon diode efficiency for a diode of dead layer thickness, L. The diode efficiency PIN(E,L) is:

$$PIN(E,L) = \exp[-\mu_2(E,Si) \cdot R \cdot L] \cdot \frac{\mu_2(E,Si)}{\mu_1(E,Si)} \cdot \{1.0 - \exp[-\mu_1(E,Si) \cdot R \cdot M]\}. \tag{6}$$

In Equation (6), L is the dead layer thickness; M is the depletion depth, R is the volume density of silicon; $\mu_1(E,Si)$ and $\mu_2(E,Si)$ are the total photon interaction cross-section and the photon energy-absorption cross-section, respectively, in silicon at energy, E.

RESULTS

Equation (4) was evaluated for various dead layer thicknesses, L, assuming a detector depletion depth of 250 µm. The ratio of the measured detector current, i_m, to the theoretical detector current, $i_t(L)$ for different values of L was calculated; these ratios are shown for a representative detector in Figure 4.

For each detector and for each dead layer thickness, L, the weighted, least-squares average of the ratios, $i_m/i_t(L)$, was calculated. In this calculation, the statistical weight assigned to each ratio was directly proportional to the average total photon cross section for the energies of the measurement and inversely proportional to the standard deviation of the measured diode currents about their mean. Then, for each detector the measured dead layer thickness was taken to be that dead layer thickness for which for "best", i.e., minimal associated fractional error, weighted least-squares fit to the ratio data was obtained. The limits of error in the measured dead layer thicknesses were taken to be those dead layer thicknesses to which correspond increased in the associated fractional error of the average ratios of 30 percent from the minimum associated fractional error. The measured diode dead layers together with the averaged ratios,

$$\frac{i_m}{i_t(L)},$$

are shown in Table 2.

The silicon diodes' efficiencies are characterized by Equation (6) using the respective dead layers listed in Table 2. A diode efficiency curve is shown in Figure 5.

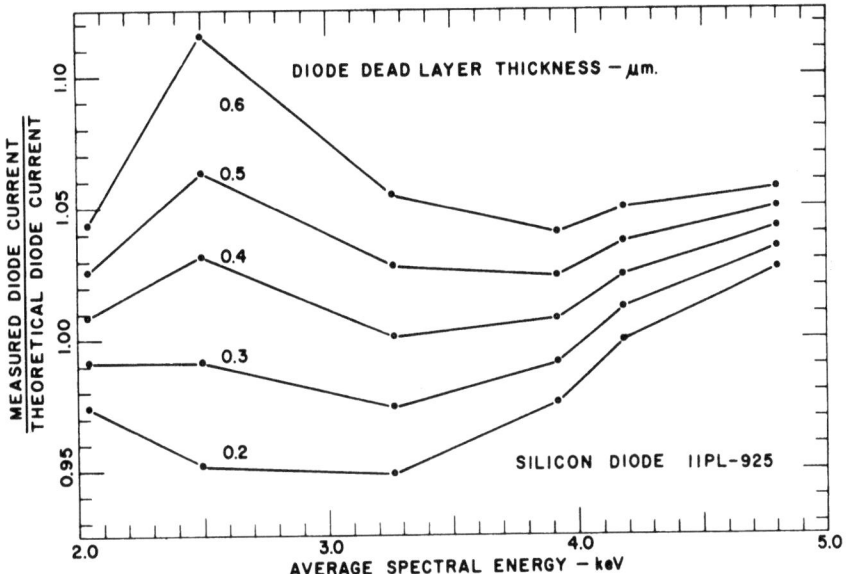

FIGURE 4. Silicon Diode Current Ratios

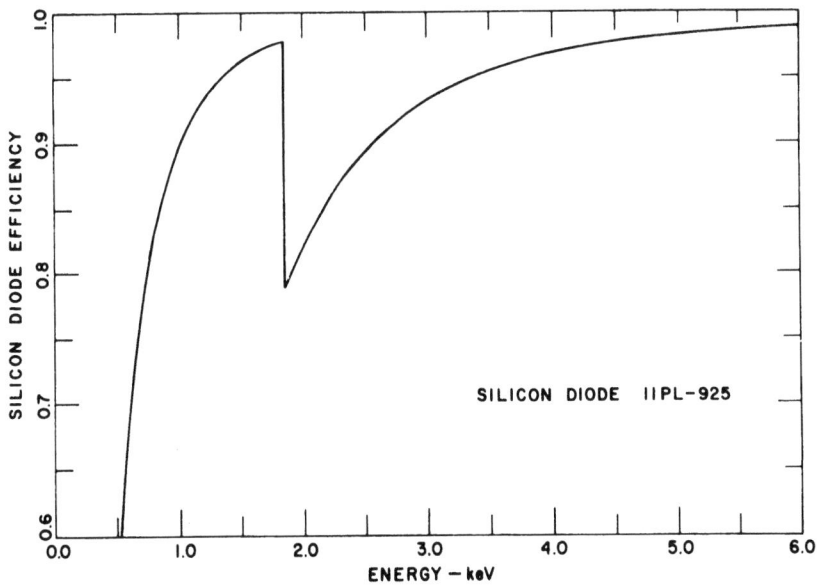

FIGURE 5. Silicon Diode Efficiency

TABLE 2

SILICON DIODE DEAD LAYER THICKNESSES

Diode Serial Number	Dead Layer Thickness Silicon – µm	Dead Layer Thickness SiO_2 – µm	Best Fit Ratio of Currents $\frac{i_m \text{(MEASURED)}}{i_t \text{(THEORETICAL)}}$
2PL-925	0.34 + 0.21, -0.06	–	1.066
5PL-925	0.40 + 0.17, -0.08	–	1.060
6PL-925	0.31 + 0.05, -0.05	–	1.056
8PL-925	0.32 + 0.24, -0.08	–	1.032
11PL-925	0.30 + 0.11, - 0.04	–	1.035
21PL-925	0.34 + 0.20, -0.18	–	1.071
18PL-649	0.98	0.44	–
10PL-878	0.60	0.44	–

DISCUSSION

The author ascribes the five percent mean deviation of the ratios shown in Table 2 from unity to a systematic experimental error. A source of such a systematic error may be the parameter, DC/RMS; the discrepancy between its theoretical value and its measured value has not been resolved. An error in the parameter, DC/RMS, would not, however, affect the measured diode dead layer thicknesses, and these measured thicknesses are consistent with the manufacturer's dead layer specifications.

Similar diodes have been calibrated for Sandia Laboratories under a contract with Science Applications, Inc. (6). Two, PIN type, planar silicon diodes with specified dead layers of 1.0 μm were found to have both silicon and oxygen in the dead layer. The measured areal densities have been converted to effective thicknesses of SiO_2, the most probable compound, and silicon and are shown in Table 2. The dead layer thicknesses inferred from these analyses, 1.0 μm and 1.4 μm are consistent with the detectors' performance.

ACKNOWLEDGEMENTS

The author would like to thank L. W. Morrison, R. Villegas, and J. Harris for their assistance in the design and fabrication of the equipment as well as in the acquisition and analysis of the data. This work was supported by the U. S. Atomic Energy Commission.

REFERENCES

1. Available from Solid State Radiations, Inc., 2261 South Carmelina Avenue, Los Angeles, California 90064.

2. L. W. Morrison and K. M. Glibert, "Sandia Laboratories' Henke X-Ray System," Sandia Laboratories, Albuquerque, New Mexico, SLA-73-0059.

3. B. L. Henke, "Microanalysis with Ultrasoft X-Radiations," in W. H. Mueller, Editor, Advances in X-Ray Analysis, Vol. 5, p. 285-305, Plenum Press (1962).

4. P. B. Lyons, J. A. Baron and J. H. McCrary, "A Total Absorption Ionization Chamber for 1.5-10 keV X Rays," Nucl. Instru. Methods, 95, p. 571-583 (1971).

5. Science Applications, Inc. 1261 Birchwood Drive, Sunnyvale, California, under contract with Sandia Laboratories, Albuquerque, New Mexico.

DEVELOPMENT OF THE HIGH PERFORMANCE 'SOLFA' ON-LINE ANALYSER TO MEASURE TOTAL SULPHUR IN PETROLEUM DISTILLATES AND RESIDUAL FUELS USING NON-DISPERSIVE X-RAY FLUORESCENCE

C.F. Gamage and W.H. Topham

Applied Research Labs. Ltd.   The British Petroleum Co. Ltd.

Wingate Road, Luton   BP House

Bedfordshire, England   London, England

INTRODUCTION

In any petroleum product, a high sulphur content is considered undesirable. Sulphur in diesel fuels is claimed to aggravate problems of wear and deposit. With residual oils used in some metallurgical operations, the sulphur combustion products can have an adverse effect on the metal. In combustion equipment, the oxides of sulphur together with water vapour can lead to undesirable acid condensation on cool metal surfaces.

The more general problem is air pollution. Local and national legislation present the oil industry with a continuing demand for fuels of lower and lower sulphur content. Best use must therefore be made of crude oils of inherently low sulphur content. Desulphurising processes are commonly used for distillates and are available for residual products. Such processes however tend to add significantly to production costs.

Means of measuring the sulphur content accurately and quickly are vital to refinery operations, to minimise quality give-away and delays in product testing. Routine laboratory tests have been successfully developed to the point where analyses may take only a few minutes and be accurate to typically 0.01 to 0.10% wt sulphur. Laboratory testing has limitations for controlling operations like in-line blending, especially where direct product dispatch may be required. For these situations the logical approach is to use on-line measurement for sulphur. This paper describes the most recent attempt by BP in conjunction with A.R.L. to make available an accurate and reliable sulphur monitor.

BP's interest in on-line sulphur measurement originated in the late 1950's from the widespread adoption of in-line blending and gas oil desulphurisation. Between 1960 and 1966, BP spent a great deal of effort developing and testing monitors based on x-ray absorption. This was partially successful for distillates, but not satisfactory for residual fuels because of viscosity problems and metals content. It was therefore decided to study methods based on continuous combustion but this could not achieve the required accuracy on gas oils and was not usable for fuel oils on a laboratory scale.

In 1970, it was estimated that the accuracy required from any new monitors should be 200 ppm sulphur for gas oils and 500 ppm sulphur for fuel oils (based on a 95% confidence level) and that fuel oil applications were the more important because sulphur in fuel oils would become a major restriction and quality give-away would therefore be more costly. As neither absorption nor combustion appeared capable of achieving the specified performance, BP examined the possibilities of x-ray fluorescence as the basis for a new monitor. Tests with the A.R.L. non-dispersive N900 XRF laboratory instrument indicated that this offered the possibility of achieving the accuracy required. Therefore, it was agreed to develop a sulphur monitor based on XRF techniques and by April 1971 agreement was reached between BP and A.R.L. to develop the prototype SOLFA monitor described in this paper.

MONITOR SPECIFICATION

The specification for the sulphur monitor is summarised as follows:

1. Applicable to the whole range of middle distillates (domestic heating oils, diesel fuels) and residual fuels having the properties shown in Table 1.

2. Absolute accuracy (95% confidence limits) to be 0.02% wt sulphur on distillates and 0.05% wt sulphur on residuals, under routine service conditions and including any effects from density, C/H ratio and the presence of other elements.

3. Conform to the stringent regulations covering electrical and

radiation safety, for unattended operation in potentially hazardous areas (Division 1). Be suitable for the service and ambient conditions in almost any refinery in the world (i.e. operable from -20°C to +40°C); specified accuracy to be maintained for variations of 10°C either side of the calibration temperature; for barometric pressures from 745 to 775 mm Hg and for power supply variations of ±10% on voltage and ±4% on frequency.

	Middle Distillates	Residual Fuels
Sulphur, % weight	0 to 2	0 to 4
Density, gm/ml at 15°C	0.77 to 0.89	0.90 to 1.00
Carbon to hydrogen weight ratio	6.00 to 7.00	6.80 to 8.00
Viscosity, cSt at 50°C	0.50 to 5.00	20 to 500
Other constituents (ppm max.)	200 Water	200 Vanadium 100 Nickel 200 Sodium 500 Water

TABLE 1

4. Direct readout in percent weight sulphur, with linear calibration if possible.

5. Response time compatible with its use for automatic quality control of in-line blending systems. (Up to 5 minutes would be considered reasonable and somewhat longer times might be acceptable).

6. Reliable enough to operate with less than 1% down-time with attention for maintenance or calibration only once per week if possible and certainly not more than daily.

7. The monitor to be available as a total package including sample conditioning to suit each particular duty, and calibration/checking facilities.

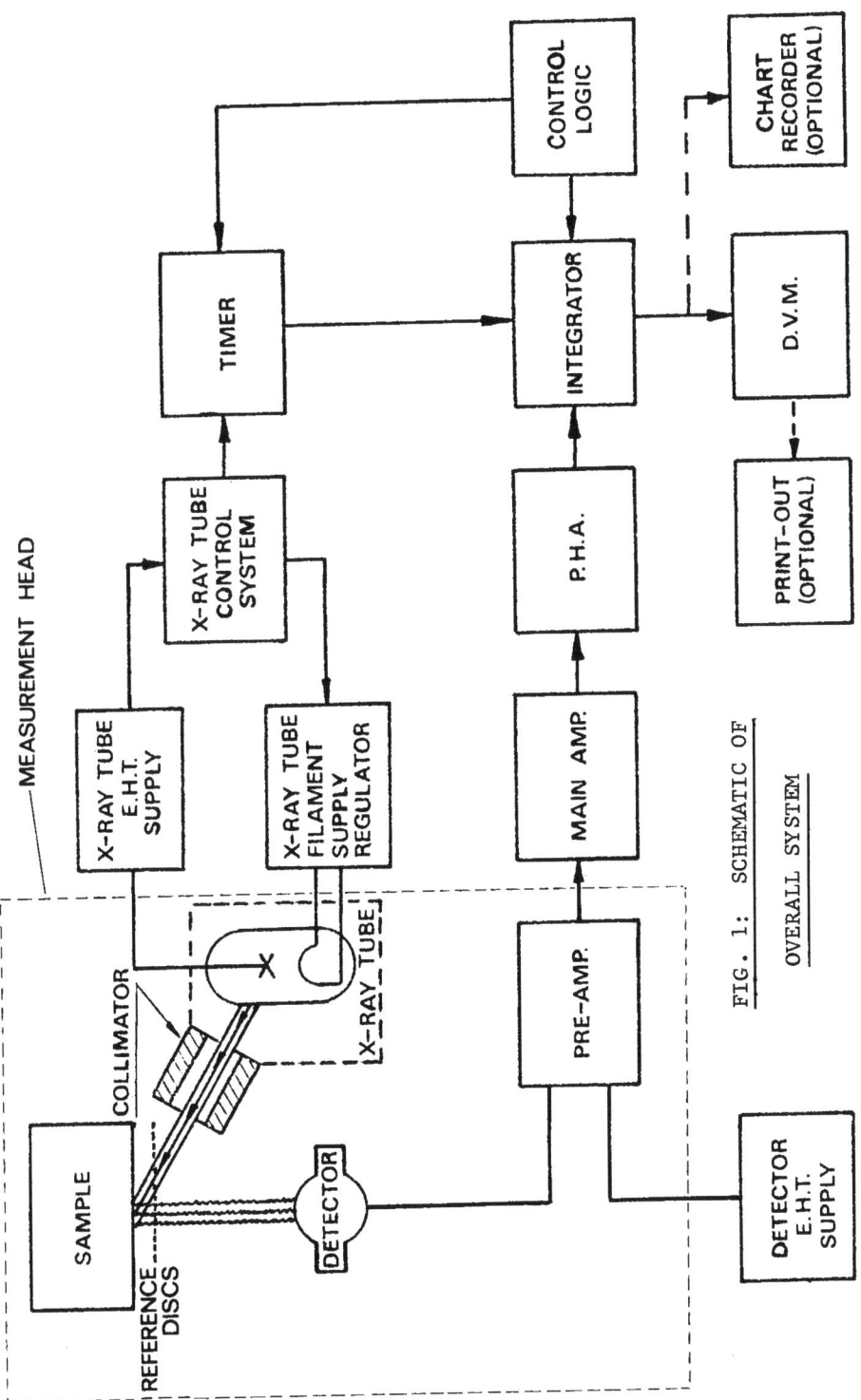

FIG. 1: SCHEMATIC OF OVERALL SYSTEM

PRELIMINARY WORK

Phase 1 of the project had to show that the agreed analytical specification was likely to be achievable. The main aspects to be studied were: Choice of excitation source and the source/sample/detector geometry; flow cell design; effect of C/H ratio and density on signal; effect of other elements; precision of sulphur measurement.

Choice of Excitation Source and System Geometry

The standard N900 instrument has an isotope source in annular form around the detector. However, A.R.L. had developed a miniature air-cooled x-ray tube and because of difficulties with acceptability of radio-active sources on the earlier x-ray absorption type monitors, and to obtain the maximum precision within a restricted analysis time, it was decided to use the x-ray tube as radiation source. The tube is entirely passive when switched off, whereas radio-isotopes are permanently active.

Using the x-ray tube makes the geometry somewhat different from the annular radio-isotope case. To reduce the background signal from scattered radiation and to restrict any radiological hazards, the exciting radiation is collimated so that the beam strikes the centre of the flow cell. A schematic of the overall system is in Fig. 1, whilst Fig. 2 shows the measurement head which includes the flow cell window and window failure detector, beam collimator, reference discs shutter, x-ray detector and pre-amplifier.

Design of Flow Cell

For on-line service, it is necessary to have a cell through which the sample flows continuously. The main problem was to find a window to withstand the operating pressure without distorting at the temperature involved (up to 95°C), yet to be sufficiently transparent to sulphur fluorescent x-rays for the system to have adequate sensitivity. We tested unsupported Mylar, Melinex and Kapton for their temperature and pressure behaviour and from these tests selected 8 μm Kapton polyimide film. This could withstand up to 20 psig at 95°C without bursting and at pressures below 2 psig did not appear to exceed its elastic limit, but it did bow sufficiently under small pressures to affect the source/window detector geometry and hence measurement accuracy. Therefore, the Kapton window was supported outside by a photo-etched nickel grid. This combination reduced the effective transparency of the window by 40% but had all the other essential properties in relation to pressure and temper-

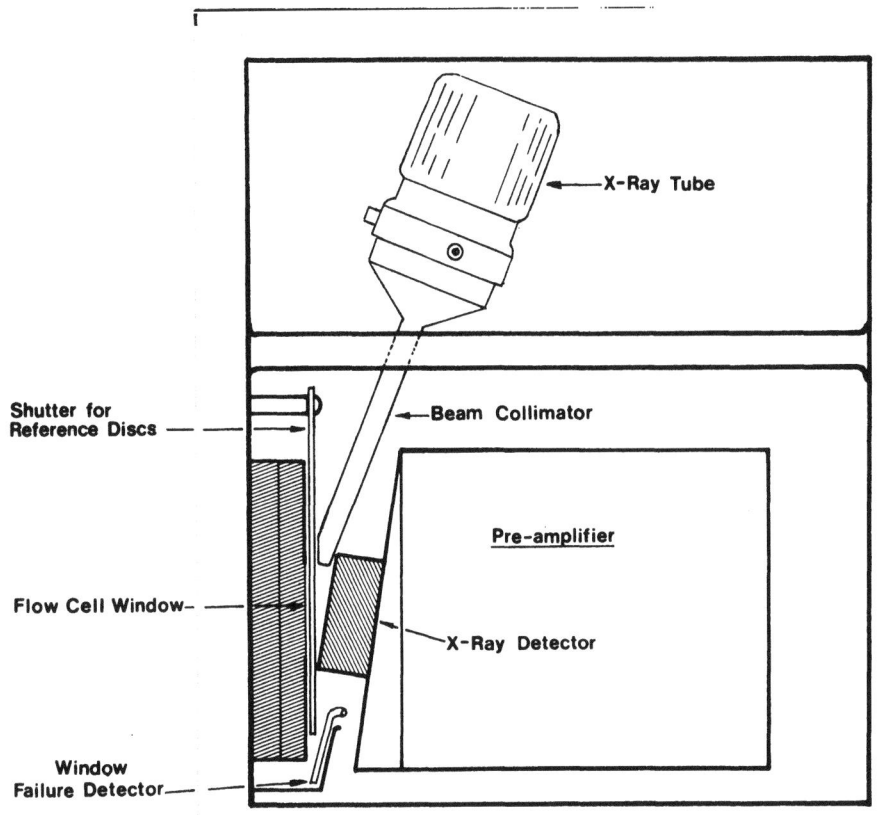

FIG. 2: THE MEASUREMENT HEAD

ature, ease of fitting etc. Fig. 3 shows the present flow cell for which a patent application has been filed.

The cell entries and measurement chamber are so shaped to provide good scouring of the window with minimum dead spots at a flow of about 20 litres/hour. The cell could be operated at pressures up to 30 psig but in practice is run at atmospheric pressure.

In practice, the amount of bowing that occurs when even the most viscous oil specified is pumped through the cell gives rise to a change of signal equivalent to less than 50 ppm sulphur.

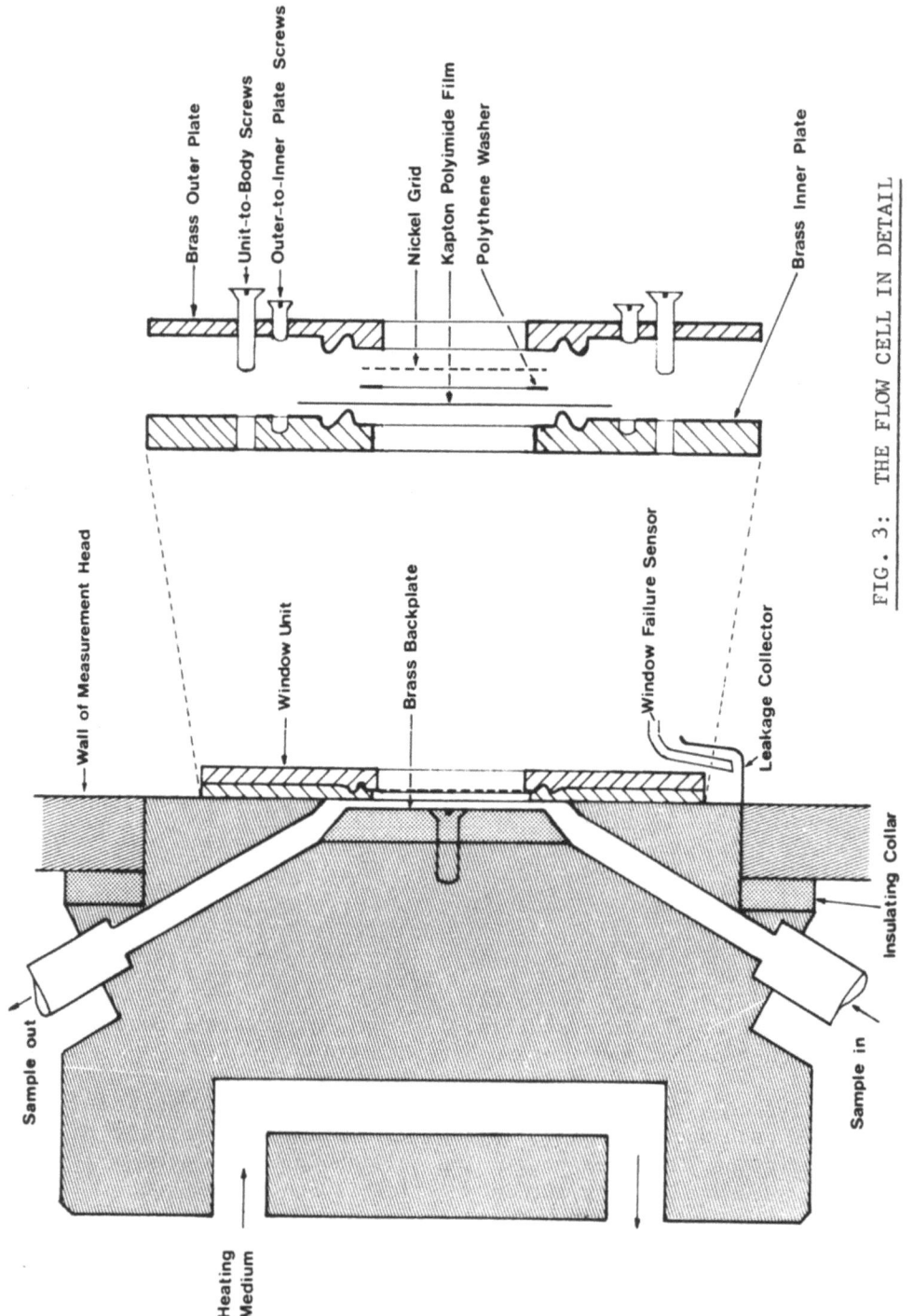

FIG. 3: THE FLOW CELL IN DETAIL

The monitor incorporates a special pneumatic system to detect oil leakage through the cell window. Should a leak be detected, the power is switched off and the oil flow stopped.

Density and C/H Ratio Effects

Tests were done on a range of samples prepared by dissolving accurately known proportions of t-butyl disulphide in solvents of widely differing C/H ratio and density. The results from the N900 instrument (isotope and tube source) showed that the density effect was negligible and that the C/H ratio effect varied with the depth of sample in the cell. At optimum depth on the test equipment, the change in apparent sulphur for a C/H change of 1 was 110 ppm at 1% sulphur and 200 ppm at 4% sulphur. These effects were significant in relation to the accuracy levels being aimed for, but were considered acceptable.

Effect of Other Elements

Residual fuels contain naturally occurring vanadium and nickel, and sodium from caustic soda injected to reduce crude tower corrosion. The monitor is required to achieve the accuracy specified despite the effects of these elements.

Tests with the N900 instrument showed no effect from more than 100 ppm nickel and 200 ppm sodium. Vanadium interferes as a result of the resolution of gas filled proportional counters. From many tests at the 500 ppm level, it was concluded that the maximum effect of the specified 200 ppm max. vanadium would be equivalent to only +100 ppm sulphur.

It it worth noting that x-ray absorption methods are much more sensitive to the specified contaminating elements. The apparent increase in sulphur content for (200 ppm V + 100 ppm Ni + 200 ppm Na) is estimated to be 1200 ppm sulphur for an absorption monitor, i.e. ten times the effect with the XRF system. This means that the effect of these elements alone would be 2½ times the maximum permissible error on fuel oils, ignoring all other sources of error. This appears to rule out categorically any possibility of an absorption monitor meeting the BP specification for sulphur monitors for residual applications.

Precision

Tests with the flow cell on a range of gas oils and residual fuels, hot and cold, static and flowing, showed a linear calibration and a standard deviation for cycle to cycle repeatability of typically 100 ppm sulphur. Considering the breadboard nature of the apparatus and the un-optimised conditions, this performance was considered an adequate basis to proceed to Phase 2.

MONITOR DESIGN

Having decided that the specification appeared to be achievable, the production prototype was designed. The resulting equipment shown in Fig. 4 comprises six main sections:

1. Flow cell, mounted through the wall of the measurement head with external connections for sample and for heating medium if necessary.

2. Measurement head containing the x-ray tube with collimator, flow cell window and leakage detection unit, sealed neon-filled detector, signal pre-amplifier and temperature sensor for air density compensation. There are external facilities for manually inserting two solid reference sample discs, mounted on a shutter, for rapid checking.

3. Power supplies unit containing the x-ray tube and detector power supplies.

4. Measurement electronics unit. This unit includes a digital voltmeter to display both the output and various test voltages.

5. Input distribution box containing main power switch, trip relay, fuses, etc.

6. Sample conditioning unit with filter, pressure reducer, pumps for flow regulation and sample disposal, heating facilities and test sample pot.

The electrical safety relies on using explosionproof-type enclosures, but all with air purging and pressure switches to shut off power on loss of purge pressure. The system is so arranged that it trips out if air flow falls below that necessary for cooling the two main electrical enclosures. There is no detectable radiation outside the x-ray tube enclosure, and a key interlock arrangement ensures that the tube cannot be exposed with power on.

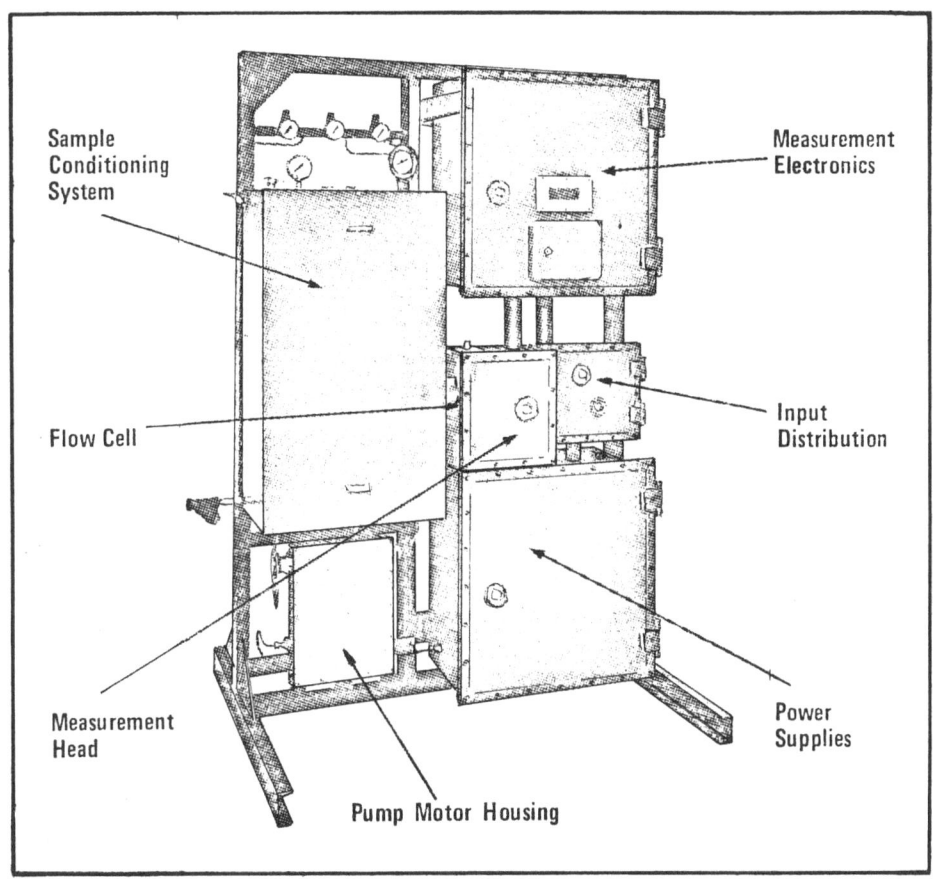

FIG. 4: THE COMPLETE SOLFA ANALYSER

TEST PROGRAMME AND RESULTS

The following items summarise the results achieved to date in the laboratory:

Reliability

This was assessed initially by running the monitor continuously day and night for 2000 hours (12 weeks). Apart from an initial faulty meter on the EHT circuit and a DVM that had to be replaced, there were no other failures. Since that time the monitor has run for several thousand hours more and the only further troubles have been a noisy switch and wear on the sample pumps.

The x-ray tube system has proved very satisfactory. The tube is rated for 30 kV and 1000 microamps but is run conservatively at 27 kV and 200 microamps. The specified drift on the EHT supply is less than 0.008 kV per day, corresponding to 40 ppm sulphur. The tube current appears never to vary by more than ±1 microamp but the monitor can compensate for very much larger variations.

Flow Cell

The flow cell performance has exceeded all expectations. One window was in almost continuous use for a year and was then only changed to inspect its condition. There has been no evidence of leakage or of change of transmission with age. The precision tests were performed with a 6 months old window. Tests with cold viscous lubricating oil (1.3% sulphur) and extract (5.9% sulphur) showed that it takes less than 2 minutes to effect a 99% sample change in the flow cell at the normal flow rate.

Effects of Barometric Pressure and Temperature

The relatively low energy sulphur fluorescent x-ray (2.3 keV) is absorbed by the air between the window and the detector. Variations of the air density change the amount of absorption. Air density depends upon temperature and pressure and the effects for liquid samples are +30 ppm sulphur/1% sulphur/$°C$ and -12 ppm sulphur/1% sulphur/mm Hg. These effects are small but significant at the levels of accuracy required and the ambient conditions specified, i.e. ±10$°C$ and ±15 mm Hg. The prototype monitor compensates automatically for air gap temperature effects and production instruments will also be able to compensate automatically for barometric effects.

Analytical Performance

Accuracy and stability were assessed by a statistically-designed programme covering five consecutive days and nights. The monitor ran continuously on hot circulating samples and no adjustment was permitted to any of the conditions, other than sample changes.

The test samples were six gas oils and five residual oils all accurately tested for sulphur content, vanadium, nickel, C/H ratio,

SAMPLE	TYPE	DENSITY gm/ml AT 15°C	C/H WT RATIO	Ni PPM	V PPM	VISC cSt AT 50°C	SULPHUR % WT*
A	Special Gas Oil	0.825	6.16				<0.01
B	Hydrofined Gas Oil	0.840	6.71				0.16
C	Sales Gas Oil	0.822	6.14				0.67
D	Kuwait Gas Oil	0.820	6.27				0.85
E	Vacuum Gas Oil	0.870	6.71				1.74
F	Cracked Gas Oil	0.900	7.90				2.16
G	Libyan Residue	0.918	7.37	12	11	160	0.62
H	Mixed Residue	0.931	7.33	15	31	170	1.37
I	Residual Fuel Oil	0.931	7.51	25	60	62	2.38
J	Residual Fuel Oil	0.948	7.65	15	38	110	2.97
K	Residual Fuel Oil	0.957	7.01	26	50	111	3.57
L	Lub. Oil Extract	1.020	9.26	3	7	36	5.85

TABLE 2

* The sulphur contents are the means of 8 analyses on a Philips PW 1540 dispersive instrument using the internal standard procedure.

density, plus viscosity and pour point where appropriate. Sample details are given in Table 2. They were chosen to cover the full monitor range and be as representative as possible of refinery products.

Each of the 11 test samples was run in random order on each of the five days. Ten consecutive 3 minute readings were taken on each sample on each run. During each sample changeover 3 readings were taken on the sulphur-free disc and three times each day 3 readings were taken on the sulphur reference disc. Overnight, the monitor was allowed to run unattended on one of the test samples. The results derived from these 800-odd readings were very good with the exception of zero drift (+0.075% sulphur over five days), so the monitor was modified and appropriate parts of the test programme were repeated over an 8-day period. The laboratory performance of the monitor in its present form can be summarised as follows: (Unless otherwise stated, results are those from the first 5-day tests)

(a) <u>Zero Stability</u>. Measurements on the sulphur-free disc are an excellent means of assessing overall monitor stability, excluding any effect in the sample cell. After modification, the zero was measured over eight consecutive days and the maximum drift over seven days was +230 ppm sulphur. The standard deviation for daily changes in zero reading was 96 ppm sulphur, the same obtained for distillate samples.

(b) <u>Sulphur Reference Disc</u>. The disc produces a reading corresponding to 4.62% sulphur. Over eight days, the daily readings after correction for barometric effects has a maximum spread of only ±127 ppm sulphur and the standard deviation for long-term precision was 88 ppm sulphur. These figures include any drift and any errors in the positioning of the disc.

(c) <u>Span</u>. The 'effective span' is defined as the difference between the readings for the two reference discs and has been shown to remain constant even when the zero drifted, as in the early tests. Both the five and eight day formal tests showed span to be constant to within ±175 ppm sulphur. This represents the maximum deviation in the slope of the calibration curve at the 4.6% sulphur level, indicating that in routine service the monitor could be adequately checked at a single point only.

(d) <u>Repeatability</u>. The standard deviation (sigma) was calculated for every set of ten consecutive 3-minute analyses for each disc and liquid sample. The average results for sigma expressed in ppm sulphur were as follows:

Zero sulphur disc = 50
Sulphur disc (4.6% sulphur) = 66
Six gas oils (0 to 2.2% sulphur) = 62
Five residual oils (0.6 to 3.6% sulphur) = 64

(Note: Repeatability as defined under Table 4 is 2.8 x sigma)

(e) **Long Term Precision.** Each of the eleven samples was tested daily, and the differences between the daily mean of ten results gives a measure of the long-term precision. Correcting all results for barometric pressure, then the standard deviation for all 30 tests on gas oils about their respective mean values was 96 ppm sulphur. The corresponding figures for the 25 residual oil results was 160 ppm sulphur, for the sulphur disc was 120 ppm (8 days) for the zero disc 96 ppm. These figures include any effect from incomplete flushing at sample changes, and any positioning errors for the discs.

(f) **Accuracy.** The mean value of every measurement on each sample corrected for barometric effect and any zero drift represents the monitor's best estimate for the sulphur content, i.e. monitor accuracy. The mean values (in millivolts) were calculated and plotted against the sulphur contents listed in Table 2. The deviation of mean monitor results from these best fit lines were as shown in Table 3. (Note: There is one calibration line for gas oils and a different line for residuals) The maximum deviation is 0.015% sulphur for gas oils and 0.008% sulphur for fuel oils.

(g) **Overall Quantitative Performance.** Interpreting the results in several ways showed the overall absolute accuracy for a single 3-minute analysis with 95% confidence limits (±2 sigma) was 0.03% sulphur, for gas oils and fuel oils. These results were obtained under favourable laboratory conditions and some deterioration might have been expected under field service conditions.

Drift is perhaps the most significant factor in overall accuracy, over a long period. Zero drift over 7 days was 0.02% sulphur. If this occurred, then the absolute accuracy with weekly attention only would be about ± 0.05% sulphur at the 95% confidence level, for a single analysis. Yet the drift is about the same magnitude as the repeatability so it would not be meaningful to consider automatic zero correction based on a single measurement. Auto correction based on the mean of, say, ten analyses would be much too complex.

	GAS OILS						FUEL OILS			
	A	B	C	D	E	F	G	H	I	J	K
% wt S	<0.01	0.16	0.67	0.85	1.74	2.16	0.62	1.37	2.38	2.97	3.57
Difference ppm S	+125	-122	+152	+70	+152	-150	+80	+80	-53	-80	+80

TABLE 3

PERCENT WEIGHT SULPHUR		1.0	2.0	5.0
Repeatability	SOLFA Monitor	.02	.02	.02
	ASTM D 1552	.07	.10	.24
	ASTM D 129	.06	.10	.18
	Promethium Source	.05	.08	.08
	PW 1540 Int. Std.	.04	.06	.12
	PW 1540 Direct	.01	.03	.05
	N900	.02	.04	.09
Long-Term Precision	SOLFA Monitor	.03	.04	.05
	Promethium Source	.07	.13	.13
	PW 1540 Int. Std.	.05	.08	.17
	PW 1540 Direct	.03	.05	.10
	N900	.02	.04	.10

TABLE 4. - For definitions see end of DISCUSSION

DISCUSSION

The target specification requires "monitor accuracy to be such that in the long run, without any correction other than the routine calibration checks, nineteen indications out of twenty shall be within ±0.02% sulphur (for gas oils) and 0.05% sulphur (for fuel oils) of the true values as determined using agreed test samples."

The most rigorous interpretation of this specification would require the accuracy criteria to apply to every single 3-minute analysis, during a week's unattended operation on-line. The results calculated for overall weekly performance (i.e. ±0.05% sulphur) shows that the monitor meets this most rigorous requirement in respect of residuals, but it is most unlikely that it could be made to do so for a single analysis on gas oils.

However, the primary object of the monitor would be to allow a line blending system to produce a total blend to the accuracy specified. Thus in practice the mean value recorded over a period could be what mattered, so repeatability would be ignored in calculating overall performance. This would also allow using shorter analysis times, say 1 minute for better automatic process control.

For relatively short production runs (say 12 hours) it might be practicable to check the monitor before each blend. For runs over several days, overall accuracy of blending should be a function only of calibration accuracy and any systematic long-term drift on the monitor. In either case, the calculated monitor performance would then be very close to 0.02% sulphur for either gas oils or fuel oils.

It was recognised at the start of the joint project that the specification was extremely demanding. Subsequent work by BP on the precision of laboratory methods for sulphur determination (1) certainly confirms that view. The best technique available in sulphur measurement is dispersive x-ray fluorescence spectrometry, and the best likely to be available in refinery laboratories is non-dispersive XRF. The results shown in Table 4 are the precision data for other methods commonly used in laboratories (2, 3) together with the SOLFA results expressed in the same terms. The SOLFA results confirm the initial decision that x-ray fluorescence could produce better results than alternative techniques.

It will be appreciated that the data in Table 4 gives no indication of absolute accuracy for the methods. However, if the PW 1540 is accurately calibrated then the long-term precision of the direct method could be taken as corresponding to absolute accuracy limits for a single test. Otherwise the internal standard method using lead must be used and this is less precise.

The PW 1540 is a relatively old spectrometer, but it is estimated that the best accuracy of the latest x-ray spectrometers would be about 0.01 to 0.02% at the 2% sulphur level and 0.05% sulphur at the 5% sulphur, the same as the monitor specification yet with more frequent checking of calibration.

In the light of the above data for laboratory methods, the monitor's performance is remarkably good. Calibration is completely linear over the full range for both types of products, even though slight non-linearity was expected on theoretical grounds (1). Reliability is excellent and speed of analysis compatible with automatic reset control. Overall accuracy is very high, better than most laboratory methods, and almost certainly much better than could be achieved with any other monitor.

On-Line Experience

At the time of writing the monitor has just finished a month's trial on-line on a gas-oil hydrofiner. This presented only minor, readily soluble operational problems and has led to results only marginally inferior to those obtained in the laboratory, e.g. long-term precision of 0.033% sulphur at the 95% confidence level. The monitor will shortly be evaluated on a residual fuel application after completing temperature tests over the ambient temperature range of -20°C to + 40°C.

Definitions (Table 4)

Repeatability: Duplicate results by the same operator should be considered suspect if they differ by more than the amount stated.

Long Term Precision: Results obtained within the same laboratory over a period of about a week should be considered suspect if they differ by more than the amount stated.

ACKNOWLEDGEMENT

Permission to publish this paper has been given by The British Petroleum Company Limited and by Applied Research Laboratories Limited.

REFERENCES

1. R.W. WILLIAMS (BP) 'The Determination of Sulphur in Petroleum

Fractions by Dispersive & Non-Dispersive XRF Analysis.'
Proceedings of the 8th X-Ray Analytical Conference, University
of Birmingham, England, 15 September 1972.

2. 'Promethium Source X-Ray Apparatus for Sulphur.' Paper by
   R.L. GILPIN and M.C. FRANKS (BP Sunbury) to API Mid Year
   Meeting, 15 May 1963.

3. 1971 Book of ASTM Standards - American Society for Testing
   and Materials.

AUTOMATIC DATA ACQUISITION AND REDUCTION FOR ELEMENTAL
ANALYSIS OF AEROSOL SAMPLES *

J. F. Harrison and R. A. Eldred

Crocker Nuclear Laboratory

University of California, Davis

ABSTRACT

A PDP 15/40 computer with ADC and CAMAC interfaces is used to control data collection apparatus, acquire data, and reduce data to determine the elemental composition of aerosol samples. The background is subtracted from each energy spectrum, peak centers are located automatically using a Gaussian correlation technique, peak multiplets are resolved with Gaussian fits, peak energies are compared with entries in a table of x-ray lines for possible identification, multiple identification of peaks and line interferences are resolved, and the elemental amounts are determined from the areas of the Gaussian fits.

INTRODUCTION

In order to analyze a large number of samples for elemental composition economically an automated system is necessary. Using the 76" cyclotron of the Crocker Nuclear Laboratory of the University of California, Davis, an ion-excited x-ray emission system has been developed in which the data acquisition and reduction are under the control of a PDP 15/40 computer. 1000 aerosol samples can be analyzed by this system in approximately 36 hours.

The most important criterion for an automated system is reliability. Based on our experience with a large number of aero-

---

*Work supported by the National Science Foundation/RANN and the California Air Resources Board.

sol samples, we have found the present system very reliable. The few visible peaks not located or correctly identified were of low intensity.

SAMPLE ANALYSIS FACILITY

A rapid and efficient sample handling system has been designed around standard 35 mm plastic slide mounts arranged in linear trays. The samples, which are generally on thin substrates as mylar or Nuclepore, are mounted on the slides prior to analysis. Figure 1 is a schematic of the facility to handle the samples and detect the x-rays. The slide trays are inserted in the tray chute, which is 45° from the beam and five cm off center. The slide to be analyzed is centered in the beam by the sample changer, which consists of a commercial slide changer, drive motors and position sensors for the slide and tray, and appropriate electronics. The changer is controlled by the computer.

A diffusion foil and current-readable collimators before and after the sample ensure a well collimated and nearly uniform beam spot at the sample. The excitation beam of 18 MeV alpha particles passes through the sample and is collected by the Faraday cup. The emitted x-rays are observed with an LN cooled Si(Li) detector with pulsed optical feedback. The Kevex 4500P linear analysis system includes pulse pile-up rejection and dead time correction. When an x-ray is detected a signal triggers the cyclotron beam

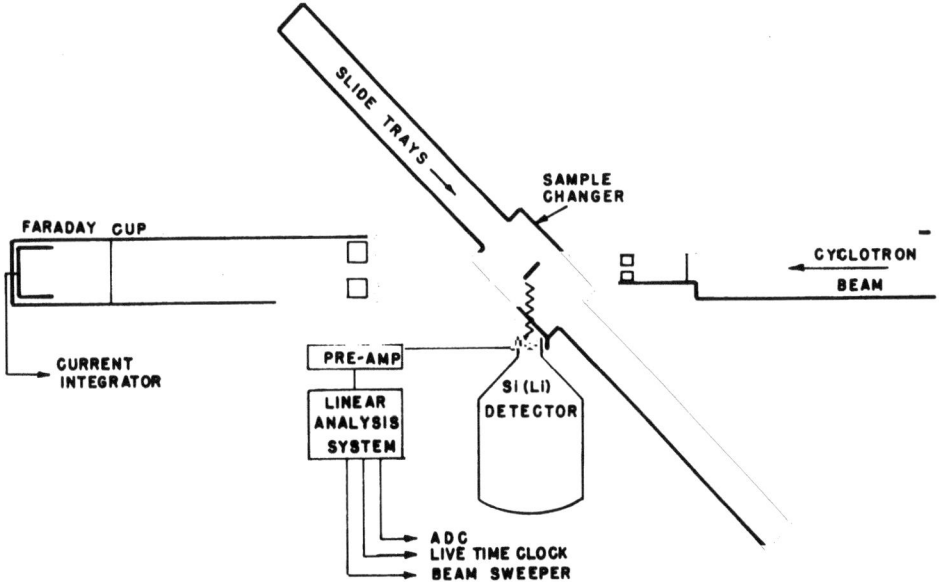

Fig. 1. Sample analysis facility.

sweeper (1). The sweeper consists of parallel plates about the beam with a normal voltage of 3000 volts. The supply voltage is turned off in about 150 nsec, causing the beam to go off axis and strike a slit. The beam is kept off for 40 μsec while the event is being processed. This permits much higher count rates for a given dead time correction, lowers the high energy background, and improves the detector resolution.

The sensitivity of the system has been enhanced by the use of a special filter between the sample and the detector. Filters are useful for count rate limited systems when most of the counts for a typical spectrum are in the low energy region. When the low energy x-rays are attenuated the beam can be increased, improving the sensitivity per unit time for a given count rate for the higher lines. However the lowest energy lines are completely lost. This problem is solved by making a small hole in the center of the filter. The lowest energy x-rays reach the detector, but with a smaller solid angle than that for the higher lines. The cross section times solid angle times efficiency for the three configurations is shown in Figure 2. With this filter the detectable limit for medium and heavy elements is one-third that with no filter, for the same time and count rate.

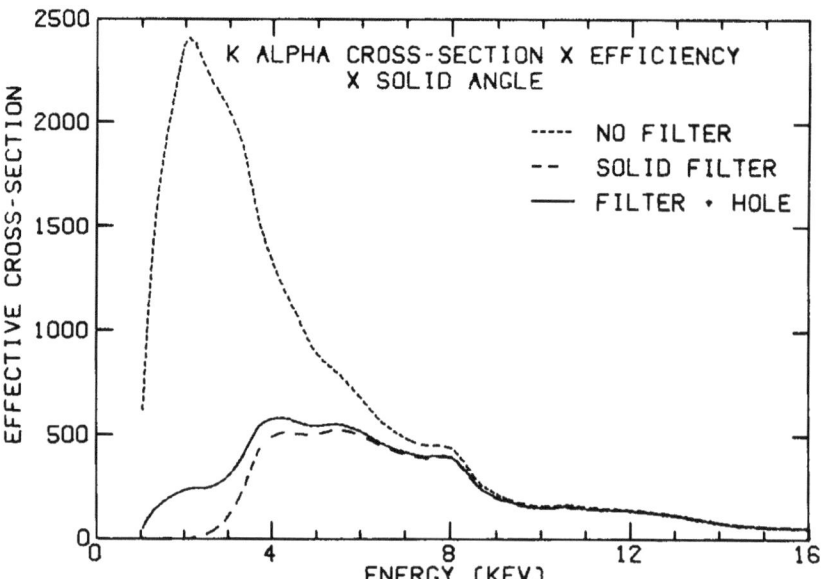

Fig. 2. Effective cross section for Kα lines. Below 3.5 keV the β lines are included. The filter material is 30 mg/cm² Kapton. The area of the hole is 0.1 times that of the detector.

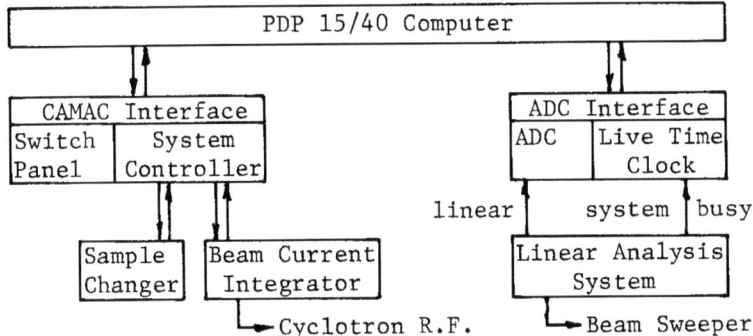

Fig. 3. Schematic of automatic data acquisition system.

AUTOMATIC DATA ACQUISITION

A schematic of the automatic data acquisition system is shown in Figure 3. The energy spectra are accumulated in the computer through the ADC interface. The operator issues commands to the acquisition program through a switch panel via the CAMAC interface. The program controls the data accumulation apparatus through the system controller via the CAMAC interface.

The sequence of acquisition steps is:
1. Insert slide.
2. Turn on current integrator, enable ADC, and turn on beam.
3. Count until preset charge has accumulated.
4. Disable ADC, turn off beam, and record data.
5. Reset current integrator and remove slide
6. Advance tray.

The energy spectrum accumulated during step 3 is stored in 500 channels, and various acquisition parameters are stored in the following 12 channels. These include live time, real time, charge, tray number, slide number, and the sample's conversion factor from square cm to cubic meters. In step 4 these data are recorded on magnetic tape for permanent storage and on magnetic disk for the reduction routine.

DATA REDUCTION

The data reduction is performed while the spectra for following samples are being acquired, using the data files written on the magnetic disk. A flow chart for the reduction is shown in Figure 4. Up to 90 seconds are needed to reduce the data from each sample. The various segments are chained so that the total program, including the acquisition segment but excluding the monitor and input/output handlers, fits into 9K words of computer memory.

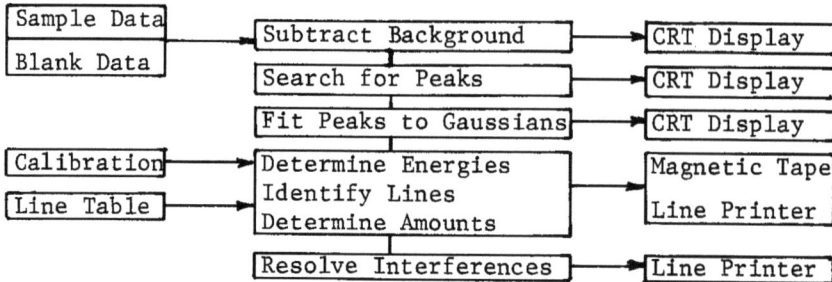

Fig. 4. Flow chart of data reduction

In addition to the sample spectrum, the disk contains the spectrum from a blank (same substrate but without loading), a linear energy calibration, and a table of energies and 'cross sections' for all reasonable x-ray lines. (The 'cross section' includes detector efficiency and solid angle.)

The 'cross sections' were measured using 30 commercially obtained foils of known areal density (2). The same geometry and reduction program were used to ensure consistency. Smooth curves of the 'cross sections' were made in order to reduce the statistical uncertainty and to obtain the values for other elements. The table used for the aerosol studies contains approximately 70 entries. The table does not include elements whose typical values in urban aerosols are well below our detectable limit (3). Selected foils are rerun at the beginning of every cyclotron run to check the entire system.

Background Subtraction

The background of the x-ray spectrum is subtracted before further analysis since 1) the peak search is more reliable with the background removed, especially in the region where the background is largest, 2) considerably less time is required in the Gaussian fits since there are no background parameters to be fit and fewer channels need to be included in the fit, and 3) problems encountered with a simultaneous background and peak fit are eliminated. Some of these problems are: 1) The functional form of the background under low energy multiplets is difficult to determine. 2) Simultaneous background and peak fits are unreliable if there are inadequate data points on either side of the peak(s) due to unresolved low intensity peaks in the background region. 3) The background region on the sides of a poor statistics peak is difficult to determine.

The background shape is obtained from a blank, which has the same substrate but no loading. The spectra of an aerosol sample

and corresponding blank are shown in the upper plot of Figure 5.*
The primary source of background is the bremsstrahlung produced by
electron scattering. For an 18 MeV alpha beam the predicted cut-
off is near 9 keV (5). Above this level the background is due to
the gamma Compton tail, which extends down to this energy region.
This also depends on the amount of material intercepted by the
beam. For a sample the background is produced by the loading as
well as by the substrate. The background is estimated by multiplying the blank spectrum by a variable which allows the blank spectrum
to come up to the valleys in the sample spectrum. For the spectrum
shown the value is 1.04.

For heavily loaded samples a small amount of background may
remain. This is estimated by a second method, modified from one
used for gamma spectra (6). This consists of averaging the data
over a number of channels, which is five to ten times the peak
widths; subtracting a constant (less than one) times the square
root of the average; and choosing the smaller of the data or the
adjusted average. This is repeated several times, each time
decreasing the size of the bump under a peak. The results are
shown in the second plot of Figure 5. Since the first background
estimate was good, the generated background removed very little.
The resultant spectrum, with both estimates used, is employed for
all subsequent steps. This is shown in the third plot.

Peak Search

To avoid finding insignificant peaks, twice the square root of
the background is subtracted from the resultant spectrum. This is
smoothed twice using a five-point smooth (7). A correlation spectrum is produced using the Gaussian cross-correlation technique of
Connelly and Black (8). This spectrum consists of the sum of two
spectra generated with correlation Gaussians of widths W and W/2.
Using two widths allows the search to locate peaks of different
widths, including narrow peaks on the side of large ones. The data
peak center is located at the center of the correlation peak. The
cut points are located in the valley between the peaks or at 2W
from the center, whichever is closer. A first estimate of the area
of the peak is obtained by a linear sum between the cut points of
the unsmoothed data. The correlation method is more reliable than
methods involving derivatives, which agrees with results found
elsewhere for x-ray spectra (9).

---

*The square root of the data is displayed since this is the most
appropriate representation (4).

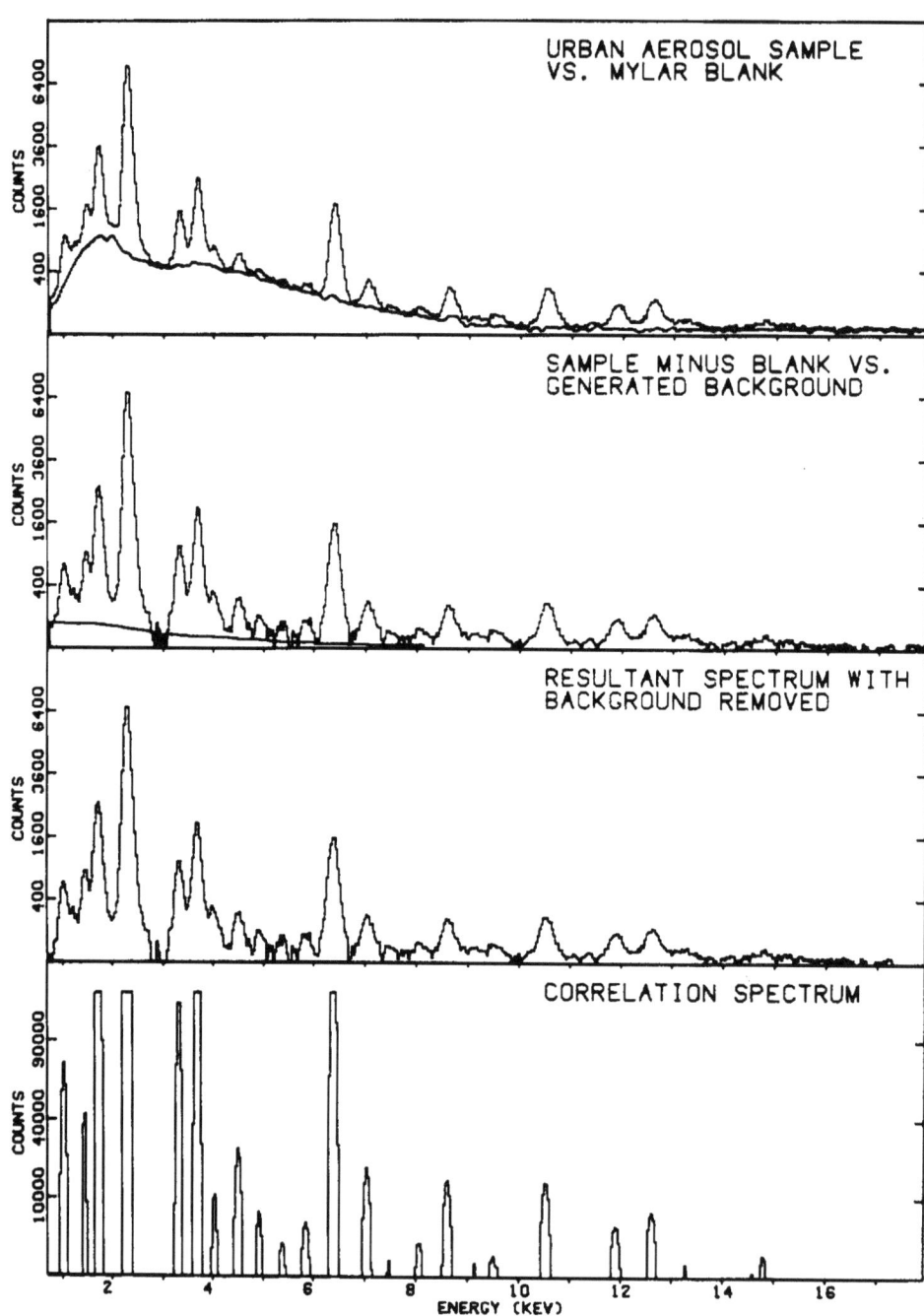

Fig. 5. Urban aerosol sample on mylar substrate at various stages of data reduction. Square root of counts per channel vs. energy.

Gaussian Fit

The peaks are fitted to Gaussians in order to obtain more reliable values for the area and center. If peaks are less than 2W channels apart, they are fitted together as a multiplet. Each peak is chi-square fit to three parameters, namely position, width, and area, using the linearization of function technique (10). The initial parameters are taken from the peak search step and the Gaussian fit is taken from the center of the first peak minus 3W/5 channels to the center of the last peak plus 3W/5. For a singlet the fit is over 6W/5 channels, which is approximately equal to the full width at half maximum. This is done to avoid the unresolved low intensity peaks which often appear at the edges of the located peaks. Two fits are shown in Figure 6.

The uncertainty in the peak position is much less than the resolution of the system. A survey was made of the deviation of the energy corresponding to the peak center from the actual energy of the line for a large number of spectra, using peaks of good and poor statistics. The result was that the average deviation was only 12 eV, which corresponds to one-third of a channel.

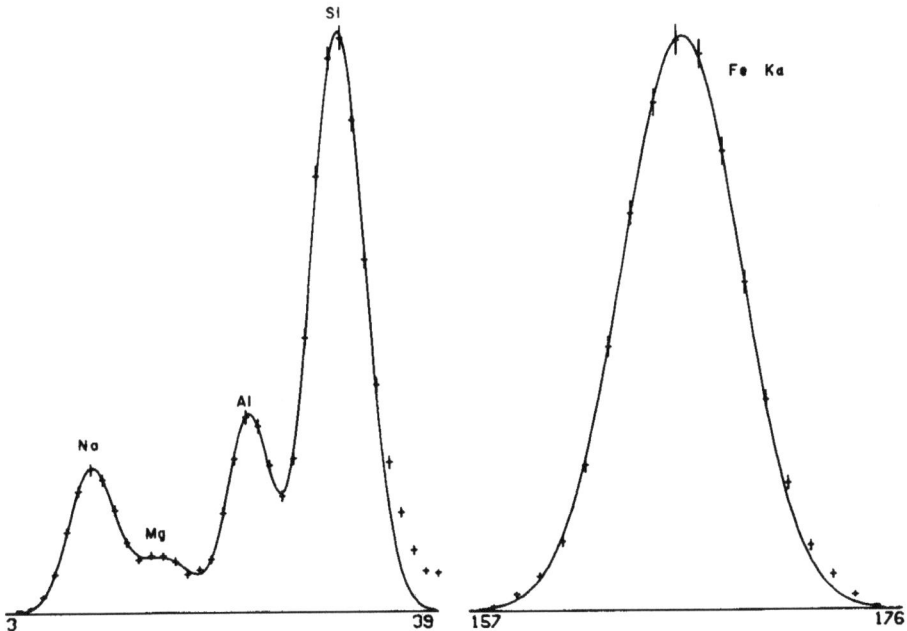

Fig. 6. Gaussian fits for quadruplet and singlet. Counts per channel vs. channel.

Peak Identification and Elemental Amounts

Using the predetermined energy calibration and the well-defined peak center, the energy of each peak is determined. Using the table of line energies, each peak is identified with all lines which lie within ± 70 eV. The amount in nanograms per square cm can be calculated using the 'cross section', the Gaussian area, and the amount of charge passing through the sample, which is corrected for dead time. The uncertainty is calculated from the quadratic sum of the Poisson uncertainty, the error from the Gaussian fit, plus 10% for uncertainties in the 'cross section' and charge integration. The results for the sample are written on magnetic tape and, if desired, on the line printer. In order to have a complete record of the analysis, all relevant information is included in this output.

Resolution of Interferences

The analysis is not complete since there is ambiguity in the identification of peaks when more than one line is near the peak. Other interferences lead to erroneous elemental amounts. The resolution of these interferences will be discussed according to three regions of the spectrum.

The region below four keV contains the K lines of elements up to calcium and the L and M lines of heavier elements. In several cases these lines interfere: zinc with sodium, bromine with aluminum, and lead with sulfur. In all of these the heavy element can be identified from higher energy lines in the same spectrum. The contributions of the heavy elements are removed by using the amounts obtained from the higher lines. The appropriate ratios of cross sections are obtained from foil standards.

The most difficult region for interferences is between four and eight keV. The biggest single problem is the double interference of the K lines of titanium and the L lines of barium, plus the additional interference of the Kα line of vanadium with the beta lines. The separation between the alpha lines is 49 eV and between the beta lines is 103 eV. Although the detector cannot resolve these, the uncertainty in the peak position is small enough to permit a reasonable identification. The first step is to compare the peak energy with the various line energies. Use is also made of the fact that the β/α intensity ratio is five times greater for barium than for titanium. Further, the Lβ2 and Lγ peaks of barium are separately identifiable. These are all included into probabilities for barium and titanium. The program is able to make a clear resolution when one of the two elements is dominant. In addition, this region has a series of interferences between the beta line of one element and the alpha line of the next higher element. If the beta of element X and the alpha of element Y interfere, the amount for element X is obtained from the previous alpha peak of X and is used to subtract the proper number of counts from the peak in question. In this way a corrected alpha amount for Y is obtained.

Above eight keV the only significant interference is between the Kα line of arsenic and the Lα line of lead. This is resolved by considering the secondary line of each. If both have secondary lines, the amounts are taken from them. If neither has any secondary lines, the peak is assumed to be lead, since the typical value in urban aerosols for lead is two orders of magnitude greater than that for arsenic.

Matrix Effects

No correction is made for matrix effects. Scanning electron microscope pictures show that all of the particles remain in the surface of mylar and Nuclepore, indicating that these substrates do not attenuate the x-rays. However, matrix effects for the low energy lines are produced by the finite size of the particles, primarily in the largest size region (above six microns). The correction can be estimated from samples of aerosols containing known elemental ratios, as marine air with NaCl. It can also be estimated by comparing the x-ray results with those obtained by methods not involving x-rays, as elastic alpha scattering and neutron activation.

ACKNOWLEDGMENTS

Parts of the data reduction program were developed by R. J. Sommerville and R. K. Wyrick. The sample analysis facility was developed by T. A. Cahill, R. G. Flocchini, P. J. Feeney and D. J. Shadoan, with the structural design by C. D. Goodart. The system controller and the electronics for the sample changer were designed by J. W. Cline.

REFERENCES

1. H. Thibeau, J. Stadle, W. Cline and T. A. Cahill, "On-Demand Beam Pulsing for an Accelerator," submitted to Nucl. Instr. and Meth.

2. Micromatter Corporation, Seattle, Washington.

3. J. A. Cooper, "Review of a Workshop on X-Ray Fluorescence Analysis of Aerosols," Battelle Northwest Publication BNWL-SA-4690, p. C2 (1973).

4. I. N. Hooton, "The Display of Statistical Data from Nuclear Counting Experiments," Nucl. Instr. and Meth. **56**, 277-283 (1967).

5. R. G. Flocchini, P. J. Feeney, R. J. Sommerville and T. A. Cahill, "Sensitivity Versus Target Backings for Elemental Analysis by Alpha Excited X-Ray Emission," Nucl. Instr. and Meth. **100**, 397-402 (1972).

6. H. R. Ralston and G. E. Wilcox, "A Computer Method of Peak Area Determination from Ge(Li) Gamma Spectra," in J. R. DeVoe, Editor, *Modern Trends in Activation Analysis*, Vol. 2, p. 1238-1243, National Bureau of Standards (1969).

7. A. Savitzky and M. J. E. Golay, "Smoothing and Differentiation of Data by Simplified Least Squares Procedures," Anal. Chem. **36**, 1627-1639 (1964).

8. A. L. Connelly and W. W. Black, "Automatic Location and Area Determination of Photopeaks," Nucl. Instr. and Meth. **82**, 141-148 (1970).

9. R. L. Heath, R. S. Frankel, R. J. Gehrke and J. Barstow, "Energy Dispersive X-Ray Spectrometry with Dedicated Computer Data Reduction," in *Analysis Instrumentation*, Vol. 9, Sec. F2, p. 1-10, Instrument Society of America (1971).

10. P. R. Bevington, *Data Reduction and Error Analysis for the Physical Sciences*, p. 232-245, McGraw-Hill (1969).

A SECONDARY-SOURCE, ENERGY-DISPERSIVE X-RAY SPECTROMETER AND

ITS APPLICATION TO QUANTITATIVE ANALYTICAL CHEMISTRY

R. P. Larsen and J. O. Karttunen

Argonne National Laboratory

Argonne, Illinois 60439

ABSTRACT

An energy-dispersive X-ray spectrometer that (1) uses as the primary excitation source the power supply and tungsten X-ray tube from a conventional crystal spectrometer (General Electric XRD-6) and (2) uses as the secondary excitation source elemental metal foils that are readily interchangeable has been built and operated. The use of an X-ray tube with a high-voltage capability, 75 kilovolts max, enables the determination of elements with atomic numbers as high as 66 (terbium) to be based on the K series of X-rays; the high-power capability, 3.7 kilowatts max, enables a particularly intense beam of X-rays to be generated by the secondary source and hence, provides a particularly high detection capability for trace elements in a sample. An instrument that uses interchangeable secondary sources to irradiate the samples has several advantages over those instruments in which excitation is accomplished by direct irradiation with an X-ray tube: (1) the background radiation in the energy range where the X-rays of interest are measured is several orders of magnitude lower and is very uniform and (2) the energy of the excitation radiation can be closely matched to the absorption edges of the elements of interest in the sample.

In the application of the instrument, particular emphasis has been placed on the development of techniques that will enable an energy-dispersive X-ray spectrometer to be used as the detection instrument for quantitative elemental analysis. Methods for the determination of the individual rare earths, plutonium and uranium at the microgram level with an accuracy of ± 1% are outlined and for the determination of plutonium and uranium at the milligram level with an accuracy of ± 0.1% are proposed.

INTRODUCTION

In an energy-dispersive X-ray spectrometer that uses an X-ray tube to irradiate the sample directly, the factor that limits the detection of a particular element to a particular value is the background radiation, radiation from the source that is scattered by the sample, the sample holder, and the adjacent walls and that has the same energy as the characteristic X-ray of the element being determined. If the beam from the tube is collimated, most of this background is due to (1) the X-rays from the tube that are isoenergetic and are coherently scattered by the sample, and (2) the X-rays that are slightly hyperenergetic and are incoherently scattered. There is some background radiation that is due to the incoherent scattering of the X-rays that are significantly hyperenergetic but it is several orders of magnitude less than that from the aforementioned sources. Coherent and incoherent scattering of the X-rays from the tube that are hypoenergetic relative to the characteristic X-ray being measured do not affect the background, but if these X-rays constitute a significant fraction of the beam, the assay time will be extended due to the fact that the detector will be occupied with pulses that are of no interest.

From the above, it is apparent that if the intensities of the X-rays that are responsible for the background can be markedly reduced while the intensities of the X-rays that are sufficiently energetic to excite the characteristic X-rays can be maintained at a high level, the detection limit can be markedly improved. This can be accomplished in two ways: (1) by interposing between the X-ray tube and the sample an absorber that will remove the low-energy X-rays and allow the high-energy X-rays to pass and (2) by irradiating the sample with the X-rays from a secondary source, a massive amount of an element which is irradiated by an X-ray tube and whose characteristic X-rays are above the energy of the absorption edge of the element being determined. The principal advantage of the absorber approach is that a low-current X-ray tube can be used; the principal disadvantage is that the average energy of the X-rays passed by the absorber is much higher than that of the absorption edge of the element being determined and is hence, less than optimum for exciting the characteristic X-rays. The principal advantage of the secondary-source approach is that the energy of the excitation radiation can be matched to the absorption edge of the element being determined and hence, the detection sensitivity is optimized; the principal disadvantage is that the X-ray generator and tube must both have a particularly high power rating.

Jaklevic et al. (1), Cooper (2), and Porter and Waldseth (3) have discussed the relative merits of direct-tube and secondary-source excitation spectrometers. The conclusions drawn by these authors is that the improvement in the signal-to-noise ratio derived from the use of a secondary-source system makes it, by far, the irradiation method of choice. This paper describes the secondary-source instrument that the authors have built and operated, the instrument's operational characteristics are discussed and the merits of energy-dispersive X-ray spectrometers in general, and tube-excited secondary-source instruments in particular, for both trace and macro elemental analysis are compared with those of other analytical methodologies.

THE SPECTROMETER

A cross-section of the spectrometer dog-house is shown in Figure 1. The outer walls are fabricated from 0.625 inch thick steel plate, the inside dimensions are 5 x 5 inches. A lead shielding wall 0.25 inch thick divides the dog-house into secondary-source and sample chambers. Silver is used for the collimators and as a liner for both the sample drawer and the lead wall of the sample chamber when the elements being determined have characteristic X-rays of about 14 keV (strontium) and less. Copper is used for these components when the X-ray energies are greater than 14 keV. When copper is used, a thin aluminum absorber is placed ahead of the detector to absorb the copper X-rays.

The tungsten target X-ray tube and the power-supply are from a General Electric XRD-6 crystal spectrometer. The maximum power rating is 3750 watts, the maximum voltage is 75 kV, and the maximum current is 100 mA.

The Si(Li) detector which has a 185 eV resolution for the manganese K_α X-ray was manufactured by the Kevex Corporation. The associated amplifiers and other electronic circuitry are capable of handling 5000 pulses per second without serious loss in resolution and pulse pile-up.

Performance Characteristics

The performance characteristics of an X-ray spectrometer are best described by the results obtained under optimum sensitivity conditions, a microgram or so of an element mounted on an essentially weightless backing. Under these conditions, the detection sensitivity is limited by only the operational characteristics of the instrument itself: there is no absorption of the X-rays and no scattering by the sample, the background is due primarily to the scattering of radiation by the instrument's structural materials. The decrease in the intensity of the element's characteristic X-rays in the analysis of a real sample due to mass absorption and the increase in background due to scattering of the excitation radiation are limitations imposed by the sample, not the instrument.

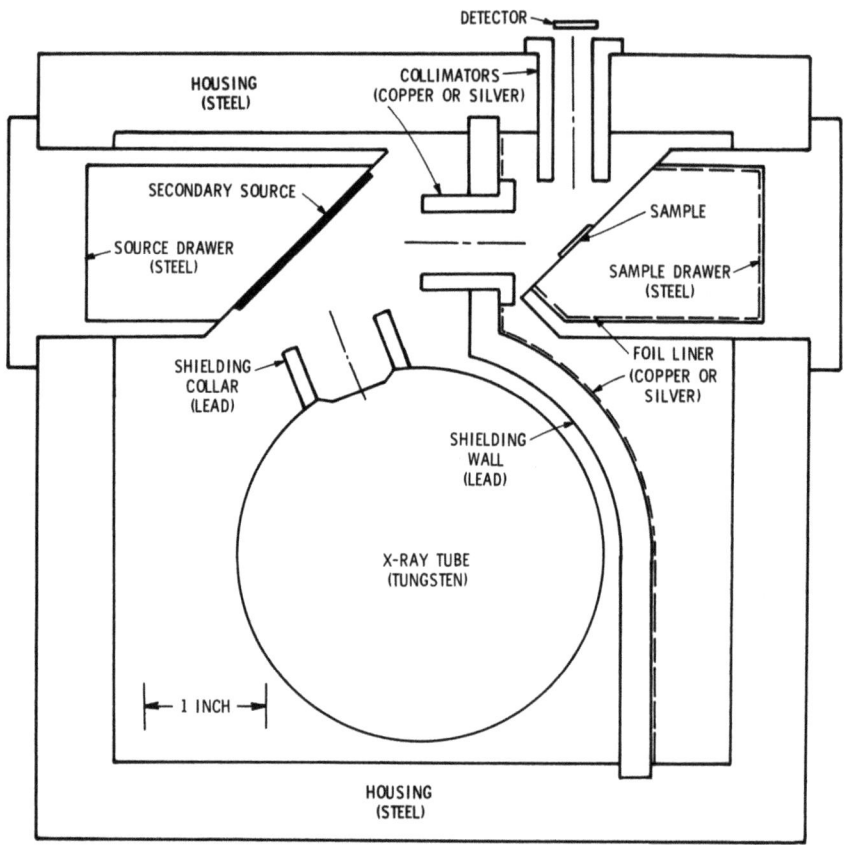

Figure 1. Schematic of secondary-source, energy-dispersive X-ray spectrometer.

Typical performance characteristics of the spectrometer are shown by the results obtained when 1.0 μg of lead mounted on a one mil Mylar sheet was assayed for 60 minutes. The secondary source was molybdenum, the X-ray tube was operated at 50 kV and 40 mA, and a 1.0 mil aluminum absorber was placed in front of the detector to absorb the X-rays having energies of about 5 keV and less. The total counts were 6×10^5 of which 2×10^4 were due to lead L_α and 3×10^5 were scattered molybdenum K_α and K_β above 15 keV. The total background radiation in the energy range 5 to 15 keV was 2×10^5 and was constant (not energy dependent), the background for the lead L_α was 1×10^4 counts, the background above 20 keV was zero. Using the criterion for detection limit of three times the standard deviation of the background, the detection limit for lead is 7 ng. For elements with K absorption edges and emission lines in the same energy region as those for lead, e.g., bromine and rubidium, the detection limit is about 3 ng.

From the above and the fact that detectors can be operated at about 3×10^7 counts/hr (10^4 counts/sec), it is apparent that the sensitivity of this instrument is limited by the intensity of the irradiation source. The detection limits could be lowered by improving the geometry of the system, that is by reducing the distance on each leg of the radiation path, tube-to-secondary source-to-sample-to-detector. However, this would appear to be difficult. These distances were recognized to be important in the design of this instrument and were, therefore, minimized. Each of these distances is now about 5 cm.

The detection limit could also be lowered by reducing the background. However, efforts in this direction would appear to be more of an academic interest than a practical one. About 50% of the background is due to the radiation scattered by the sample support, one mil Mylar. As the mass of this material viewed by the detector is about one mg and this appears to be about the minimum amount of material that will be present in such samples as those obtained from air filters and impactors, endeavors to reduce the background that arises from scattering off the structural materials, e.g., sample holder and walls, would appear to be of only marginal value. It has been stated by Porter and Waldseth (3) that the residual background is due to bremsstrahlung radiation from the tube that is scattered first by the secondary source and then by the materials in and around the sample. It is our observation that the residual background is due to incoherent scattering of the characteristic X-rays from the secondary source. There is no background in our spectrometer above the energy of the coherently scattered K_β radiation of the secondary source, i.e., no scattering of the high-energy bremsstrahlung. It follows, therefore, that the scattering of low-energy bremsstrahlung is a negligible part of the background in the 5 to 15 keV energy range.

Because the photoelectric cross-section of an element is highly energy dependent, it has been suggested that maximum detection sensitivity for the determination of a particular element, A, would be obtained by using as the secondary source an element, S, whose K_α X-ray energy is just above the absorption edge of A. However, this is not always the case, the intensity of the excitation radiation must also be considered. By using as the secondary source an element whose atomic number is much higher than that of the element that is optimum energywise, the increase in the intensity of excitation radiation can be sufficient to more than offset the disadvantage of a lower photoelectric cross-section. An example of this is the determination of manganese, K_{ab} = 6.54 keV, using copper, K_α = 8.05 keV, or molybdenum, K_α = 17.48 keV, as the secondary source. The cross-sections for manganese at the two energies are 4×10^4 and 4×10^3 barns/atom, respectively. For the same input to the tube, 50 kV and 20 mA, the detection sensitivity is a factor of two better when molybdenum is used. The mass absorption coefficient of copper for its K_α X-ray is significantly higher than that of molybdenum for its K_α X-ray and the intensity of the K_α X-rays from copper is, therefore, significantly lower than that from molybdenum. Because the intensity of the X-rays from the secondary source is also a function of the intensity versus energy spectral characteristics of the radiation from the tube, this factor will also affect the detection sensitivity for element A. For a tungsten tube, the primary source of radiation is bremsstrahlung, and the intensity of X-rays from the secondary source will, with increasing atomic number, reach a maximum and then decrease.

If the excitation energy, $S_{K\alpha}$, is too closely matched to the absorption edge, A_{Kab}, there will be a decrease in the detection sensitivity due to increased background, the $A_{K\alpha}$ will appear on the side of the incoherently scattered $S_{K\alpha}$ peak. It is, therefore, necessary to choose as a secondary source an element whose atomic number is higher than that of the element that is optimum energywise. How much higher will depend on the instrument since the average energy of the source's incoherently scattered K_α depends on the source-to-sample-to-detector geometry and the breadth of the peak depends on the degree of collimation between the source and the sample. In our instrument, for example, it is necessary to use molybdenum (Z = 42) rather than zirconium (Z = 40) to control this background problem in the determination of rubidium (Z = 37).

An important advantage can be realized in the analysis of certain samples by the judicious choice of a secondary source. If the constituent of interest is a minor one and if its atomic number is less than that of a major constituent, the secondary source can be chosen such that the characteristic X-rays of the minor component will be generated but not those of the major one. The secondary source in this case will be an element whose K_β X-rays are less energetic than the absorption edge of the element whose X-rays are being suppressed.

From the above discussion, it is apparent that by carefully considering all the factors relating to the operational characteristics of a secondary-source spectrometer, one can select the element to be used as the secondary source which provides maximum detection sensitivity for the determination of a particular element in a sample. However, in comparing secondary source with direct-tube-excitation instruments, this advantage is heavily overshadowed by the decrease in background that is obtained in the secondary-source instrument.

APPLICATION TO QUANTITATIVE ANALYSIS

In the publications describing the merits of tube-excited, energy-dispersive X-ray spectrometry, the emphasis to date has been placed primarily on the high detection sensitivity and the capability for detecting a large number of elements in a single assay. From the spectra presented for such diverse materials as blood plasma, orchard leaves, and paint samples, the high potential of this type of instrumentation for the solution of complex analytical problems is apparent. However, little has been said about how the data obtained, the number of counts in the principal photopeak of each element observed to be present, are to be used to obtain the information about the sample that is of prime interest, namely, the concentration of each of these elements. In nearly all the cases where a direct assay is made of a sample, a significant fraction of the X-rays of some, or all of the elements in the sample are absorbed by the matrix and the detection of trace elements is either inhibited by this absorption or the presence in the spectrum of the X-rays of the major constituents. The only types of samples where the mass absorption effects are negligible are those obtained in aerosol analysis. The application of X-ray spectrometry to this area of analytical chemistry is obviously an extremely important one, but it is, nevertheless, a specialized one.

Although the general solution to the problem of deriving concentrations from count-rate data is the same for both energy- and wavelength-dispersive X-ray spectrometers, the fact that the energy-dispersive instrument provides X-ray intensity data over a range of energies in a single assay expedites the solution to this problem and provides the means for establishing a higher level of confidence in the values obtained with no additional expenditure of time. The general solution can be accomplished in two ways: (1) the sample and a standard whose composition is known and whose matrix and geometric configuration are the same as the sample are assayed and the concentration of the element of interest is calculated from the measured X-ray intensities and the composition of the standard and

(2) the sample is dissolved, the element of interest is separated from the major constituents of the sample, the separated material is incorporated into a small amount of an "assay" matrix, the sample is mounted, and the assay made. A comparison of the measured X-ray intensity with those obtained with standards, known amounts of the element in the same matrix and mounted in the same way, enables the amount of the element in the sample to be calculated. The latter approach we have chosen to term "matrix transfer".

The particular advantages of the matrix transfer-approach are improved specificity and marked increases in sensitivity (often several orders of magnitude) and, hence, the applicability of X-ray spectrometry to the solution of a much broader range of analytical problems. For the analytical laboratory that is faced with the problem of analyzing samples that vary widely in composition, the matrix-transfer approach is also the most expeditous. The preparation of standards that are identical to the original matrix is frequently a time consuming task and is subject to many pitfalls. Unless there are a large number of samples to be analyzed that have the same matrix, the time required to prepare the standards and establish confidence in them can be excessive.

In a paper that the authors consider to be a classic in the field of applied X-ray spectrometry, Luke (4) has set forth the principles and discussed the advantages of the matrix-transfer approach and then outlined the procedures that can be used for the coprecipitation of trace amounts of a very large number of elements with submilligram amounts of others. His method of detection was wavelength-dispersive spectrometry. By using such separation procedures as solvent extraction and ion exchange in conjunction with Luke's coprecipitation methods, by incorporating an internal standard into the assay matrix, and by using energy-dispersive X-ray detection, X-ray spectrochemical analysis will more closely approach the goal which Luke has envisioned it to be: "An almost universally applicable method for trace analysis."

The incorporation of an internal standard into a sample and then measuring the X-ray intensity of the element of interest relative to that of the internal standard is a technique that is commonly used in X-ray spectrometry to overcome the problem of X-ray attenuation by the sample matrix and/or variations in sample geometry. The X-ray intensity of the element of interest is measured relative to that of the internal standard and the concentration of the element is calculated from this ratio and the ratios obtained from material of known composition. If the element chosen as an internal standard is one whose characteristic X-ray has very nearly the same energy as the X-ray of the element being determined, wide variations in the sample geometry and matrix can be tolerated without affecting the accuracy of the determination.

The applicability of Luke's matrix transfer approach to the solution of the matrix problem in X-ray spectrometry is markedly improved by the incorporation of an internal standard into the assay matrix. In the determination of all elements, save those with very low X-ray energies, this eliminates the need for geometric reproducibility of sample and standard mounts. The incorporation of the internal standard into the assay matrix is a particularly attractive approach when detection is made with an energy-dispersive spectrometer. From a single assay the data required to calculate the net X-ray intensities of both the element of interest and the internal standard are obtained. In a wavelength spectrometer, it would be necessary to make six individual intensity measurements, one each for the element and the internal standard and four to establish their respective backgrounds. Examples of the matrix-transfer, internal-standard approach to the solution of what have in the past been rather difficult analytical chemical problems are given below.

Rare Earth Analysis

Determinations of the concentrations of the individual rare earths, particularly when these elements are present at trace levels, have traditionally been some of the most difficult in the field of analytical chemistry. Of the methodologies that have the requisite sensitivity, mass spectrometric isotopic dilution (MSID) and neutron activation (NA), the interelemental interferences and/or differences in sensitivity are such that separation of the rare earths from each other must be a part of the analytical procedures. MSID, cannot be used to determine praseodymium, holmium, and thulium as each of these elements has only one stable isotope and no long-lived one. Application of these methods is quite limited due to the high cost and specialized nature of the equipment required.

Energy-dispersive X-ray spectrometry is applicable to the determination of the rare earths and particularly so if the energy of the excitation source is sufficient to generate the K series. The energy difference between the K_α X-rays of adjacent rare earths is large (the average difference is about 1.5 keV), there is essentially no difference in sensitivity in going from lanthanum to lutetium, and the rare earth group can be easily separated from all the other elements by either ion exchange or precipitation. If there is only one rare earth present in the sample, a known amount of another rare earth, a "spike", is added to the sample, the chemical separation is carried out, the separated rare earths are coprecipitated with a small amount of ferric hydroxide, and the precipitate is assayed. The recovery in the separation procedure need not be quantitative as the percent recovery of the rare earth being determined will be the same as that of the spike. From the X-ray intensity ratios obtained from the assays of the sample and the standards, the concentration of the rare earth in the sample can be calculated. If there is more than one rare earth present in the sample, the rare earth spike can, if necessary, be one of those present. In this case, two portions of the sample are carried through the separation procedure, one spiked and one unspiked. The concentrations of all the rare earths in the sample can be calculated from the X-ray intensity ratios obtained for the two samples and the standards and the amount of spike added.

An X-ray spectrogram of lanthanum, cerium, praseodymium and neodymium is shown in Figure 2. The secondary source was a 10 mil thick holmium metal foil; the X-ray tube was operated at 75 kV and 2 mA; the assay time was 30 minutes. The sample was prepared by coprecipitating the amounts of each rare earth shown on the figure with one mg of ferric hydroxide, centrifuging the solution, transferring the precipitate onto a 1 mil thick sheet of Mylar, and evaporating to dryness.

The quantitative aspects of an X-ray spectrometric rare earth determination have been demonstrated using neodymium as the analyte and cerium as the internal standard. Solutions containing 5 to 40 µg of neodymium were each spiked with 5 µg of cerium, the rare earths were coprecipitated with ferric hydroxide, and the precipitates assayed under the same conditions used to obtain the spectrogram shown in Figure 2. A plot of the net neodymium to net cerium count ratio versus weight of neodymium is given in Figure 3.

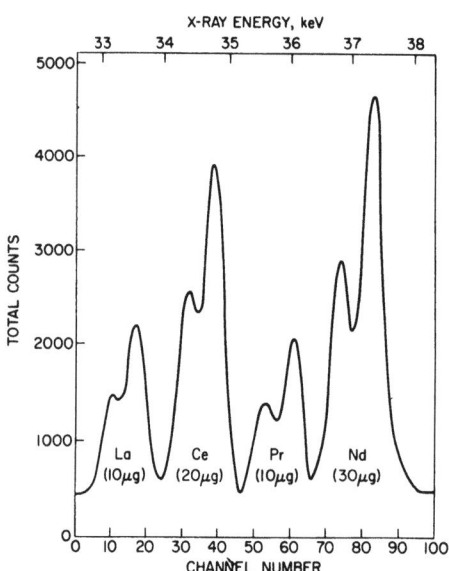

Figure 2. X-ray spectrogram of the light rare-earth elements.

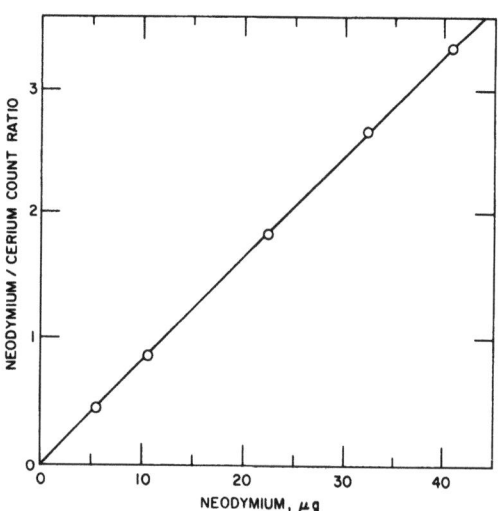

Figure 3. Calibration curve for the determination of neodymium using cerium as the internal standard.

An X-ray spectrometric analysis of the rare earths based on the K series rather than the L series has important advantages. The K series of X-rays are much simpler and the energy difference between the K_α X-rays of adjacent rare earths is much larger than that between the L_α X-rays, about 1.5 as compared to 0.2 keV. The low value of the mass absorption coefficients of iron for rare earth K X-rays enables the deposition of the ferric hydroxide on Mylar to be made with no regard for its geometric configuration. If the analysis was based on the L X-rays, it would be necessary to coprecipitate the rare earth hydroxides with some hydroxide other than iron, e.g., aluminum. (The K_α and K_β of iron occur at approximately the same energy as the L X-rays of several rare earths and would, therefore, interfere.) The precipitate would also have to be spread in a fixed area and have a uniform thickness, e.g., by filtration with a Millipore filter in a filter-chimney assembly. The absorption of the rare earth L X-rays by one milligram of aluminum is significant and varies markedly in going from lanthanum to lutetium.

An energy-dispersive spectrometer is quite superior to a wavelength instrument for the determination of rare earths when the analysis is based on the K series of X-rays. The sensitivities are much higher and the resolution is much better.

The application of the secondary-source instrument described in the first section of this paper is limited to those rare earths whose K absorption edges are at a lower energy than the K_α of holmium, i.e., lanthanum to terbium. The K_α X-rays of the rare earths above holmium cannot be generated by the tungsten K_α. For the determination of the heavy rare earths, it would be necessary to irradiate the separated rare earths directly with the bremsstrahlung radiation from a platinum or gold target tube operated at 75 kV, at least, and preferably at 100 kV. The detection sensitivity would, of course, be somewhat lower than that for the light rare earths due to the higher background.

Uranium and Plutonium

In the nuclear power industry analytical methods for the determination of uranium and plutonium in a wide variety of materials are required. The accuracy requirements vary from ±0.2% (R.S.D.) for samples having high concentrations of these elements to ±5% for samples having concentrations at the ppm level. It is particularly important that the confidence level of the analyses be high due to the high monetary value of these materials and to guard against their diversion for clandestine use. There is, at present, no simple method for the determination of trace amounts of elemental plutonium.

Because plutonium can be readily separated from other elements by simple ion exchange and solvent extraction procedures and can be coprecipitated as the hydroxide with milligram amounts of other hydroxides, energy-dispersive X-ray spectrometry is particularly well suited to its determination. Thorium would be used as the internal standard and aluminum hydroxide would be the assay matrix. Both thorium and plutonium are quantitatively and selectively adsorbed onto Dowex-1 anion exchange resin from strong nitric acid, eluted with dilute hydrochloric acid, and coprecipitated with aluminum hydroxide.

The quantitative aspects of an X-ray spectrometric plutonium analysis have been demonstrated. Uranium was used as the analyte rather than plutonium to avoid the radioactivity handling problem. Silver was used as the secondary source. A plot of uranium to thorium count ratio against micrograms of uranium comparable to that shown in Figure 3 was obtained.

For the determination of uranium and plutonium at high concentrations, the high accuracy methods used most frequently are controlled-potential coulometry and amperometric titration. Although these methods are highly specific and have the potential for determining both uranium and plutonium with accuracies of ±0.05%, the accuracies obtained in the round-robin evaluations of the plutonium methods (5) have not approached this. Repetitive analysis of the same material, UO_2 - 20% PuO_2, by a number of laboratories has shown that the average value obtained by a laboratory is frequently biased, from 0.3 to 0.5% of the amount known to be present, and the precision is often poor, ±0.5% for a single determination.

It would appear that energy-dispersive X-ray spectrometry could be used to determine uranium and plutonium in this type of material with a precision of ±0.1% to 0.2% and with very little chance of bias. The UO_2-PuO_2 would be dissolved, yttrium or zirconium would be added as the internal standard, a portion of the solution would be assayed, and the concentrations calculated from the relative intensity ratios obtained from this solution and standards. The precision of the analysis would be limited by the counting statistics, the bias would be limited by the care exercised in preparing the standards and the ability to reproduce the conditions of the standard and sample assays. Porter and Waldseth (3) have reported a precision of ±0.1% for the determination of silicon to germanium ratios for assay times of only 20 minutes. In the analysis of uranium and plutonium, it is estimated that the analysis time would have to be about 40 minutes, since the detector would be accumulating L_α plutonium counts only 5% of the time.

CONCLUSION

It is the authors' hope that from the above examples the analytical chemist who is responsible for carrying out a wide variety of elemental analysis will recognize that tube-excited, secondary-source, energy-dispersive X-ray spectrometry can, when used in conjunction with the vast literature of separations chemistry, provide a highly specific and sensitive approach to the determination of a very large number of elements in a wide variety of matrices.

Probably the most important quality of energy-dispersive X-ray spectrometric elemental analysis is the assurance, which the analyst derives from the features of the spectrogram obtained in the assay, that the counts in the photopeak upon which the analysis is based are due primarily to the element whose concentration is being determined. The fact that the major and minor photopeaks obtained in the assays of the sample and the standard occur in the same channels of the analyzer, that the shape of each sample peak is the same as the corresponding peak of the standard, and that the major-to-minor photopeak net count ratios are the same establishes unequivocally that the contribution of some other element to the photopeak upon which the analysis is based is, at most, only a small fraction of the counts in that photopeak.

REFERENCES

1. J. M. Jaklevic, R. D. Giauque, D. F. Malone and W. L. Searles, "Small X-Ray Tubes for Energy Dispersive Analysis Using Semi-Conductor Spectrometers," K. F. J. Heinrich, C. S. Barrett, J. B. Newkirk, and Clayton O. Ruud, Editors, Advances in X-Ray Analysis, Vol. 15, p. 266-275, Plenum Press (1972).

2. J. A. Cooper, "Comparison of Particle and Photon Excited X-Ray Fluorescence Applied to Trace Element Measurements of Environmental Samples," Nucl. Inst. and Method 106, 525-538 (1973).

3. D. E. Porter and R. Waldseth, "X-Ray Energy Spectrometry," Anal. Chem. 45, 604A-614A (1973).

4. C. L. Luke, "Determination of Trace Elements in Inorganic and Organic Materials by X-Ray Fluorescence Spectroscopy," Anal. Chim. Acta 41, 237-250 (1968).

5. J. E. Rein, R. K. Zeigler and C. F. Metz, "LMFBR/FFTF Fuel Development Analytical Program - Phase II," Los Alamos Scientific Laboratory Report, LA-4407 (1970).

Author Index

A

Ammons, A. M.	371
Anderson, C. H.	214
Artz, B. E.	225

B

Baucum, W. E.	371
Biederman, R. R.	139
Birks, L. S.	302, 436
Boehm, H.	354
Bosworth, T. J.	116
Bourgault, R. F.	139
Brower, I.	116
Brown, D. B.	436
Brown, J. D.	479
Burkhalter, P. G.	423

C

Campbell, J. L.	457
Campbell, W. J.	247, 279
Carpenter, J. A., Jr.	395
Cate, J. L., Jr.	337
Chaturvedi, R. P.	445
Cheng, E. L.	269
Chessin, H.	225
Chung, F. H.	106
Clark, B. C.	258
Clark, R. R.	75

D

Douglas, L. A.	88
Dozier, C. M.	423
Duggan, J. L.	445

E

Ebisu, E. S.	150
Eldred, R. A.	560
Elsheimer, H. N.	236

F

Fabbi, B. P.	236
Fatemi, M.	302, 436
Ferran, G.	416
Feuerbacher, D. G.	75
Fitzpatrick, R. L.	467

G

Gamage, C. F.	542
Gianelos, J.	325
Gillieson, A. H.	16
Gould, R. W.	384
Grant, C. L.	44
Gray, T. J.	445

H

Harrison, J. F.	560
Hayakawa, K.	498
Heinrich, K. F. J.	309
Henke, B. L.	150
Herglotz, H. K.	509
Herman, A. W.	457
Hill, M. S.	384
Hohlfelder, J. J.	531
Hruska, S. J.	487

J

Jatczak, C. F.	354
Jenkins, R.	32

K

Karttunen, J. O.	571
Kawase, S.	498
Klein, B. M.	423

L

Larsen, R. P.	571
Law, S. L.	279
Leitner, J. W.	214
Leyden, D. L.	293
Liedl, G. L.	487
Lin, J.	445
Lynch, D. R.	509

M

Mander, J. E.	214
Marr, H, E. III	247
Mathieson, J. M.	318
McMurdie, H. F.	20
McNelles, L. A.	457

N

Nagel, D. J.	423
Neylan, D. L.	247

O

Okano, H.	498
Ong, P. S.	269
Orr, B. H.	457

P

Parker, D. L.	521
Parobek, L.	479
Parrish, W.	97
Pelton, P. A.	44
Pickles, W. L.	337

R

Rasberry, S. D.	309
Roberts, E. C.	116
Roof, R. B.	348

S

Sachtleben, C. C.	445
Skogerboe, R. K.	68
Smith, R. W.	139
Sroka, G.	269
Stine, P. A.	487

T

Tenney, D. R.	395
Topham, W. H.	542

V

Viana, C. S.	416
Voskamp, A. P.	124

W

Whitlock. R. R.	423
Willoughby, R. A.	457

Y

Yamamoto, S.	498
Yolken, H. T.	1

SUBJECT INDEX

A

Absorption, 107
  Coefficient, 160
  Correction, 479
  Correction, for V and Pu Analysis, 337
  Correction, in Energy Dispersive Analysis, 275
  Effects, 310
  Effects in XRF, 302
  In Diffused Layers, 395
  In XRF, 264
  Spectrometry for Oils, 549
Accuracy, 2
Aerosol Analysis, Automatic, 560
Air Pollution Particulates, 288
$Al\ Cr\ Be_4$, 122
$Al\ Fe\ Be_4$, 116
Al-K alpha, 158
$Al\ Mu\ Be_4$, 122
$Al\ Ni\ Be_4$, 122

Al_2O_3, 112
Aluminum, 254
  In Nonferrous Alloys, 252
  In Steel, 249
Alkaline Earth, 281
Alloy Films, Analysis of, 487
Amorphous, 112
  Scattering, from Glasses, 384
Analysis,
  Fluorescent, of Portland Cement, 214
  Of Recycled Metal, 247
  Of Plated Wires, 325
Arkansas Stone, As Standard, 33
"Artificial" Homogenization, 70
Arsenic, Analysis of, 465
Atmospheric Aerosols, 225
Atomic Number Effect, 479
Austenite, Retained, Analyzed, 124
Auger Electrons, 158
  Spectroscopy, 150, 498
  Emission Microanalysis, 498
Automatic Data Acquisition, 560

B

Ba F_2, 158
Be, 116
Bentonite, Patterns of, 88
Berg-Barrett and Lang
    Topographic, 468
Beta Rays, Discrimination
    against, 344
Binary Systems, 114
Biological Sample Analysis, 269
Biomedical Analyses, 457
Biopsy, Analysis of, 273
Bismuth, 102, 104
Blood, Analysis of, 459
Boemite, Transformations, 37
Bragg-Brentamo, 419
Bragg and Packers Method, 416
Brass, Analysis of, 325

C

Ca CO_3, 113
Cadminum, in Liver, 7
Cadminum (109) Excitation for
    XRF, 258
Calcite, 77
  Transformations, 37
Calculators, Programmable,
    Use of, 224
Calibration,
  For Brass Analysis, 325
  For Portland Cement Analysis,
    214
  For Steel Analysis, 249
  For X-ray Fluorescent Spectrometry,
    225
Carbon in Steel, 8
Cd CO_3 Transformations, 37
Cd O, 112
Centroid Calculations, 375
Certification of Standards, 17
Chelating Ion Exchange Resins, 293
Chemical Analysis by Combined
    Instrument, 509
Chlorites, Patterns of, 88
Chopper Wheel, 537
Chromium, 297
  In Steel, 249
  In Superalloys, 309
Clay,
  Minerals, 75, 102
  Patterns of, 88
Coating, Analysis of 509
Cobalt in Superalloys, 309
Coefficient of Integral Reflection,
    436
Collimator, for Microdot Samples,
    318
Compton,
  Energy Shift, 261
  Scattering, Use in Quantitative
    Analysis, 272
  Scattering, 575
Copper, 281, 282
  Alloys, 18
  Films, Analysis of, 487
  In Nonferrous Alloys, 252

Subject Index

In Scrap Steel Analyses, 255
In Steel, 249, 251
Spectrum, Continuous, 129
Copper-base Alloys, 255
Copresc Precipitation
  Technique, 321
Corrections in X-ray
  Spectrometry, 479
Correction Factors, for X-ray
  Fluorescent Analysis, 214
Criss-Birks Computer Program, 333
Cross-sections, X-ray, 258
Crystal Data, Book, 20
Cyclotron, Use in Analysis, 560

D

Data Reduction,
  Spectrometric, 563
Delta Adaptation, 313
Density, 55
Detection Limit for Sulfur, 244
Detector,
  Efficiency, 262
  Si(Li), 531
Dickite, 102
Diffraction Analysis, 106
  of Minerals, 88
  Quantitative, 139
Diffractometer,
  Choice of Crystals for, 436
  Geometry, 97
  Used at N.B.S., 27
Diffractometry for Stress Measurement, 354
Diffusion,
  Determinations, Review of, 395
  Welding, 116
  X-ray Analysis of, 395
Distribution, 45
Drift in XRF System, 555
Drilling Mud Solids, 75
Duncumb-Reed Corrections, 484

E

Electron-Ionization Cross Section, 152
Electron Energy Analyzer, 510
Electron Excitation for XRF, 338
Electron Microprobe, 68
  Analysis, 152
  Analyzer, 498
Electron Spectroscopy for Chemical
  Analysis ESCA, 509
Emission from Thin Films, 487
Empirical Correction Method, 309
Energy Dispersive Analysis,
  Compared with Wavelength Dispersive, 124,
  For Biological Samples, 269
  Microdot Samples, 318
  Of Oils, 542
  Of Retained Austenite, 124
  Of U and Pu, 337
Energy Dispersive Spectrometry,
  Resin-loaded Filters for, 279

Energy Dispersion Spectrometer,
    Combined with ESCA, 509
Energy Electron Spectroscopy,
    152
Energy Shift, 448
Enhancement in XRF, 266
Error, 44, 253
  In Angle Measurements, 100
  In Diffractometry, 97
  In Lattice Parameter
    Determination, 100
  In XRF, 302
Exchange Resins, 293
Excitation of X-ray by
    Electrons in Films, 487
Extinction in LiF and KAP, 436

F

Fe Be, 116
Feldspar, 77
Ferrous Alloys, 19
Filter Materials, 227
Filtering,
  For Aerosol Samples, 226
  Samples for Analysis, 279
Filters in Spectrometry, 562
FINAGLE Program, 302
Fixed Angle Diffraction
    Method (see Energy
    Dispersion), 124
Flash X-ray, 423
Flow Bell for Oil, 546, 552
Fluorescence,
  Probability, 260
  Proton Induced, 457
  Spectrometer, 509
  Spectrometry of Aerosols, 225
  Yield, 450
Fluorescence Analysis, 152
  Corrections, 302
  For Plated Wires, 325
  For U and Pu, 337
  Of Liquids, 518
  Of Portland Cement, 214
  With Ion Exchange, 293
Fluorescent,
  Photons, 158
  Radiation Induced Energy
    Dispersive Analysis (FRIEDA),
    269
  X-ray Analysis of Scrap, 247
  X-ray Analysis of Oils, 542
  X-ray Analysis of Resin-loaded
    Filters for, 279
  X-ray Analysis of Submicrogram
    Samples, 318
  X-ray Sources in Spectrometry, 57
Fluorescent Effects, 310
  In XRF, 309
Flux Measurement, 535
Focusing,
  For Stress Measurement, 354
  Spectrometer, 521
Fogel'son Method, 407

G

Gamma Rays, Discrimination against,
    344

Ga P, 473
Ge I, 113
Glass, 288
  Radial Distribution
    Functions in, 384
Gold, 281, 287, 298
  Films, Analysis of, 487
Graphite, 473
Grinding, Effect on Analysis, 32
Guiner Camera, 100
  Geometry, 102
Gun Shot Residue, 322

H
Half Width of PuO_2, 348
Hematite, 77
High-temperature Superalloys, 309
Houska Method, 408

I
Identification by Diffraction, 20
Illite, Patterns of, 88
Indexing of Diffraction File, 29
Intensities in Powder Diffraction, 28
Intensity,
  From Crystal Reflection, 436
  Relations in Energy
    Dispersion System, 124

Interatomic Distances in Glass, 388
Interelement Corrections, 254
Interelement Effects, 479
  Determination of, 302
  In Emission from Films, 487
  In X-ray Fluorescence, 214
Interferences in Spectrometry, 568
Interferometer, 4
Internal Standards, 75, 578
  For Diffraction, 27
Interplanar Spacings, 28
Ion Exchange,
  In Spectrometry, 279
  Papers for, 294
Ion Microprobe, 68
Ionization,
  By Ions, 445
  Potential, 482
Iron, (55) 282
  Excitation for XRF, 258
  In Glass, 253
  In Nonferrous Alloys, 252
  In Superalloys, 309
Irradiation of PuO_2, 348
Isotopic Source, Compared with X-ray, 513

J
Jones-type Riffler, 54
Jost Method, 398

K
Kaolinite, Transformations, 37

KAP Crystals for Diffractometer, 436
K Cl, 112
Kidney, Analysis of, 459
Kossel Pattern, 467

L

Laser,
  Microprobe, 68
  Plasma, 423
  Pulses, 423
Lattice Constant of PuO_2, 348
Leached Glass, Structure of, 384
Lead,
  Analysis of, 462, 575
  In Glass, 6
  In Nonferrous Alloys, 252
Least Squares in XRF Corrections, 305
Levitskaya and Vodop Yanova Method, 405
Li F, 112
  Crystals for Diffractometer, 436
Light Element, 150
Litharge Transformations, 37
Liver, Analysis of, 459
Log-Normal Distribution, 49
Lorentz-polarization Factor, 107

M

Macres-Wolf Correction, 479

Magnesium in Nonferrous Alloys, 252
Magnetic Alloy, 313
Manganese,
  In Nonferrous Alloys, 252
  In Steel, 249
Maraging Steel, 313
Mass Absorption Coefficients, 150
Massicot Transformations, 37
Matamo Method, 398
Matrix,
  Effects, 72, 106
  Effects in XRF, 299
  Flushing, 106
  Transfer, 578
Mercury, 281, 282, 285, 286
Metal Ions, 295
Micas, Patterns of, 88
Microanalysis, Electron Corrections, 479
Microanalyzer, Using Auger Electrons, 498
Microdot Samples, 318
Microprobe, 68
  Techniques, 73
  Use with Thin Films, 487
Minerals, 16, 288
  Analysis, 236
  Diffraction Standards of, 88
Molybdenum, 281, 283
Monochromator, 99
Montmorillonite Transformations, 37
Multicomponent Analysis, 106
Multiplicity Factor, 416

Subject Index

N

N.B.S., 1
Nd: Glass Laser, 426
Nickel in,
   Nonferrous Alloys, 252
   Steel, 249
   Superalloys, 309
Ni O, 112
Nondispersive (see Energy
    Dispersive), 337
Nonferrous,
   Alloys, 252
   Metals, 255

O

Oil Well Drilling Fluids, 75
Orchard Leaves, 9
Oxidation State Effects, 236
Oxides of Steel, Analysis of, 139
Oxygen,
   Distribution on Iron
    Surface, 498
   In Impact, 448, 452

P

Pair Function Distribution
   Curves, 384
Paladium, 287
Partial Photoionization and
   Electron Ionization Cross
   Sections, 158
Particle,
   Blender, 56
   Size Effect in ESCA, 515
Peak,
   As Parabola, 358
   Location, 358
   Shifts Amorphous, 389
Petroleum, 288
   Analysis of, 542
Phase Sensitive Detector, 536
Philiberts' Correction, 483
Phosphorus in Steel, 249
Photoelectron, 150, 158
   Spectrometer, 509
Photoionization, 152
   Cross Sections, 150
Pigments, 288
Pines Method, 400
Plasmas, 423
Platelet Particles, 62
Plating Thickness Analyzed, 325
Platinum, 281, 287
Plutonium, 337, 581
Pole,
   From Iron, 416
   Figure, 419
   Plotting, 416
Polutants, Standards for, 225
Poly/Vinyl Chloride, 41
Portland Cement, Analysis of, 214
Potassium-Silicate Glasses, 384
Powder Diffraction
   File, 20
   Standards, 32
Precipitation on Filter Paper, 289
Precision, 2

Preferred Orientation, 102
  Effect of on Analysis, 142
  In Powders, 28
Preparation of Samples for XRF, 344
Production Cross Sections, 445
Proton,
  Excitation of X-rays, 457
  Stopping Power, 454
PuO_2 Lattice of, 348

Q

Quantitative Analysis,
  Of Thin Films, 487
  Powder Samples for, 32
Quantitative X-ray Diffraction Analysis, 139
Quantometers, Use of, 217, 219
Quartz, 14, 77

R

Radiation Effects on Samples, 40
Radioactive Sample Analysis, 337
Radioisotopic,
  Standardization, 287
  X-ray Analyzer, 256
Rasberry-Heinrich Method, 315, 316
Rare Earth Analyses, 579
Reciprocal Lattice, 417, 418
Recycling, 247
Reflection Coefficient, 436
Reference Materials, 1, 16
Regression Equations, 255, 302
Residual Stress, 371
  Measurement, 354
Resin, 113
  -Loaded Filters, 279
Resolution, in Energy Dispersive Analysis, 124
Rocks
  Bismuth in, 296
  Silicate, 296
Rudman Method, 399
Rutherford Scattering, 448

S

Sample,
  Mounts for Automatic Systems, 561
  Utilization, 68
Sampling,
  Of Recycled Metals, 247
  Theory, 45
Sapphire, 475
Scanning Electron Microscopy, 152
Scattering Coefficients, 259
Secondary Materials, 247
Secondary Source (see Fluorescent X-ray Analysis)
Seeman-Bohlin, 100
  Diffractometer, 102
Segregation, 45
Self-Irradiation of PuO_2, 348
Semiconductor Materials, 467
Serum, Analysis of, 274
Sharpening Factor, 386

Subject Index

Si(Li),
  Dead Layers in, 531
  Detector, 459
  Sensitivity for X-rays, 531
  Use of, 560
Silicon, 470
  Crystal, 4
  Diode Efficiencies, 539
  In Steel, 249
Silver, 285
  Films, Analysis of, 487
SiO_2, 113
Slits, Choice of in Stress Work, 360
Smectites, Patterns of, 88
Soft X-ray Shift, 236
SOLFA Analyser, 542
Specimen Preparation, 102
  For Diffractometry, 97
Spectrometer,
  Design, 571
  Spherically Bent Crystal, 521
  X-ray Bent Crystal, 521
Spectrometry,
  Energy Dispersive, 560, 571
  Of Aerosols, 225
  Proton Excitation in, 457
  Using Alpha Particles, 560
Spherical Particles, 60
Spinning Riffer, 45
Standard,
  Deviations, 44
  Powder Diffraction Data, 20
  Reference Materials, 1
Standardizing Biological Sample Analysis, 269
Standards,
  For Analysis of Aerosols, 225
  For Minerals, 88
  For Powder Diffractometry, 32
  In Analysis, 578, 579, 582
  In Trace Analysis, 462
  Secondary Metal XRF, 247
  Stainless Steel, 305
Steel,
  From Ferrous Scrap, 248
  Oxides of, 139
  Stress Measurement on, 354
  Wire, Analysis of Plating on, 325
Stress,
  Analysis, 379
  Diffractometer, 379
Structure Factor, 107
Sulfate Analysis, 236
Sulfide Analysis, 236
Sulphur,
  Analysis, 236
  In Oils, Analysis of, 542
Superalloys, 309
Surfaces, Analysis of with Electrons, 498
Synthetic Powder Standards, 44

T

Tertiary Fluorescent Excitation, 516
Tetra-ethyl Lead, 40

Textureless Sample, 416
Thermal Vibration, 107
Thin Film Techniques in XRF, 279
Thin Target Excitation for XRF, 258
Three-point Parabola Technique, 371
Tin in,
  Nonferrous Alloys, 252
  Steel, 249, 251
TiO_2, 114
Titanium,
  In Superalloys, 309
  Monitor, 337
Total Electron Ionization Cross Sections, 158
Trace Analysis, 571
  By Novel Spectrometer, 521
Trace Element, 48, 68
  Analysis, 293, 318, 454, 457
Trace Metal Analysis, 279
Transformations from Grinding, 32
Tube Transmission Anode, 338
Tungsten, 283

U
Ultrasoft X-ray Region, 152
"Urban Ores," 247
Uranium, 337, 581
  Alloy, 371
  Impurities, 288
  -Titanium Alloy, 371

V
Variances, 44
Vermiculites, Patterns of, 88
Visman's Method, 47

W
Water, 288
  Analysis, 322
Wavelength Dispersive, Compared with Energy Dispersion, 581
Welded Be, 116
Wurtzite Transformations, 37

Z
Zinc, 282
  Analysis of, 462
  Implanted Semiconductor, 453
  In Nonferrous Alloys, 252
$ZnCO_3$ Transformations, 37
ZnO, 112

MIX
Papier aus verantwortungsvollen Quellen
Paper from responsible sources
FSC® C105338

If you have any concerns about our products,
you can contact us on
ProductSafety@springernature.com

In case Publisher is established outside the EU,
the EU authorized representative is:
**Springer Nature Customer Service Center GmbH
Europaplatz 3, 69115 Heidelberg, Germany**

Printed by Libri Plureos GmbH
in Hamburg, Germany